The projective, Möbius, Laguerre, and Minkowski planes over the real numbers are just a few examples of a host of fundamental classical topological geometries on surfaces that satisfy an axiom of joining. This book summarises all known major results and open problems related to these classical geometries and their close (non-classical) relatives.

Topics covered include: classical geometries; methods for constructing non-classical geometries; classifications and characterisations of geometries. This work is related to a host of other fields including interpolation theory, convexity, differential geometry, topology, the theory of Lie groups and many more. The authors detail these connections, some of which are well-known, but many much less so.

Acting both as a referee for experts and as an accessible introduction for beginners, this book will interest anyone wishing to know more about incidence geometries and the way they interact.

Burkard Polster is Logan Research Fellow at Monash University.
Günter Steinke is a Senior Lecturer at the University of Canterbury.

Volume 84

Geometries on Surfaces

ENCYCLOPEDIA OF MATHEMATICS AND ITS APPLICATIONS

4 W. Miller, Jr. *Symmetry and separation of variables*
6 H. Minc *Permanents*
11 W. B. Jones and W. J. Thron *Continued fractions*
12 N. F. G. Martin and J. W. England *Mathematical theory of entropy*
18 H. O. Fattorini *The Cauchy problem*
19 G. G. Lorentz, K. Jetter and S. D. Riemenschneider *Birkhoff interpolation*
21 W. T. Tutte *Graph theory*
22 J. R. Bastida *Field extensions and Galois theory*
23 J. R. Cannon *The one-dimensional heat equation*
25 A. Salomaa *Computation and automata*
26 N. White (ed.) *Theory of matroids*
27 N. H. Bingham, C. M. Goldie and J. L. Teugels *Regular variation*
28 P. P. Petrushev and V. A. Popov *Rational approximation of real functions*
29 N. White (ed.) *Combinatorial geometrics*
30 M. Pohst and H. Zassenhaus *Algorithmic algebraic number theory*
31 J. Aczel and J. Dhombres *Functional equations containing several variables*
32 M. Kuczma, B. Chozewski and R. Ger *Iterative functional equations*
33 R. V. Ambartzumian *Factorization calculus and geometric probability*
34 G. Gripenberg, S.-O. Londen and O. Staffans *Volterra integral and functional equations*
35 G. Gasper and M. Rahman *Basic hypergeometric series*
36 E. Torgersen *Comparison of statistical experiments*
37 A. Neumaier *Intervals methods for systems of equations*
38 N. Korneichuk *Exact constants in approximation theory*
39 R. A. Brualdi and H. J. Ryser *Combinatorial matrix theory*
40 N. White (ed.) *Matroid applications*
41 S. Sakai *Operator algebras in dynamical systems*
42 W. Hodges *Model theory*
43 H. Stahl and V. Totik *General orthogonal polynomials*
44 R. Schneider *Convex bodies*
45 G. Da Prato and J. Zabczyk *Stochastic equations in infinite dimensions*
46 A. Bjorner, M. Las Vergnas, B. Sturmfels, N. White and G. Ziegler *Oriented matroids*
47 E. A. Edgar and L. Sucheston *Stopping times and directed processes*
48 C. Sims *Computation with finitely presented groups*
49 T. Palmer *Banach algebras and the general theory of *-algebras*
50 F. Borceux *Handbook of categorical algebra I*
51 F. Borceux *Handbook of categorical algebra II*
52 F. Borceux *Handbook of categorical algebra III*
54 A. Katok and B. Hassleblatt *Introduction to the modern theory of dynamical systems*
55 V. N. Sachkov *Combinatorial methods in discrete mathematics*
56 V. N. Sachkov *Probabilistic methods in discrete mathematics*
57 P. M. Cohn *Skew Fields*
58 Richard J. Gardner *Geometric tomography*
59 George A. Baker, Jr. and Peter Graves-Morris *Padé approximants*
60 Jan Krajicek *Bounded arithmetic, propositional logic, and complex theory*
61 H. Gromer *Geometric applications of Fourier series and spherical harmonics*
62 H. O. Fattorini *Infinite dimensional optimization and control theory*
63 A. C. Thompson *Minkowski geometry*
64 R. B. Bapat and T. E. S. Raghavan *Nonnegative matrices and applications*
65 K. Engel *Sperner theory*
66 D. Cvetkovic, P. Rowlinson and S. Simic *Eigenspaces of graphs*
67 F. Bergeron, G. Labelle and P. Leroux *Combinatorial species and tree-like structures*
68 R. Goodman and N. Wallach *Representations of the classical groups*
69 T. Beth, D. Jungnickel and H. Lenz *Design Theory volume I 2 ed.*
70 A. Pietsch and J. Wenzel *Orthonormal systems and Banach space geometry*
71 George E. Andrews, Richard Askey and Ranjan Roy *Special Functions*
72 R. Ticciati *Quantum field theory for mathematicians*
73 M. Stern *Semimodular lattices*
74 I. Lasiecka and R. Triggiani *Control theory for partial differential equations I*
75 I. Lasiecka and R. Triggiani *Control theory for partial differential equations II*
76 A. A. Ivanov *Geometry of sporadic groups I*
77 A. Schinzel *Polynomials with special regard to reducibility*
78 T. Beth, D. Jungnickel and H. Lenz *Design Theory volume II 2 ed.*
79 T. Palmer *Banach algebras and the general theory of *-algebras II*
80 O. Stormark *Lie's Structural Approach to PDE Systems*
81 C. F. Dunkl and Y. Xu *Orthogonal polynomials of several variables*
82 J. Mayberry *The foundations of mathematics in the theory of sets*

ENCYCLOPEDIA OF MATHEMATICS AND ITS APPLICATIONS

Geometries on Surfaces

BURKARD POLSTER
Monash University

and

GÜNTER STEINKE
University of Canterbury

PUBLISHED BY THE PRESS SYNDICATE OF THE UNIVERSITY OF CAMBRIDGE
The Pitt Building, Trumpington Street, Cambridge, United Kingdom

CAMBRIDGE UNIVERSITY PRESS
The Edinburgh Building, Cambridge, CB2 2RU, UK
40 West 20th Street, New York, NY 10011–4211, USA
10 Stamford Road, Oakleigh, VIC 3166, Australia
Ruiz de Alarcón 13, 28014 Madrid, Spain
Dock House, The Waterfront, Cape Town 8001, South Africa

http://www.cambridge.org

First published 2001

Printed in the United Kingdom at the University Press, Cambridge

Typeface Computer Modern 10/13pt *System* LaTeX 2ε [UPH]

A catalogue record of this book is available from the British Library

ISBN 0 521 66058 0 hardback

To Anu and Marina

Contents

Preface

What This Book Is All About

'Geometries on surfaces'—what do you think of when you read such a title? Whatever it is will depend to a large extent on your background in mathematics. Our background is in incidence geometry, and, even if we were not the authors of this book, we would first think of examples such as the Euclidean plane and the geometry of circles on a sphere. These two geometries have a number of features in common. For example, the point sets of both geometries are surfaces, the lines or circles are curves that are nicely embedded in these surfaces, and both geometries satisfy an 'axiom of joining'—in the Euclidean plane two points are contained in exactly one line and in the geometry on the sphere three points are contained in exactly one circle.

The Euclidean plane and the geometry of circles on a sphere are just two examples of a host of classical examples of geometries on surfaces. This book is about these classical geometries and their close relatives which live on the same surfaces, have the same kinds of lines, and satisfy the same axioms as their classical counterparts.

The history of our geometries on surfaces starts with Hilbert constructing a first example of a nonclassical $\mathbf{R}^2$-plane, that is, a close relative of the Euclidean plane. Today, one century of research later, our book tries to summarize all major results about geometries on surfaces. This includes the following topics.

- Detailed descriptions of the classical geometries.
- The main methods of constructing nonclassical geometries.
- Classifications and characterizations of the geometries that are 'most homogeneous'. In particular, classifications of the various kinds of geometries with respect to the dimensions of their groups of auto-

morphisms which, in most cases, are known to be Lie groups of finite dimensions.
- Descriptions of the various geometries associated with a given geometry such as subgeometries, fix-geometries of automorphisms, and Lie geometries.
- Connections with other fields such as the theories of interpolation and approximation, convexity, differential geometry, topology, Lie groups, differential equations, oriented matroids, finite geometries, and higher-dimensional topological geometries.
- Open problems.

Putting Things into Perspective

Real Geometries: Most of the 'classical' types of geometries investigated in incidence geometry, such as affine and projective planes and circle planes, can be defined over all fields. The classical examples of geometries on surfaces are geometries associated with the real numbers. Just as the real numbers occupy a central position in the theory of fields, the classical geometries on surfaces occupy a central position in incidence geometry.

Topological Geometries: Our geometries on surfaces are topological geometries in that both the point and line/circle sets of our geometries carry natural topologies such that the geometric operations in the axioms that these geometries satisfy are continuous. As a result, the connecting line of two points in the Euclidean plane and the connecting circle of three points on a sphere depend continuously on the positions of the points on the surface. Among the topological incidence geometries, the geometries on surfaces are the ones that are most easily described and the ones that are best understood. A thorough understanding of these geometries is necessary for everybody interested in working in topological geometry.

Networks of Geometries: One of the most appealing features of geometries on surfaces is the tightly knit network of relationships that has been established between them. This network of relationships plus the wealth of our knowledge about the individual geometries identifies this cluster of geometries as one of the most well-charted sectors in incidence geometry. In other sectors of incidence geometry, in particular finite geometry, the most popular branch of incidence geometry today, the counterparts of our geometries are tied together into highly complex networks, the overall structure of which is very hard to discern because of the presence of a lot of 'noise', that is, sporadic geometries and links.

By stepping back and looking at the network of relationships between finite geometries of odd order through our 'topological filter', it is possible also to discern many of the fundamental links and results in our model cluster in the finite setting that would otherwise be very hard to see. We only remark that finite geometries of even order behave and are linked in ways that are very different from their odd order counterparts. Therefore it is not surprising that the correspondence between geometries on surfaces and finite geometries of even order is not very strong.

Geometric Combinatorics: The similarity of results in the finite and topological settings establishes a first close link between the theory of geometries on surfaces and combinatorics. There are two further strong links between these two disciplines. First, once a basic toolbox of techniques has been developed, these techniques can be employed in a very combinatorial manner to construct and derive results about geometries on surfaces. Second, finite subsets of circles of rank n geometries correspond to rank n pseudocircle arrangements that are fundamental objects in geometric combinatorics. In particular, finite subsets of lines in flat projective planes are pseudoline arrangements that have interpretations as important combinatorial structures such as oriented matroids, switching sequences, and primitive sorting networks.

Interpolation: The classical examples of our geometries also correspond to the classical objects around which other seemingly unrelated theories are built. The so-called classical tubular circle plane of rank n, for example, corresponds to the set of polynomials of degree at most $n-1$ over the reals, and the axiom of joining that this circle plane satisfies translates into the fact that this set of polynomials solves the Lagrange interpolation problem of order n. Similarly, every tubular circle plane of rank n corresponds to a set of continuous functions that solves the Lagrange interpolation problem of order n. In developing a theory of geometries on surfaces, we and our colleagues have also tried to provide a unifying topological foundation for the different kinds of disciplines that deal, in whatever guise, with geometries on surfaces.

Lie Groups: For many of the geometries on surfaces it is known that all their automorphisms are homeomorphisms of the surfaces, and their groups of automorphisms are finite-dimensional Lie transformation groups. This establishes an important connection between the theory of geometries on surfaces and the theory of Lie groups. In fact, the classification of finite-dimensional Lie groups is one of the most important tools used in the classification of geometries on surfaces.

Intended Audience

First and foremost our book is a reference book and people who will be interested in every single aspect of the book will most likely be mathematicians specializing in incidence geometry. Nevertheless, given that most of the geometries we are dealing with can be described, understood, visualized, and appreciated with only an advanced undergraduate's knowledge of topology and analysis, we have made every effort to keep large parts of the book accessible and appealing to as broad an audience as possible. Therefore, if you are a student interested in incidence geometry, try to develop a feeling for the way incidence geometries behave and interact by first studying the classical geometries over fields and geometries on surfaces. If you are a lecturer looking for new, beautiful, and easily accessible topics to spice up your introductory course on topology, analysis, the theory Lie groups, geometric combinatorics, or even interpolation theory, you will find that our book of geometries on surfaces has a lot to offer to you.

Contents and How to Read and Use This Book

We have tried to make this book as self-contained as possible. However, to get the most out of it and to arrive at a full understanding of every aspect of the theory of geometries on surfaces and topological geometry in general, we recommend that you use it side by side with a number of excellent books and survey articles. In particular, *Compact Projective Planes*, Salzmann et al. [1995], is the most comprehensive collection of results on topological projective planes and is a must-read for everybody interested in topological geometry. We also recommend the *Handbook of Incidence Geometry*, Buekenhout ed. [1995]. This book is an excellent up-to-date survey of modern incidence geometry. In particular, the last two chapters of this book deal with topological geometries. For a pictorial guide to geometries on surfaces see *A Geometrical Picture Book*, Polster [1998g]. For information about finite geometries also check out *Finite Geometries*, Dembowski [1968].

You should start by reading Chapter 1. In it we define the most important terms that we will be using in the rest of the book, describe the different kinds of geometries we are dealing with, and list many of their most important features in simple language.

In Chapter 2 we concentrate on flat linear spaces. These include the close relatives of the Euclidean plane and the projective plane over the real numbers. The results in this chapter have counterparts in the following chapters and many of these counterparts are based on the results in this chapter. In particular, we show that the line set of a flat linear space carries a natural topology, that automorphisms of such a geometry are continuous, and that the automorphism group of such a geometry is a finite-dimensional Lie transformation group. The logical continuation of this is to give an account of the classification of the flat linear spaces with respect to the dimensions of their automorphism groups. An excellent exposition of the classification of the flat projective planes whose automorphism groups have dimension at least 3 can be found in the book Salzmann et al. [1995]. Rather than copying all the relevant material in that book, we only summarize the main results derived in that book and extend them by many results, constructions, and classifications outside its scope. Highlights include the description of all flat projective planes whose automorphism groups are at least 2-dimensional, and all Möbius strip planes and $\mathbf{R}^2$-planes whose automorphism groups are at least 3-dimensional.

In Chapter 3 we focus on the spherical circle planes, that is, the relatives of the geometry of circles on a sphere. In it we demonstrate how to use the results about $\mathbf{R}^2$-planes developed in Chapter 2 to prove similar results for higher-rank flat circle planes. We also give a more detailed account of the group-dimension classification of spherical circle planes than we did for flat linear spaces. So, if you are interested in familiarizing yourself with typical arguments that are used in group-dimension classifications of topological geometries, here is a good spot to start. Some of the highlights in this chapter are the classification of the spherical circle planes whose automorphism groups are at least 4-dimensional, the Hering classification of these planes, and many new constructions of flat linear spaces from spherical circle planes.

In Chapters 4 and 5 we concentrate on the toroidal circle planes and the cylindrical circle planes. These include the flat Minkowski and Laguerre planes, respectively. Just like the spherical circle planes, the two types of circle planes we focus on in these chapters are of rank 3 and the results in these chapters are closely related to the results in Chapter 3. On the other hand, some new considerations enter the scene as both toroidal and cylindrical circle planes have nontrivial parallelisms. Included in these chapters are the classification of the flat Minkowski planes of group dimension at least 4 and the classification of the flat Laguerre planes of group dimension at least 5.

Chapter 6 is devoted to the theory of antiregular 3-dimensional generalized quadrangles as developed in *Topological Circle Planes and Topo-*

logical Quadrangles, Schroth [1995a]. This theory ties the flat Möbius, Laguerre, and Minkowski planes into a tight knot full of beautiful connections and unexpected relationships. In this chapter we give a summary of Schroth's theory and state the most important results about flat Laguerre, Möbius, and Minkowski planes that are best expressed in the language of generalized quadrangles. Also included in this chapter are the recently discovered connections between antiregular generalized quadrangles, circle planes, and semibiplanes.

In Chapter 7 we give an account of the connection between classical Lagrange and Hermite interpolation and higher-rank circle planes on surfaces.

In two appendices we summarize some basic results from analysis, topology, and the theory of Lie transformation groups that are frequently used in this book.

Acknowledgements

This book is based on the work of, and support by, many mathematicians: in particular, Helmut Salzmann under whose guidance topological geometry was created over the past 50 years, and Karl Strambach who introduced us to this beautiful branch of mathematics and encouraged us to pursue its study.

Particular thanks are due to Rainer Löwen and Andreas Schroth who read parts of the book and offered lots of helpful advice. Thanks are also due to Markus Stroppel who provided us with some urgently needed manuscripts.

The first author wishes to thank the Australian Research Council for its support through an Australian Postdoctoral Research Fellowship and Monash University for its support through a Logan Research Fellowship.

Finally, special thanks are due to Anu, Marina, and the kids for their support and patience throughout the writing of this book.

Burkard Polster
Melbourne, December 2000

Günter F. Steinke
Christchurch, December 2000

1
Geometries for Pedestrians

1.1 Geometries of Points and Lines

In this book a *geometry* will usually consists of a nonempty *point set* and a nonempty *line set*, where a line is a subset of the point set containing at least three points and every point is contained in at least three lines. For example, the *Euclidean plane* is a geometry whose points are the points of the coordinate plane $\mathbf{R}^2$ and whose lines are the Euclidean lines. For historical reasons, lines are sometimes called *blocks* or *circles*. The lines of the geometry of circles on a sphere, for example, are really called circles. All geometries that we are interested in also satisfy a number of structuring *axioms*. In particular, both the Euclidean plane and the geometry of circles satisfy an 'axiom of joining'—in the Euclidean plane any two points are contained in exactly one line, and in the geometry of circles any three points are contained in exactly one circle. In fact, virtually all the geometries considered in this book satisfy an axiom of joining.

In most textbooks in our subject area lines and the sets of lines through points are required to contain only at least two elements. By restricting ourselves to *thick* geometries, it is possible to omit one axiom from each of the systems of axioms that we will encounter in this book.

Geometries Are Thick

All geometries in this book are *thick*, that is, every line contains at least three points and every point is contained in at least three lines.

The main disadvantage with our approach is that certain important graphs that are usually also regarded as geometries get excluded. In particular, the complete graph on four vertices, which is the smallest finite counterpart of the Euclidean plane, is not a geometry according to our definition. However, since we are concentrating on geometries on surfaces, we feel that the advantages of our approach outweigh its disadvantages.

By dealing with such basic geometries as the Euclidean plane, you will already have acquired a certain familiarity with many terms used in this book that should enable you to understand this chapter. For precise definitions of the most important terms used in this chapter and the rest of the book see Section 1.3 at the end of this chapter.

In the following we define some of the most fundamental types of geometries incidence geometers are dealing with: linear spaces, in particular projective planes and affine planes; the three types of Benz planes, that is, Möbius planes, Laguerre planes, and Minkowski planes; and orthogonal arrays. The classical examples of these geometries can be defined over many fields. The different geometries that correspond to the real numbers are the most important classical geometries on surfaces. We will also refer to projective planes, Benz planes, and certain maximal orthogonal arrays as *circle planes* since the lines/circles in the classical real examples of these planes are really topological circles.

A *linear space* is a geometry in which every two distinct points are contained in exactly one line. We start by defining projective and affine planes, two special types of linear spaces.

1.1.1 Projective Planes

A *projective plane* is a geometry that satisfies the following two axioms.

Axioms for Projective Planes

(P1) Two distinct points are contained in a unique line.

(P2) Two distinct lines intersect in a unique point.

The *classical examples* of projective planes are the projective planes over fields. Given a field F, this projective plane can be constructed as follows. The point set is the set of all 1-dimensional subspaces and the line set is the set of all 2-dimensional subspaces of the 3-dimensional vector space over F. In this model the two axioms are easily verified. Axiom P1 translates into the fact that two distinct 1-dimensional subspaces of a 3-dimensional vector space are contained in a uniquely determined 2-dimensional subspace. Similarly, Axiom P2 translates into the fact that two distinct 2-dimensional subspaces of a 3-dimensional vector space intersect in a 1-dimensional subspace. The classical projective plane over F is denoted by $\mathrm{PG}(2, F)$.

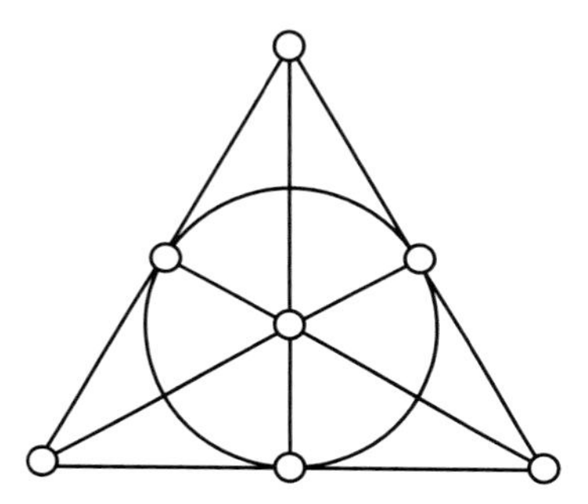

Fig. 1.1. The Fano plane

Figure 1.1 is a picture of the projective plane over the field with two elements. It has seven points and seven lines, every line contains three points and every point is contained in three lines. This projective plane is called the *Fano plane*. It is the unique smallest projective plane. A finite projective plane is of *order* n if and only if every one of its lines contains $n+1$ points and every one of its points is contained in $n+1$ lines. For example, the Fano plane is of order 2 and, in general, the classical projective plane over the field with q elements is of order $q+1$.

There are many ways in which the classical projective planes are characterized among the projective planes. One of the most famous such characterizations is via *Pappus' configuration*. Pappus' configuration is a geometry with nine points and nine lines, every line of which contains three points and every point of which is contained in three lines. Figure 1.2 is a picture of this configuration. Note that this diagram does not capture all the symmetries of this configuration. In fact, contrary to what the diagram suggests, none of the points and lines of the configuration is distinguished in any way.

It turns out that a projective plane is classical if and only if Pappus'

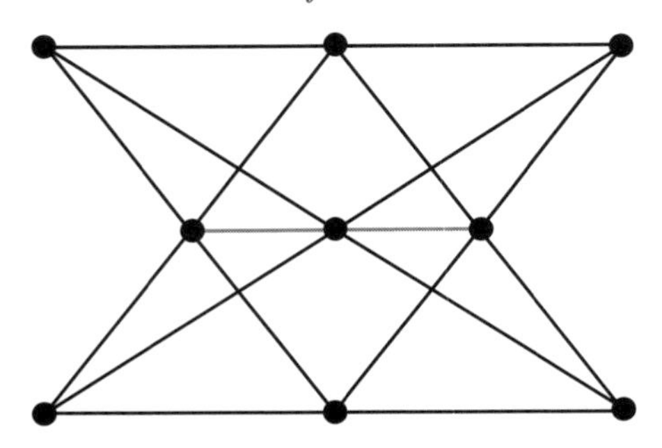

Fig. 1.2. Pappus' configuration

configuration *closes* in the projective plane; see p. 22 for a precise definition of the term 'closes'. In particular, for Pappus' configuration to close means that no matter how one draws the solid black parts of the configuration in Figure 1.2 using points and lines of the geometry, the three points that are contained in only two black lines each are always contained in a line of the geometry (the grey line). We only remark that the term 'Pappus' configuration closes' also encompasses some degenerate cases that arise when certain points of the configuration get identified.

Note that projective planes can also be defined over skewfields and that the projective planes constructed like this are precisely the ones in which *Desargues' configuration* closes; see Figure 1.3.

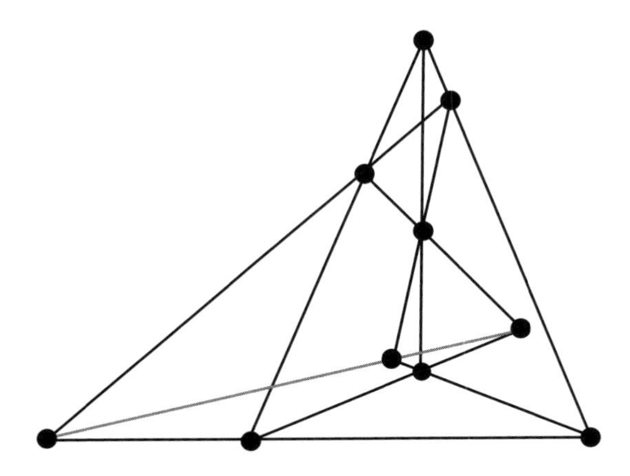

Fig. 1.3. Desargues' configuration

A projective plane is called a *Pappian* or *Desarguesian* projective plane if and only if Pappus' configuration or Desargues' configuration closes in it, respectively. Since every field is also a skewfield this implies that every Pappian projective plane is also Desarguesian. The converse is not true. However, for the projective planes that we are concentrating on in this book, namely the close relatives of the classical projective plane over the real numbers, the two notions coincide.

1.1.2 Affine Planes

By removing a line and all the points contained in it from a projective plane with at least four points on a line, we arrive at a geometry that satisfies the following axioms.

Axioms for Affine Planes

(A1) Two distinct points are contained in a unique line.

(A2) Given a line and a point not on this line, there is a unique line through the point that does not intersect the given line.

Every geometry that satisfies these axioms is called an *affine plane*. Two lines in an affine plane are called *parallel* if they coincide or do not intersect in a point. Axiom A2 implies that being parallel defines an equivalence relation on the line set. The equivalence classes of this equivalence relation are called *parallel classes*.

The affine plane we arrive at by removing a line from the classical projective plane associated with the real numbers is the Euclidean plane. All affine planes obtained from a classical projective plane by removing a line are isomorphic. Every affine plane has a unique *projective extension* to a projective plane. Starting with an affine plane, this projective plane can be constructed as follows. The points of the projective plane are the points of the affine plane plus its parallel classes of lines. The set of parallel classes is one of the lines of the projective plane. We construct the remaining lines by extending every line in the affine plane by the parallel class it is contained in.

A finite affine plane is of order n if its projective extension is of order n. Note that by removing a line and all its points from the projective plane of order 2, we arrive at the complete graph on four vertices. For completeness' sake, we call this graph the affine plane of order 2.

The affine plane associated with the classical projective plane over the field F can also be constructed as follows. The point set is $F \times F$, the lines are the verticals in the point set plus the graphs of all linear functions $F \to F$. In this plane the lines of a given slope form a parallel class. The verticals also form a parallel class.

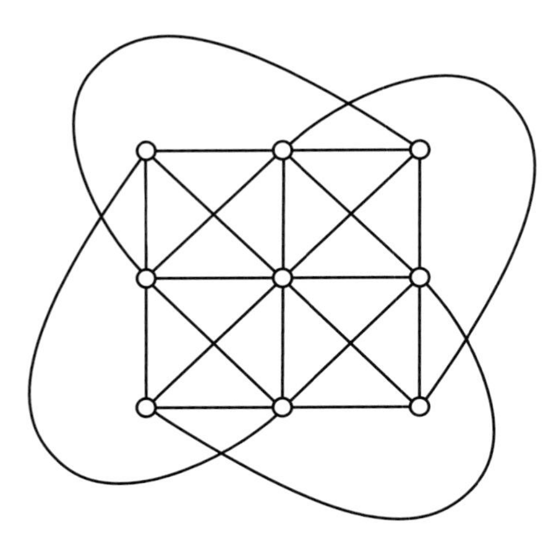

Fig. 1.4. The affine plane of order 3

Figure 1.4 shows a picture of the affine plane of order 3. In this plane every line contains three points and every point is contained in four lines. There are four parallel classes consisting of three lines each.

For comprehensive introductions to projective planes and their associated affine planes see Hughes–Piper [1973], Dembowski [1968], and Salzmann et al. [1995].

1.1.3 Benz Planes—the Original Circle Planes

When geometers talk about circle planes, they usually have the three different types of Benz planes in mind. In this book the term circle plane comprises infinitely many different types of geometries, not only the Benz planes. The following definition of Benz planes can be found in Steinke [1995].

A *Benz plane* is a geometry whose point set is equipped with one or two equivalence relations called *parallelisms*. Two points of a Benz plane are *parallel* if they are in relation with respect to one of the parallelisms. A parallelism is called *trivial* if two points are parallel if and only if they coincide. Furthermore, Benz planes satisfy the following axioms.

Axioms for Benz Planes

(B1) Three pairwise nonparallel points are contained in a uniquely determined circle.

(B2) Given a point p on a circle C and a point q not parallel to p, there is a uniquely determined circle that contains both points and *touches C geometrically*, that is, intersects C only in p or coincides with C.

(B3) Parallel classes with respect to a nontrivial parallelism and circles intersect in a unique point.

(B4) Parallel classes with respect to different nontrivial parallelisms intersect in a unique point.

If a Benz plane has two different nontrivial parallelisms, it is a *Minkowski plane*. If it has only one nontrivial parallelism, it is a *Laguerre plane*. In this case Axiom B4 does not apply. If it has only a trivial parallelism, both Axioms B3 and B4 do not apply, 'nonparallel' translates into 'distinct', and the Benz plane is a *Möbius plane*.

The classical examples of Möbius, Laguerre, and Minkowski planes arise as the geometries of nontrivial plane sections of elliptic quadrics, elliptic cones, and hyperbolic quadrics, respectively, in the 3-dimensional projective spaces over fields.

In the real case that we are most interested in two of the classical geometries live in a real Euclidean 3-space. The classical real Möbius plane is just the geometry of circles on a sphere, and the classical real Laguerre plane is the geometry of nontrivial plane sections of a cylinder. The set of generator lines on this cylinder are the parallel classes of points. The classical real Minkowski plane minus one of its circles and all points on this circle is the geometry of nontrivial plane sections of a (one-sheeted) hyperboloid. The two sets of parallel classes correspond to the disjoint partitions of the hyperboloid into Euclidean lines embedded in the hyperboloid.

The *derived plane* at a point p of a geometry $\mathcal{G}$ whose point set carries one or more parallelisms has all the points of $\mathcal{G}$ not parallel to p as points. Its lines are all lines of $\mathcal{G}$ through p that have been punctured at p plus all nontrivial parallel classes not containing p that have been punctured at all points parallel to p. It turns out that a geometry having one or two parallelisms and satisfying Axioms B3 and B4 is a Benz plane if and only if the derived planes at all its points are affine planes.

A finite Benz plane is of order n if all its circles contain $n+1$ points. The derived planes of a Benz plane of order n are affine planes of order n. A derived plane of a classical Benz plane over the field F is isomorphic to the classical affine plane over F. The derived planes of the geometry of circles on a sphere are isomorphic to the Euclidean plane.

The picture of the affine plane of order 3 in Figure 1.4 is also a picture of the unique Minkowski plane of order 2. Here the parallel classes are the verticals and horizontals in the grid and the circles are the remaining six lines in the affine plane.

1.1.4 Orthogonal Arrays

An *orthogonal array of rank n* is a geometry whose point set carries one nontrivial parallelism. Its point set is of the form $C \times E$, where C and E are disjoint sets, with $|C| \geq n$ and $|E| \geq 2$. Furthermore, two of its points $(a, b), (c, d) \in C \times E$ are parallel if and only if $a = c$, and it satisfies the following axioms.

Axioms for Orthogonal Arrays

(O1) Any n pairwise nonparallel points are contained in a uniquely determined circle.

(O2) A parallel class and a circle intersect in a unique point.

Given an orthogonal array O of rank n, we can identify C and E with a circle and a parallel class, respectively. Because of Axiom O2, circles are graphs of functions $C \to E$.

Starting with the polynomials of degree at most $n-1$ over a field F with at least $n \geq 3$ elements, we define a geometry $\mathrm{Poly}(n, F)$ whose point set is $F \times F$ and whose circles are the graphs of the polynomials. Keeping in mind the interpolating property of the set of polynomials under consideration, it is clear that this geometry is an orthogonal array of rank n. The parallel classes are the verticals in the point set. For example, the geometry $\mathrm{Poly}(2, \mathbf{R})$ is the Euclidean plane in which the verticals lines have been turned into parallel classes.

By removing one parallel class together with all the points contained in it from an orthogonal array of rank n, we are left with an orthogonal array of the same rank as long as $|C| \geq \max\{4, n+1\}$. An orthogonal array of rank n is called *maximal* if it does not arise from an orthogonal array of the same rank in this manner. The polynomial geometry $\mathrm{Poly}(n, F)$ is not maximal since it can be extended by a *parallel class at infinity* $\{\infty\} \times F$ to a larger orthogonal array $\overline{\mathrm{Poly}}(n, F)$. Here the point (∞, a) extends the circles that correspond to polynomials in which a is the coefficient of x^{n-1}.

The geometries $\mathrm{Poly}(n, F)$ and $\overline{\mathrm{Poly}}(n, F)$ are the *classical orthogonal arrays*. For example, the geometry $\overline{\mathrm{Poly}}(3, \mathbf{R})$ is isomorphic to the Laguerre plane over the real numbers. Clearly, every Laguerre plane is an orthogonal array of rank 3.

We remark that some of the terminology in the literature on finite orthogonal arrays differs from ours. For example, the word 'strength' is used instead of 'rank' and the orthogonal arrays we defined here are further said to be of index 1; see Rao [1947], Bush [1952], and Colbourn–Dinitz [1996] Part 2.

1.2 Geometries on Surfaces

In this section we first have a quick look at flat linear spaces, that is, geometries that are closely related to the Euclidean plane. Following this, we give an overview of the most important geometries on surfaces. These are geometries whose lines or circles are topological circles. We call these geometries *flat circle planes*. They form the backbone of the theory of geometries on surfaces, and most other types of geometries that we will come across in this book have representatives that can be easily derived from flat circle planes.

We also try to give you a feel for what geometries on surfaces are all about by describing some of their basic features, connections with the theories of interpolation and convexity, and parts of the network of relationships that turns the different types of geometries on surfaces into a larger whole.

1.2.1 (Ideal) Flat Linear Spaces

Ideally, we would like to develop a general theory of the *linear spaces on surfaces*, that is, linear spaces whose point sets are surfaces and all of whose lines are **R**-lines and/or **S**-lines, that is, closed subsets of the point sets homeomorphic to **R** and/or the circle $\mathbf{S}^1$, respectively. All known linear spaces on surfaces are *flat linear spaces*, that is, linear spaces on surfaces whose line sets can be equipped with topologies such that the operations of joining two points by a line and intersecting two lines in a point are continuous on their domains of definition. Furthermore, in a flat linear space the set of pairs of distinct intersecting lines is required to be open; see Section 2.5. Flat linear spaces live on surfaces homeomorphic to $\mathbf{R}^2$, the real projective plane, and the Möbius strip and are referred to as $\mathbf{R}^2$-planes, flat projective planes, and Möbius strip planes, respectively. We suspect that the flat linear spaces may well encompass all linear spaces on surfaces. However, it has not even been proved that the three surfaces mentioned above are the only point sets that a linear space on a surface can live on; see Problem 2.11.2.

We do know that the $\mathbf{R}^2$-planes and flat projective planes are precisely the linear spaces on surfaces that are homeomorphic to $\mathbf{R}^2$ and the real projective plane, respectively. In $\mathbf{R}^2$-planes all lines are **R**-lines and in flat projective planes all lines are **S**-lines. Möbius strip planes are of mixed type, that is, have both **R**- and **S**-lines. We do not know whether the Möbius strip planes are the only linear spaces on surfaces homeomorphic to the Möbius strip; see Problem 2.11.1.

Given a flat projective plane with point set P, examples of $\mathbf{R}^2$-planes and Möbius strip planes arise as the restrictions of this flat projective plane to certain open subsets of P. However, not all $\mathbf{R}^2$-planes and Möbius strip planes arise in this manner. The restriction to the complement of a line is called a flat affine plane and the restriction to the complement of a point is called a point Möbius strip plane. Both flat affine planes and point Möbius strip planes can be considered as special representations of the flat projective plane they are associated with since this flat projective plane can be reconstructed from both geometries in a unique manner.

Apart from being models of flat projective planes, point Möbius strip planes also play an important role in the general theory of tubular circle planes as they are in one-to-one correspondence with the tubular circle planes of rank 2; see Subsection 1.2.2 and Chapter 7.

We do not want to start our investigations with the above abstract definition of flat linear spaces as is usually done, and rather develop the theory as far as possible starting with the simple concept of linear spaces on surfaces. In view of the results mentioned above, we will therefore first develop the theory of $\mathbf{R}^2$-planes and flat projective planes. We will also include point Möbius strip planes in our considerations because of their importance for the general theory of geometries on surfaces and because they are just special representations of flat projective planes and the most well-behaved Möbius strip planes. We will jointly refer to these three 'ideal' types of linear spaces as *ideal flat linear spaces.* We formally introduce general flat linear spaces in Section 2.5.

In the following we have a closer look at the three different types of ideal flat linear spaces.

1.2.1.1 $\mathbf{R}^2$*-Planes*

We define an $\mathbf{R}^2$*-plane* to be a linear space on a surface homeomorphic to $\mathbf{R}^2$. Any $\mathbf{R}$- or $\mathbf{S}$-line in $\mathbf{R}^2$ separates $\mathbf{R}^2$ into two connected components. Clearly, given an $\mathbf{R}$- or $\mathbf{S}$-line L and an $\mathbf{S}$-line K in $\mathbf{R}^2$ such that L contains points in the two different connected components of $\mathbf{R}^2 \setminus K$, L and K have at least two points in common. Since two lines in a linear space intersect in at most one point, we see that an $\mathbf{R}^2$-plane cannot contain $\mathbf{S}$-lines.

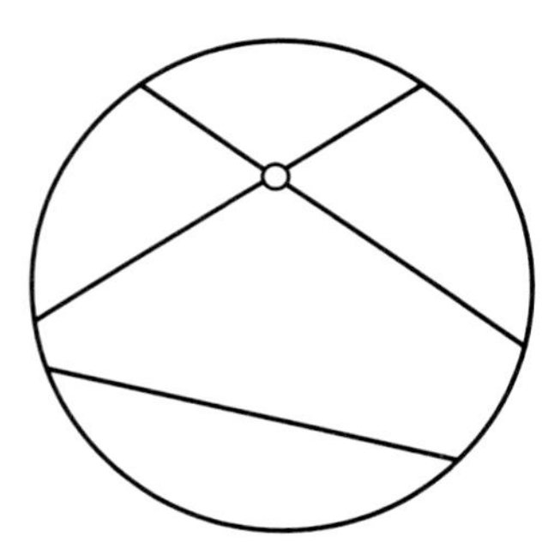

Fig. 1.5. More than one parallel line

The Euclidean plane is the classical example of an $\mathbf{R}^2$-plane. In fact, the Euclidean plane is a *flat affine plane*, that is, an $\mathbf{R}^2$-plane that also satisfies Axiom A2. It is easy to construct $\mathbf{R}^2$-planes that are not flat affine planes. Consider, for example, the restriction of the Euclidean plane to an open strictly convex region. Then we are still dealing with an $\mathbf{R}^2$-plane. Nevertheless, unless this region is the whole plane, Axiom A2 will never be satisfied. As an example, let us have a look at the restriction of the Euclidean plane to the open unit disk. Figure 1.5 shows two lines through a point that are both parallel to a third line, that is, do not intersect this third line.

Important $\mathbf{R}^2$-planes that are embedded in the Euclidean plane are the restrictions of the Euclidean plane to the open upper half-plane and the open unit disk. These two geometries are also called the *real cylinder plane* $\mathrm{C}(\mathbf{R})$ and the *real hyperbolic plane* $\mathrm{H}(\mathbf{R})$, respectively.

1.2.1.2 Flat Projective Planes

A *flat projective plane* is a linear space on a surface homeomorphic to the *real projective plane*. The classical flat projective plane is the projective plane over the real numbers. The surface it lives on is a compact surface which, just like the Klein bottle, cannot be embedded in Euclidean 3-dimensional space without self-intersections. We arrive at a visually appealing model of the classical flat projective plane from the geometry of great circles on a 2-dimensional sphere by topologically identifying antipodal points on this sphere. Under this identification the sphere turns into the real projective plane and every single one of the great circles into an **S**-line on this surface. Based on this model it is easy to deduce many topological properties of this flat projective plane such as the fact that the point set is compact and that circles are nonseparating.

Since there are no **R**-lines in the real projective plane, all lines of a flat projective plane are **S**-lines. We remark that flat projective planes are really projective planes as we defined them above. In particular, two distinct lines in such a plane automatically intersect in a unique point; see Theorem 2.2.7.

The restriction of any flat projective plane to the complement of a line is a flat affine plane and every flat affine plane arises like this from a flat projective plane.

1.2.1.3 Point Möbius Strip Planes

Remember that after removing a point from the real projective plane we are left with a Möbius strip. A linear space on a surface homeomorphic to the Möbius strip is called a *point Möbius strip plane* if it is the restriction of a flat projective plane to the complement of a point p.

A line in such a plane is an **R**- or **S**-line depending on whether it arises from a line in the projective plane through p or not through p, respectively. From the remarks we made about flat projective planes it is clear that two **R**-lines never intersect and, more strongly, that the set of **R**-lines of a point Möbius strip plane forms a partition of its point set. Furthermore, since two lines in a flat projective plane intersect in exactly one point, every line in a point Möbius strip plane intersects an **S**-line in exactly one point.

Conversely, any linear space on a Möbius strip whose **R**-lines form a partition of the Möbius strip is a point Möbius strip plane. The point set of the corresponding flat projective plane is the one-point compactification of the Möbius strip by an additional point ∞. Its lines are the **S**-lines of the linear space plus the **R**-lines that have been extended by ∞.

The classical example of a point Möbius strip plane is the restriction of the classical flat projective plane to the complement of a point.

1.2.2 Flat Circle Planes

The *flat circle planes* are geometries living on surfaces whose circles are topological circles, that is, closed subsets of the point sets homeomorphic to $\mathbf{S}^1$. In Figure 1.6 the known types of flat circle planes are represented by one icon each. You have to think of this diagram as being continued to the right as indicated. This means that we are really dealing with an infinite number of different types of geometries.

As you can see, the surfaces these geometries live on are the Möbius strip, the cylinder, the sphere, the torus, and the real projective plane. Depending on the type of surface the geometry lives on, it only has the trivial, one nontrival, or two nontrivial parallelisms of points. We describe these parallelisms by specifying their parallel classes.

- A flat circle plane living on the real projective plane or the sphere has only the trivial parallelism.
- A flat circle plane living on the Möbius strip or the cylinder has one nontrivial parallelism whose parallel classes are the verticals on this surface.
- A flat circle plane living on the torus has two parallelisms. The parallel classes are the horizontal and vertical circles on the torus.

The number n of solid black dots in one of the icons signifies that the corresponding geometry satisfies the following axiom of joining.

Axiom of Joining for Flat Circle Planes

(J) Every n pairwise nonparallel points are contained in a unique circle.

We call n the *rank* of the geometry. Note that parallel points are never contained in a circle, whereas points that are not parallel are always connected by a circle.

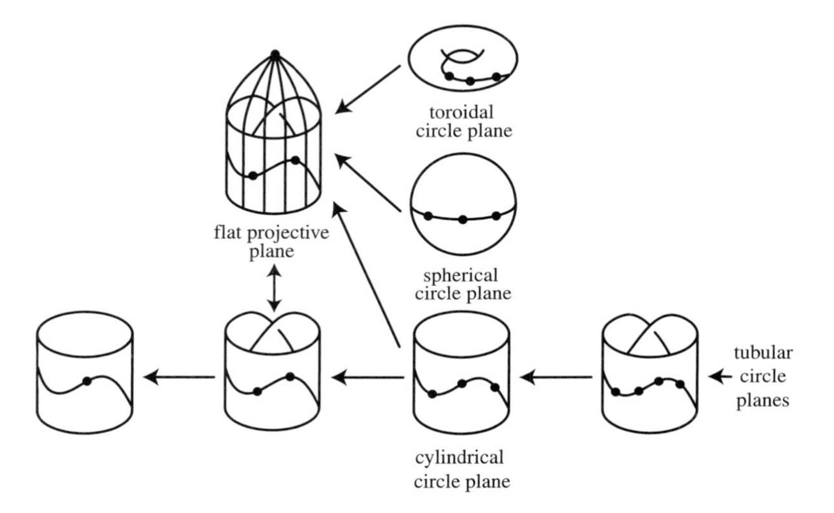

Fig. 1.6. The different types of flat circle planes

The different types of geometries are grouped around the following *classical examples.*

- *Flat projective planes:* The classical projective plane over the real numbers; see Subsection 1.2.1.
- *Spherical circle planes:* The classical flat Möbius plane, that is, the geometry of Euclidean circles on a sphere.
- *Toroidal circle planes:* The classical flat Minkowski plane, that is, the geometry of nontrivial plane sections of a hyperbolic quadric in real projective 3-space. Remember that a (projective) plane intersects a hyperbolic quadric in either a circle or a pair of intersecting projective lines. In the second case we say that the plane intersects the

quadric trivially, otherwise nontrivially. The set of lines on the hyperbolic quadric that correspond to trivial plane sections is the disjoint union of two *reguli*, that is, sets of lines that partition the set of points of the quadric. If you have problems visualizing all this, remember that a hyperbolic quadric minus one of the nontrivial plane sections corresponds to a hyperboloid in Euclidean 3-space. Also, as a topological space a hyperbolic quadric is a torus, that is, homeomorphic to $\mathbf{S}^1 \times \mathbf{S}^1$, and we can identify the two spaces via a homeomorphism in such a way that the two reguli correspond to the horizontals and verticals on the torus. Under this identification nontrivial plane sections correspond to graphs of homeomorphisms $\mathbf{S}^1 \to \mathbf{S}^1$ considered as subsets of the torus $\mathbf{S}^1 \times \mathbf{S}^1$.

- *Tubular circle planes of rank n:* The (topological) projective extension of the geometry that corresponds to the set of polynomials of degree at most $n-1$ over the real numbers. As an abstract geometry (without any topology) this is the geometry $\overline{\text{Poly}}(n, \mathbf{R})$ that we introduced in the section on orthogonal arrays; see p. 8. The point set of this (topological) geometry is the Möbius strip or the cylinder, depending on whether the rank n is even or odd. For example, $\overline{\text{Poly}}(2, \mathbf{R})$ is the projective plane over the reals from which a point and all lines through this point have been removed. In fact, considered as a surface only, the real projective plane minus a point is a Möbius strip. On the other hand, $\overline{\text{Poly}}(3, \mathbf{R})$ is the classical flat Laguerre plane and really lives on the cylinder, as it can be represented as the geometry of nontrivial plane sections of a cylinder over a circle in the xy-plane. Similarly, the classical tubular circle plane of rank 1 can be considered as the geometry of horizontal plane sections of a vertical cylinder.

It is the three different types of circle planes of rank 3 that are usually referred to as *the* flat circle planes. Their classical examples arise as the geometries of nontrivial plane sections of the three different kinds of nondegenerate quadrics in real projective 3-space and are representatives of the three different kinds of Benz planes.

1.2.3 A Network of Relationships

We already mentioned that there exists a tightly knit network of relationships between the different types of geometries on surfaces. Let us have a look at some of these relationships.

1.2.3.1 Nested Flat Circle Planes

It is possible that a flat circle plane has a much richer local structure than implied by Axiom J. To every point of one of the classical circle planes of rank $n > 2$, for example, there is associated a classical circle plane of rank $n-1$. The single arrows in Figure 1.6 indicate what type of circle plane this associated plane is. Consider, as a concrete example, the geometry of circles on the sphere. The derived plane at one of its points is the Euclidean plane and the flat circle plane associated with the point is the projective extension of the Euclidean plane, that is, the classical flat projective plane.

All this means that the classical flat circle planes have a nested structure. If a flat circle plane has a local structure that is similar to that of its classical counterpart, we call it *nested*. All flat circle planes of rank at most 2 turn out to be nested. Not all flat circle planes of higher rank are nested and the ones of rank 3 that are have special names. Nested toroidal circle planes are called *flat Minkowski planes*, nested spherical circle planes are called *flat Möbius planes*, and nested cylindrical circle planes are called *flat Laguerre planes*. These geometries are really Benz planes as we defined them in Subsection 1.1.3.

1.2.3.2 Other Connections

Apart from the connections that we just described, there exist a multitude of other connections between flat circle planes and between flat circle planes and other topological geometries. For example, the derived plane of a spherical circle plane at a point is an $\mathbf{R}^2$-plane and the geometry of great circles on the unit sphere is the 'double cover' of the classical flat projective plane and a subgeometry of the geometry of circles on the unit sphere. Further geometries associated with spherical circle planes include certain generalized quadrangles, semibiplanes, flat Laguerre planes, and flat Minkowski planes.

1.2.4 Interpolation and the Axiom of Joining

Clearly, Axiom J has a lot to do with Lagrange interpolation. Depending on whether n is odd or even, the circle set of a tubular circle plane of rank n can be interpreted as a system of continuous periodic or half-periodic functions defined over some interval that solves the Lagrange interpolation problem of order n over the interval. For example, after identifying the unit circle with the interval $[0, 2\pi)$ in the usual manner, the classical tubular circle plane of rank 3 can be easily seen to correspond to the set of periodic trigonometric polynomials

$$[0, 2\pi) \to \mathbf{R} : x \mapsto a + b \sin x + c \cos x,$$

where $a, b, c \in \mathbf{R}$. Also, for a tubular circle plane to be nested just means that it solves a topological version of the Hermite interpolation problem. In particular, a system of $n-1$ times continuously differentiable periodic or half-periodic functions that solves the Hermite interpolation problem of order n corresponds to a nested tubular circle plane of rank n; see Theorem 7.2.7.

1.2.5 Convexity

There are a number of equivalent definitions of convexity in the Euclidean plane. For example, define a set to be convex if the connecting line segment of any two of its points is completely contained in the set. The whole theory of convexity in the Euclidean plane can be derived from this definition.

Starting with the same definition of convex sets in any $\mathbf{R}^2$-plane, it is possible to derive a theory of convexity that is basically the same as the one that has been derived for the classical geometry.

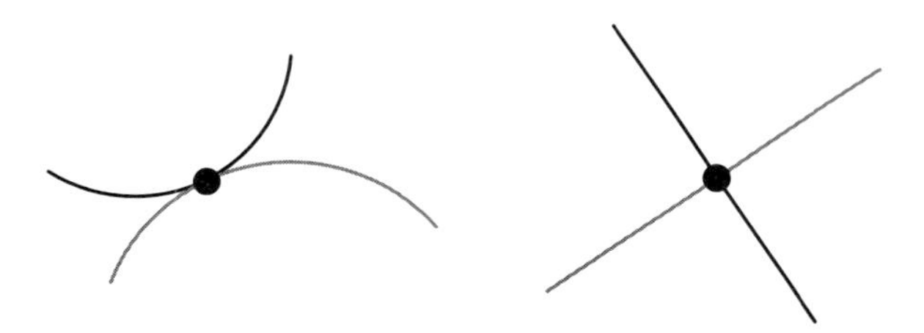

Fig. 1.7. Touching curves and curves that intersect transversally

One of the results that follow in the course of developing such a theory is that two lines in an $\mathbf{R}^2$-plane *intersect transversally* just like lines in the Euclidean plane. They never just *touch topologically*; see Figure 1.7 for the picture to keep in mind and p. 22 for the precise definitions of the two terms.

Remember also that the axiom of joining satisfied by $\mathbf{R}^2$-planes implies that two lines never intersect in more than one point. More generally, in one of our rank n geometries two lines or circles intersect in at most $n-1$ points, and, if they intersect in this maximal number of points, then they intersect transversally in every single one of the points.

If they intersect in less than the maximal number of points, they may also touch topologically in some of these points. For example, the rank of the geometry of circles on the sphere is 3. Therefore, if two circles intersect in two points, they intersect as shown in Figure 1.8 on the left. If two circles intersect in only one point, then they touch topologically at this point.

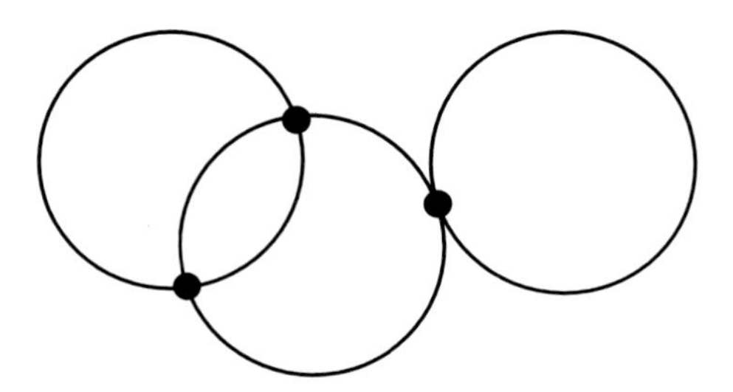

Fig. 1.8. Two circles that intersect in the maximal number of points intersect transversally. If they intersect in less than the maximal number of points, they may touch topologically.

Various notions of convexity with respect to polynomials have been explored that generalize the notion of convexity with respect to the linear functions, or equivalently, the Euclidean plane. These notions of convexity can be generalized even further to provide notions of convexity for higher-rank circle planes. Useful results that can be derived in this context include the above result about the ways lines intersect in such planes, and important cut-and-paste construction principles that allow combining two or more planes of the same type into new planes.

1.2.6 Topological Geometries on Surfaces

The interplay between geometry and topology is one of the most attractive and important features of the theory of the geometries on surfaces that we focus on in this book.

1.2.6.1 Continuity of the Geometric Operations

All our geometries are *topological geometries* in that the geometric operations in the axioms that they satisfy are continuous. For example, in a flat circle plane of rank n the connecting circle of n pairwise nonparallel points changes its shape and position continuously as the n points are moved about continuously.

1.2.6.2 Line Sets Are Line Spaces

Our geometries are also topological in the sense that it is possible to turn the line or circle sets of a given geometry into a topological space in a natural way using the notion of convergence in the sense of Hausdorff. The spaces we arrive at in this manner from geometries of the same type can often be shown to be homeomorphic. For example, when provided with the natural topology, the line set of any flat projective plane is homeomorphic to the real projective plane and the circle set of a tubular circle plane of rank n is homeomorphic to $\mathbf{R}^n$.

Remember that the classical geometries of the classes of geometries we are dealing with in this book can be defined over many fields. Of course, the classical examples over the complex numbers are also 'topological' geometries, and it makes sense to investigate close relatives of these geometries. We also know a lot about these kinds of topological geometries. See Salzmann et al. [1995] for exhaustive information about compact projective planes, that is, the topological projective planes that are close relatives of the projective planes defined over the real numbers, the complex numbers, the quaternions, and the octonions. See Steinke [1995] and Grundhöfer–Löwen [1995] for surveys dealing with different types of topological geometries.

1.2.7 Classification with Respect to the Group Dimension

It is natural to try to classify the 'most homogeneous' geometries of a certain type. Of course, we first have to make up our minds what we mean by 'most homogeneous'. When we are dealing with finite geometries it is natural to classify the geometries with large automorphism groups, that is, geometries that admit a large number of symmetries. In the case of our geometries something similar is possible. For many of our geometries it has been shown that automorphisms are homeomorphisms of the point spaces and that the groups of automorphisms of the geometries are *Lie transformation groups* of finite dimensions. We call the dimension of the automorphism group of a geometry its *group dimension.* Both the Euclidean plane and the geometry of circles on a sphere have group dimension 6. For example, the automorphisms of the Euclidean plane are the maps of the form

$$\mathbf{R}^2 \to \mathbf{R}^2 : x \mapsto Ax + t,$$

where A is an invertible 2×2 matrix with real entries and $t \in \mathbf{R}^2$.

It has also been shown for many of the different types of geometries that the classical geometries are characterized among their close relatives by the fact that they have maximal group dimension. Furthermore, since all relevant Lie groups of finite dimensions are classified, it is possible to classify all geometries of the same type having a large group dimension by first figuring out how the higher-dimensional Lie groups can act as automorphism groups of the given type of geometry, and then reconstructing all possible geometries from the possible group actions.

1.3 Definitions of Frequently Used Terms

In the following we define the most important geometrical terms that we will use in this book. Frequently used terms and results from topology, analysis, and the theory of Lie groups are summarized in the appendices.

Two or more points of a geometry are *collinear* or *concircular* if they are all contained in some line or circle, respectively. A *flag* or *antiflag* of a geometry is a pair (p, L) consisting of a point p and a line L such that $p \in L$ or $p \notin L$, respectively.

The point or line sets of a geometry may be equipped with some special equivalence relations, called *parallelisms.* Two points or lines are *parallel* if they are in relation for such an equivalence relation. A parallelism is called *trivial* if two points or lines are parallel with respect to it if and only if they coincide. A nontrivial parallelism on the point set has the property that two distinct points that are *parallel* with respect to it are not connected by a line. A nontrivial parallelism on the line set has the property that two distinct lines that are *parallel* with respect to it do not intersect in a point. The equivalence classes of a parallelism are called *parallel classes.* The usual parallelism of lines in the Euclidean plane is an example of a nontrivial parallelism on the line set of a geometry. Its parallel classes are the usual parallel classes of Euclidean lines.

A *pencil* of lines associated with a set of one or more points of a geometry is the set of all lines containing all the points in the given set. The points in the set are called the *carriers* of the pencil. Pencils of circles are usually referred to as *bundles* of circles.

Given a point p in a geometry $\mathcal{G}$, the *derived geometry of* $\mathcal{G}$ *at the point* p has a point set that consists of all points of $\mathcal{G}$ that are not parallel to p or different from p depending on whether the point set of $\mathcal{G}$ is equipped with a parallelism or not. The lines of the derived geometry are the lines in $\mathcal{G}$ through p that have been punctured at p. If the point set of $\mathcal{G}$ is not equipped with a parallelism, the *derived plane of* $\mathcal{G}$ *at the point* p coincides with the derived geometry at this point. If the point set of $\mathcal{G}$ is equipped with a parallelism, the *derived plane of* $\mathcal{G}$ *at the point* p is the derived geometry at p whose line set has been augmented by the nontrivial parallel classes not containing p that have been punctured at all points parallel to p.

A map from the point set of one geometry to the point set of a second geometry is *collineation-preserving* if it maps sets of collinear points to sets of collinear points. It is an *isomorphism* if, in addition, it is bijective, maps lines to lines, and induces a bijection between the line sets. An isomorphism between two geometries with parallelisms is also required to map parallel classes to parallel classes. An *automorphism* of a geometry is an isomorphism from the geometry to itself. Automorphisms are also called *collineations*. The set of automorphisms of a geometry is a group. It is called the *automorphism* or *collineation group* of the geometry. Most of the geometries in this book have point sets that are topological spaces. We call two such geometries *topologically isomorphic* if there is a homeomorphism between the point sets of the two geometries that is an isomorphism of the geometries.

Sometimes geometries will be given in a more abstract form, where the point and line sets are disjoint sets, together with an *incidence relation*, that is, a rule which defines when a point is *incident with*, or equivalently, *contained in* a line. Of course, in a geometry that is given as above, with lines being subsets of the point set, this incidence relation is just containment. Usually, a geometry given in the abstract form can be turned into the more elementary form by replacing every abstract line with the set of points incident with it. When we execute this replacement, we only have to make sure that there are no repeated lines, that is, two or more lines that are incident with exactly the same points. Clearly, if there are repeated lines, then we will have lost some information about the geometry after having turned it into the elementary form. However, there will not be any repeated lines in any of our geometries, so let us not worry about this any longer.

Given a geometry, we construct its *dual* by exchanging the roles of points and lines, that is, the points and lines of the dual geometry are the lines and points of the original geometry with (new) points and lines being incident if they were incident in the original geometry. A geometry is *self-dual* if it is isomorphic to its dual and an isomorphism between a geometry and its dual is called a *duality*. A duality of a geometry with point set P and line set $\mathcal{L}$ can be described as a bijection π of the disjoint union of P and $\mathcal{L}$ to itself that exchanges P and $\mathcal{L}$. If π is an involution, that is, if π^2 is the identity, then the duality that corresponds to π is called a *polarity*.

A geometry is called *finite* or *infinite* depending on whether its point set contains a finite or infinite number of points. The geometries we are focusing on in this book are probably the most well-behaved infinite geometries. Their point sets are well-known surfaces such as the Euclidean plane, the real projective plane, the sphere, the torus, the Möbius strip, and the cylinder. Their lines, blocks, or circles are curves homeomorphic to the 1-sphere or some interval and are nicely embedded in these surfaces.

Often we will simultaneously introduce a geometry $\mathcal{G}$, its point set P, and its line set $\mathcal{L}$ by saying: 'Let $\mathcal{G} = (P, \mathcal{L})$ be ....' Also, if in $\mathcal{G}$ incidence I between points and lines is not just containment, then we will often introduce $\mathcal{G}$ by saying: 'Let $\mathcal{G} = (P, \mathcal{L}, I)$ be ....'

Let $\mathcal{G}$ be a geometry with a point- and line-transitive collineation group. This just means that none of its points or lines plays a distinguished role. Let $\mathcal{G}'$ be the geometry that shares all points and lines with $\mathcal{G}$ except for one line. We say that $\mathcal{G}$ *closes* in a geometry $\mathcal{H}$ if and only if all collineation-preserving maps from the point set of $\mathcal{G}'$ to the point set of $\mathcal{H}$ extend to collineation-preserving maps from $\mathcal{G}$ to $\mathcal{H}$. Note that this may also involve maps that are not injective. Many of the classical geometries are characterized by the fact that certain small geometries close in them. For example, the classical projective planes are characterized by the fact that Pappus' configuration closes in them; see Subsection 1.1.1.

Here are some more definitions that we will need in the following. A *topological circle* or *topological line* on a surface is a closed subset of the surface homeomorphic to $\mathbf{S}^1$ or $\mathbf{R}$, respectively. We also refer to topological circles and lines as $\mathbf{S}$- and $\mathbf{R}$-lines, respectively. A *topological sphere* and a *topological disk* are topological spaces homeomorphic to a 2-sphere and $\mathbf{R}^2$, respectively.

Let C_1, C_2 be two Jordan curves on a surface that have the point p in common. We say that the curves *touch each other topologically* at the point p if there are a neighbourhood U of p and a homeomorphism between U and $\mathbf{R}^2$ that maps $C_1 \cap U$ to the x-axis and $(C_2 \cap U) \setminus \{p\}$ to a subset of the upper half-plane. We say that the curves *intersect each other transversally* at p if there are a neighbourhood U of p and a homeomorphism between U and $\mathbf{R}^2$ that maps $C_1 \cap U$ to the x-axis and $C_2 \cap U$ to the y-axis.

2
Flat Linear Spaces

We start this chapter by deriving the most important results about the ideal flat linear spaces, that is, the flat projective planes, $\mathbf{R}^2$-planes, and point Möbius strip planes. In Section 2.5 we then describe how these results generalize to general flat linear spaces.

To provide full proofs of all results listed below is beyond the scope of an Encyclopedia volume such as this. Also, Salzmann et al. [1995] is an excellent source for many of these results and proofs. Therefore, it was very tempting to restrict ourselves to just listing results and to referring the reader to that book and the literature about flat linear spaces for proofs. However, to make our book as self-contained as possible and to give the reader a good feel for why flat linear spaces behave in the way they do, we have included sketches of proofs for some of the most important results that illustrate the main arguments used to prove them. In particular, we show that the line set of an ideal flat linear space carries a natural topology, that automorphisms of such a geometry are continuous, and that the automorphism group of such a geometry is a finite-dimensional Lie transformation group. On the other hand, our exposition of the group-dimension classification of the flat linear spaces is purely descriptive and we do refer to the relevant literature for proofs. Highlights include the description of all flat projective planes whose automorphism groups are at least 2-dimensional, and all Möbius strip planes and $\mathbf{R}^2$-planes whose automorphism groups are at least 3-dimensional.

In the last part of this chapter we deal with flat linear spaces that are special in one way or another, the Lenz–Barlotti classification of flat projective planes, and geometries closely related to flat linear spaces.

Very good books and articles dealing with various aspects of flat linear spaces are available. In particular, Salzmann et al. [1995] is a must-read for everybody interested in flat affine and projective planes and their higher-dimensional relatives. We also recommend Grundhöfer–Löwen [1995] for an overview and Salzmann [1967b] for the most important results about $\mathbf{R}^2$-planes. Many of the proofs in this chapter are adapted from Salzmann [1967b] and Salzmann et al. [1995].

Conventions: Unless specified otherwise, throughout this chapter we will assume that all $\mathbf{R}^2$-planes live on $\mathbf{R}^2$, that all flat projective planes live on the real projective plane, and that all point Möbius strip planes live on a punctured real projective plane.

If A is a flat affine plane, $\overline{A}$ denotes the projective extension of A. Note that by Theorem 2.3.8 this projective plane is isomorphic to a uniquely determined flat projective plane.

When we speak about the *real projective plane* we have the topological space in mind. On the other hand, the *classical flat projective plane* refers to the point–line geometry $\mathrm{PG}(2, \mathbf{R})$.

Given two points p and q in a linear space, the line connecting p and q is denoted by pq. If K and L are two lines that intersect, their point of intersection is denoted by $K \wedge L$.

2.1 Models of the Classical Flat Projective Plane

All classical ideal flat linear spaces are subgeometries of the classical flat projective plane $\mathrm{PG}(2, \mathbf{R})$. The Euclidean plane arises by restricting the flat projective plane $\mathrm{PG}(2, \mathbf{R})$ to the complement of one of its lines and the classical point Möbius strip plane is the restriction of $\mathrm{PG}(2, \mathbf{R})$ to the complement of a point.

The $\mathbf{R}^2$-planes, point Möbius strip planes, and flat projective planes have point sets that are homeomorphic to $\mathbf{R}^2$, the Möbius strip, and the real projective plane, respectively. While the first two surfaces are embeddable in $\mathbf{R}^3$ without self-intersections, the third is not. This means that it is relatively easy to visualize $\mathbf{R}^2$-planes and point Möbius strip planes, whereas visualizing flat projective planes requires a little bit more effort.

We start this chapter by introducing a number of simple models of $\mathrm{PG}(2, \mathbf{R})$ that are easy to visualize. We also indicate some ways in which these models can be modified to give models of nonclassical flat projective planes.

In Subsection 1.1.1 we encountered a first model of the classical flat projective plane. The points and lines in this model are the Euclidean lines and Euclidean planes of the 3-dimensional Euclidean space containing the origin. This model is equivalent to the usual description of this projective plane in terms of homogeneous coordinates.

Given a surface S in $\mathbf{R}^3$ that does not contain the origin, we define two geometries. The points and lines of the first geometry are the points of S and the intersections of S with the Euclidean planes containing the origin, respectively. If some of the Euclidean lines through the origin intersect S in more than one point, we define a second geometry by identifying all points on S that are contained in the same Euclidean line through the origin.

The most popular representations of the classical flat projective plane arise in this manner and are described in the following three subsections. We begin every subsection by specifying which surface S we want to consider.

2.1.1 The Euclidean Plane Plus Its Line at Infinity

The surface S is a plane in $\mathbf{R}^3$ that does not contain the origin.

The points and lines of the first geometry associated with S are the points of S and the Euclidean lines contained in S, respectively.

Of course, this is just a copy of the Euclidean plane. Its line at infinity corresponds to the plane through the origin of $\mathbf{R}^3$ that is parallel to S and the points at infinity correspond to the lines through the origin in this plane.

On the other hand, by restricting a flat projective plane to the complement of one of its lines, we arrive at a flat affine plane. It turns out that two flat projective planes that yield topologically isomorphic flat affine planes in this manner are topologically isomorphic themselves; see Theorem 2.3.8. Furthermore, all flat affine planes arise from flat projective planes and contain all the information about these flat projective planes necessary to reconstruct them.

2.1.2 The Geometry of Great Circles and the Disk Model

The surface S is the unit sphere $\mathbf{S}^2$ centred at the origin.

The points and lines of the first geometry are the points of S and the great circles. Two points on S are contained in the same line through the origin if and only if they are antipodal.

We construct the second geometry by identifying antipodal points. Of course, this geometry is isomorphic to the classical flat projective plane.

A model of this geometry lives on the upper closed hemisphere of $\mathbf{S}^2$. Topologically speaking this hemisphere is a closed disk. The points of the model are the points of the upper open hemisphere plus pairs of antipodal points on the equator.

We stereographically project the upper hemisphere of $\mathbf{S}^2$ from its south pole onto the unit disk in the xy-plane to arrive at a representation of the classical flat projective plane on the closed unit disk. Its points are the interior points of the disk plus all pairs of antipodal points on the unit circle $\mathbf{S}^1$. Its lines are the unit circle plus the Euclidean line and circle segments that connect antipodal points of the unit circle.

The restriction of this geometry to the interior of the disk is a representation of the Euclidean plane. The line at infinity of this Euclidean plane is the unit circle. A parallel class of lines corresponds to the set of Euclidean line and circle segments through two antipodal points on the unit circle. Figure 2.1 shows an embedding of Desargues' configuration in this 'disk' model of $\mathrm{PG}(2, \mathbf{R})$.

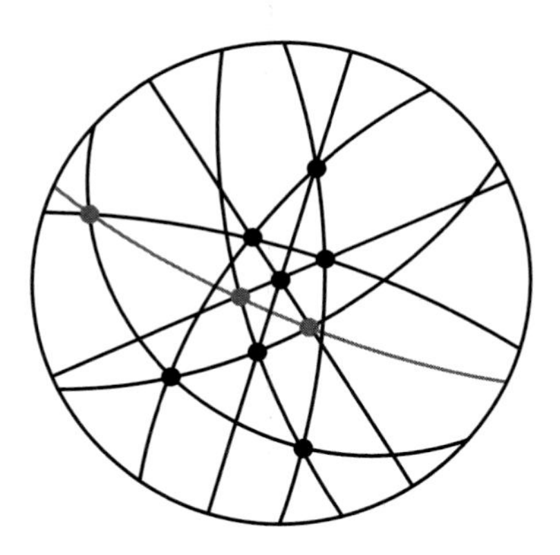

Fig. 2.1. The disk model

In general, a *disk model of a flat projective plane* is a geometry on a closed topological disk D together with a fixed-point-free involutory homeomorphism γ of the boundary of D to itself. Its points are the interior points of the disk together with the set of pairs of boundary points that get exchanged by γ. Its line set consists of the boundary of D and a collection of Jordan curves in D every single one of which has two endpoints which get exchanged by γ. Furthermore, the geometry satisfies Axiom P1. The topological space we arrive at by identifying the boundary of the topological disk via γ is a surface homeomorphic to the real projective plane. The line set of our geometry turns into a set of topological circles on this surface which is the line set of a flat projective plane. In this way a disk model is a special representation of this particular flat projective plane. Also, every flat projective plane has a disk model. See Subsection 3.8.1 for more information about disk models.

If we replace the antipodal map in the construction of the disk model of the classical plane by an arbitrary fixed-point-free involutory homeomorphism of the unit circle to itself, then the constructed geometry is the disk model of a flat projective plane; see Subsection 3.8.1. In general, projective planes constructed like this will not be classical. Consider again Desargues' configuration in Figure 2.1. It is easy to construct an involutory fixed-point-free homeomorphism γ' of the unit circle to itself which, just like the antipodal map, exchanges the ends of the solid black segments in the diagram but does not exchange the end points of the grey segment. As before, the lines in the disk model associated with γ' are the restrictions to the unit disk of all those Euclidean lines and circles that contain pairs of points that get exchanged by γ'. This means that the associated flat projective plane still contains the solid black part of the configuration, but does not contain the grey segment. Since the three grey points are contained in exactly one Euclidean line or circle, Desargues' configuration does not close in the new flat projective plane. Hence this plane is not classical.

2.1.3 The Classical Point Möbius Strip Plane

The surface S is the vertical cylinder over the unit circle $\mathbf{S}^1$ in the xy-plane.

The points of the first geometry are the points of the cylinder. Its lines are pairs of (vertical) Euclidean lines contained in the cylinder plus a selection of ellipses on the cylinder. This geometry is naturally isomorphic to the restriction of the geometry of great circles on the unit sphere $\mathbf{S}^2$ to the complement of the set consisting of the north and south poles.

If we parametrize $\mathbf{S}^1$ by $[0, 2\pi)$ in the usual way, then the ellipses on the cylinder are the graphs of the periodic functions

$$[0, 2\pi) \to \mathbf{R} : t \mapsto a \sin t + b \cos t,$$

where $a, b \in \mathbf{R}$. Two points on the cylinder are *antipodal* if their connecting line contains the origin. By identifying antipodal points on the cylinder, we arrive at a model of the classical point Möbius strip plane. Its point set is the Möbius strip M, which is the strip $[0,\pi] \times \mathbf{R}$, whose left and right boundaries have been identified via the (antipodal) map

$$\{0\} \times \mathbf{R} \to \{\pi\} \times \mathbf{R} : (0, y) \mapsto (\pi, -y).$$

The lines of the point Möbius strip plane are the verticals and the graphs of the functions

$$[0, \pi) \to \mathbf{R} : t \mapsto a \sin t + b \cos t,$$

where $a, b \in \mathbf{R}$.

The set of continuous half-periodic functions above solves the Lagrange interpolation problem of order 2. We will see in Chapter 7 that any such set of functions gives rise to a point Möbius strip plane whose point set is M and whose lines are the verticals in M plus all graphs of functions in the interpolating set.

Note also that, in general, sets of continuous functions that solve the Lagrange interpolation problem of some order correspond to geometries that we are focusing on in this book.

2.2 Convexity Theory

In this section we summarize the basics of a general theory of convexity for $\mathbf{R}^2$-planes and related results for ideal flat linear spaces.

We first note that lines in an $\mathbf{R}^2$-plane are embedded in $\mathbf{R}^2$ just like a Euclidean line, that is, there are no 'wild' embeddings. More precisely, given a line in an $\mathbf{R}^2$-plane, there is a homeomorphism of $\mathbf{R}^2$ to itself that maps this line to a Euclidean line. This is an easy corollary of the Jordan–Schoenflies Theorem; see Theorem A1.2.4.

The complement of every line L in an $\mathbf{R}^2$-plane $\mathcal{R}$ has two open components. These components are called the *open half-planes* defined by the line and the topological closure $\overline{H} = H \cup L$ of such a half-plane H is called a *closed half-plane*.

For distinct points a and b in the plane $\mathcal{R}$, the *closed interval* $[a, b]$ is the intersection of all connected subsets of the line ab in the plane $\mathcal{R}$ containing a and b. *Open intervals* are defined by $(a, b) = [a, b] \setminus \{a, b\}$.

A subset S of $\mathbf{R}^2$ is *convex* (with respect to $\mathcal{R}$) if $[a, b] \subset S$ for any two distinct points $a, b \in S$. The *convex hull* $H(D)$ (with respect to $\mathcal{R}$) of a subset D of $\mathbf{R}^2$ is the intersection of all convex sets containing D.

Let p_1, p_2, p_3 be three noncollinear points and let H_i be the open half-plane defined by the line $p_j p_k$ such that $\{i, j, k\} = \{1, 2, 3\}$ and $p_i \in H_i$.

Then the *open convex triangle* $T(p_1, p_2, p_3)$ is the intersection of the three half-planes H_i, $i = 1, 2, 3$, and the *closed convex triangle* $\overline{T}(p_1, p_2, p_3)$ is the intersection of the three closed half-planes $\overline{H}_i$, $i = 1, 2, 3$. It can be shown that $\overline{T}(p_1, p_2, p_3)$ is the convex hull of the set consisting of the three points p_1, p_2, p_3, that the boundary of this triangle is the union of the intervals $[p_1, p_2]$, $[p_2, p_3]$, $[p_1, p_3]$, and that the convex open triangle $T(p_1, p_2, p_3)$ is the set of topological interior points of $\overline{T}(p_1, p_2, p_3)$.

In general, if the boundary of the convex hull of n points contains all n points and no three among the n points are collinear, then this convex hull is called a *closed convex n-gon.* The set of interior points of a closed convex n-gon is a *open convex n-gon.*

Given a point in a flat projective plane, the affine plane associated with any line not through this point is an $\mathbf{R}^2$-plane that coincides with the flat projective plane 'close to' the point. This means that locally every flat projective plane behaves like an $\mathbf{R}^2$-plane. Similarly, given a point in a point Möbius strip plane, the restriction of this plane to the complement of an $\mathbf{R}$-line in its point set is an $\mathbf{R}^2$-plane that contains the point p. Also, we want to stress again that point Möbius strip planes are really just flat projective planes minus a point. Keeping these observations in mind, it is easy to translate the various facts about $\mathbf{R}^2$-planes listed below into local results about flat projective planes and point Möbius strip planes.

We say that a set of points in an ideal flat linear space $\mathcal{R}$ has a convexity-related property, such as being convex or being a convex triangle, if it has this property with respect to a certain *sub-$\mathbf{R}^2$-plane* of $\mathcal{R}$, that is, the restriction of $\mathcal{R}$ to an open subset of its point set that is an $\mathbf{R}^2$-plane.

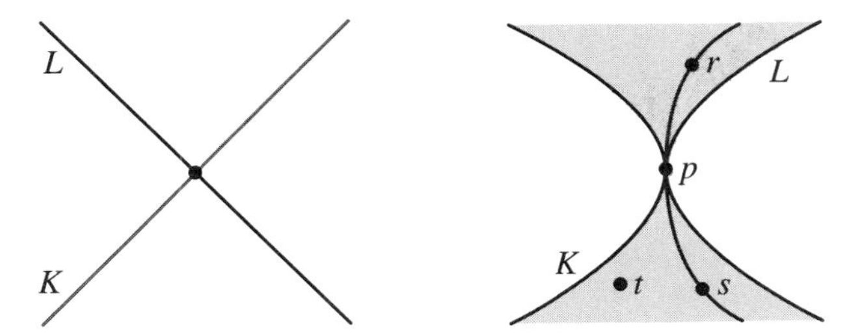

Fig. 2.2. Lines in an $\mathbf{R}^2$-plane intersect transversally as in the left diagram. They never touch as on the right.

We now list a number of important properties of $\mathbf{R}^2$-planes that illustrate that $\mathbf{R}^2$-planes are really very closely related to the Euclidean plane. Perhaps the most important such property is the fact that lines intersect 'transversally' and never 'touch'; see p. 22 for precise definitions of these terms.

THEOREM 2.2.1 (Lines Intersect Transversally) *Let K and L be two distinct lines in an $\mathbf{R}^2$-plane that intersect in the point p. Then K and L intersect transversally.*

Note that this implies that lines in point Möbius strip planes and flat projective planes also intersect transversally.

Proof of theorem. Assume that K and L do not intersect transversally. Then the two lines touch. If H_L^K is the open half-plane determined by L that contains $K\backslash p$ and H_K^L is the open half-plane determined by K that contains $L\backslash p$, let r, s, and t be three noncollinear points in $H_L^K \cap H_K^L$ such that not all three points are contained in the same connected component of $H_L^K \cap H_K^L$ as in Figure 2.2. Then the connecting line of r and s has to be contained in the set $(H_L^K \cap H_K^L) \cup \{p\}$, that is, the shaded region in the diagram (otherwise it would intersect either K or L in at least two points, which is impossible). Since the same has to be true for the line connecting r and t, we conclude that the lines rs and rt have the two points r and p in common, which is a contradiction. $\square$

THEOREM 2.2.2 (Further Basic Results about Convexity) *If $\mathcal{R}$ is an $\mathbf{R}^2$-plane, then the following hold.*

(i) *Intervals and half-planes are convex sets.*

(ii) *Convex sets are connected. The closed intervals contained in a line are exactly the compact, connected subsets of the line.*

(iii) *A line L and a closed interval $[p, q]$ with $p, q \notin L$ are disjoint if and only if p and q lie on the same side of L.*

(iv) *(The Pasch Axiom) Let p, q, r be three noncollinear points. Every line L meeting the set $[p, q] \cup [q, r] \cup [p, r]$ intersects at least two of the intervals in this union. A line $L \neq pq, qr, pr$ meets the open convex triangle $T(p, q, r)$ if and only if L meets one of the open intervals (p, q), (q, r), or (p, r).*

(v) *Every point has a neighbourhood basis consisting of finite intersections of open half-planes.*

(vi) *Every point has a neighbourhood basis consisting of open convex triangles (quadrangles).*

For proofs of these simple results see Salzmann et al. [1995] Proposition 31.5, Corollary 31.7, Proposition 31.10, and Corollary 31.11.

2.2.1 Convex Curves, Arcs, and Ovals

Let S be a set of points in a linear space. A line is an *exterior line*, a *tangent*, or a *secant* of S if it intersects the set in no, 1, or at least two points, respectively. A point not contained in S is a *geometrical interior point* of S if every line containing it is a secant of S. A point not in S is a *geometrical exterior point* if it is contained in an exterior line. An *arc* in a linear space is a nonempty set of points such that no three points in the set are collinear. An arc is called *maximal* if it is not contained in a larger arc. An arc is called an *oval* if every one of its points is contained in exactly one tangent line.

A Jordan curve in an ideal flat linear space is called a *convex curve* if every line intersects the curve in no, 1, or two points, or in an interval. A convex curve is *closed* if it is a simply closed Jordan curve. A convex curve is *strictly convex* or a *topological arc* if none of the lines intersect it in an interval that contains more than one point. A closed topological arc is a *topological oval* if it is an oval. In the Euclidean plane a topological arc is just a strictly convex Jordan curve and a topological oval is just a differentiable, strictly convex, simply closed Jordan curve.

Let A be a topological arc in an ideal flat linear space that is closed or homeomorphic to $\mathbf{R}$ and let p be a point on this arc. Then a line through p is called a *supporting line* of A if it is a tangent of this arc that touches A topologically at p.

Judging from the available results, topological arcs and ovals in ideal flat linear spaces seem to behave exactly like their Euclidean counterparts. We summarize some important results in the following theorem.

THEOREM 2.2.3 (Boundaries of Compact Convex Sets) *Let $\mathcal{R} = (\mathbf{R}^2, \mathcal{L})$ be an $\mathbf{R}^2$-plane, let C be a closed topological arc, and let D be a compact convex set in $\mathbf{R}^2$ whose topological interior is nonempty (this implies that D is not fully contained in a line of $\mathcal{R}$). Then the following hold.*

(i) *The convex hull $H(C)$ of C is the compact convex set consisting of the points of C and the geometrical interior points of C. The topological boundary of $H(C)$ is C.*
(ii) *The topological boundary of D is a closed topological arc B. Furthermore, $H(B) = D$.*

(iii) *The topological interior and exterior points of D are exactly the geometrical interior and exterior points of B, respectively. Every exterior point is contained in exactly two tangent lines.*

(iv) *The arc B has secant lines and exterior lines, and every point in B is contained in a tangent line. Apart from its point of intersection with B every point of such a tangent is an exterior point of B. Every tangent touches B topologically at its point of intersection with the curve. A secant intersects B transversally in its two points of intersection.*

(v) *The arc B is, and, in general, every closed topological arc is, a maximal arc.*

(vi) *Let K and L be secant lines of C. Let k_1, k_2 be the points in which K intersects C and let l_1, l_2 be the points in which L intersects C. Assume that the four points of intersection are distinct. Then K intersects L in an interior point of C if and only if k_1 and k_2 are contained in different connected components of the set $C \setminus \{l_1, l_2\}$. Furthermore, K and L do not intersect or intersect in an exterior point if and only if k_1 and k_2 are contained in the same connected components of $C \setminus \{l_1, l_2\}$.*

All these results are proved in Drandell [1952] or are easy corollaries of the theorems in that paper.

THEOREM 2.2.4 (Topological Ovals) *An ideal flat linear space $\mathcal{R}$ has the following properties.*

(i) *It contains topological ovals.*

(ii) *The complement of a topological oval in $\mathcal{R}$ has two connected components, one of which is homeomorphic to an open disk. The points of the open disk are the geometrical interior points of the oval and the points of the second component are the geometrical exterior points of the oval. Except for their point of intersection with the oval, tangents are made up of exterior points. Every exterior point is contained in two tangents. If $\mathcal{R}$ is a flat projective plane, the second connected component is homeomorphic to a Möbius strip.*

(iii) *If $\mathcal{R}$ is a flat projective plane, the set of tangents of a topological oval is a topological oval in the dual of $\mathcal{R}$ (the dual of $\mathcal{R}$ is also a flat projective plane; see Corollary 2.3.7 below). The exterior lines of the oval are the interior points of the dual oval and the secant lines are the exterior points.*

Proof. (i) Let T be a closed convex triangle in a flat projective plane, or, more precisely, in one of the sub-$\mathbf{R}^2$-planes of this flat projective plane. Then the main result in Polster–Rosehr–Steinke [1997] guarantees the existence of a topological oval in the projective plane that is contained in T. As a consequence of Proposition 2.7.13, given any closed convex triangle T in an ideal flat linear space $\mathcal{R}_1$ there is a flat projective plane $\mathcal{R}_2$ that also contains T such that the restrictions of $\mathcal{R}_i$, $i = 1, 2$, to T coincide. Now it is easy to see that any topological oval in $\mathcal{R}_2$ that is contained in T is also a topological oval in $\mathcal{R}_1$.

Part (ii) is a corollary of Theorem 2.2.3. For a proof of (iii) see Buchanan–Hähl–Löwen [1980]. □

Topological ovals play an important role in the study of flat rank 3 circle planes. Consider, for example, the derived affine plane of a flat Möbius plane at a point p. Every circle in the Möbius plane not passing through this point induces a topological oval in the derived plane. This means that a flat Möbius plane has a representation in terms of a special flat affine plane plus a set of topological ovals in this affine plane. In the classical case this is the geometry of Euclidean lines and circles in $\mathbf{R}^2$. See Chapter 3 for more details about this kind of representation. Note also that most of the facts about topological ovals listed above have counterparts for ovals in finite projective planes of odd order; see Dembowski [1968].

How many arcs can two different ideal flat linear spaces share?

THEOREM 2.2.5 (Many Arcs Determine a Flat Linear Space) *Let $\mathcal{R} = (P, \mathcal{L})$ be an ideal flat linear space and let K be a set of arcs of $\mathcal{R}$. If every three distinct points of P that are not contained in a line of $\mathcal{R}$ are contained in one of the arcs in K, then $\mathcal{R}$ is the only ideal flat linear space on P in which all elements of K are arcs.*

Proof. Let L be some line in $\mathcal{L}$ and let p and q be two points on L. The arcs in K that contain these two points cover all the points of the set $P \setminus (L \setminus \{p, q\})$. Hence L has to be contained in any ideal flat linear space with point set P in which every element of K is an arc. Hence any such ideal flat linear space on P contains all the lines of $\mathcal{L}$, which of course implies that $\mathcal{R}$ is the only such plane. □

Here are two examples for sets of arcs having the property stated in this theorem.

- The set of all topological ovals in a flat projective plane (this is a corollary of the main result in Polster–Rosehr–Steinke [1997]).
- The set of all those circles in the affine part of a spherical circle plane that are topological circles; see Chapter 3.

We conclude this subsection by exhibiting three examples of maximal arcs in the Euclidean plane that are not topological ovals; see Figure 2.3.

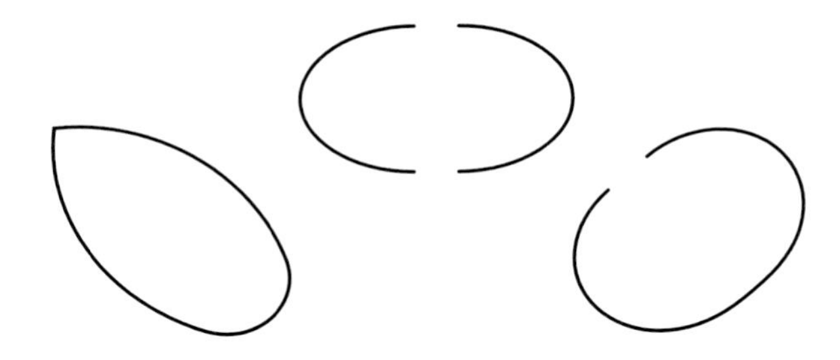

Fig. 2.3. Maximal arcs that are not ovals

Note that the first set is a topological circle and no three of its points are collinear. However, it is not an oval since there are many lines that only contain the distinguished corner. The set on the right is homeomorphic to a closed interval and the connecting line of its two endpoints touches the curve in both points in the analytical sense. The example in the middle consists of two topological arcs that are both homeomorphic to a closed interval.

For more information about convexity in $\mathbf{R}^2$-planes see the references mentioned above, as well as Cantwell [1974], [1978], and Cantwell–Kay [1978].

2.2.2 Dependence of the Axioms of Joining and Intersection

In our definition of flat projective planes we only required that these geometries satisfy Axiom P1; see p. 2. We now show that Axiom P2 is automatically satisfied in a flat projective plane. This means that flat projective planes are really projective planes as we defined them in Chapter 1.

LEMMA 2.2.6 (Complements of Lines) *Let $\mathcal{P} = (P, \mathcal{L})$ be a flat projective plane. If L is a line of $\mathcal{P}$, then $P \setminus L$ is homeomorphic to $\mathbf{R}^2$.*

Proof. The set $P \setminus L$ is connected because, otherwise, the connecting line of two points contained in different components would intersect L in at least two points, which is impossible. Let $\kappa : \mathbf{S}^2 \to P$ be the map that identifies antipodal points on $\mathbf{S}^2$ with the corresponding point of P and let $\overline{L} = \kappa^{-1}(L)$. Then $\overline{L}$ is either the union of two disjoint topological circles or a topological circle. In the first case, as a consequence of Theorem A1.2.4, the set $\mathbf{S}^2 \setminus \overline{L}$ has three connected components, two homeomorphic to $\mathbf{R}^2$ and one homeomorphic to a cylinder. This implies that $P \setminus L$ is not connected, which is contradiction. In the second case the set $\mathbf{S}^2 \setminus \overline{L}$ has two connected components homeomorphic to $\mathbf{R}^2$ and κ defines a homeomorphism between one of them and $P \setminus L$. □

THEOREM 2.2.7 (Joining Implies Intersection) *Let $\mathcal{P} = (P, \mathcal{L})$ be a flat projective plane. Then $\mathcal{P}$ satisfies Axiom P2, that is, any two lines in $\mathcal{P}$ intersect in exactly one point.*

See also Salzmann [1967b] 2.5.

Proof of theorem. Let K and L be two lines in $\mathcal{P}$. Since $\mathcal{P}$ satisfies Axiom P1, the two lines intersect in at most one point. Assume they do not intersect. Then L is completely contained in the set $P \setminus K$ which, by Lemma 2.2.6, is homeomorphic to $\mathbf{R}^2$. We conclude that $P \setminus L$ has two connected components, which is a contradiction to Lemma 2.2.6. □

A geometry of topological circles on the real projective plane that satisfies Axiom P2 is not automatically a flat projective plane (just consider a flat projective plane from which some lines have been removed).

THEOREM 2.2.8 (Local Implies Global) *Let $\mathcal{G} = (P, \mathcal{L})$ be a geometry whose point set P is the real projective plane and whose lines are topological circles in P. Then $\mathcal{G}$ is a flat projective plane if and only if it satisfies the following two conditions.*

(i) *Two distinct lines intersect in exactly one point.*
(ii) *Every point has an open neighbourhood such that G restricted to that neighbourhood is an $\mathbf{R}^2$-plane.*

Proof. If $\mathcal{G}$ is a flat projective plane, then (i) and (ii) are clearly satisfied.

To prove the converse we need to use some terminology and results that will only be introduced and proved in the following sections. Since P is compact, there are finitely many sub-$\mathbf{R}^2$-planes whose point sets form a cover of the point set P. Let p be a point. We have to show that every point q different from p is contained in exactly one of the lines through p. Let $\mathcal{R}$ be one of the $\mathbf{R}^2$-planes in the finite cover that contains p and let r be a point in $\mathcal{R}$ with $r \neq p$. Then there is a unique connecting line $K \in \mathcal{L}$ of p and r. Repeated application of Theorem 2.3.4 yields that this unique connecting line and its point of intersection with a fixed line L, with $p \notin L$ and $q \in L$, depend continuously on r (in the sense of Hausdorff; see the next section for a definition). Equipped with its natural topology the pencil $\mathcal{L}_p$ of lines through p is homeomorphic to $\mathbf{S}^1$. Hence the function $\gamma : \mathcal{L}_p \to L$ which maps a line through p to its unique point of intersection with L is injective and continuous. In fact, since both $\mathcal{L}_p$ and L are homeomorphic to $\mathbf{S}^1$, the function γ is a homeomorphism. Since $q \in L$, there is a line through p and q. As a consequence of (i), this line is unique. We conclude that $\mathcal{G}$ is a flat projective plane. □

An ideal flat linear space $\mathcal{R}$ is *locally classical* if all its points have open neighbourhoods such that $\mathcal{R}$ restricted to such a neighbourhood is an $\mathbf{R}^2$-plane isomorphic to a sub-$\mathbf{R}^2$-plane of the Euclidean plane.

THEOREM 2.2.9 (Locally Classical Equals Classical) *Let $\mathcal{R}$ be a locally classical ideal flat linear space. Then $\mathcal{R}$ is isomorphic to the restriction of the classical flat projective plane to one of its subsets.*

See Polley [1968] for a proof of this result.

2.3 Continuity of Geometric Operations and the Line Space

In this section we show that in an ideal flat linear space the shape and position of the connecting line of two distinct points depend continuously on the positions of the two points. Furthermore, the line space of an ideal flat linear space carries a natural topology that makes the ideal flat linear space into a topological geometry in which the operations of joining and intersection are continuous.

In the proofs we will focus on the case of $\mathbf{R}^2$-planes and sometimes only sketch how the proofs have to be modified or used to arrive at the respective results for point Möbius strip planes and flat projective planes. In many cases results about such linear spaces can be derived by restricting to suitable $\mathbf{R}^2$-planes and then applying the respective result for $\mathbf{R}^2$-planes. Also, since, by definition, point Möbius strip planes are really just 'punctured' flat projective planes, results about point Möbius strip planes are usually straightforward corollaries of the respective results for flat projective planes.

Most of the proofs of the results in this section are modelled after the proofs of the respective results in Salzmann [1967b] and Salzmann et al. [1995].

LEMMA 2.3.1 (Collinearity Is Preserved under Limits) *In an ideal flat linear space $\mathcal{R}$ collinearity and the order of collinear point triples are preserved under limits as follows.*

(i) *If the sequences of points (a_i), (b_i), (c_i) have mutually distinct limits a, b, c, respectively, and if a_i, b_i, c_i are collinear for infinitely many $i \in \mathbf{N}$, then a, b, c are also collinear.*

(ii) *If $\mathcal{R}$ is an $\mathbf{R}^2$-plane and $b_i \in (a_i, c_i)$ for infinitely many $i \in \mathbf{N}$, then also $b \in (a, c)$.*

Proof. Without loss of generality, we may assume that the ideal flat linear space under consideration is an $\mathbf{R}^2$-plane and that $b_i \in (a_i, c_i)$ for $i \in \mathbf{N}$. Now we can follow Salzmann et al. [1995] Proposition 31.12 to prove this result.

(i) Assume that a, b, c are not collinear and choose a line L that intersects the intervals (a, b) and (b, c). Then for sufficiently large i the point b_i and the interval (a_i, c_i) lie on different sides of L, which is a contradiction to Theorem 2.2.2(iii).

(ii) Assume that $b \notin (a, c)$. Then we may also assume that $c \in (a, b)$. If $L \neq ac$ is a line that intersects (b, c), then for sufficiently large i the point b_i and the interval (a_i, c_i) lie on different sides of L, which is again a contradiction to Theorem 2.2.2(iii). □

Let (L_i) be a sequence of lines in an ideal flat linear space. Furthermore, let $\liminf_i L_i$ be the set of all limits of convergent sequences (p_i) with $p_i \in L_i$, and let $\limsup_i L_i$ be the set of all accumulation points of sequences (q_i) with $q_i \in L_i$. We say that (L_i) *converges to* L *in the sense of Hausdorff* if $L = \liminf_i L_i = \limsup_i L_i$. This means that, roughly speaking, the points of L_i approximate precisely the points of L, and that no parts of L_i stay away from L as i tends to infinity.

LEMMA 2.3.2 (Hausdorff Limits of Lines) *In an ideal flat linear space let (a_i) and (b_i) be two convergent sequences of points with limits a and b such that $a_i \neq b_i$ and such that $a \neq b$. Then the sequence of lines (L_i) with $L_i = a_i b_i$ converges to $L = ab$ in the sense of Hausdorff.*

Proof. We prove this result for $\mathbf{R}^2$-planes. The proof for flat projective planes is a variation of the following proof in Salzmann et al. [1995] Proposition 31.14.

Clearly, $\liminf_i L_i \subseteq \limsup_i L_i$. It remains to show that

$$\limsup_i L_i \subseteq L \subseteq \liminf_i L_i.$$

If x is an accumulation point of a sequence of points (x_i) with $x_i \in L_i$, then Lemma 2.3.1(i) guarantees that $x \in L$.

We proceed to prove the second inclusion. Let $c \in L$. We distinguish three cases depending on the position of c on L relative to a and b. First, if $c \in \{a, b\}$, then, clearly, $c \in \liminf_i L_i$.

Second, let $c \in (a, b)$, let $K \neq L$ be a line through c, and let D be a compact convex set that contains (a, b) in its interior. Then, for sufficiently large i, the points a_i and b_i lie on different sides of K and are interior points of D. Hence for sufficiently large i the points $t_i = a_i b_i \wedge K$ exist and are contained in D. We conclude that the sequence (t_i) has an accumulation point and that, by Lemma 2.3.1(i), this accumulation point has to be c. Hence $c \in \liminf_i L_i$.

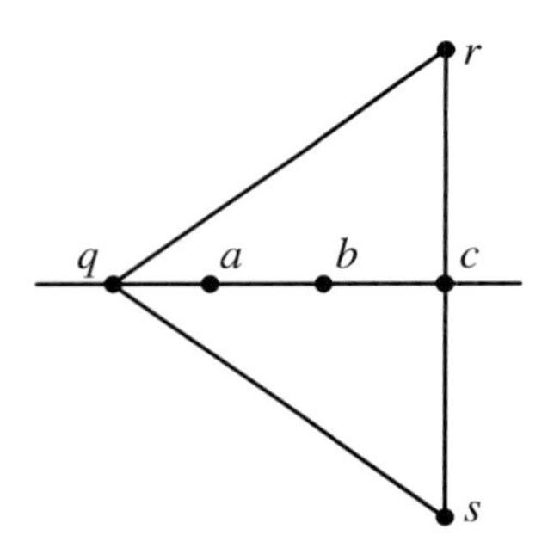

Fig. 2.4.

Third, let $c \in ab \setminus [a, b]$. This final case is dealt with using a configuration of points and lines as in Figure 2.4. Here we may assume, without loss of generality, that $b \in (a, c)$, $L_i \neq L$ for all $i \in \mathbf{N}$, $q \notin L_i$ for all $i \in \mathbf{N}$. If for large enough i the intersection points $t_i = L_i \wedge [r, s]$ exist, we conclude, as above, that (t_i) converges to c. Now assume that for large i the line L_i does not intersect $[r, s]$. For large i the interval (a_i, b_i) is contained in the open convex triangle $T(r, s, q)$ and, by Theorem 2.2.2(iv), L_i intersects the two sides $[q, r]$ and $[q, s]$ of this triangle. Hence we can choose points $t_i \in [q, r] \cup [q, s]$ such that $b_i \in (a_i, t_i)$ (note that $t_i \neq q$, since $q \notin L_i$). The sequence (t_i) accumulates at a point t of the compact set $[q,r] \cup [q,s]$, and Lemma 2.3.1(i) implies that the points t, a, b are collinear. Consequently, $t = q$ and Lemma 2.3.1(ii) implies that $b \in (a, q)$, which is a contradiction. □

The *natural topology* H defined on the line set $\mathcal{L}$ of a given ideal flat linear space is defined as follows. A set A of lines in $\mathcal{L}$ is H-closed if and only if A contains the Hausdorff limit of any convergent sequence in A. The open sets of the topology H are just the complements of the H-closed subsets of $\mathcal{L}$.

Here are a number of other ways to define 'natural' topologies on the line set $\mathcal{L}$ of an ideal flat linear space $(P, \mathcal{L})$. All these topologies can be shown to coincide with the topology H.

The *final topology* F on $\mathcal{L}$ is the finest topology F on $\mathcal{L}$ such that the join map

$$\{(p,q) \in P \times P \mid p \neq q\} \to \mathcal{L} : (p,q) \mapsto pq$$

is continuous. The *open join topology* OJ is generated by the subbasis elements $O_1O_2 = \{pq \mid p \in O_1, q \in O_2\}$, where O_1 and O_2 are disjoint open subsets of the point set P. The *interval join topology* IJ has a subbasis consisting of the sets I_1I_2, where I_1 and I_2 are disjoint open intervals in a sub-$\mathbf{R}^2$-plane that span a convex quadrangle. Finally, the *open meet topology* OM is defined by the subbasis whose elements are the sets $M_O = \{L \in \mathcal{L} \mid L \cap O \neq \emptyset\}$, where O is an open set in P.

THEOREM 2.3.3 (Topology of the Line Space) *Let $(P, \mathcal{L})$ be an ideal flat linear space. Then the following hold.*

(i) *The topologies* H, F, OJ, IJ, OM *for $\mathcal{L}$ coincide.*
(ii) *Equipped with its natural topology $\mathcal{L}$ is a Hausdorff space.*
(iii) *Convergence in $\mathcal{L}$ is equivalent to convergence in the sense of Hausdorff.*

See Salzmann et al. [1995] Proposition 31.19 for a proof of this result for $\mathbf{R}^2$-planes that easily generalizes to flat projective planes.

We mention one more way of generating the natural topology on the line set $\mathcal{L}$ of an $\mathbf{R}^2$-plane; see Busemann [1955] Section 3. See also Section A1.1. For $K, L \in \mathcal{L}$ let

$$d(K,L) = \sup\left\{ |\delta(x,K) - \delta(x,L)| e^{-|x|} \mid x \in \mathbf{R}^2 \right\},$$

where

$$\delta(x, L) = \inf \{ \, |x - y| \mid y \in L \} \, .$$

Then d is a metric on $\mathcal{L}$ whose associated topology is the natural topology on $\mathcal{L}$.

THEOREM 2.3.4 (Continuity of Joining and Intersection) *Let $(P, \mathcal{L})$ be an ideal flat linear space and let $\mathcal{L}$ be endowed with the natural topology. Then the following hold.*

(i) *The join map*

$$\{(p, q) \in P \times P \mid p \neq q\} \to \mathcal{L} : (p, q) \mapsto pq$$

is continuous.

(ii) *The pairs of distinct intersecting lines form an open subset I of $\mathcal{L}^2$, and the intersection map*

$$\wedge : I \to P : (K, L) \mapsto K \wedge L$$

is continuous.

Proof. The following proof is a combination of the proofs of Salzmann et al. [1995] Propositions 31.16 and 31.21.

(i) Assume that the join map is not continuous. Let p and q be distinct points with connecting line L, let O be an open neighbourhood of L, and let $\{P_i \mid i \in \mathbf{N}\}$ and $\{Q_i \mid i \in \mathbf{N}\}$ be neighbourhood bases of p and q, respectively. Without loss of generality, we may assume that P_i and Q_i are disjoint sets for all $i \in \mathbf{N}$ and that the set of lines P_iQ_i is not contained in O for all $i \in \mathbf{N}$. Then there are points $p_i \in P_i$ and $q_i \in Q_i$ such that the line p_iq_i is not contained in O. By Lemma 2.3.2, the sequence of lines (p_iq_i) converges to L in the sense of Hausdorff. Consequently, L belongs to the H-closed set $\mathcal{L} \setminus O$, which is a contradiction.

(ii) To prove that the pairs of distinct intersecting lines form an open subset of $\mathcal{L}^2$ it suffices to show that any two intersecting lines K and L have open neighbourhoods N and M such that every line in M intersects every line in N. Choose a convex quadrangle (p, q, r, s) such that K intersects the disjoint open intervals $K_1 = (p, q)$ and $K_2 = (r, s)$ and L intersects the disjoint open intervals $L_1 = (p, s)$ and $L_2 = (q, r)$. Using the description of the natural topology on $\mathcal{L}$ as the interval join topology IJ it is clear that the sets $N = K_1K_2$ and $M = L_1L_2$ are open neighbourhoods of K and L with the desired property. By Theorem 2.2.2(vi), the open convex quadrangles containing a point form a neighbourhood basis for this point. This implies that the intersection map is continuous on its domain of definition. □

Let $\mathcal{R} = (P, \mathcal{L})$ be an ideal flat linear space whose line set is equipped with some topology, and let I be the set of pairs of distinct intersecting lines. We say that intersection is *stable* if I is an open subset of $\mathcal{L}^2$ and the intersection map is continuous on this subset.

THEOREM 2.3.5 (Characterization of the Natural Topology) *Let $\mathcal{R} = (P, \mathcal{L})$ be an ideal flat linear space. Then the natural topology* H *on $\mathcal{L}$ is the only topology such that the join map is continuous and the intersection map is stable.*

For a proof see Salzmann et al. [1995] Theorem 31.22.

THEOREM 2.3.6 (Line Space and Flag Space) *Let $\mathcal{R} = (P, \mathcal{L})$ be an ideal flat linear space, let L be a line, and let p be a point that is not contained in L. Then $\mathcal{R}$ has the following properties.*

(i) *The pencil of lines through p is homeomorphic to $\mathbf{S}^1$.*

(ii) *The set of all lines through p that do not intersect L is homeomorphic to a nonempty closed interval if L is an* **R***-line, and empty if L is an* **S***-line.*

(iii) *If $\mathcal{R}$ is a flat affine plane, then the set of lines parallel to L is a closed subset of $\mathcal{L}$ homeomorphic to* **R**.

(iv) *If $\mathcal{R}$ is a point Möbius strip plane, then the set of all* **R***-lines is homeomorphic to $\mathbf{S}^1$.*

(v) *If $\mathcal{R}$ is a point Möbius strip plane, or a flat projective plane, then $\mathcal{L}$ is homeomorphic to the real projective plane. If it is a point Möbius strip plane, the set of all* **S***-lines is homeomorphic to $\mathbf{R}^2$.*

(vi) *The set of lines in an $\mathbf{R}^2$-plane is homeomorphic to a Möbius strip.*

(vii) *If $\mathcal{R}$ is a flat projective plane, then the space of all flags of $\mathcal{R}$ is homeomorphic to the flag space of the classical flat projective plane.*

Note that (ii) implies that $\mathbf{R}^2$-planes and point Möbius strip planes cannot be projective planes.

Proof of theorem. (i) Consider an open convex triangle $T(a,b,c)$ that contains the point p and let $B = [a,b] \cup [b,c] \cup [a,c]$ be its boundary. Note that B is homeomorphic to $\mathbf{S}^1$. Then every line through p intersects B in exactly two points. We define an involutory permutation γ of B that exchanges the two points of intersection of lines through p with B. This involution is a homeomorphism that is topologically equivalent to the antipodal map of $\mathbf{S}^1$. We identify points that get exchanged by γ and arrive at a topological space B' homeomorphic to $\mathbf{S}^1$. The map that assigns to every line through p the unique point in B' that corresponds to its points of intersection with B defines a homeomorphism.

(ii) Using the same setup as in (i), we define a map that assigns to every point q of the line L the unique point in B' that corresponds to the points of intersection of pq with B. This map is injective and continuous. Hence the complement of its image is homeomorphic to a nonempty closed interval if L is homeomorphic to $\mathbf{R}$. The map is a homeomorphism if L is a topological circle.

(iii) Let K be a line not in the parallel class E under consideration. Then the function $E \to K$ that maps elements of E to their respective points of intersection with K is a homeomorphism. Using the IJ description of the natural topology on $\mathcal{L}$ it is easy to construct a neighbourhood of K that does not contain any line in the parallel class. This shows that E is a closed subset of $\mathcal{L}$.

(iv) A similar argument as under (iii) yields the homeomorphism between the set of $\mathbf{R}$-lines and the circle $\mathbf{S}^1$.

The following arguments are adapted from Salzmann et al. [1995] Proposition 31.23, Lemma 31.24, and Proposition 32.3.

(v) We start by showing that $\mathcal{L}$ is a compact surface. To begin with let a and b be two distinct points of L. By Theorem 2.3.3(i), the sets IJ, where I and J are disjoint open intervals containing a and b, respectively, and $I, J \not\subset L$, form a neighbourhood basis of L. This together with the continuity of joining points and intersecting lines implies that $\mathcal{L}$ is locally homeomorphic to $\mathbf{R}^2$. Since $\mathbf{R}^2$ has a countable basis, it is possible, using the OM-description of the natural topology, to construct a countable basis for $\mathcal{L}$. Since, by Theorem 2.3.3, the line set $\mathcal{L}$ is a Hausdorff space, it is a surface.

To show that this surface is compact, let (L_i) be a sequence of lines and assume that L is an $\mathbf{S}$-line and different from all L_is. Then the sequence of points of intersection $(L_i \wedge L)$ has an accumulation point c and we may assume that this sequence tends to c as i tends to infinity. Let K be an $\mathbf{S}$-line whose point of intersection with L is distinct from c

and which is different from all L_is. Then the sequence $(L_i \wedge K)$ has an accumulation point d. This implies that the sequence of lines we started with has a subsequence that converges to the line cd. Hence the surface under consideration is compact.

Let K, L, M be three distinct **S**-lines. Let a, b, c, d be distinct points of K such that a and b separate c and d, and such that $c = K \wedge L$ and $d = K \wedge M$. We proceed to show that $\mathcal{L}' = \mathcal{L} \setminus \{K\}$ is a Möbius strip. The line K is the union of two closed intervals I and J whose common boundary points are a and b. Let $c \in I$ and $d \in J$. This implies that $\mathcal{L}'$ is the union of the set $\mathcal{I}$ of all those lines that meet I and the set $\mathcal{J}$ of all lines that meet J in exactly one point. We identify the sets $L \setminus \{c\}$ and $M \setminus \{d\}$ with $\mathbf{R}$, I with the closed interval $[0, 1]$, and J with the closed interval $[2, 3]$ such that 0 and 3 correspond to a and 1 and 2 to b. Now the function that maps a line N in $\mathcal{I}$ to the point $(N \wedge K, N \wedge M)$ is a homeomorphism $\mathcal{I} \to [0, 1] \times \mathbf{R}$. Likewise, the function that assigns a line N in $\mathcal{J}$ the point $(N \wedge K, N \wedge L)$ is a homeomorphism $\mathcal{J} \to [2, 3] \times \mathbf{R}$. This means that $\mathcal{L}'$ can be represented as a quotient space

$$([0, 1] \times \mathbf{R} \cup [2, 3] \times \mathbf{R}) / \sim,$$

where $(1, t) \sim (2, f(t))$ and $(3, t) \sim (0, g(t))$ for all $t \in \mathbf{R}$ with two homeomorphisms $f, g : \mathbf{R} \to \mathbf{R}$. If f and g have the same orientation, then $\mathcal{L}'$ is a cylinder, otherwise it is a Möbius strip. Since $\mathcal{L}$ is the one-point compactification of $\mathcal{L}'$, we can exclude the first possibility (the one-point compactification of a cylinder is a sphere with two points identified, and the compactifying point does not have a neighbourhood homeomorphic to $\mathbf{R}^2$). We conclude that $\mathcal{L}'$ is a Möbius strip and $\mathcal{L}$ a real projective plane.

(vi) The proof of this fact is harder than that of (v) and requires deeper topological results; see Löwen [1995] 3.2. In the special case that the $\mathbf{R}^2$-plane $\mathcal{R}$ under consideration is the restriction of a flat projective plane to one of its convex open disks D, we may argue as follows. Let B be the boundary of D. Then every line of $\mathcal{R}$ arises from a line in the projective plane that intersects B in exactly two points. On the other hand every two points on B determine one such line. Since B is homeomorphic to $\mathbf{S}^1$, this means that the line set of $\mathcal{R}$ is homeomorphic to the topological space of unordered pairs of distinct points on $\mathbf{S}^1$, which can easily be shown to be isomorphic to the Möbius strip.

(vii) See Breitsprecher [1972] and Salzmann et al. [1995] Proposition 32.3. □

COROLLARY 2.3.7 (The Dual of a Flat Projective Plane) *The dual of a flat projective plane $\mathcal{P}$ is a flat projective plane and the dual of the dual of $\mathcal{P}$ is (topologically isomorphic to) $\mathcal{P}$.*

Proof. This follows immediately from Theorem 2.3.6 (i), (v) and Theorem 2.3.5. □

The duals of $\mathbf{R}^2$-planes and point Möbius strip planes are not linear spaces. However, these duals are very close to being point Möbius strip planes and flat projective planes, respectively. See also Section 2.5 for information about 'opposite planes' of flat linear spaces.

Let $\mathcal{M} = (M, \mathcal{L})$ be a point Möbius strip plane. We call the flat projective plane it arises from the *projective extension of* $\mathcal{M}$. To reconstruct this flat projective plane, simply one-point-compactify the Möbius strip M this geometry is living on by a point ∞, and extend all lines that correspond to $\mathbf{R}$-lines of $\mathcal{L}$ by ∞.

Let $\mathcal{A}$ be a flat affine plane. As a consequence of Theorem 2.3.6 (i), (iii), and (vi), the dual of $\mathcal{A}$ whose line set has been augmented by the parallel classes of lines in $\mathcal{A}$ is a point Möbius strip plane. It is clear that the dual of the projective extension of this geometry is a flat projective plane that is isomorphic to the (geometric) projective extension of $\mathcal{A}$ and we will refer to this flat projective plane as the (geometric and topological) projective extension of $\mathcal{A}$.

THEOREM 2.3.8 (Projective Extension) *Let $\mathcal{R}$ be a flat affine plane or a point Möbius strip plane. Let $\mathcal{P}$ be a flat projective plane whose restriction to the complement of a line or point, respectively, is topologically isomorphic to $\mathcal{R}$. Then $\mathcal{P}$ is isomorphic to the projective extension of $\mathcal{R}$.*

See Salzmann et al. [1995] Section 32, for a detailed discussion of this result.

These considerations also show that flat projective planes, flat affine planes, and point Möbius strip planes are basically identical geometrical objects.

2.4 Isomorphisms, Automorphism Groups, and Polarities

In this section we prove that isomorphisms between ideal flat linear spaces are automatically topological isomorphisms and that the automorphism group of an ideal flat linear space carries a natural topology that makes it into a Lie group of dimension at most 8.

Let $\mathcal{R}$ be an ideal flat linear space and let A be a set of its points. We define the subplane $\langle A \rangle$ generated by A to be the smallest subgeometry of $\mathcal{R}$ that is closed under the operations of joining and intersection. Its point set F and line set G can also be defined inductively as follows.

(i) $F_0 = A$.
(ii) $F_{n+1} = F_n \cup \{p_1p_2 \wedge p_3p_4 \mid p_i \in F_n, p_1 \neq p_2, p_3 \neq p_4, p_1p_2 \neq p_3p_4\}$.
(iii) $F = \bigcup_i F_i$.
(iv) G is the set of all intersections of F with all those lines in $\mathcal{R}$ that have at least two points in common with F.

Note that every automorphism of $\mathcal{R}$ that leaves the set A pointwise fixed also fixes every element of the subplane generated by A.

We call the set consisting of the vertices and one interior point of a closed convex triangle a *tripod.*

THEOREM 2.4.1 (Tripods Generate Dense Subplanes) *In an ideal flat linear space $\mathcal{R} = (P, \mathcal{L})$ a tripod generates a dense subplane of $\mathcal{R}$.*

Proof. We prove this result only for $\mathbf{R}^2$-planes with $P = \mathbf{R}^2$ following the exposition in Salzmann [1967b] Theorem 3.4. Given a tripod in a flat projective plane or point Möbius strip plane $\mathcal{S}$, just observe that there is a certain $\mathbf{S}$- or $\mathbf{R}$-line, respectively, such that the tripod is contained in the sub-$\mathbf{R}^2$-plane that is the restriction of $\mathcal{S}$ to the complement of this line in the point set. The tripod then generates a dense subplane of this sub-$\mathbf{R}^2$-plane that is also dense in $\mathcal{S}$.

Let A consist of the vertices of a convex triangle $T = \overline{T}(a_1, a_2, a_3)$ and one of its interior points a_0 (together these four points form a tripod). We want to show that the set of points F of $\langle A \rangle$ is a dense subset of $\mathbf{R}^2$. Each point in $\mathbf{R}^2$ is the point of intersection of two lines intersecting the boundary $[a_1, a_2] \cup [a_2, a_3] \cup [a_1, a_3]$ of T nontrivially. Hence it suffices to show that the intersection of F with this boundary is dense in the boundary.

Without loss of generality, we only show that the set $F \cap [a_1, a_2]$ is dense in $[a_1, a_2]$, or equivalently, that $[a_1, a_2] \setminus \overline{F \cap [a_1, a_2]}$ is empty. Assume that this is not the case. Then this set contains a nonempty open interval $(b_1, b_2) \subset (a_1, a_2)$ whose boundary points are contained in $\overline{F}$.

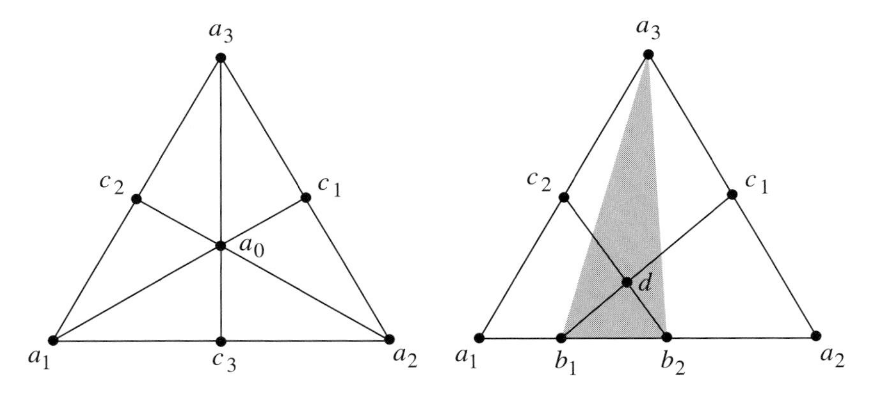

Fig. 2.5.

Let $c_k = a_i a_j \wedge a_0 a_k$ and $d = b_1 c_1 \wedge b_2 c_2$; see Figure 2.5. Then d is contained in the open convex triangle $T(b_1, b_2, a_3)$. By continuity and because b_1 and b_2 are contained in $\overline{F}$, there exist points b_1' and b_2' in F such that $d' = b_1' c_1 \wedge b_2' c_2$ is also contained in the triangle $T(b_1, b_2, a_3)$. But then the point $a_3 d' \wedge [a_1, a_2]$ is contained in both F and (b_1, b_2), which is a contradiction. □

THEOREM 2.4.2 (Isomorphisms Are Continuous) *Isomorphisms between two ideal flat linear spaces $\mathcal{R} = (P_1, \mathcal{L}_1)$ and $\mathcal{S} = (P_2, \mathcal{L}_2)$ are topological, that is, induced by homeomorphisms $P_1 \to P_2$.*

Proof. We prove this result only in the case that both $\mathcal{R}$ and $\mathcal{S}$ are $\mathbf{R}^2$-planes. Without loss of generality, we may assume that $P_1 = P_2 = \mathbf{R}^2$. We follow the exposition in Salzmann [1967b] Theorem 3.5. Given an isomorphism between two point Möbius strip planes or two flat projective planes $\mathcal{T}$ and $\mathcal{U}$, we can then use this result to conclude that the restriction of this isomorphism to a sub-$\mathbf{R}^2$-plane of $\mathcal{T}$ that is the restriction of $\mathcal{T}$ to the complement of a line is induced by a homeomorphism. Since the point sets of these sub-$\mathbf{R}^2$-planes cover the point set of $\mathcal{T}$, it then follows easily that the isomorphism we started with is itself induced by a homeomorphism between the point sets of $\mathcal{T}$ and $\mathcal{U}$.

Since, by Theorem 2.2.2(vi), the open convex triangles of $\mathcal{R}$ form a basis for the topology on $\mathbf{R}^2$, we can restrict ourselves to showing that, given an isomorphism $\gamma : \mathcal{R} \to \mathcal{S}$, we have

$$\gamma(T(a_1, a_2, a_3)) = T(\gamma(a_1), \gamma(a_2), \gamma(a_3))$$

for all open convex triangles in $\mathcal{R}$. Without loss or generality, it suffices to show that $\gamma[a_1, a_2] = [\gamma(a_1), \gamma(a_2)]$.

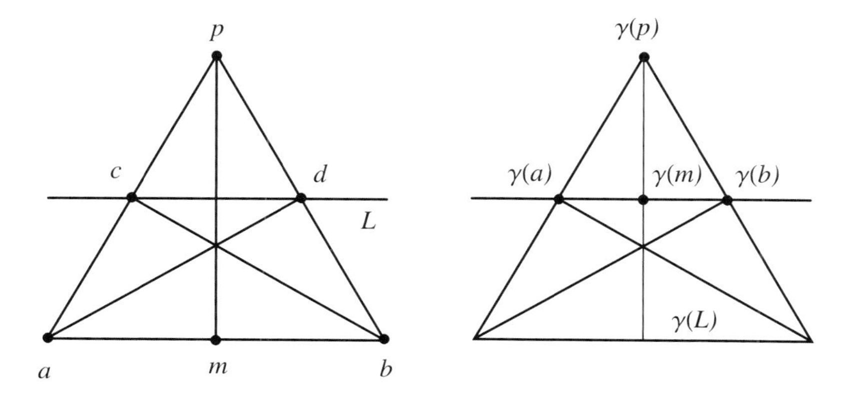

Fig. 2.6.

Given a point p, a line ab, and a line L that separates p from ab, the points $c = ap \wedge L$ and $d = bp \wedge L$ exist, and $m = (ad \wedge bc)p \wedge ab$ is contained in the interval $[a, b]$; see the left diagram in Figure 2.6. We call a point that can be constructed like this a midpoint of a and b. There always exists at least one midpoint and there may be many midpoints (for different choices of L). Note that, no matter what the relative positions of $\gamma(p)$, $\gamma(ab)$, and $\gamma(L)$ are, a midpoint of a and b is always mapped to a point of the interval $(\gamma(a), \gamma(b))$; the right diagram shows $\gamma(m)$ with respect to $\gamma(p)$, $\gamma(ab)$, and $\gamma(L)$ in one of the possible essentially different three configurations of these three objects. Note, in particular, that $\gamma(L)$ does not separate $\gamma(p)$ from $\gamma(ab)$ in this diagram.

Let c_i and c_j be midpoints of $[a_j, a_k]$ and $[a_i, a_k]$ for some choice of $\{i, j, k\} = \{1, 2, 3\}$, and let $a_0 = a_i c_i \wedge a_j c_j$. Then $a_0 \in T(a_1, a_2, a_3)$ and $\gamma(a_0) \in \gamma(T(a_1, a_2, a_3))$.

For $x \in a_1 a_2$ construct

$$\tau(x) = ((x a_3 \wedge a_0 a_1) a_2 \wedge (x a_3 \wedge a_0 a_2) a_1) a_3 \wedge a_1 a_2$$

if this is possible; see Figure 2.7. Let X be the domain of definition of the map τ thus defined and let $I = [a_1, a_2]$. Then $a_0 \in T(a_1, a_2, a_3)$ implies that $I \subseteq X$ and $\tau(I) \subseteq \tau(X) \subseteq I$. The continuous function τ takes on all intermediate values between a_1 and a_2. Hence, $I \subseteq \tau(I)$ and, finally, $\tau(X) = I$. This purely geometric characterization of the inter-

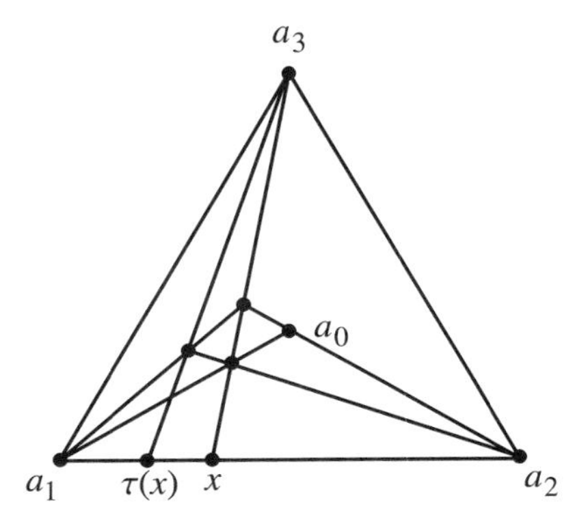

Fig. 2.7.

val I together with the fact that $\gamma(a_0) \in \gamma(T(a_0, a_1, a_2))$ demonstrates that $\gamma([a_1, a_2]) = [\gamma(a_1), \gamma(a_2)]$. □

COROLLARY 2.4.3 (Tripods Determine Isomorphisms) *An isomorphism between two ideal flat linear spaces is uniquely determined by four points of a tripod and their images.*

In the literature tripods in flat projective planes are usually referred to as *nondegenerate quadrangles.* Because of its importance within the theory of flat projective planes we restate the above result as follows.

COROLLARY 2.4.4 (Nondegenerate Quadrangles) *An automorphism of a flat projective plane that fixes each point of a nondegenerate quadrangle is the identity.*

Using these results we can identify the group of automorphisms Γ of an ideal flat linear space $\mathcal{R} = (P, \mathcal{L})$ in a natural way with a subset of P^4 and thereby turn it into a topological space. Here the set of all tripods is an open subset of P^4 which is an 8-dimensional manifold. We fix one tripod t and identify every automorphism γ of $\mathcal{R}$ with the point of P^4 that corresponds to the image of the tripod t under γ. We call the corresponding topology on Γ the *topology of convergence on t.*

Another natural topology on Γ is the *compact-open topology* which is generated by the subbasis consisting of all sets $\{\gamma \in \Gamma \mid \gamma(C) \subseteq U\}$, where C is a compact subset and U is an open subset of the point set of the geometry $\mathcal{R}$.

THEOREM 2.4.5 (Topology of the Automorphism Groups) *Let $\mathcal{R} = (P, \mathcal{L})$ be an ideal flat linear space and Γ its automorphism group. Then the compact-open topology on Γ coincides with the topology of convergence on any of the tripods of $\mathcal{R}$.*

With this topology Γ is a locally compact topological transformation group on P. Even stronger, Γ is a Lie group of dimension at most 8.

No topology on Γ that is finer than the compact-open topology or coarser than the topology of convergence on a tripod adequately relates the topological properties of Γ to those of the underlying geometry. This means that the compact-open topology is the 'natural' topology on Γ.

The first part of the following proof is modelled after the proofs of the respective results in Salzmann [1967b] 3.7–3.11, and the second part after Salzmann [1962a] 3.2, and Salzmann et al. [1995] 32.21.

Proof of theorem. To prove the first statement we only need to show that, given a tripod t, a homeomorphism γ and a sequence of automorphisms (γ_i) such that

$$\lim_i \gamma_i(t) = \gamma(t),$$

we have

$$\lim_i |\gamma_i(x) - \gamma(x)| = 0$$

uniformly for $x \in C$, where C is any compact set in P.

Since γ is a homeomorphism, we may assume that it is the identity. It is also clear that the sequence of homeomorphisms converges on all the points of the dense subplane generated by the tripod t.

Let us assume that there are a positive ε and a sequence (x_i) of points in a compact subset of P such that $|\gamma_i(x_i) - x_i| \geq 3\varepsilon$ for infinitely many i. Then this sequence accumulates at a point a and we may assume that the sequence converges to this point. Let U be the Euclidean open disk with radius ε and centre a. Since the open convex triangles form a neighbourhood basis for the topology of P, there is such a triangle T contained in U whose vertices are points of the dense subplane generated by the tripod t. Hence, for large i, $x_i \in T \subset U$ and, consequently, $\gamma_i(x_i) \in \gamma_i(T) \subseteq U$ and $|\gamma_i(x_i) - x_i| \leq 2\varepsilon$, which is a contradiction.

The characterization of the compact-open topology by uniform convergence on compact sets implies that Γ provided with this topology acts as a topological transformation group on P, that is, the group operations and the map $P \times \Gamma \to P : (x, \gamma) \mapsto \gamma(x)$ are continuous; see Section A2.3.

We prove that Γ is locally compact. The group Γ considered as a subset of P^4 is the same as $\Gamma(s)$ for any tripod s. Hence it suffices to show that $\Gamma(s)$ is a closed subset of the collection of all tripods (this is an open subset of P^4), or equivalently: given tripods s and t and a sequence (γ_i) of collineations such that $\lim_i \gamma_i(s) = t$, the γ_i converge to a collineation γ with $\gamma(s) = t$. Because $\langle s \rangle$ is countable, we can construct a subsequence (γ'_i) that converges pointwise in the dense subplane $\langle s \rangle$. Then

$$\alpha(p) = \lim_i \gamma'_i(p)$$

defines a collineation preserving map from $\langle s \rangle$ to $\langle t \rangle$. We need to show that α is a bijection. Let us first assume that $\mathcal{R}$ is a flat projective plane. Let v be a point in $\langle s \rangle$. Since the image of α contains the tripod t, it is not degenerate and we can choose points o, u, e such that $\alpha(o), \alpha(e), \alpha(u)$, and $\alpha(v)$ form a tripod, and o is contained in some line of $\langle s \rangle$. We coordinatize $\mathcal{R}$ with respect to the tripod o, e, v, u in the usual manner (see, for example, Dembowksi [1968] p. 127) and use the addition in the associated ternary ring to define on the line ov the points

$$c_0 = o, \;\; c_{k+1} = 1 + c_k$$

for all $k \in \mathbf{Z}$. Now the union of all intervals $[c_k, c_{k+1}]$ equals $ov \setminus \{v\}$. From the way addition is defined, the fact that α preserves collinearity, and the order properties of $\mathcal{R}$, we conclude that

$$c_k \in \langle s \rangle, \alpha(c_k) \neq \alpha(c_{k+1}) \text{ and } \alpha(v) \notin [\alpha(c_k), \alpha(c_{k+1})]$$

for all $k \in \mathbf{Z}$. If x is an arbitrary point in the interval $[c_k, c_{k+1}]$, then all accumulation points of the sequence $(\gamma'_i(x))$ are contained in the interval $[\alpha(c_k), \alpha(c_{k+1})]$. In particular, for $x \in [c_k, c_{k+1}] \cap \langle s \rangle$, the accumulation points of the sequence $(\gamma'_i(x))$ are contained in $[\alpha(c_k), \alpha(c_{k+1})]$. Hence $\alpha(x) \neq \alpha(v)$, which implies that α is bijective. We have also proved that for any two distinct points $p, q \in \langle s \rangle$ and a point $x \in [p, q]$, the set of all accumulation points of $(\gamma'_i(x))$ is contained in $[\alpha(p), \alpha(q)]$. Since $\langle t \rangle$ is dense in P (see Theorem 2.4.1), $\langle t \rangle$ is also dense in all lines of $\mathcal{R}$ that have at least two points in common with $\langle t \rangle$. Hence the intersection of all intervals $[\alpha(p), \alpha(q)]$ with $x \in [p, q]$, $p, q \in \langle s \rangle$ can contain only one point. Hence $\lim_i \gamma'_i(x)$ exists for every $x \in P$ that is collinear with two points in $\langle s \rangle$. Consequently, $\lim_i \gamma'_i(L)$ exists for every line L in $\mathcal{R}$. Finally, $\lim_i \gamma'_i(x)$ exists for every point in $\mathcal{R}$ and, therefore, $\gamma(x) = \lim_i \gamma'_i(x)$ defines a collineation-preserving map of $\mathcal{R}$ to itself that extends α. Since $\gamma(s) = t$ this map is the collineation we have been looking for. This proves that Γ is locally compact if $\mathcal{R}$ is a flat projective plane.

For the other types of ideal flat linear spaces similar (more complicated) arguments lead to the desired conclusion; see Salzmann [1967b] Theorem 3.11 for the case that $\mathcal{R}$ is an $\mathbf{R}^2$-plane, and Löwen [1976] Section 2 for the general case. Löwen's paper deals with stable planes, a class of geometries that comprise the flat linear spaces. He proves that the automorphism group of a locally compact, locally connected stable plane is a locally compact, second countable topological transformation group of both the point and the line set of such a plane.

Theorem A2.3.5 guarantees that a locally compact transformation group that acts effectively on a surface is a Lie group. We conclude that Γ is a Lie group. That it has dimension at most 8 follows from the fact that it can be considered as a closed subset of $\mathbf{R}^8$. □

We say that an ideal flat linear space has *group dimension* n if its automorphism group is a Lie group of dimension n.

THEOREM 2.4.6 (Maximal Group Dimension) *An* $\mathbf{R}^2$*-plane or a point Möbius strip plane has group dimension at most* 6*. A flat projective plane has group dimension at most* 8*.*

The classical flat projective plane, the Euclidean plane, and the classical point Möbius strip plane, have group dimension 8, 6, *and* 6, *respectively.*

For a proof of this refinement of the last part of Theorem 2.4.5 see Salzmann [1967b] Theorem 3.19 and Betten [1968]. Note that in the case of the point Möbius strip planes this refinement is an easy consequence of the dimension formula; see Theorem A2.3.6.

An ideal flat linear space is called *flexible* if its automorphism group has an open orbit in the space of all flags. Since the flag space is 3-dimensional (see Theorem 2.3.6(vii)), a flexible ideal flat linear space must have an automorphism group of dimension at least 3.

THEOREM 2.4.7 (Flexible Planes) *A flat projective plane is flexible if and only if it has group dimension at least* 3*. A flat projective plane has group dimension at least* 2 *if its automorphism group has an open orbit in the point space.*

For proofs see Salzmann [1967b] Section 4, Salzmann et al. [1995] Theorem 38.4 and Theorem 38.5.

THEOREM 2.4.8 (Transitive Automorphism Group) *A flat projective plane has an automorphism group that acts transitively on its point set if and only if it is classical.*

See Salzmann [1975a] and Löwen [1981a] for proofs of this fact.

It is a long-standing open conjecture in finite geometry that the corresponding statement is also true for finite projective planes. Note that there are many flat affine planes whose groups act transitively on their point sets; the shift planes that we will introduce in Section 2.7 are examples of such planes. However, none of these flat affine planes gives rise to a nonclassical flat projective plane that is a *translation plane*; see Salzmann et al. [1995] 24.6 for a definition of this term.

PROPOSITION 2.4.9 (Translation Planes Are Classical) *A flat projective plane is a translation plane if and only if it is isomorphic to the classical flat projective plane.*

See Salzmann et al. [1995] Corollary 64.7 for a proof of this result.

THEOREM 2.4.10 (Automorphisms of Flat Projective Planes) *Let $\mathcal{P}$ be a flat projective plane and γ one of its automorphisms. Then the following hold.*

(i) *The automorphism γ fixes a point and a line.*

(ii) *If γ is an involution, it is a reflection which is uniquely determined by its centre and its axis.*

For a proof see Salzmann et al. [1995] Theorem 32.11 and Lemma 32.12.

THEOREM 2.4.11 (Involutions of $\mathbf{R}^2$-Planes) *An involutory automorphism of an $\mathbf{R}^2$-plane is a reflection about a point or about a line. In the first case it is an orientation-preserving homeomorphism and it is orientation-reversing in the second case.*

For a proof of this result see Salzmann [1967b] 3.25.

The only ideal flat linear spaces that can admit polarities are flat projective planes. An *absolute point* of a polarity is a point that is contained in its image under the polarity. Similarly, an *absolute line* of a polarity is a line whose image under the polarity is contained in the line. If a polarity of a flat projective plane has no absolute points, it is called *elliptic*, otherwise it is called *hyperbolic*. The classical flat projective plane admits both kinds of polarities; see, for example, Salzmann et al. [1995] 13.12.

THEOREM 2.4.12 (Polarities of Flat Projective Planes) *Let π be a polarity of a flat projective plane. Then π is continuous. If π is hyperbolic, then the set of absolute points is a topological oval and a line is absolute if and only if it is a tangent of this oval.*

For a proof of this result see Bedürftig [1974a]. Similar results hold for 4-dimensional compact projective planes, and finite projective planes of odd order; see Buchanan–Hähl–Löwen [1980] Satz 4.3 and Hughes–Piper [1973] Chapter 12, respectively.

In the classical flat projective plane a topological oval is the set of absolute points of a polarity if and only if it is a nondegenerate conic section.

A hyperbolic polarity in a flat projective plane is completely determined by its associated topological oval consisting of the absolute points. It maps the points of the oval to the tangents, the exterior points to the secants and the interior points to the exterior lines of the oval as follows. A point on the oval is mapped to the tangent at this point. An exterior point is contained in two tangents of the oval. The point is mapped to the line that intersects the oval in the two points at which the tangents touch the oval. Given an interior point, we already know the images of any two secants through this point under the polarity. The connecting line of the images is the image of the interior point under the polarity.

Finally, we also mention the following result about topological ovals in ideal flat linear spaces.

THEOREM 2.4.13 (Topological Ovals, Secants $\rightarrow$ Tangent) *Let O be a topological oval in an ideal flat linear space, let $p \in O$, and let (p_n) be a sequence of points of O distinct from p that converge to p as n goes to infinity. Then the sequence of lines (pp_n) converges to the tangent line of O at p.*

This general result is a corollary of the restriction of this result to flat projective planes, a proof of which can be found in Salzmann et al. [1995] Sections 55.9–55.18 or Buchanan–Hähl–Löwen [1980].

2.5 Topological Planes and Flat Linear Spaces

A *topological geometry* is an abstract geometry satisfying a number of axioms whose sets of points and lines carry nonindiscrete topologies such that the geometric operations in the axioms are continuous maps on their domains of definition. For example, a projective plane is topological if both its point and line sets carry topologies such that the operations of joining two points by a line and intersecting two lines in a point are continuous operations. A topological geometry is an X *geometry* if its point space is a topological space of type X.

Not much can be said about general topological geometries. However, it can be shown that the finite-dimensional, locally compact, connected projective planes are closely related to the projective planes over the real numbers, the complex numbers, the quaternions, and the octonions. In fact, the 2-dimensional locally compact, connected projective planes are exactly the flat projective planes; see Salzmann et al. [1995] Section 42.

The first part of Theorem 2.3.4 says that with respect to the natural topologies, ideal flat linear spaces are topological linear spaces. The second part highlights a special property of these topological linear spaces which is usually referred to as *stability*. This means that the ideal flat linear spaces are 2-dimensional, locally compact stable linear spaces. The converse of this result is not true. For example, the restriction of the Euclidean plane to the union of two disjoint open disks is a stable plane not all of whose lines are connected.

The *flat linear spaces* are the 2-dimensional, locally compact stable linear spaces with connected lines; see also Subsection 1.2.1 for some background information. Let $\mathcal{P} = (P, \mathcal{L})$ be a flat projective plane and let $C \subset P$ be either P, an open convex topological disk, the complement of a point, the complement of a closed convex topological disk, or the complement of a proper closed interval of a line. It is easy to verify that the restriction of $\mathcal{P}$ to C yields a flat linear space. It turns out that every flat linear space 'looks like' one of the five essentially different types of geometries arising in this manner. In particular, the possible point sets of the flat linear spaces are the same as those of the ideal flat linear spaces and lines are either **R**- or **S**-lines. Furthermore, the classical examples of flat linear spaces that are not ideal flat linear spaces are the restrictions of the classical flat projective plane to the complement of the closed unit disk and the closed unit interval, respectively. There are flat linear spaces on both $\mathbf{R}^2$ and the Möbius strip that do not arise from flat projective planes as described above. It is not known whether every linear space on a Möbius strip is a flat linear space; see Problem 2.11.1.

An **R**-line L in a flat linear space is called *rigid* if every one of its points is contained in only one **R**-line, namely L itself. It is called *variable* if every one of its points contains more than one **R**-line. Clearly, all **R**-lines in an $\mathbf{R}^2$-plane are variable and all **R**-lines in a point Möbius strip plane are rigid.

It turns out that most of the results of this chapter stay valid if we replace the term 'ideal flat linear space' by 'flat linear space'. In particular, all results about the way lines intersect, the topology of the line sets, line pencils, and automorphism groups remain valid. We can complement our results about ideal flat linear spaces as follows.

THEOREM 2.5.1 (Flat Stable Planes) *Let $\mathcal{R} = (\mathcal{P}, \mathcal{L})$ be a flat linear space that is not ideal and let $p \in P$. Then the following hold.*

(i) *The point set P is homeomorphic to the Möbius strip.*
(ii) *The line set $\mathcal{L}$ contains both $\mathbf{R}$- and $\mathbf{S}$-lines.*
(iii) *Every $\mathbf{R}$-line is either rigid or variable.*
(iv) *The sets of all lines through p and $\mathbf{R}$-lines through p are homeomorphic to $\mathbf{S}^1$ and a nonempty closed interval, respectively. This implies that at least one of the lines through p is an $\mathbf{R}$-line and that there are infinitely many lines through p that are $\mathbf{S}$-lines.*
(v) *The line set $\mathcal{L}$ is homeomorphic to the real projective plane and the set of $\mathbf{S}$-lines is homeomorphic to $\mathbf{R}^2$.*
(vi) *The Möbius strip plane $\mathcal{R}$ has group dimension at most* 4.

The *Möbius strip planes* are the flat linear spaces whose point set is the Möbius strip. Depending on whether a Möbius strip plane has only rigid $\mathbf{R}$-lines, both rigid and variable $\mathbf{R}$-lines, or only variable $\mathbf{R}$-lines, we call it a *point Möbius strip plane*, an *interval Möbius strip plane*, or a *disk Möbius strip plane*, respectively. Note that the restriction of a flat projective plane to the complement of a point, a proper closed interval of a line, or a proper closed convex disk, is a point, interval, or a disk Möbius strip plane, respectively.

Let $\mathcal{R}$ be a flat linear space that has $\mathbf{S}$-lines, that is, a flat linear space that is not an $\mathbf{R}^2$-plane. The *opposite plane* of $\mathcal{R}$ has as point set the set of $\mathbf{S}$-lines of $\mathcal{R}$. Associated with every point of $\mathcal{R}$ is a line of the opposite plane consisting of all $\mathbf{S}$-lines through the point.

THEOREM 2.5.2 (Opposite Planes) *Let $\mathcal{R}$ be a flat linear space that is not an $\mathbf{R}^2$-plane. Then the opposite plane of $\mathcal{R}$ is an $\mathbf{R}^2$-plane. If $\mathcal{R}$ is a point Möbius strip plane, then the opposite plane is a flat affine plane. If $\mathcal{R}$ is a flat projective plane, then its opposite plane coincides with the dual of $\mathcal{R}$.*

The fact that two lines in an opposite plane are connected by exactly one line follows from the fact that given an $\mathbf{S}$-line in a flat linear space, every other line intersects this $\mathbf{S}$-line in exactly one point.

Salzmann [1969] and Löwen [1972] derived the possible point and line spaces of flat linear spaces. The best reference for the results concerning the topology of the objects and object spaces associated with flat linear spaces is Löwen [1995]. For more details about flat linear spaces and topological geometries, and a detailed list of references see Grundhöfer–Löwen [1995]. For detailed information about opposite planes see Löwen [1981c], [1995].

2.6 Classification with Respect to the Group Dimension

In the following we summarize the classifications of the different types of flat linear spaces in terms of the group dimension. For descriptions of the different kinds of flat linear spaces in these classifications see Subsection 1.2.1 and Section 2.7.

The overall strategy for classifying flat linear spaces (and other topological geometries) is as follows. First, we have to derive the maximal group dimension of the geometries under consideration. Second, we use various classifications of Lie groups and their actions on the point sets of the geometries under consideration to come up with a first list of the (connected) Lie groups that may act as automorphism groups of our geometries. Usually there are only a few possibilities for the automorphism groups of the geometries with large group dimension and their possible actions. Third, for every single one of the actions of groups in our list we determine all geometries admitting this action. The line joining distinct points is fixed by the stabilizer S of the two points, hence the line can often be found by studying the orbits of S. Once all possible geometries have been found it remains to determine the isomorphisms between them and their full automorphism groups.

It turns out that among the flat linear spaces of a given type, the classical plane has maximal group dimension and that every nonclassical plane has a group dimension that is less than that of the classical plane.

An excellent thorough exposition of the classification of the flat projective planes of group dimension at least 3 can be found in the book Salzmann et al. [1995] Chapter 3. Rather than copying all the relevant material in that book, we only summarize the main results and extend them by many results and constructions outside its scope. Most of the other classification results are scattered throughout the literature, although some good summaries are available; see, in particular, Grundhöfer–Löwen [1995] Theorems 5.14 and 5.15.

2.6.1 Flat Projective Planes

If $\mathcal{P}$ is a flat projective plane of group dimension $n \geq 2$, then its automorphism group contains a closed connected subgroup Δ of the same dimension. A 1-dimensional orbit (of points) of Δ is *nontrivial* if it is not contained in a line. Such an orbit is either a topological oval, or a topological oval minus a point, or a *topological semioval* (see Subsection 2.7.12) whose endpoints are distinct. Note that the possible nontrivial 1-dimensional orbits correspond to the three different ways in which a line can intersect a topological oval. The *fix-configuration of* Δ is the union of all its fixed points, fixed lines, and nontrivial 1-dimensional orbits. If F is the set of all points contained in one of the elements of the fix-configuration, its complement in the point set is the union of the 2-dimensional orbits of the group Δ.

THEOREM 2.6.1 (Closed Connected Automorphism Groups) *Let Δ be a closed connected subgroup of the automorphism group of a nonclassical flat projective plane.*

(i) *If* $\dim \Delta = 4$, *then the commutator subgroup* Δ' *is isomorphic to the universal covering group of* $\mathrm{SL}_2(\mathbf{R})$. *Its fix-configuration consists of an antiflag.*

(ii) *If* $\dim \Delta = 3$, *then the fix-configuration consists of* t *points and* t *lines, where* t *is either* 0, 1, *or* 2. *The possible fix-configurations are illustrated by the different icons in Table 2.1, column 'fix'.*

 (a) *If* $t = 0$, *then* Δ *is isomorphic to* $\mathrm{PSL}_2(\mathbf{R})$.

 (b) *If* $t = 2$, *then* Δ *is isomorphic to* $\mathbf{R} \times \mathrm{L}_2$.

(iii) *If* $\dim \Delta = 2$, *then* Δ *is isomorphic to* $\mathbf{R}^2$, L_2, *or* $\mathbf{R} \times \mathbf{S}^1$. *Up to duality, the corresponding possible fix-configurations are those illustrated by the different icons in Table 2.1.*

See Salzmann et al. [1995] Theorem 33.9 and Theorem 34.9 for (i), Theorem 33.9 and Proposition 37.2 for (ii). Finally, for (iii), Groh [1976] Corollary 2.6 determines the possible groups, Groh [1977] determines the fix-configurations in the case $\Delta = \mathbf{R}^2$, Groh [1976] 2.7 proves that the fix-configuration in the case $\Delta = \mathbf{R} \times \mathbf{S}^1$ is an antiflag, and, based on results in Groh [1976], Lippert [1986] Theorem F in Section 4 determines the possible fix-configurations in the case $\Delta = \mathrm{L}_2$ up to duality.

Table 2.1 summarizes the classification of the flat projective planes of group dimension at least 2. This means that, given a plane like this, the plane itself or its dual is isomorphic to one of the planes in this table.

n	Δ	fix	name	page no.
8	$\mathrm{PGL}_3(\mathbf{R})$		Classical plane	
4			Moulton plane	63, 66
3	$\mathrm{PSL}_2(\mathbf{R})$		Skew hyperbolic plane	76
			Skew parabola plane	69
	$\mathbf{R} \times \mathrm{L}_2$		Cartesian plane	77
2	$\mathbf{R} \times \mathbf{S}^1$		Radial plane	66
	$\mathbf{R}^2$		Shift plane	67
			Pasted plane	91
			Triangle plane	94
	L_2		Stretchshift plane	73
			Stretchshift plane	73
			Pasted plane	92
			Pasted plane	93
			Semioval plane	98
			Semioval plane	100

Table 2.1. The flat projective planes of group dimension at least 2 up to duality

In fact, if the group dimension is at least 3, dualizing a plane of a given type yields a plane of the same type. For example, the dual of a Moulton plane is again a Moulton plane. The way to extract information from this table is by reading from left to right. For example, if the group dimension is 2, Δ is isomorphic to $\mathbf{R}^2$, and Δ fixes exactly one flag, then the plane under consideration is a shift plane.

The classification of the flat projective planes of group dimension at least 3 is due to Salzmann and marks the beginning of topological incidence geometry. The main contributors to the classification of the flat projective planes of group dimension at least 2 are Salzmann, Groh, Lippert, Pohl, and Schellhammer. For a concise exposition of Salzmann's classification we refer to Salzmann et al. [1995] Chapter 3 instead of referring to Salzmann's original papers. However, all his papers dealing with flat linear spaces are listed in the Bibliography. The relevant references for the classification of the planes of group dimension 2 are provided together with a description of these planes in Section 2.7. The main results beyond Salzmann's classification are contained in Groh [1976], [1977], [1979], [1981], Lippert [1986], Pohl [1990], and Schellhammer [1981].

THEOREM 2.6.2 (Homogeneous Flat Projective Planes) *Let $\mathcal{P}$ be a flat projective plane.*

(i) *The maximal possible group dimension of $\mathcal{P}$ is that of the classical flat projective plane. Its group dimension is* 8.

(ii) *The plane $\mathcal{P}$ has group dimension at least* 5 *if and only if it is classical.*

(iii) *The plane $\mathcal{P}$ has group dimension* 4 *if and only if it is isomorphic to a nonclassical projective Moulton plane $\overline{\mathcal{M}}(k)$, that is, a projective Moulton plane with parameter $0 < k \neq 1$; see p. 66.*

(iv) *The plane $\mathcal{P}$ has group dimension* 3 *if and only if it is isomorphic to one of the following planes.*

  (a) *A nonclassical skew hyperbolic plane $\mathcal{H}_t$, that is, a skew hyperbolic plane with parameter $t > 0$; see p. 76.*

  (b) *A nonclassical projective skew parabola plane $\overline{\mathcal{E}}_{c,d}$, that is, a projective skew parabola plane with parameters $0 < c \leq 1 < d$, $(c,d) \neq (1,2)$; see p. 69.*

  (c) *A proper projective Cartesian plane $\overline{\mathcal{P}}_{\alpha,\beta,c}$, that is, a projective Cartesian plane with parameters $0 < \alpha, \beta \leq 1$, $(\alpha, \beta) \neq (1,1)$, $0 < c$, and $c \leq 1$ if $1 \in \{\alpha, \beta\}$; see p. 77.*

For this result in particular see Salzmann et al. [1995] Theorem 38.1.

2.6.2 $\mathbb{R}^2$*-Planes*

Let $\mathcal{A}$ be a flat affine plane, Δ its group of automorphisms, and $\overline{\Delta}$ the group of automorphisms of the projective extension of $\mathcal{A}$. Since every automorphism of $\mathcal{A}$ extends to an automorphism of its projective extension, the group Δ is the stabilizer of the line at infinity of $\mathcal{A}$ in $\overline{\Delta}$. This implies that the classification of the flat projective planes translates in a straightforward way into a classification of the flat affine planes. The following theorem summarizes the classification of the planes with group dimension at least 3.

THEOREM 2.6.3 (Homogeneous Flat Affine Planes) *Let $\mathcal{A}$ be a flat affine plane.*

(i) *The maximal possible group dimension of $\mathcal{A}$ is that of the classical flat affine plane. Its group dimension is* 6.

(ii) *The plane $\mathcal{A}$ has group dimension at least* 5 *if and only if it is classical.*

(iii) *The plane $\mathcal{A}$ has group dimension* 4 *if and only if it is isomorphic to one of the nonclassical radial Moulton planes $\mathcal{M}(s)$; see p. 66.*

(iv) *The plane $\mathcal{A}$ has group dimension* 3 *if and only if it is isomorphic to one of the following planes.*

  (a) *A nonclassical skew parabola plane $\mathcal{E}_{c,d}$; see p. 69.*

  (b) *A proper Cartesian plane $\mathcal{P}_{\alpha,\beta,c}$; see p. 77.*

  (c) *An affine plane that we arrive at by removing from a proper projective Cartesian plane $\overline{\mathcal{P}}_{\alpha,\beta,c}$ the fixed line of its automorphism group that contains only one fixed point; see p. 77.*

  (d) *An affine plane that we arrive at by removing from a non-classical projective Moulton plane $\overline{\mathcal{M}}(s)$ a line through the fixed point of its automorphism group; see p. 66.*

The most homogeneous $\mathbf{R}^2$-planes that are not flat affine planes have also been classified.

THEOREM 2.6.4 (Homogeneous Proper $\mathbf{R}^2$-Planes) *Let $\mathcal{R}$ be an $\mathbf{R}^2$-plane that is not a flat affine plane.*

(i) *The maximal possible group dimension of $\mathcal{R}$ is* 4. *The only plane that has this group dimension is the real cylinder plane* C(**R**), *that is, the restriction of the Euclidean plane to an open half-plane; see p. 12.*

(ii) *The plane $\mathcal{R}$ has group dimension* 3 *if and only if it is isomorphic to one of the following planes.*

  (a) *The real hyperbolic plane* H(**R**), *that is, the restriction of the Euclidean plane to the interior of the unit circle.*

(b) *Strambach's* $\mathrm{SL}_2(\mathbf{R})$*-plane;* *see p. 78.*

(c) *The following arc planes of Type* $\mathbf{R}^2$.

1. *The exponential arc planes;* *see p. 72.*
2. *The hyperbolic arc planes;* *see p. 72.*

(d) *The following pasted planes made up of two arc planes each.*

1. *A three-parameter family of pasted planes made up of exponential arc planes;* *see p. 89.*
2. *A four-parameter family of pasted planes made up of hyperbolic arc planes;* *see p. 90.*

For (i) see Salzmann [1967b] Section 4. The remaining results were proved by Salzmann, Strambach, Ostmann, Betten, Löwen, Groh, Lippert, and Pohl. See Strambach [1970b], Groh [1982a], Groh–Lippert–Pohl [1983], and the references given in these papers.

2.6.3 Möbius Strip Planes

As in the case of flat affine planes, the group-dimension classification of the point Möbius strip planes is a corollary of the group-dimension classification of the flat projective planes. Also, apart from a few exceptions (see Theorem 2.6.5(iv)(d)) the most homogeneous interval and disk Möbius strip planes arise as subgeometries of the most homogeneous flat projective planes.

THEOREM 2.6.5 (Homogeneous Möbius Strip Planes) *Let* $\mathcal{M}$ *be a Möbius strip plane.*

(i) *The maximal possible group dimension of* $\mathcal{M}$ *is that of the classical point Möbius strip plane;* *see p. 12.* *This plane has group dimension* 6.

(ii) *The plane* $\mathcal{M}$ *has group dimension at least* 5 *if and only if it is isomorphic to the classical point Möbius strip plane.*

(iii) *The plane* $\mathcal{M}$ *has group dimension* 4 *if and only if it is isomorphic to one of the following two planes.*

(a) *The point Möbius strip plane that is obtained from a nonclassical projective Moulton plane* $\overline{\mathcal{M}}(s)$ *(see p. 66) by removing the fixed point of its automorphism group.*

(b) *The classical interval Möbius strip plane;* *see p. 55.*

(iv) *The plane $\mathcal{M}$ has group dimension 3 if and only if it is isomorphic to one of the following planes.*

(a) *A point Möbius strip plane obtained from a flat projective plane $\mathcal{P}$ by removing a point p as follows.*

1. *$\mathcal{P}$ is a nonclassical projective Moulton plane $\overline{\mathcal{M}}(k)$ (see p. 66) and p is one of the points on the fixed line of its automorphism group.*
2. *$\mathcal{P}$ is one of the nonclassical projective skew parabola planes $\overline{\mathcal{E}}_{c,d}$ (see p. 69) and p is the fixed point of its automorphism group.*
3. *$\mathcal{P}$ is one of the proper projective Cartesian planes (see p. 77) and p is the fixed point of its automorphism group that is the intersection of the two fixed lines.*
4. *$\mathcal{P}$ is one of the proper projective Cartesian planes (see p. 77) and p is the fixed point of its automorphism group that is not the intersection of the two fixed lines.*

(b) *An interval Möbius strip plane obtained from a flat projective plane $\mathcal{P}$ by removing a closed interval I as follows.*

1. *$\mathcal{P}$ is a nonclassical projective Moulton plane $\overline{\mathcal{M}}(k)$ (see p. 66) and I is one of the two intervals that connects the fixed point of the automorphism group with a point on the fixed line.*
2. *$\mathcal{P}$ is one of the proper projective Cartesian planes $\overline{\mathcal{P}}_{\alpha,\beta,c}$ (see p. 77) and I is one of the two intervals that connect the two fixed points of its automorphism group.*

(c) *The disk Möbius strip plane that is obtained by restricting a skew hyperbolic plane $\mathcal{H}_t$ (see p. 76) to the complement of the closed unit disk (note that this family includes the classical disk Möbius strip plane; see p. 55).*

(d) *The exponential and hyperbolic arc planes generated by one arc each (see p. 72) with points at infinity added as described in the special case* MDC($\mathbf{R}$)*; see p. 103. These planes are also disk Möbius strip planes.*

The proof can be found in Betten [1968]. The disk Möbius planes arising from exponential arc planes (iv)(d) have been overlooked in case 1 of the proof of his result 5.8; see Grundhöfer–Löwen [1995], Proof of Theorem 5.15.

2.7 Constructions

In this section we present some of the most famous, important, and appealing constructions for flat linear spaces. In particular, we construct the flat linear spaces with group dimension at least 3 and list many important characterizations and properties of these geometries.

We have already encountered a number of constructions that allow us to construct one type of flat linear space from another as follows.

- The restriction of a flat linear space to an open convex topological disk is an $\mathbf{R}^2$-plane.
- The restriction of a flat projective plane to the complement of a proper closed convex subset is a Möbius strip plane.
- Any point Möbius strip plane or flat affine plane can be extended to a flat projective plane.
- The opposite plane of a Möbius strip plane is an $\mathbf{R}^2$-plane. The opposite plane or, equivalently, the dual of a flat projective plane is also a flat projective plane.

A number of other constructions of flat projective planes are mentioned in Subsection 1.1.3 (flat affine planes as derived planes of flat Möbius, Laguerre, and Minkowski planes) and Section 2.1 (disk models of flat projective planes, point Möbius strip planes from interpolating sets of functions). Further constructions of flat linear spaces from higher-rank circle planes will be investigated in the following chapters.

Many of the geometries in this section are constructed by specifying the point set of a geometry, some *generating lines*, and a generating group of automorphisms of the geometry. The lines of the geometry are the images of the generating lines under elements of the group. Many of the generating lines can be given as graphs of real-valued functions. Given such a function f, the *slope supply of* f, denoted by slp f, is the set of all slopes of secants of the graph of f. For example, the slope supply of the exponential function $\mathbf{R} \to \mathbf{R} : x \mapsto e^x$ is the interval $(0, \infty)$.

2.7.1 Original Moulton Planes

The (original) Moulton planes are flat affine planes. They were introduced by Moulton [1902] and are some of the earliest examples of nonclassical flat affine planes. The most homogeneous nonclassical flat projective planes are the projective extensions of these flat affine planes.

We fix a positive real number k. We arrive at the Moulton plane $\mathcal{M}_k$ by replacing every line in the Euclidean plane with positive slope m by a line that starts out as this Euclidean line in the right half-plane and continues as a line of slope km in the left half-plane. This gives the following 'bent' lines.

$$\{(x, mx + t) \in \mathbf{R}^2 \mid x \geq 0\} \cup \{(x, kmx + t) \in \mathbf{R}^2 \mid x \leq 0\},$$

where $m, t \in \mathbf{R}, m > 0$; see Figure 2.8.

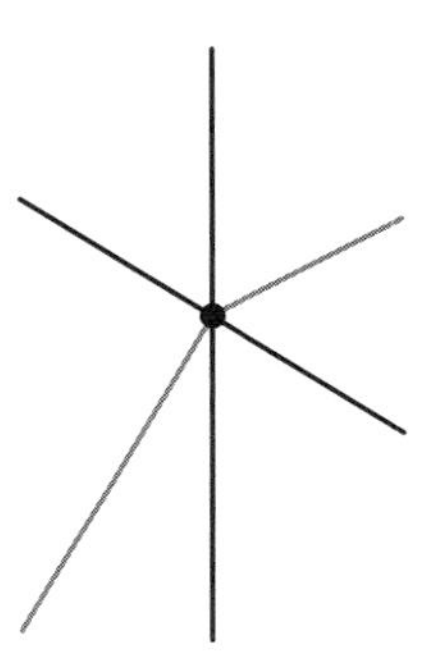

Fig. 2.8. Three lines through the origin in an original Moulton plane

You can also think of this plane as being glued together along the y-axis from two Euclidean halves. The number k is the 'glue' factor. Two such planes $\mathcal{M}_k$ and $\mathcal{M}_{k'}$ are isomorphic if and only if $k' = k$ or $k' = 1/k$. In particular, we obtain the Euclidean plane for $k = 1$. For $k \neq 1$ the resulting planes are nonclassical.

We already mentioned that the projective extensions of the Moulton planes play a very important role in the theory of flat projective planes. It is quite cumbersome to extract information about the full automorphism group of these planes from the above model. We only observe that the maps of the form

$$\mathbf{R}^2 \to \mathbf{R}^2 : (x, y) \mapsto (ax, by + c)$$

where $a, b, c \in \mathbf{R}$, $a, b > 0$, form a 3-dimensional group of automorphisms of $\mathcal{M}_k$. The radial model of the projective Moulton planes that we will introduce in Subsection 2.7.3 is more suitable for describing the full automorphism group of the projective extension of $\mathcal{M}_k$. For comprehensive information about Moulton planes see Salzmann et al. [1995] Example 31.25(b) and Section 34.

2.7.2 Semi-classical Planes and Generalized Moulton Planes

The Moulton planes have been generalized in various ways in the topological, field, and finite settings; see Steinke [1985a] and the references given there. In this subsection we describe all those flat affine planes that are, just like the Moulton planes, glued together from two Euclidean half-planes; see Steinke's paper for details. Let $h, g : \mathbf{R} \to \mathbf{R}$ be two orientation-preserving homeomorphisms. Then the *semi-classical flat affine plane* $\mathcal{A}_{h,g}$ is constructed as follows. The point set is $\mathbf{R}^2$. Lines are the vertical Euclidean lines and the sets

$$\{(x, mx + t) \in \mathbf{R}^2 \mid x \geq 0\} \cup \{(x, g^{-1}(h(m)x + g(t))) \in \mathbf{R}^2 \mid x < 0\},$$

where $m, t \in \mathbf{R}$.

We call these planes semi-classical because the geometries and topologies on $A_+ = \mathbf{R}^+ \times \mathbf{R}$ and $A_- = \mathbf{R}^- \times \mathbf{R}$ are the same as on the corresponding subsets of the (topological) Euclidean plane. One can think of these semi-classical planes as being obtained from the Euclidean plane by cutting the plane along a line—the y-axis—and gluing the two pieces together in a new way.

Let $\mathcal{H}_{0,1}$ be the set of all homeomorphisms $\mathbf{R} \to \mathbf{R}$ that fix both 0 and 1.

PROPOSITION 2.7.1 (Semi-classical Planes) *Let $\mathcal{A} = (P, \mathcal{L})$ be a flat affine plane and let $L \in \mathcal{L}$. If the restrictions of $\mathcal{A}$ to the connected components of $P \setminus L$ are both isomorphic to Euclidean half-planes, then there exist $g, h \in \mathcal{H}_{0,1}$ such that $\mathcal{A}$ is isomorphic to the semi-classical flat affine plane $\mathcal{A}_{h,g}$.*

A semi-classical plane $\mathcal{A}_{h,g}$ with $g, h \in \mathcal{H}_{0,1}$ is classical if and only if $g = h = id$.

For a proof of this result see Steinke [1985a] Proposition 2.2 and Corollary 3.2. The semi-classical planes comprise the *generalized Moulton planes* which are obtained for $g = id$ and the original Moulton planes which are obtained for $g = id$ and

$$h : \mathbf{R} \to \mathbf{R} : x \mapsto \begin{cases} x & \text{for } x \leq 0, \\ kx & \text{for } x \geq 0. \end{cases}$$

See Subsections 2.7.10 and 2.7.11 for a number of different cut-and-paste constructions of ideal flat linear spaces that generalize the construction of the semi-classical planes.

2.7.3 Radial Planes and Radial Moulton Planes

The whole Euclidean plane can be generated as follows. Start with a (generating) Euclidean line that does not contain the origin and consider the images of this line under all possible rotations around the origin and dilatations with all possible positive dilatation factors. Finally, add the Euclidean lines that pass through the origin to get all lines of the Euclidean plane. We can construct nonclassical flat affine planes by replacing the generator in the above construction by a topological line that is not too different from a Euclidean line.

Schellhammer [1981] Satz 7.9 proves that up to rotation around the origin every topological line L that generates a flat affine plane in this manner has a description in polar coordinates as follows:

$$\{(\varphi, (f(\varphi))^{-1}) \mid -\pi/2 < \varphi < \pi/2\},$$

where $f : [-\pi/2, \pi/2] \to \mathbf{R}$ is a continuous function such that

(i) f is positive when restricted to $(-\pi/2, \pi/2)$,
(ii) $f(-\pi/2) = f(\pi/2) = 0$,
(iii) $\ln f : (-\pi/2, \pi/2) \to \mathbf{R} : \varphi \mapsto \ln f(\varphi)$ is strictly concave.

Conversely, every function such as this generates a flat affine plane.

Consider the function

$$f_s : [-\pi/2, \pi/2] \to [0, \infty) : \varphi \mapsto e^{-s\varphi} \cos\varphi,$$

where $s \geq 0$. Then f_s satisfies the three conditions above. Let $\mathcal{M}(s)$ denote the resulting flat affine plane. It turns out that the flat projective plane $\overline{\mathcal{M}}(s)$ which corresponds to this function is isomorphic to the projective original Moulton plane $\overline{\mathcal{M}}_k$ with 'glue factor' $k = e^{2\pi s}$ that we introduced in Subsection 2.7.1. In particular, $\mathcal{M}(0)$ is the Euclidean plane and $\overline{\mathcal{M}}(0)$ is the classical flat projective plane.

Identify $\mathbf{R}^2$ with $\mathbf{C}$ in the usual manner. Then an isomorphism between $\overline{\mathcal{M}}(s)$ and $\overline{\mathcal{M}}_k$ is given by the extension of the homeomorphism

$$\mathbf{C} \setminus i\mathbf{R} \to (\mathbf{R} \setminus \{0\}) \times \mathbf{R} : re^{i\varphi} \mapsto \left(\frac{e^{s\varphi}}{r\cos\varphi}, \tan\varphi\right),$$

where $r > 0$, $\pi/2 \neq \varphi \in (-\pi/2, \pi/2)$.

Note that this homeomorphism extends uniquely to an isomorphism between $\overline{\mathcal{M}}(s)$ and $\overline{\mathcal{M}}_k$ which maps the line at infinity of $\mathcal{M}(s)$ to the line in $\overline{\mathcal{M}}_k$ that corresponds to the y-axis (the 'bending line'). The main advantage of the radial model is that none of the automorphisms of a nonclassical projective Moulton plane $\overline{\mathcal{M}}(s)$ moves this special line. This implies that the automorphism group of the projective Moulton plane $\overline{\mathcal{M}}(s)$ coincides with the automorphism group of the flat affine plane $\mathcal{M}(s)$.

The rotations around the origin form a 1-dimensional group of automorphisms of $\mathcal{M}(s)$. Together with the 3-dimensional group of automorphisms apparent in the flat affine plane $\mathcal{M}_k$ this 1-dimensional group generates a 4-dimensional group of automorphisms of the common projective extension of these two flat affine planes.

The only point and line fixed by all automorphisms of a nonclassical projective Moulton plane are the origin and the line at infinity of $\mathcal{M}(s)$ ('=' the infinite point of the x-axis and the line that extends the y-axis, respectively, in $\overline{\mathcal{M}}_k$).

THEOREM 2.7.2 (Group Dimension 4**)** *A flat projective plane has group dimension* 4 *if and only if it is isomorphic to one of the mutually nonisomorphic projective Moulton planes* $\overline{\mathcal{M}}(s)$ *for* $s > 0$*.*

As we already mentioned in Theorem 2.6.1, if Δ is the automorphism group of the plane $\overline{\mathcal{M}}(s)$, $s > 0$, then Δ' is isomorphic to the universal covering group of $\mathrm{SL}_2(\mathbf{R})$. See Salzmann et al. [1995] Section 34 and Schellhammer [1981] for more detailed information about these results and constructions.

2.7.4 Shift Planes and Planar Functions

We deform the Euclidean plane by the homeomorphism

$$\mathbf{R}^2 \to \mathbf{R}^2 : (x, y) \mapsto (x, y + x^2).$$

Under this deformation the verticals stay unchanged and the nonvertical line that is the graph of the linear function $\mathbf{R} \to \mathbf{R} : x \mapsto bx + c$ turns into the parabola given by the quadratic function

$$\mathbf{R} \to \mathbf{R} : x \mapsto x^2 + bx + c.$$

This model of the Euclidean plane is called the *parabola model.*

A second method of constructing the same plane from the Euclidean plane is to keep all vertical lines and to replace all nonvertical lines by the parabola $\{(x, x^2) \in \mathbf{R}^2 \mid x \in \mathbf{R}\}$ and its 'translates in all possible directions', that is, the nonvertical lines are the images of the parabola under all Euclidean translations

$$\mathbf{R}^2 \to \mathbf{R}^2 : (x, y) \mapsto (x + a, y + b),$$

where $a, b \in \mathbf{R}$; see Figure 2.9.

We generalize this construction by replacing the parabola by the graph of a function $f : \mathbf{R} \to \mathbf{R}$ and consider the geometry $\mathcal{G}(f)$ whose lines are the verticals in $\mathbf{R}^2$ plus all translates of the graph of f. We say that f is a *planar function* if and only if $\mathcal{G}(f)$ is an affine plane.

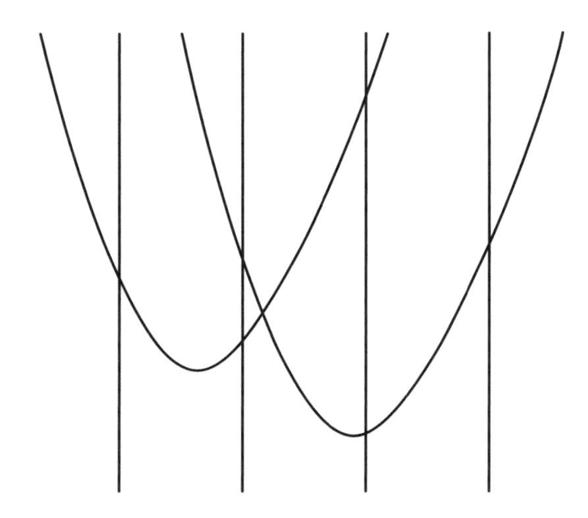

Fig. 2.9. A shift plane

THEOREM 2.7.3 (Continuous Planar Functions) *Let $f : \mathbf{R} \to \mathbf{R}$ be a continuous function. Then $\mathcal{G}(f)$ is a flat affine plane if and only if f has the following properties:*

(i) *f is strictly convex (or concave);*
(ii) *$\lim_{x \to \pm\infty} f(x) - ax = +\infty$ (or $-\infty$) for all $a \in \mathbf{R}$.*

COROLLARY 2.7.4 (Differentiable Planar Functions) *Given a differentiable function $f : \mathbf{R} \to \mathbf{R}$, the geometry $\mathcal{G}(f)$ is a flat affine plane if and only if the derivative of f is a homeomorphism $\mathbf{R} \to \mathbf{R}$.*

Continuous or differentiable planar functions $\mathbf{R} \to \mathbf{R}$ are also called *parabolic* or *strictly parabolic* functions. The flat affine plane generated by a parabolic function is called a *shift plane* (see Figure 2.9) and we call the projective extension of such a plane a *projective shift plane.*

Note that all Euclidean translations are automorphisms of all shift planes. However, only the translations in the vertical direction are translations of the shift plane in the geometric sense. Because of this fact the Euclidean translations are called shifts of the plane.

THEOREM 2.7.5 (Classical Shift Planes) *Let f be a parabolic function. Then the shift plane $\mathcal{G}(f)$ is classical if and only if f is a quadratic polynomial.*

THEOREM 2.7.6 (Fix-Configuration) *A flat projective plane is isomorphic to a projective shift plane if and only if its automorphism group contains a subgroup Δ isomorphic to $\mathbf{R}^2$ that fixes precisely one line.*

See Groh [1976] 3.6 A1, B1 and Groh [1982b] for proofs of these results.

The above construction generalizes as follows. Let A and B be two abelian groups and let $f : A \to B$. As above, we define a geometry with point set $A \times B$ whose lines are the verticals in $A \times B$ and the graphs of the functions $A \to B : x \mapsto f(x-a)+b$, where $(a,b) \in A \times B$. Then the function f is called planar if and only if this geometry is an affine plane. It has been shown that continuous planar functions $\mathbf{R}^n \to \mathbf{R}^n$ exist only for $n = 1, 2$. See Salzmann et al. [1995] Section 74 for more information about continuous planar functions.

2.7.4.1 Skew Parabola Planes

We concentrate on a particularly nice class of continuous planar functions to construct the so-called *skew parabola planes.*

For $c, d \in \mathbf{R}$ with $c > 0$ and $d > 1$ let

$$f : \mathbf{R} \to \mathbf{R} : x \mapsto \begin{cases} x^d & \text{for } x \geq 0, \\ c|x|^d & \text{for } x \leq 0. \end{cases}$$

We call f a *skew parabola function* and the graph of f a *skew parabola.*

Clearly, the derivative of such a function f is a homeomorphism. Therefore f generates a shift plane which we denote by $\mathcal{E}_{c,d}$. Besides the shifts these skew parabola planes also admit the automorphisms

$$\mathbf{R}^2 \to \mathbf{R}^2 : (x,y) \mapsto (rx, r^d y),$$

where $r \in \mathbf{R}^+$.

Clearly, $\mathcal{E}_{1,2}$ is the Euclidean plane and $\mathcal{E}_{c,d}$ is isomorphic to $\mathcal{E}_{c^{-1},d}$ via the homeomorphism $\mathbf{R}^2 \to \mathbf{R}^2 : (x,y) \mapsto (-x, c^{-1}y)$. Only the point at infinity of the verticals and the line at infinity are fixed by all automorphisms of a nonclassical projective skew parabola plane.

The following theorem characterizes the nonclassical projective skew parabola planes among the flat projective planes.

THEOREM 2.7.7 (Characterization of Skew Parabola Planes) *Let Σ be the full automorphism group of a flat projective plane. Assume that the connected component Σ^1 has dimension 3 and fixes precisely one flag of the plane. Then the plane is isomorphic to one of the mutually nonisomorphic projective shift planes $\overline{\mathcal{E}}_{c,d}$, where $0 < c \le 1 < d$ and $(c,d) \ne (1,2)$. Conversely, all these planes satisfy our hypotheses.*

Let $(c,d) \ne (1,2)$. Then the full automorphism group Σ of $\overline{\mathcal{E}}_{c,d}$ has dimension 3, and the connected component Σ^1, acting on the points of the flat affine plane $\mathcal{E}_{c,d}$, is

$$\{\mathbf{R}^2 \to \mathbf{R}^2 : (x,y) \mapsto (rx + a, r^d y + b) \mid a, b, r \in \mathbf{R}, r > 0\}.$$

Furthermore, $\Sigma = \Sigma^1$ for $c \ne 1$, and

$$\Sigma = \Sigma^1 \langle \mathbf{R}^2 \to \mathbf{R}^2 : (x,y) \mapsto (-x, y) \rangle$$

for $c = 1$.

For these and other results about the skew parabola planes see Salzmann et al. [1995] Chapter 36.

2.7.5 Arc Planes

Generalizing shift planes leads to $\mathbf{R}^2$-planes which are usually referred to as *arc planes*. The main results about these planes were proved in Groh [1976], [1979], [1982a], and [1982b].

Just as in the case of shift planes, the automorphism group of an arc plane has a subgroup that acts sharply transitively on the point set of the plane. This subgroup is homeomorphic either to the group $\mathbf{R}^2$ consisting of all Euclidean translations

$$t_{a,b} : \mathbf{R}^2 \to \mathbf{R}^2 : (x,y) \mapsto (x + a, y + b),$$

where $a, b \in \mathbf{R}^2$, or to the group L_2 consisting of all homeomorphisms

$$h_{a,b} : H \to H : (x,y) \mapsto (ax, ay + b),$$

where H is the half-plane $\{(x, y) \in \mathbf{R}^2 \mid x > 0\}$ and $a, b \in \mathbf{R}$, $a > 0$. Accordingly, we distinguish between arc planes of type $\mathbf{R}^2$ and arc planes of type L_2.

Let P be the coordinate plane $\mathbf{R}^2$ or the half-plane H, let G be the group $\mathbf{R}^2$ or the group L_2, and let $\mathcal{R}$ be the Euclidean plane or the restriction of the Euclidean plane to the half-plane H, respectively. We say that a line in $\mathcal{R}$ has *slope* $s \in \mathbf{R} \cup \{\infty\}$ if it is part of a Euclidean line that has this slope. The *slope supply* of a set of points in P is the set of all slopes of secant lines of the set. Two sets of points are *slope disjoint* if their slope supplies do not have any elements in common. If $P = \mathbf{R}^2$, then a set of points is called *projectable* if the following condition is satisfied.

- There is a parallel class of Euclidean lines such that every one of these lines intersects the set in at most one point and the set itself is not contained in any strip bounded by two of these parallel lines.

If $P = H$, then a set of points is projectable if one of the following two conditions is satisfied.

- Every vertical Euclidean line meets the set in at most one point and the set is not contained in a strip bounded by two of these lines.
- There is a point p on the y-axis such that all the lines in $\mathcal{R}$ that have p as their boundary point intersect the set in at most one point and the set is not contained in a sector bounded by two of these lines.

Let A be a collection of topological lines in P such that the following hold.

(i) All elements of A are topological arcs in $\mathcal{R}$.
(ii) None of the elements of A has a slope supply equal to $\mathbf{R} \cup \{\infty\}$.
(iii) Any two distinct elements of A are slope disjoint.
(iv) All elements of A are projectable.

Let S be the set of all lines in $\mathcal{R}$ that have a slope that is not contained in the slope supply of any arc in A. Finally, let $\mathcal{L}$ consist of all elements of S and all images of the elements of A under elements of the group G. The geometry $(P, \mathcal{L})$ is called the *arc plane* generated by A.

THEOREM 2.7.8 (General Arc Planes) *Arc planes are* $\mathbf{R}^2$*-planes. Given an* $\mathbf{R}^2$*-plane whose automorphism group contains a 2-dimensional, closed, point-transitive subgroup, this* $\mathbf{R}^2$*-plane is isomorphic to an arc plane.*

For a proof of this theorem see Groh [1979] Theorem 7.5.

2.7.5.1 Arc Planes of Type $\mathbf{R}^2$

Figure 2.10 shows some examples for sets A of arcs that generate arc planes of type $\mathbf{R}^2$ and the corresponding sets S.

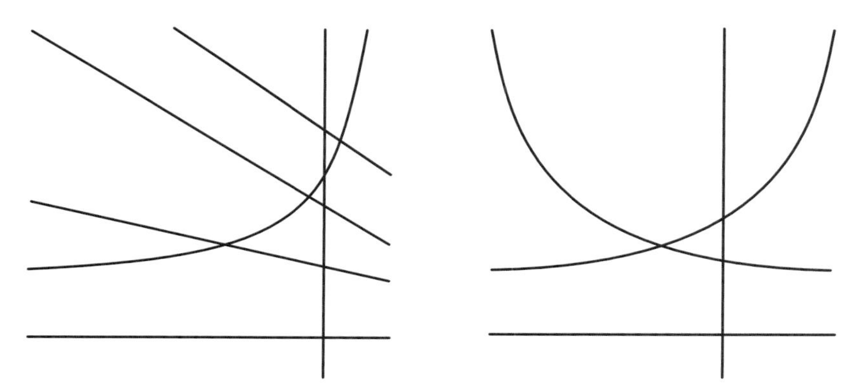

Fig. 2.10. Two arc planes

In the first example A contains only the graph of the exponential function $\mathbf{R} \to \mathbf{R} : x \mapsto e^x$ and S contains all Euclidean lines with nonpositive slope. In the second example A contains both the graphs of the exponential function and the function $\mathbf{R} \to \mathbf{R} : x \mapsto e^{-x}$. The set S consists of all the horizontals and verticals. We remark that this last example is isomorphic to the restriction of the Euclidean plane to an open half-plane; see also the end of Subsubsection 2.7.5.2.

The following result characterizes the shift planes among the arc planes of type $\mathbf{R}^2$; see Groh [1976] Theorem 3.6 A1, B1 and Groh [1979] Theorem 6.3.

THEOREM 2.7.9 (Arc Planes and Shift Planes) *An arc plane of type* $\mathbf{R}^2$ *is a flat affine plane if and only if it is the image of a shift plane under a rotation of its point set.*

Apart from the skew parabola planes there are two further classes of arc planes of group dimension 3. These are the *hyperbolic arc planes* and the *exponential arc planes*; see Groh–Lippert–Pohl [1983].

There are two types of hyperbolic arc planes. A plane of the first type is generated by the graph of one of the functions

$$\mathbf{R}^+ \to \mathbf{R} : x \mapsto x^s,$$

where $s \leq -1$. A plane of the second type is generated by the graphs of the two functions

$$\mathbf{R}^+ \to \mathbf{R} : x \mapsto x^s \text{ and } \mathbf{R}^+ \to \mathbf{R} : x \mapsto rx^s,$$

where $r \leq -1$ and $s < 0$. Apart from the Euclidean translations the maps

$$\mathbf{R}^2 \to \mathbf{R}^2 : (x, y) \mapsto (tx, t^s y),$$

where $t > 0$, are also automorphisms of these geometries.

There are also two types of exponential arc planes. There is only one plane of the first type. It is generated by the graph of the exponential function $\mathbf{R} \to \mathbf{R} : x \mapsto e^x$. A plane of the second type is generated by the graph of the exponential function and the graph of one of the functions

$$\mathbf{R} \to \mathbf{R} : x \mapsto -\text{sgn}(s)e^{sx},$$

where $|s| \geq 1$. Apart from the Euclidean translations the geometries also admit the automorphisms

$$\mathbf{R}^2 \to \mathbf{R}^2 : (x, y) \mapsto (x, ty),$$

where $t > 0$.

Convince yourself that none of the hyperbolic or exponential arc planes is a flat affine plane.

2.7.5.2 Arc Planes of Type L_2

Figure 2.11 shows some examples for sets A of arcs that generate arc planes of type L_2 and the corresponding sets S. In the picture on the

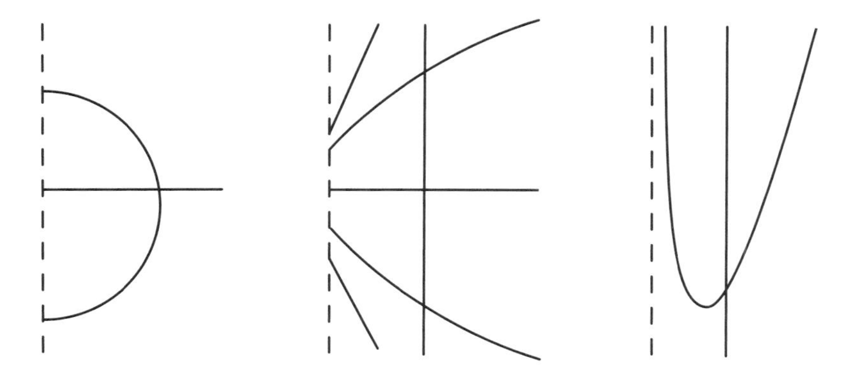

Fig. 2.11. Three arc planes of Type L_2

left the set A consists of the intersection of the unit circle with H and the set S consists of all horizontals. This is the well-known Poincaré model of the real hyperbolic plane on H. The set A corresponding to the diagram in the middle consists of the graphs of the functions

$$\mathbf{R}^+ \to \mathbf{R} : x \mapsto \ln(x+1) + 1 \text{ and } \mathbf{R}^+ \to \mathbf{R} : x \mapsto -\ln(x+1) - 1.$$

The corresponding set S consists of the lines of slope greater than 1, those of slope less than -1, and those of slope 0. The set A corresponding to the diagram on the right consists of the graph of the function

$$\mathbf{R}^+ \to \mathbf{R} : x \mapsto x^2 + 1/x.$$

The corresponding set S consists of all the vertical lines. The topological arcs in Figure 2.12 do not generate arc planes of type L_2.

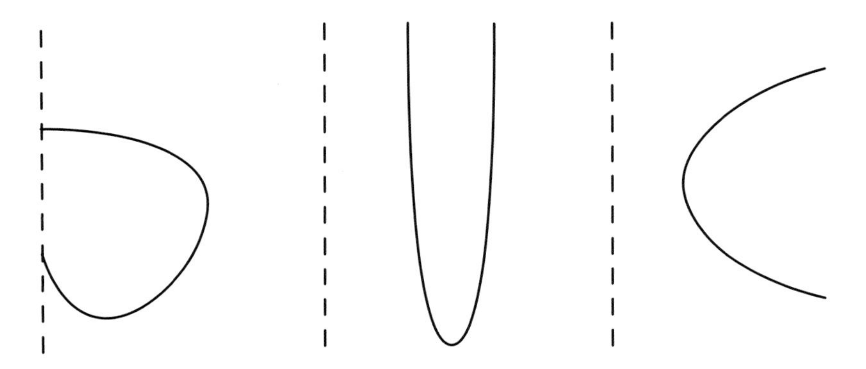

Fig. 2.12. Arcs that do not generate arc planes of type L_2

Groh [1976] 3.6 A2, A3, B2, B3 characterizes the arc planes of type L_2 that are affine planes; see also Groh [1979] Theorem 6.4. There are two different kinds of such planes. The planes of the first kind are generated by one topological arc and the planes of the second kind are generated by two topological arcs. We call these affine planes *stretchshift planes.*

Stretchshift planes generated by one arc: Let L be the graph of a strictly convex continuous function $f : \mathbf{R}^+ \to \mathbf{R}$ such that

- $\lim_{x\to 0,+\infty} f(x) - ax = +\infty$, for all $a \in \mathbf{R}$.

For example, the function $\mathbf{R}^+ \to \mathbf{R} : x \mapsto x^2 + 1/x$ satisfies this condition. The set S consists of all vertical lines in the half-plane H.

In the projective extension of such a flat affine plane the group L_2

fixes exactly one point and one line (the infinite point of the verticals and the line at infinity).

Stretchshift planes generated by two arcs: Let L be the graph of a strictly convex continuous function $f : \mathbf{R}^+ \to \mathbf{R}$ that satisfies the following conditions:

- $\lim_{x\to 0} f(x) = \infty$;
- $\lim_{x\to+\infty} f(x) = -\infty$;
- $\lim_{x\to+\infty} f'(x) = 0$ (note that since f is convex, it is differentiable everywhere, with the possible exception of a set of Lebesgue measure 0).

For example, the function $\mathbf{R}^+ \to \mathbf{R} : x \mapsto -\ln x$ satisfies these conditions. Furthermore, let M be the graph of a continuous strictly concave function g such that $-g$ also satisfies the three conditions above. Then the set A consists of L and M and the set S consists of all horizontal and vertical lines in H.

In the projective extension of such a flat affine plane the group L_2 fixes exactly two points and one line (the infinite points of the horizontals and verticals, and the line at infinity).

THEOREM 2.7.10 (Arc Planes and Stretchshift Planes) *An arc plane of type* L_2 *is a flat affine plane if and only if it is a stretchshift plane generated by one arc or the image of a stretchshift plane generated by two arcs under one of the homeomorphisms* $H \to H : (x, y) \mapsto (x, \pm y + ax)$, *where* $a \in \mathbf{R}$.

If $\mathcal{P} = (P, \mathcal{L})$ *is a flat projective plane whose automorphism group contains a subgroup* Δ *isomorphic to* L_2 *that fixes precisely one line* L, *and acts transitively on* $P \setminus L$, *then* Δ *fixes one or two points on this line and* $\mathcal{P}$ *is a projective stretchshift plane.*

In Groh [1982b] the question when two arc planes are isomorphic is answered completely. In particular, the arc planes of type $\mathbf{R}^2$ that are isomorphic to arc planes of type L_2 are described explicitly; see Groh [1982b] 5.3:

- The Euclidean plane is isomorphic to the arc plane of type L_2 generated by the graphs of the functions

$$\mathbf{R}^+ \to \mathbf{R} : x \to \ln x \text{ and } \mathbf{R}^+ \to \mathbf{R} : x \to -\ln x.$$

- The arc plane of type $\mathbf{R}^2$ generated by the functions

$$\mathbf{R} \to \mathbf{R} : x \to e^x \text{ and } \mathbf{R} \to \mathbf{R} : x \to -e^x$$

is isomorphic to the restriction of the Euclidean plane to the half-plane H.

- The arc plane of type $\mathbf{R}^2$ generated by the function

$$\mathbf{R} \to \mathbf{R} : x \to e^x$$

is isomorphic to the arc plane of type L_2 generated by the function

$$\mathbf{R}^+ \to \mathbf{R} : x \to \ln x.$$

- The arc planes of type $\mathbf{R}^2$ generated by the functions

$$\mathbf{R} \to \mathbf{R} : x \mapsto e^x \text{ and } \mathbf{R} \to \mathbf{R} : x \mapsto -\mathrm{sgn}(s)e^{sx},$$

where $s \neq 0, 1$ is isomorphic to the arc plane of type L_2 generated by the function

$$\mathbf{R}^+ \to \mathbf{R} : x \mapsto x^s.$$

See Groh [1976], [1979], [1982a], and [1982b] for more information about arc planes.

2.7.6 Skew Hyperbolic Planes

The family of flat projective planes described in the following is one of the exceptional families of flat projective planes of group dimension 3.

For $t \in \mathbf{R}$, define the generating curve

$$\{(x,0) \in \mathbf{R}^2 \mid -1 \leq x \leq 1\} \cup \{(x,y) \in \mathbf{R}^2 \mid txy \geq 0, y^2 = t^2(x^2-1)\}.$$

For $t \neq 0$ this curve consists of the part of the x-axis contained in the unit circle and two halves of two branches of a certain hyperbola. See Figure 2.13. For $t = 0$ the curve coincides with the x-axis. In either case the curve together with the infinite point of its asymptote, the Euclidean line of slope t through the origin, is a topological circle L_t embedded in the real projective plane.

We construct a flat projective plane as follows. Let Δ be the group of automorphisms of the classical flat projective plane that fix the unit circle and induce orientation-preserving homeomorphisms of the circle to itself. This group is the so-called hyperbolic motion group $\mathrm{PO}_3(\mathbf{R}, 1)$ which is isomorphic to $\mathrm{PSL}_2(\mathbf{R})$. The lines of the plane are the exterior lines and tangents of the unit circle in the classical plane plus the images of L_t under elements of the group Δ. We call the resulting flat projective plane the *skew hyperbolic plane* $\mathcal{H}_t$.

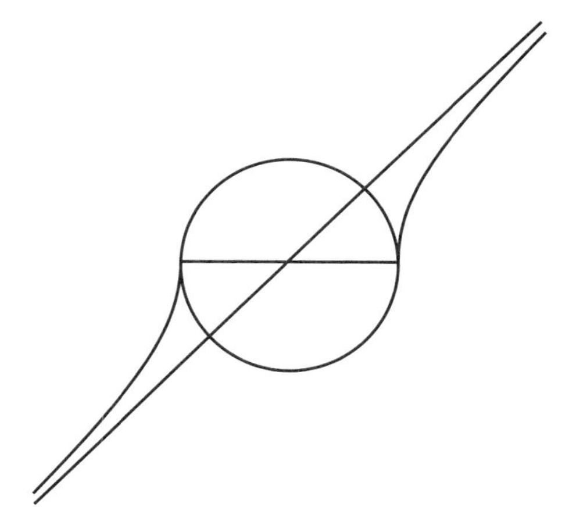

Fig. 2.13. The generating curve L_t

THEOREM 2.7.11 (Characterization Skew Hyperbolic Planes) *A flat projective plane is isomorphic to a skew hyperbolic plane if and only if its automorphism group contains a closed subgroup isomorphic to the 3-dimensional Lie group* $\mathrm{PSL}_2(\mathbf{R})$.

Furthermore, $\mathcal{H}_t$ *is isomorphic to* $\mathcal{H}_{-t}$ *for all* $t \in \mathbf{R}$ *and* $\mathcal{H}_0$ *is the classical flat projective plane. For* $t > 0$ *the planes are nonclassical and pairwise nonisomorphic. Their full automorphism group is isomorphic to* $\mathrm{PSL}_2(\mathbf{R})$. *This automorphism group fixes the unit circle, but has neither fixed points, nor fixed lines.*

See Salzmann et al. [1995] Section 35 for this result and further information about this class of flat projective planes.

2.7.7 Cartesian Planes

Let $\alpha, \beta, \gamma, \bar{\gamma}, c$ be positive parameters. We define a new multiplication on $\mathbf{R}$ as follows:

$$a * x = \begin{cases} ax & \text{for } a, x \geq 0, \\ a^{\alpha}x & \text{for } a \geq 0, x \leq 0, \\ ax^{\beta} & \text{for } a \leq 0, x \geq 0, \\ |a|^{\gamma}c|x|^{\bar{\gamma}} & \text{for } a, x \leq 0. \end{cases}$$

Then $\mathbf{R}$ provided with the new multiplication and the usual addition is a *Cartesian field.* This implies that the set consisting of all vertical

Euclidean lines plus all graphs of the 'linear' functions

$$\mathbf{R} \to \mathbf{R} : x \mapsto a * x + b,$$

where $a, b \in \mathbf{R}$, is the line set of a flat affine plane on $\mathbf{R}^2$.

Figure 2.14 depicts a line pencil in one of these flat affine planes. The other lines in this plane are vertical translates of the lines in this pencil.

Let $\alpha^{-1} + \beta^{-1} > 1, \gamma = \alpha k$ and $\bar{\gamma} = \beta k$ with $k = (\alpha + \beta - \alpha\beta)^{-1} > 0$. The corresponding flat affine plane is denoted by $\mathcal{P}_{\alpha,\beta,c}$ and is called a *Cartesian plane.* It admits the automorphisms

$$\mathbf{R}^2 \to \mathbf{R}^2 : (x, y) \mapsto (a(b * x), ay + t),$$

where $a, b, t \in \mathbf{R}$, $a, b > 0$.

These automorphisms form a group that is isomorphic to $\mathbf{R} \times \mathrm{L}_2$.

THEOREM 2.7.12 (Characterization of Cartesian Planes) *A flat projective plane is isomorphic to one of the pairwise nonisomorphic nonclassical projective planes $\overline{\mathcal{P}}_{\alpha,\beta,c}$ with parameters $0 < \alpha, \beta \leq 1$, $(\alpha, \beta) \neq (1, 1)$, $0 < c$, and $c \leq 1$ if $1 \in \{\alpha, \beta\}$ if and only if its full collineation group has dimension 3 and fixes precisely two points and two lines. The fixed points are the infinite points of the horizontals and verticals of $\mathcal{P}_{\alpha,\beta,c}$ and the fixed lines are the line at infinity of $\mathcal{P}_{\alpha,\beta,c}$ and the line that corresponds to the y-axis.*

Furthermore, $\overline{\mathcal{P}}_{1,1,1}$ is the classical flat projective plane and $\mathcal{P}_{1,1,c}$ is the Moulton plane $\mathcal{M}_c$.

See Salzmann et al. [1995] Section 37 and, in particular, Theorem 37.4 for more detailed information about the Cartesian planes. We call a (projective) Cartesian plane ($\overline{\mathcal{P}}_{\alpha,\beta,c}$) $\mathcal{P}_{\alpha,\beta,c}$ *proper* if its parameters α, β, c are as in this theorem, that is, $0 < \alpha, \beta \leq 1$, $(\alpha, \beta) \neq (1, 1)$, $0 < c$, and $c \leq 1$ if $1 \in \{\alpha, \beta\}$.

2.7.8 Strambach's $\mathrm{SL}_2(\mathbf{R})$*-Plane*

This proper $\mathbf{R}^2$-plane of group dimension 3 was constructed in Strambach [1968]. We start with the Euclidean plane and replace all lines that do not pass through the origin by all images of the topological arc $\{(x, x^{-1}) \in \mathbf{R}^2 \mid x \in \mathbf{R}^+\}$ under the group $\mathrm{SL}_2(\mathbf{R})$, that is, the group of all real 2×2 matrices of determinant 1. Note that the lines that do not pass through the origin form an orbit under the group $\mathrm{SL}_2(\mathbf{R})$. Strambach proved that this plane does not admit a strong group-preserving

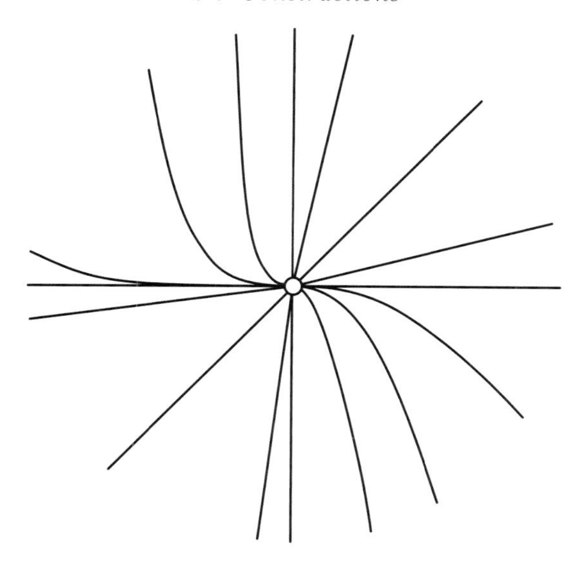

Fig. 2.14. The line pencil through the origin in a Cartesian plane

open embedding into any compact projective plane. However, Stroppel [1993b] Theorem 7 showed that it is isomorphic to the restriction of some flat projective plane to one of its open convex subsets.

2.7.9 Integrated Foliations

The nonvertical lines in the Euclidean plane are the graphs of all linear functions $\mathbf{R} \to \mathbf{R} : x \mapsto ax + b,\ a, b \in \mathbf{R}$. The linear functions are the integral functions of the constant functions $\mathbf{R} \to \mathbf{R} : x \mapsto a,\ a \in \mathbf{R}$.

In this way the Euclidean plane 'is the integral' of the parallel class consisting of the horizontal lines.

If we replace the parallel class in this construction by the parallel class consisting of all lines with slope $k \neq 0$, then the 'integral' of this parallel class is one of the parabola models of the Euclidean plane that we considered above.

More generally, let F be a 1-unisolvent set of continuous functions on $\mathbf{R}$, that is, for every $(x, y) \in \mathbf{R}^2$ there is exactly one function in the set such that $f(x) = y$. We can define a geometry whose point set is $\mathbf{R}^2$, and whose lines are the verticals plus the graphs of all integral functions of the functions in the set. As a consequence of Theorem 7.2.2, this geometry is always an $\mathbf{R}^2$-plane. It is a flat affine plane if and only if for any two distinct functions f and g in the 1-unisolvent set

$$\lim_{x\to\pm\infty}\left|\int_0^x (f(t)-g(t))dt\right| = \infty.$$

See Subsection 5.3.4 and Polster [1995b] for details about this construction and how it extends to a constructions of flat Laguerre planes. In Chapters 5 and 7 we will consider various generalizations of this construction that allow us to construct rank $n+1$ geometries by integrating rank n geometries.

2.7.10 Different Ways to Cut and Paste

In the following we describe a number of simple methods that allow us to combine flat linear spaces of the same type into new flat linear spaces. We describe various counterparts of these constructions in the following chapters.

For more information about cut-and-paste constructions in flat linear spaces see Groh [1981], Groh [1982b], Stroppel [1993a], and Polster–Steinke [1995]; see also the following subsection.

2.7.10.1 Cut and Paste in the Point Set

Let $\mathcal{R}_i = (P, \mathcal{L}_i)$, $i = 1, 2$, be two ideal flat linear spaces that share a closed topological arc or topological n-gon $\mathcal{O}$. The set $P \setminus \mathcal{O}$ has two connected components. Let C^1 be the component homeomorphic to the open unit disk and let C^2 be the other component.

We define a new geometry $\mathcal{R}$. Its point set is P and its lines are as follows.

(i) All lines in $\mathcal{R}_2$ that do not intersect $\mathcal{O}$ plus all lines that intersect $\mathcal{O}$ in an interval, including those intervals that contain only one point.

(ii) If q and r are distinct points on $\mathcal{O}$, let N_i be the restriction of the line qr in $\mathcal{R}_i$ to C^i. Then $N_1 \cup N_2 \cup \{q, r\}$ is a line in $\mathcal{R}$.

PROPOSITION 2.7.13 (Cut and Paste I) *Let $\mathcal{R}_i$, $i = 1, 2$, be two ideal flat linear spaces of the same type sharing the same point set P and a closed topological arc or topological n-gon $\mathcal{O}$. Then the geometry $\mathcal{R}$ defined above is a flat linear space of the same type as $\mathcal{R}_i$. If both $\mathcal{R}_1$ and $\mathcal{R}_2$ are flat affine planes, then so is $\mathcal{R}$.*

Proof. We only prove this in the case that $P = \mathbf{R}^2$, that is, in the case that the geometries $\mathcal{R}_i$, $i = 1, 2$, are $\mathbf{R}^2$-planes. The proofs for point Möbius strip planes and flat projective planes run along the same lines.

It suffices to check that $\mathcal{R}$ is a linear space. Consider the pencil of lines in $\mathcal{R}$ through an arbitrary point $p \in P$. First, assume that $p \in C^1$ and let $o \in \mathcal{O}$. Repeated application of Lemma 2.3.2 or Theorem 2.3.4 yields that the line op depends continuously on the point o. Furthermore, Theorem 2.2.3(vi) implies that for distinct $o_1, o_2 \in \mathcal{O}$ the lines o_1p and o_2p coincide or intersect only in the point p. Now repeated application of Theorem 2.3.4 yields that, as we move o around $\mathcal{O}$, the line op will hit every point of $\mathbf{R}^2$ different from p exactly once. This means that every point different from p is contained in exactly one line in the pencil of lines through p. For $p \in \mathcal{O} \cup C^2$ similar arguments yield the same conclusion. This proves that $\mathcal{R}$ is an $\mathbf{R}^2$-plane.

Assume that $\mathcal{R}_1$ and $\mathcal{R}_2$ are flat affine planes. Then similar arguments show that two distinct lines in $\mathcal{R}$ intersect in a point if and only if the two lines in $\mathcal{R}_2$ that gave rise to them intersect in a point. This implies that $\mathcal{R}$ is also a flat affine plane. □

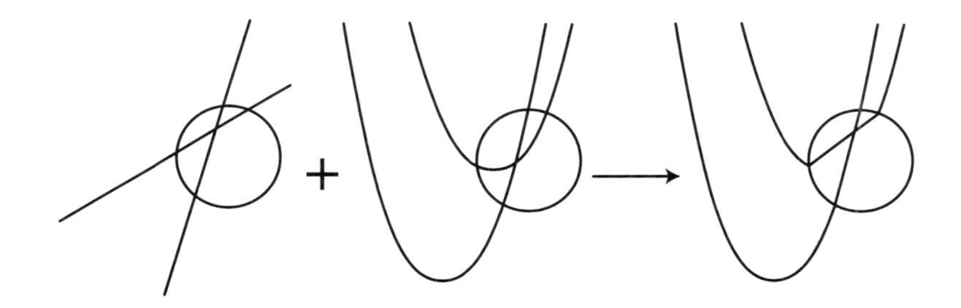

Fig. 2.15. Gluing with respect to an oval

Let $\mathcal{R}_i = (P, \mathcal{L})$, $i = 1, 2$, be two ideal flat linear spaces as above and let $\mathcal{O}_i$ be a simply closed topological arc or topological n-gon in $\mathcal{R}_i$. Then there is a homeomorphism $\gamma : P \to P$ such that $\gamma(\mathcal{O}_1) = \mathcal{O}_2$ and such that vertices are mapped to vertices if we are dealing with n-gons. This means that $\gamma(\mathcal{R}_1)$ can be combined with $\mathcal{R}_2$ into a new flat linear space using the above result.

Cutting and pasting as above between two lines is also possible. We start with two flat affine planes or two flat projective planes $\mathcal{R}_i$, $i = 1, 2$, sharing the same point set P and two distinct lines K and L.

If the two geometries are flat projective planes, then $P \setminus (K \cup L)$ has two connected components; let us call them C^1 and C^2.

If the two geometries are flat affine planes and K and L are parallel, let C^1 be the connected component of $P \setminus (K \cup L)$ bounded by these two lines and let C^2 be the union of the other two components. If K and L intersect in a point p, let J be a line through p different from K and L and let C^1 be the union of the two of the four connected components that contain parts of J. Let C^2 be the union of the other two connected components.

We define a new geometry $\mathcal{R}$. Its point set is P and its lines are as follows.

(i) All lines in $\mathcal{R}_i$ that are completely contained in $K \cup L \cup C^i$ (this includes K and L).

(ii) If q is a point on K and r is a point on L such that neither of the two points is contained in both lines, let N_i be the restriction of the line qr in $\mathcal{R}_i$ to C^i. Then $N_1 \cup N_2 \cup \{q, r\}$ is a line in $\mathcal{R}$.

(iii) (This only applies in the case that K and L intersect in the point p and we are dealing with flat affine planes.) Let $\{M, N\} = \{K, L\}$, let $r \neq p$ be a point of M, let N_i be the restriction to C^i of the parallel to N through r in $\mathcal{R}_i$. Then $N_1 \cup N_2 \cup \{r\}$ is a line in $\mathcal{R}$.

PROPOSITION 2.7.14 (Cut and Paste II) *Let $\mathcal{R}_i$, $i = 1, 2$, be two flat affine planes or two flat projective planes sharing the same point set P and two lines. Then the geometry $\mathcal{R}$ defined above is also a flat affine plane or flat projective plane, respectively.*

The proof of this result runs along the same lines as that of the previous result. Figure 2.16 illustrates the combining of two flat affine planes along two (common) parallel lines.

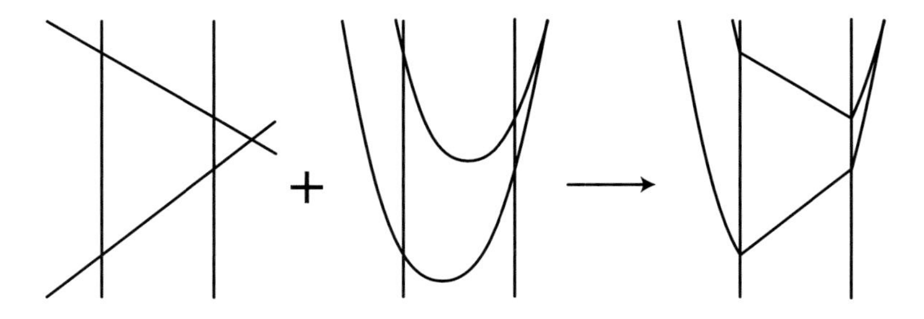

Fig. 2.16. Cut and paste with respect to two lines

Let $\mathcal{R}_i = (P, \mathcal{L}_i)$, $i = 1, 2$, be two ideal flat linear spaces of the same type as above. Choose two pairs of distinct lines $K_i, L_i \in \mathcal{R}_i$ in such a way that $|K_1 \cap L_1| = |K_2 \cap L_2|$. Then it is easy to construct a homeomorphism γ of the common point set P to itself such that $\gamma(K_1) = K_2$ and $\gamma(L_1) = L_2$. This means that $\gamma(\mathcal{R}_1)$ (which, of course, is isomorphic to $\mathcal{R}_1$) and $\mathcal{R}_2$ are two ideal flat linear spaces on P that share two lines. This implies that there are infinitely many ways to combine any two given flat affine or projective planes into new planes.

Combining two point Möbius strip planes $\mathcal{R}_i$, $i = 1, 2$, that share the same point set P and two lines K and L is also possible. Just one-point-compactify P by a point ∞ to the real projective plane P'. Then the two flat projective planes on P' that correspond to $\mathcal{R}_1$ and $\mathcal{R}_2$ can be combined into a new flat projective plane $\mathcal{R}'$ using the common lines that correspond to K and L. The restriction $\mathcal{R}$ of $\mathcal{R}'$ to P is the new point Möbius strip plane we have been looking for.

2.7.10.2 Cut and Paste in the Line Set

Dualizing Proposition 2.7.14 in the case of flat projective planes suggests also trying to cut and paste directly in the line space of flat linear spaces.

Let $\mathcal{R} = (P, \mathcal{L})$ be an ideal flat linear space and let p and q be two distinct points. Let C^+ be a connected component of $pq \setminus \{p, q\}$ that has both p and q as boundary points, let $C^- = pq \setminus (\{p, q\} \cup C^+)$, let $\mathcal{L}^{p,q}$ be the set of all lines in $\mathcal{L}$ that do not intersect pq or contain p and/or q, and let $\mathcal{L}^+$ and $\mathcal{L}^-$ be the set of lines that intersect pq in C^+ and C^-, respectively. Clearly, the set of lines $\mathcal{L}$ is the disjoint union of the three sets $\mathcal{L}^{p,q}$, $\mathcal{L}^+$, and $\mathcal{L}^-$.

PROPOSITION 2.7.15 (Cut and Paste III) *Let $\mathcal{R}_i = (P, \mathcal{L}_i)$, $i = 1, 2$, be two ideal flat linear spaces of the same type. Suppose that there are distinct points p and q such that $\mathcal{L}_1^{p,q} = \mathcal{L}_2^{p,q}$. We define a new set of lines $\mathcal{L} = \mathcal{L}_1^{p,q} \cup \mathcal{L}_1^+ \cup \mathcal{L}_2^-$. Then $\mathcal{R} = (P, \mathcal{L})$ is an ideal flat linear space of the same type as $\mathcal{R}_i$. If both $\mathcal{R}_1$ and $\mathcal{R}_2$ are flat affine planes, then so is $\mathcal{R}$.*

Proof. We first show that $\mathcal{R}$ is a flat linear space. For this it suffices to prove that, given two distinct points r and s in $\mathcal{R}_i$, it is possible to infer, by only looking at lines in the common set $\mathcal{L}_i^{p,q}$, which of the three disjoint sets $\mathcal{L}_i^{p,q}$, $\mathcal{L}_i^+$, and $\mathcal{L}_i^-$ the connecting line rs is contained in. If both points are contained in the line pq, then $rs = pq \in \mathcal{L}_i^{p,q}$. Without loss of generality we may assume that $r \notin pq$. Then Figure 2.17

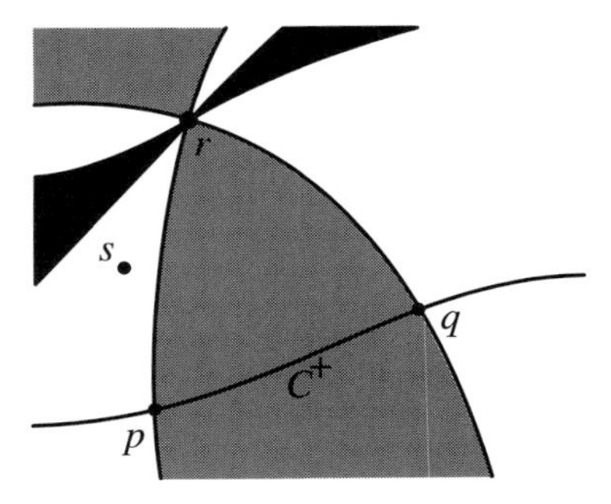

Fig. 2.17. Existence of a connecting line

illustrates that the lines through r in the common set $\mathcal{L}_i^{p,q}$ partition the point set of the plane into four disjoint sets: (1) the grey region; (2) the white region; (3) $rp \cup rq$; and (4) the black region that is exhausted by the lines in $\mathcal{L}_i^{p,q}$ that do not intersect pq. Note that this last set may be empty if pq is an **S**-line or consist of only one line (e.g., if we are dealing with affine planes); see Theorem 2.3.6. Since lines in a flat linear space intersect transversally, it is clear that, depending on whether the point s is contained in region 1, 2, or $3 \cup 4$, the connecting line rs will be contained in $\mathcal{L}_i^+$, $\mathcal{L}_i^-$, or $\mathcal{L}_i^{p,q}$, respectively. For example, if r and s are situated as in the diagram, then rs is contained in $\mathcal{L}_i^-$.

We proceed to show that $\mathcal{R}$ is a flat affine plane if both $\mathcal{R}_1$ and $\mathcal{R}_2$ are flat affine planes. Let L be a line through p. Then the set of lines E_L consisting of L and all lines in the line set of $\mathcal{R}$ that are parallel to L clearly form a partition of P. Furthermore, given a line M in $\mathcal{R}$ and a point r, there are a line M' through p parallel to M and a line through r in $E_{M'}$. This means that if $\mathcal{R}$ is a flat affine plane, then the sets of the form E_L are its parallel classes. It remains to show that lines $M \in E_{M'}$ and $N \in E_{N'}$ with $M' \neq N'$ intersect in a point. This is clearly the case if either M or N is contained in $\mathcal{L}_i^{p,q}$ or if M and N are both contained in either $\mathcal{L}_1^+$ or $\mathcal{L}_2^-$. So assume that $M \in \mathcal{L}_1^+$ and $N \in \mathcal{L}_2^-$. Then we are in the situation depicted in Figure 2.18. We want to show that the open point of intersection really exists. For this we introduce the auxiliary line O through p 'in between' M' and N'. Then the four points of intersection r, s, t, and u exist. Here t and u are the points of intersection of M and N with the line O, respectively. It is possible that t coincides with u, in which case $t = u$ is the point of intersection we are looking for. It is also possible that, in contrast to the situation

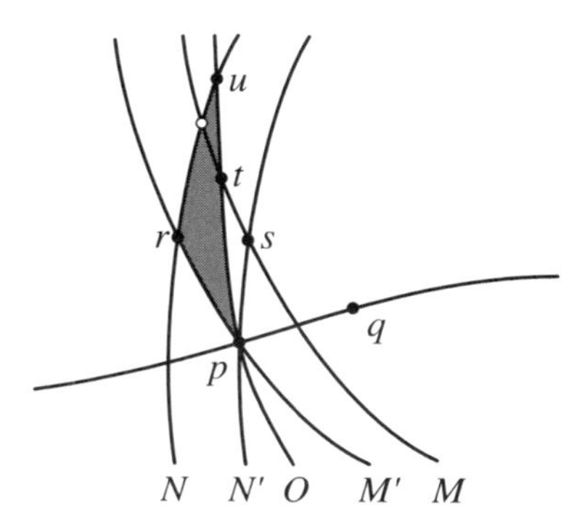

Fig. 2.18. Existence of a point of intersection

depicted in the diagram, p–u–t is the order of the three points p, t, and u on the line O, in which case the following argument can be modified in a straightforward manner.

In the situation depicted in the diagram we can be sure that, approaching from the point s, the line M crosses into the grey triangle at the point t. It has to intersect the boundary of this triangle at a second point, but this point cannot be contained in the interval $[p, u]$ because M cannot intersect O twice, nor in the interval $[p, r]$ because M is parallel to M'. This leaves us with a point of intersection in the interval $[r, u]$, which is exactly what we were looking for. □

Assume that a flat linear space $\mathcal{R} = (P, \mathcal{L})$ is embedded in a larger flat linear space $\mathcal{R}' = (P', \mathcal{L}')$. This means that $P \subset P'$ and that every line in $\mathcal{L}$ is a subset of a line in $\mathcal{L}'$. Then the restrictions to P of the lines in $\mathcal{L}'$ through a point $p \in P' \setminus P$ form a partition of P into **R**-lines that belong to $\mathcal{L}$. This suggests using a set of **R**-lines that partition the point set of a flat linear space as a 'virtual' point and trying to cut and paste with respect to virtual points.

Let $\mathcal{R} = (P, \mathcal{L})$ be an ideal flat linear space, let p be a point, and let L be an **R**-line through p. This implies that $\mathcal{R}$ is either a point Möbius strip plane or an $\mathbf{R}^2$-plane. Furthermore, let L^+ and L^- be the two connected components of $L \setminus \{p\}$, let $\mathcal{L}^{p,L}$ be the set containing all lines that do not intersect L or contain p (this set contains L), and let $\mathcal{L}^+$ and $\mathcal{L}^-$ consist of the lines that intersect L in L^+ and L^-, respectively. Clearly, $\mathcal{L}$ is the disjoint union of the sets $\mathcal{L}^{p,L}$, $\mathcal{L}^+$, and $\mathcal{L}^-$.

Just to make the connection with what we said before, note that, given an **R**-line L in an ideal flat linear space, there is always at least one partition of its point set into $\mathbf{R}$-lines that contains L. Given $p \in L$, any such partition is contained in the set $\mathcal{L}^{p,L}$. In point Möbius strip planes and flat affine planes the set $\mathcal{L}^{p,L}$ contains exactly one partition of the point set into $\mathbf{R}$-lines.

PROPOSITION 2.7.16 (Cut and Paste IV) *Let $\mathcal{R}_i = (P, \mathcal{L}_i)$, $i = 1, 2$, be two ideal flat linear spaces of the same type. Suppose that the planes have an $\mathbf{R}$-line L in common and there is a point $p \in L$ such that $\mathcal{L}_1^{p,L} = \mathcal{L}_2^{p,L}$. Let $\mathcal{L} = \mathcal{L}_1^{p,L} \cup \mathcal{L}_1^+ \cup \mathcal{L}_2^-$. Then $\mathcal{R} = (P, \mathcal{L})$ is a flat linear space of the same type as $\mathcal{R}_i$. If both $\mathcal{R}_1$ and $\mathcal{R}_2$ are flat affine planes, then so is $\mathcal{R}$*

The proof of this result is a variation of the proof of Proposition 2.7.15.

Let us consider the case of two virtual points. Let $\mathcal{R} = (P, \mathcal{L})$ be a flat linear space and let Q and R be two *transversal* partitions of P into $\mathbf{R}$-lines of $\mathcal{L}$. Here transversal means that, given a line in Q and a line in R, these two lines intersect in a point. This implies that there is a homeomorphism $\gamma : P \to \mathbf{R}^2$ that identifies the elements of Q and R with the verticals and horizontals, respectively. In particular, the geometry $\mathcal{R}$ is an $\mathbf{R}^2$-plane. Also the image of a line not in $Q \cup R$ under γ is the graph of a strictly increasing or strictly decreasing function $I \to \mathbf{R}$, where $I \subset \mathbf{R}$ is an interval. Let $\mathcal{L}^{Q,R} = Q \cup R$, and let $\mathcal{L}^+$ and $\mathcal{L}^-$ be the lines that correspond to strictly increasing and strictly decreasing functions, respectively.

PROPOSITION 2.7.17 (Cut and Paste V) *Let $\mathcal{R}_i = (P, \mathcal{L}_i)$, $i = 1, 2$, be two $\mathbf{R}^2$-planes. Suppose that there are two transversal partitions $Q, R \subset \mathcal{L}_1 \cap \mathcal{L}_2$ of P into $\mathbf{R}$-lines. Let $\mathcal{L} = \mathcal{L}_1^{Q,R} \cup \mathcal{L}_1^+ \cup \mathcal{L}_2^-$. Then $\mathcal{R} = (P, \mathcal{L})$ is also an $\mathbf{R}^2$-plane. If both $\mathcal{R}_1$ and $\mathcal{R}_2$ are flat affine planes, then so is $\mathcal{R}$.*

The proof of this result is a variation of the proof of Proposition 2.7.15.

Let $\mathcal{R}_i = (P, \mathcal{L}_i)$, $i = 1, 2$, be two $\mathbf{R}^2$-planes that both contain two transversal partitions $Q_i, R_i \subset \mathcal{L}_i$. Then it is possible to find a homeomorphism $\gamma : P \to P$ such that $\gamma(Q_1) = Q_2$ and $\gamma(R_1) = R_2$. This means that $\gamma(\mathcal{R}_1)$ (which, of course, is isomorphic to $\mathcal{R}_1$) and $\mathcal{R}_2$ are two $\mathbf{R}^2$-planes on P that share two transversal partitions. These two planes can be combined into a new plane using the above proposition.

However, a similar alignment is not always possible for the other kinds of common line sets that we considered before. For example, given two flat projective planes $\mathcal{P}_i = (P, \mathcal{L}_i)$, $i = 1, 2$, and points $p_i, q_i \in P_i$, it is not always possible to find a homeomorphism $\gamma : P \to P$ that maps the set of lines $\mathcal{L}_1^{p_1,q_1}$ (as in Proposition 2.7.15) to $\mathcal{L}_2^{p_2,q_2}$; see Rosehr [20XX] Example 8. Nevertheless, combining flat projective planes into a new flat projective plane 'along such sets' is still possible as follows. By removing the line p_iq_i and all the points contained in it from the flat projective planes $\mathcal{P}_i$, it turns into a flat affine plane and the set $\mathcal{L}_i^{p_i,q_i}$ into the union of two transversal partitions. Then the two affine planes can be aligned, combined into a new flat affine plane, and the resulting flat affine plane completed into a new flat projective plane.

Finally, we remark that Proposition 2.7.13 also 'dualizes' as follows. Let $\mathcal{R} = (P, \mathcal{L})$ be an ideal flat linear space and let $\mathcal{O}$ be a closed topological arc or topological n-gon in $\mathcal{R}$. Let $\mathcal{L}^e$, $\mathcal{L}^t$, and $\mathcal{L}^s$ be the sets of all lines that do not intersect $\mathcal{O}$, that intersect $\mathcal{O}$ in an interval (this interval may consist of one point only), and intersect $\mathcal{O}$ in exactly two points, respectively. Clearly, $\mathcal{L}$ is the disjoint union of these three sets.

PROPOSITION 2.7.18 (Cut and Paste VI) *Let $\mathcal{R}_i = (P, \mathcal{L}_i)$, $i = 1, 2$, be ideal flat linear spaces of the same type sharing a closed topological arc or topological n-gon $\mathcal{O}$ and $\mathcal{L}_1^t = \mathcal{L}_2^t$. Let $\mathcal{L} = \mathcal{L}_1^e \cup \mathcal{L}_2^t \cup \mathcal{L}^2$. Then $\mathcal{R} = (\mathcal{P}, \mathcal{L})$ is a flat linear space of the same type as $\mathcal{R}_i$. If both $\mathcal{R}_1$ and $\mathcal{R}_2$ are flat affine planes, then so is $\mathcal{R}$.*

This is a straightforward generalization of Polster–Steinke [1995] Proposition 2.

2.7.10.3 Separating Sets

In the following chapters we will see that the effortless ways of cutting and pasting in the point set do not have counterparts for flat circle planes of higher rank, whereas the cut-and-paste techniques in the line set do have such counterparts. The concept of a separating set is what seems to be behind all the different cut-and-paste constructions in line and circle sets described in this book.

In its strongest form a separating set for the flat linear spaces living on the common point set P is a set C of **R**- and **S**-lines that has the following properties.

(i) If C is the subset of a line set of a flat linear space, then it separates the line set into a finite number $n \geq 2$ of connected components.

(ii) The number n is independent of the different line sets of flat linear spaces C may be contained in.

(iii) Let $\mathcal{R}_i = (P, \mathcal{L}_i)$, $i = 1, 2$, be two flat linear spaces containing C. Then there are labellings of the connected components of the line sets of both planes with numbers from 1 to n such that replacing component number m in the line set of $\mathcal{L}_1$ by the respective component in the line set of $\mathcal{L}_2$ gives a line set of a new flat linear space.

For example, if the point set P is the real projective plane, then the sets of the form $\mathcal{L}^{p,q}$ used in Proposition 2.7.15 are clearly separating sets. Similarly, if $P = \mathbf{R}^2$, then the sets of the form $\mathcal{L}^{Q,R}$ used in Proposition 2.7.17 are examples of separating sets.

2.7.11 Pasted Planes

The $\mathbf{R}^2$-planes whose automorphism groups fix one line and act transitively on each of the two connected components of the complement of this line proved to be very important in the group-dimension classifications of flat linear spaces and, in particular, the classifications of the flat projective planes of group dimension at least 2 and the $\mathbf{R}^2$-planes of group dimension at least 3.

In the course of classifying these planes it was necessary to figure out how exactly two $\mathbf{R}^2$-planes can be pasted together along a new line to form new $\mathbf{R}^2$-planes. This was done by Groh [1981] Section 2.

Let $(P_1, \mathcal{L}_1)$ and $(P_2, \mathcal{L}_2)$ be $\mathbf{R}^2$-planes and assume that P_1 is disjoint from P_2. Let F be a subset of $\mathcal{L}_1 \times \mathcal{L}_2$ and let G be the set of all those lines in $\mathcal{L}_1 \cup \mathcal{L}_2$ that are not a first or second component of any element of F. Let $\omega : P_1 \cup P_2 \to P \setminus C$ be a homeomorphism, where P is homeomorphic to $\mathbf{R}^2$ and C is a topological line in P. Groh defines the (F, ω)-*pasting sum* to be the pair $(P, \mathcal{L})$, where

$$\mathcal{L} = \{\overline{\omega(l_1)} \cup \overline{\omega(l_2)} \mid (l_1, l_2) \in F\} \cup \{\omega(l) \mid l \in G\} \cup \{C\}.$$

We call a (F, ω)-pasting sum a *pasted plane* that is *made up of* P_1 *and* P_2 if it is an $\mathbf{R}^2$-plane. Note that every $\mathbf{R}^2$-plane can be considered as a pasted plane that is made up of the restrictions of the plane to the two connected components of the complement of a line C. Groh [1981] Section 2 gives necessary and sufficient conditions for an (F, ω)-pasting sum to be a pasted plane. These conditions ensure that the lines in an (F, ω)-pasting sum are $\mathbf{R}$-lines and that any two distinct points are joined by exactly one such line.

In the following subsubsections we describe the pasted planes made up of arc planes that are important for the classification of all flat projective planes of group dimension at least 2 and the classification of all $\mathbf{R}^2$-planes of group dimension at least 3. We also give names to the different kinds of planes to be able to refer to these planes more easily. Only if a name starts with an 'Affine' are the respective pasted planes actually flat affine planes. Usually, a plane is of some 'Type X'. Here X is a connected closed subgroup of its automorphism groups isomorphic to $\mathbf{R}^2$ or L_2.

In most of the following constructions P_1 and P_2 are the left and right open half-planes, respectively, and C is the y-axis or an interval on it. Usually we do not specify ω explicitly but only say how points on C get 'identified' when approaching from the left or right or just state the pasted lines.

Of course, it is possible to iterate the pasting process and the so-called triangle planes described in Subsubsection 2.7.11.6 are flat projective planes of group dimension at least 2 that are projective extensions of flat affine planes that are pasted together from more than two arc planes of type $\mathbf{R}^2$.

2.7.11.1 Pasted Exponential Arc Planes (Type L_2*)*

The $\mathbf{R}^2$-planes described in this subsubsection are precisely the group dimension 3 pasted planes made up of exponential arc planes; see p. 72. These pasted sums are the planes $E(v,\alpha,\beta)$, $-1 \leq v < 0$, $\alpha,\beta \in \mathbf{R}$, which are defined as follows. Let P_1 and P_2 be the left and the right open half-plane, respectively. Furthermore, let Y^+ consist of all the points of the positive y-axis and let $\overline{P}_i = P_i \cup Y^+$, where $i = 1,2$. We glue $\overline{P}_1$ to $\overline{P}_2$ by identifying the point $(0,y) \in \overline{P}_1$ with the point $(0, y^{1/v}) \in \overline{P}_2$ for all $y \in \mathbf{R}^+$. The resulting topological space is homeomorphic to $\mathbf{R}^2$ and serves as the common point set of the planes under discussion. The lines are the verticals except for the y-axis, the open half-lines emanating from the origin except for the negative y-axis, the horizontal defined by $y = 1$, and the graphs of the functions

$$\mathbf{R}^+ \to \mathbf{R} : x \mapsto \begin{cases} -(-x)^{\alpha+1} & \text{for} \quad \alpha < 0, \\ -x\ln(-x) & \text{for} \quad \alpha = 0, \\ (-x)^{\alpha+1} & \text{for} \quad \alpha > 0, \end{cases}$$

$$\mathbf{R}^- \to \mathbf{R} : x \mapsto \begin{cases} x^{v\beta+1} & \text{for} \quad \beta < 0, \\ -vx\ln x & \text{for} \quad \beta = 0, \\ -x^{v\beta+1} & \text{for} \quad \beta > 0. \end{cases}$$

Furthermore, all images of these lines under the group consisting of the following homeomorphisms of the point set P to itself:

$$(x,y) \mapsto \begin{cases} (e^{r-t}x, e^r(y+e^{-t}sx)) & \text{for} \quad (x,y) \in \overline{P}_1, \\ (e^{r/v-t}x, e^{r/v}(y-e^{-t}sx)) & \text{for} \quad (x,y) \in \overline{P}_2, \end{cases}$$

where $r, s, t \in \mathbf{R}$.

This group is isomorphic to the group $L_2 \times \mathbf{R}$. The two planes $E(v,\alpha,\beta)$ and $E(v',\alpha',\beta')$ are isomorphic if $(v,\alpha,\beta) = (v',\alpha',\beta')$ or if $v = v' = -1$ and $(\alpha,\beta) = (-\alpha',-\beta')$. For $v \neq -1$ or $\beta \neq -\alpha$ the group $L_2 \times \mathbf{R}$ is the full group of automorphisms. For $v = -1$ and $\beta = -\alpha$ the full group is generated by $L_2 \times \mathbf{R}$ and the involution $P \to P : (x,y) \mapsto (-x,y)$. None of the planes is a flat affine plane, nor is any of the planes isomorphic to a sub-$\mathbf{R}^2$-plane of the Euclidean plane. The restrictions of $E(v,\alpha,\beta)$ to P_1 and P_2 are isomorphic to exponential arc planes.

See Betten–Ostmann [1978] Satz 13 for more information about these planes. Also, by Groh et al. [1983] Theorem 7.5.B, given a proper $\mathbf{R}^2$-plane whose automorphism group has a 3-dimensional connected closed subgroup isomorphic to $L_2 \times \mathbf{R}$ which fixes precisely one line, this plane is isomorphic to one of the above planes.

2.7.11.2 Pasted Hyperbolic Arc Planes (Type $\mathbf{R}^2$)

The planes described in this subsubsection are precisely the group dimension 3 pasted planes made up of hyperbolic arc planes; see p. 72. These pasted planes are the planes $H(d,r,s,t)$, $d < 0$, $r,s,t \geq 0$. The common point set of all these pasted planes is $\mathbf{R}^2$. The arc planes they are made up of are (isomorphic to) the restrictions of these pasted planes to the left and right open half-planes of $\mathbf{R}^2$. The lines of $H(d,r,s,t)$ are the horizontals, the verticals, the graphs of the two functions

$$(-1,1) \to \mathbf{R} : x \mapsto \begin{cases} -s\big|\ln|x|\big|^d & \text{for} \quad -1 < x < 0, \\ 0 & \text{for} \quad x = 0, \\ |\ln x|^d & \text{for} \quad 0 < x < 1, \end{cases}$$

$$(-1,1) \to \mathbf{R} : x \mapsto \begin{cases} t\big|\ln|x|\big|^d & \text{for} \quad -1 < x < 0, \\ 0 & \text{for} \quad x = 0, \\ -r|\ln x|^d & \text{for} \quad 0 < x < 1, \end{cases}$$

and all images of these lines under the group consisting of the following homeomorphisms:

$$\mathbf{R}^2 \to \mathbf{R}^2 : (x,y) \mapsto (ax, y+b),$$

where $a, b \in \mathbf{R}, a > 0$. This group is homeomorphic to $\mathbf{R}^2$. The planes also admit the automorphisms

$$\mathbf{R}^2 \to \mathbf{R}^2 : (x, y) \mapsto (\operatorname{sgn}(x)|x|^c, c^d y),$$

where $c \in \mathbf{R}^+$. These automorphisms form a 1-dimensional group of automorphisms which, together with the 2-dimensional group above, generates a 3-dimensional group of automorphisms of the plane.

None of these planes is an affine plane. The two planes $H(d, r, s, t)$ and $H(d', r', s', t')$ are isomorphic if and only if $d = d'$ and either the two triples (r, s, t) and (r', s', t') coincide or both triples contain a 1 at the same position, and the remaining two numbers are interchanged. These planes were constructed in Groh [1981] 7.7. By Groh et al. [1983] Theorem 7.5, given a proper $\mathbf{R}^2$-plane whose automorphism group has a 3-dimensional connected closed subgroup not homeomorphic to $\mathrm{L}_2 \times \mathbf{R}$ which fixes precisely one line, this plane is isomorphic to one of the planes above.

2.7.11.3 Affine Pasted Planes, Two Fixed Points and Lines (Type $\mathbf{R}^2$*)*
Let $f_{i,j} : \mathbf{R} \to \mathbf{R}^+$, $i, j = 1, 2$, be four strictly convex continuous functions that satisfy the following conditions:

(i) $\lim_{x \to -\infty} f_{i,j}(x) = 0$;
(ii) $\lim_{x \to +\infty} f'_{i,j}(x) = +\infty$ (note that since $f_{i,j}$ is convex, it is differentiable everywhere, with the possible exception of a set of Lebesgue measure 0).

Furthermore, let

$$g_i : \mathbf{R} \to \mathbf{R} : x \mapsto \begin{cases} (-1)^i f_{i2}(\ln|x|) & \text{for} \quad x < 0, \\ 0 & \text{for} \quad x = 0, \\ (-1)^{i+1} f_{i1}(\ln|x|) & \text{for} \quad x > 0. \end{cases}$$

We define a flat affine plane with point set $\mathbf{R}^2$. Its lines are the horizontals and the verticals plus all the images of the graphs of g_1 and g_2 under the group Δ consisting of the homeomorphisms

$$\mathbf{R}^2 \to \mathbf{R}^2 : (x, y) \mapsto (ax, y + b),$$

where $a, b \in \mathbf{R}^+$, $a > 0$. The group Δ is isomorphic to $\mathbf{R}^2$ and the restrictions of the flat affine plane to the left and right half-planes are isomorphic to arc planes of type $\mathbf{R}^2$. Considered as a group of automorphisms of the projective extension of such a flat affine plane, the group Δ fixes exactly the line at infinity, the line that corresponds to the y-axis, the point of intersection of these two lines and the infinite point of the x-axis. Conversely, given a flat projective plane that admits a group of automorphisms isomorphic to $\mathbf{R}^2$ fixing exactly two lines, their point of intersection and one additional point on one of the two lines, this projective plane is isomorphic to one of the planes that arise as above. See Groh [1981] for more information about these planes; see in particular Theorem 5.4.

2.7.11.4 Affine Pasted Planes, Two Fixed Points (Type L_2*)*

Let $m, n \in \{-1, 1\}$ and let $f_i : [0, \infty) \to [0, \infty)$, $i = 1, \ldots, 4$, be continuous, strictly increasing, surjective functions every one of which satisfies one of the following conditions:

(i) f_i is strictly convex, $\lim_{x\to 0} \frac{f_i(x)}{x} = 0, \lim_{x\to\infty} \frac{f_i(x)}{x} = \infty$;
(ii) f_i is strictly concave, $\lim_{x\to 0} \frac{f_i(x)}{x} = \infty, \lim_{x\to\infty} \frac{f_i(x)}{x} = 0$;
(iii) the function $[0, \infty) \to [0, \infty) : x \mapsto x(f_i(ax) - f_i(x))$ is bijective for all $a > 1$.

Let

$$\begin{aligned} K &= \{(x, f_1(x)) \in \mathbf{R}^2 \mid x \leq 0\} \cup \{(x, -f_3(-x)) \in \mathbf{R}^2 \mid x \geq 0\}, \\ L &= \{(x, -f_2(x)) \in \mathbf{R}^2 \mid x \leq 0\} \cup \{(x, f_4(-x)) \in \mathbf{R}^2 \mid x \geq 0\}, \end{aligned}$$

and let the group $\Delta_{m,n}$ consist of the homeomorphisms

$$\mathbf{R}^2 \to \mathbf{R}^2 : (x, y) \mapsto \begin{cases} (a^m x, ay + b) & \text{for} \quad x \geq 0, \\ (a^n x, ay + b) & \text{for} \quad x \leq 0, \end{cases}$$

where $a, b \in \mathbf{R}$, $a > 0$. This group is isomorphic to L_2.

Define a geometry on $\mathbf{R}^2$ whose lines are the horizontals and verticals plus all images of the curves K and L under $\Delta_{m,n}$. Depending on m and n, this geometry is a flat affine plane if and only if the f_is satisfy the following conditions:

- $m = n = 1$: either all f_i satisfy condition (i) above or all satisfy (ii);
- $m = -n = 1$: f_1 and f_2 satisfy (ii) and f_3 and f_4 satisfy (iii);
- $m = -n = 1$: f_1 and f_2 satisfy (iii) and f_3 and f_4 satisfy (ii).
- $m = n = -1$: all f_i satisfy (iii).

The restrictions of this flat affine plane to the left and right half-planes are arc planes of type L_2. Considered as a group of automorphisms of the projective extension of such a flat affine plane, the group $\Delta_{m,n}$ fixes exactly the line at infinity, the line that corresponds to the y-axis, the point of intersection of these two lines and the infinite point of the x-axis. Schellhammer [1981] Section 6 derived these planes and proves that every flat projective plane that admits a group of automorphisms isomorphic to L_2 fixing exactly two lines and two points is isomorphic to one of the projective planes associated with the above flat affine planes.

2.7.11.5 Affine Pasted Planes, at Least Three Fixed Points (Type L_2*)*

Each of these planes is constructed from a triple $(\mathcal{F}_1, \mathcal{F}_2, \varphi)$ satisfying the following conditions.

- Both $\mathcal{F}_1$ and $\mathcal{F}_2$ are sets of pairwise slope disjoint strictly convex or strictly concave functions $\mathbf{R}^+ \to \mathbf{R}$ such that each $f \in \mathcal{F}_1 \cup \mathcal{F}_2$ satisfies the limit condition $\lim_{x\to 0} f(x) = 0$.
- The function $\varphi : \mathbf{R} \to \mathbf{R}$ is a decreasing homeomorphism. Furthermore, $\varphi(K_1) = K_2$, where $K_j = \mathbf{R} \setminus \bigcup_{f\in\mathcal{F}_j} \mathrm{slp}\, f$.

Define $\mathcal{F}_\varphi = \{(f_1, f_2) \in \mathcal{F}_1 \times \mathcal{F}_2 \mid \mathrm{slp}\, f_2 = \varphi(\mathrm{slp} f_1)\}$. We list a number of conditions which $(\mathcal{F}_1, \mathcal{F}_2, \varphi)$ may or may not satisfy.

(i) Let $(f_1, f_2) \in \mathcal{F}_\varphi$. Then f_2 is convex if and only if f_1 is concave.
(ii) Let $f \in \mathcal{F}_1 \cup \mathcal{F}_2$ and $\mathrm{slp}\, f = (a, b)$. Then
  (a) $\lim_{x\to\infty}(f(x) - bx) = -\infty$, if f is convex and $b < \infty$,
  (b) $\lim_{x\to\infty}(f(x) - ax) = \infty$, if f is concave and $a > -\infty$.
(iii) $|K_1| > 1$.

For $k \in K_1$ define

$$t_k : \mathbf{R} \to \mathbf{R} : x \mapsto \begin{cases} kx & \text{for} \quad x \geq 0, \\ -\varphi(k)x & \text{for} \quad x \leq 0. \end{cases}$$

For $f = (f_1, f_2) \in \mathcal{F}_\varphi$ define

$$g_f : \mathbf{R} \to \mathbf{R} : x \mapsto \begin{cases} f_1(x) & \text{for} \quad x > 0, \\ 0 & \text{for} \quad x = 0, \\ f_2(-x) & \text{for} \quad x < 0. \end{cases}$$

Define a geometry on $\mathbf{R}^2$ whose lines are the verticals and the images of the graphs of all $g_f, f \in \mathcal{F}_\varphi$ and $t_k, k \in K_1$ under the natural action $(x, y) \mapsto (ax, ay + b)$ of the group L_2 on $\mathbf{R}^2$.

Pohl [1990] Theorems 2.7, 2.9, 3.4 prove that a geometry defined like this is an $\mathbf{R}^2$-plane if and only if the triple $(\mathcal{F}_1, \mathcal{F}_2, \varphi)$ satisfies condition (i); see Pohl [1990] Theorem 2.5. It is a flat affine plane if and only if satisfies conditions (i) and (ii); see Pohl [1990] Theorem 2.7. Furthermore, every flat projective plane that admits a group of automorphisms isomorphic to L_2 fixing exactly two lines and more than two points is isomorphic to a projective plane associated with a triple $(\mathcal{F}_1, \mathcal{F}_2, \varphi)$ satisfying conditions (i), (ii), and (iii); see Pohl [1990] Theorem 3.4. In this last case the fixed lines are the line at infinity, the line that corresponds to the y-axis. The fixed points are the infinite point of the y-axis and the infinite points of the lines that are the images under L_2 of the graphs of t_k, $k \in K_1$.

2.7.11.6 Affine Triangle Planes (Type $\mathbf{R}^2$*)*

The *triangle planes* are flat projective planes that have been classified in Pohl [1990]. These planes are the flat projective planes that have an automorphism group that contains a connected 2-dimensional group isomorphic to $\mathbf{R}^2$ whose fix-configuration is a triangle. This 2-dimensional group has four 2-dimensional point orbits which are the complements of the union of the lines that make up the triangle. The restriction of a triangle plane to one of these connected components is isomorphic to an arc plane of type $\mathbf{R}^2$ and the triangle plane itself can be considered as being pasted together from the sub-$\mathbf{R}^2$-planes that are the restrictions of the triangle plane to the four connected components.

Every one of the triangle planes has six generating curves. Consequently, the description of these planes is rather complicated.

Let $d, q \in \mathbf{R}$, $v \in \mathbf{R} \cup \{\infty\}$ with $d < 1, q > d$ and either $v < 0$ and $v < d$ or $1 < v \leq \infty$ and $v > q$. Let $d', q', q'', v' \in \mathbf{R}$. Let

$$m = \frac{q-d}{1-\frac{q}{v}} \text{ and } n = \frac{1-\frac{d}{v}}{1-\frac{q}{v}} q'.$$

Then a triple of functions (f_1, f_2, f_3) is called a (d, q, v, d', q', v')-*triple* if and only if the following conditions hold; see Figure 2.19.

(i) The function $f_1 : \mathbf{R} \to \mathbf{R}$ is a strictly convex function with

$$\lim_{x \to -\infty} f_1(x) = d' \text{ and } \lim_{x \to \infty} |f_1(x) - mx - n| = 0.$$

(ii) The function $f_2 : (-\infty, v') \to \mathbf{R}$ is a strictly concave function with

$$\lim_{x \to -\infty} f_2(x) = d' \text{ and } \lim_{x \to v'} f_2(x) = -\infty.$$

(iii) The function $f_3 : (v', \infty) \to \mathbf{R}$ is a strictly concave function with

$$\lim_{x \to v'} f_3(x) = -\infty \text{ and } \lim_{x \to \infty} |f_3(x) - mx - n| = 0.$$

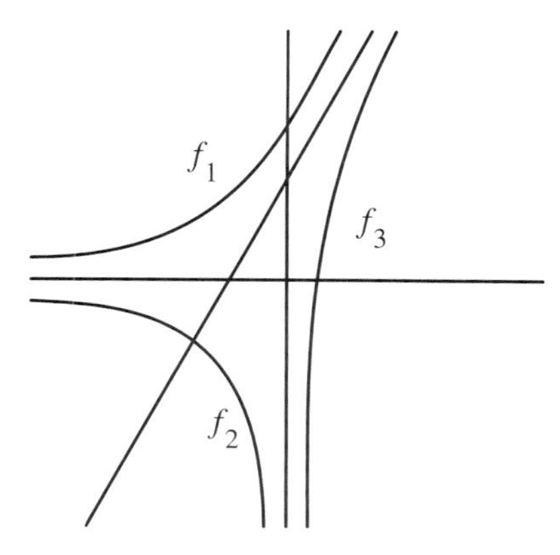

Fig. 2.19. A (d, q, v, d', q', v')-triple of functions

A $(d, q, v, d', q', v', q'')$-*system* $\mathcal{F}$ is a set consisting of twelve functions f_{ij}, with $i, j = 1, 2, 3, 4$, $i \neq j$, having the following properties:

(i) (f_{41}, f_{31}, f_{21}) is a $(0, 1, \infty, 0, q'', 0)$-triple;
(ii) (f_{32}, f_{42}, f_{12}) is a $(0, q, v, 0, q', v')$-triple;
(iii) (f_{23}, f_{13}, f_{43}) is a $(d, 1, v, d', q'', v')$-triple;
(iv) (f_{14}, f_{24}, f_{34}) is a $(d, q, \infty, d', q', 0)$-triple.

We define a 2-dimensional group Δ acting on $\mathbf{R}^2$ which is isomorphic to $\mathbf{R}^2$. It consists of the homeomorphisms

$$\mathbf{R}^2 \to \mathbf{R}^2 : (x, y) \mapsto \begin{cases} (e^a x, e^b y) & \text{for} \quad x, y \geq 0, \\ (e^{a - \frac{1}{v} b} x, e^b y) & \text{for} \quad x < 0 \leq y, \\ (e^{a - \frac{1}{v} b} x, e^{b - da} y) & \text{for} \quad x, y < 0, \\ (e^a x, e^{b - da} y) & \text{for} \quad y < 0 \leq x, \end{cases}$$

where $a, b \in \mathbf{R}$.

We proceed to define six functions $\mathbf{R} \to \mathbf{R}$ whose graphs generate a flat affine plane:

$$g_1(x) = \begin{cases} e^{f_{12}(\ln |x|)} & \text{for} \quad x < -e^{v'}, \\ 0 & \text{for} \quad x = -e^{v'}, \\ -e^{f_{13}(\ln |x|)} & \text{for} \quad -e^{v'} < x < 0, \\ -e^{q'} & \text{for} \quad x = 0, \\ -e^{f_{14}(\ln |x|)} & \text{for} \quad x > 0; \end{cases}$$

$$g_2(x) = \begin{cases} -e^{f_{23}(\ln|x|)} & \text{for} \quad x < 0, \\ -e^{d'} & \text{for} \quad x = 0, \\ -e^{f_{24}(\ln|x|)} & \text{for} \quad 0 < x < 1, \\ 0 & \text{for} \quad x = 1, \\ e^{f_{21}(\ln|x|)} & \text{for} \quad x > 1; \end{cases}$$

$$g_3(x) = \begin{cases} e^{f_{32}(\ln|x|)} & \text{for} \quad x < 0, \\ 1 & \text{for} \quad x = 0, \\ e^{f_{31}(\ln|x|)} & \text{for} \quad 0 < x < 1, \\ 0 & \text{for} \quad x = 1, \\ -e^{f_{34}(\ln|x|)} & \text{for} \quad x > 1; \end{cases}$$

$$g_4(x) = \begin{cases} -e^{f_{43}(\ln|x|)} & \text{for} \quad x < -e^{v'}, \\ 0 & \text{for} \quad x = -e^{v'}, \\ e^{f_{42}(\ln|x|)} & \text{for} \quad -e^{v'} < x < 0, \\ 1 & \text{for} \quad x = 0, \\ e^{f_{41}(\ln|x|)} & \text{for} \quad x > 0; \end{cases}$$

$$g_5(x) = \begin{cases} -|x|^{\frac{1-d}{1-\frac{q}{v}}} & \text{for} \quad x < 0, \\ x & \text{for} \quad x \geq 0; \end{cases}$$

$$g_6(x) = \begin{cases} |x|^{\frac{q}{1-\frac{q}{v}}} & \text{for} \quad x < 0, \\ -x^{q-d} & \text{for} \quad x \geq 0. \end{cases}$$

We define a geometry $\mathcal{R} = (P, \mathcal{L})$ whose lines are the horizontal and vertical Euclidean lines and the images under the group Δ of the graphs of the six functions g_i defined above.

This geometry is a flat affine plane. We call the projective extension of such a plane a *triangle plane*; see Pohl [1990] 5.13.

The fix-configuration of Δ in one of these planes is a triangle whose vertices are the origin of $\mathbf{R}^2$ and the infinite points of the horizontals and verticals. Every flat projective plane whose group of automorphisms contains a connected 2-dimensional group isomorphic to $\mathbf{R}^2$ whose fix-configuration is a triangle is isomorphic to a triangle plane; see Pohl [1990] 6.10.

It can be shown that a triangle plane has group-dimension at least 3 if and only if it is a Cartesian plane; see Pohl [1990] 7.4.

Pohl [1990] Example 3 constructs the following large family of triangle planes. Let $d, q \in \mathbf{R}$, $v \in \mathbf{R} \cup \{\infty\}$ with $d < 1, q > 0, q > d$ and either $v < 0$ and $v < d$ or $1 < v \leq \infty$ and $v > q$. Let $d', q', q'', v' \in \mathbf{R}$. Abbreviate

$$\alpha = q - d, \qquad \beta = \frac{q}{1 - \frac{q}{v}}, \qquad \text{and} \qquad \gamma = \frac{1 - \frac{d}{v}}{1 - \frac{1}{v}}.$$

Then the functions given by the equations

$$\begin{array}{ll}
f_{12}(x) = \ln(e^{\beta(x-v')} - 1) + \beta v' + q'q^{-1}, & f_{21}(x) = \ln(e^{x} - 1) + q'', \\
f_{32}(x) = \ln(e^{\beta(x+q'q^{-1})} + 1), & f_{31}(x) = \ln(1 - e^{\alpha x}), \\
f_{42}(x) = \ln(1 - e^{\gamma(x-v')}), & f_{41}(x) = \ln(e^{x+q''} + 1), \\
f_{13}(x) = \ln(1 - e^{\beta(x-v')}) + d', & f_{14}(x) = \ln(e^{\alpha x+q'} + e^{d'}), \\
f_{24}(x) = \ln(1 - e^{x}) + d', & f_{23}(x) = \ln(e^{\gamma(x+q'')}+e^{d'}), \\
f_{43}(x) = \ln(e^{\gamma(x-v')} - 1) + \gamma(v' + q''), & f_{34}(x) = \ln(e^{\alpha x} - 1) + q'
\end{array}$$

define a $(d, q, v, d', q', v', q'')$-*system.*

Let us have a closer look at what the lines of the corresponding flat affine plane look like (see Figure 2.20) if we choose

$$(d, q, v, d', q', v', q'') = (0, 1, \infty, 0, 0, 0, \ln 2).$$

In this case the lines are the vertical and horizontal Euclidean lines and the graphs of the piecewise linear functions $\mathbf{R} \to \mathbf{R}$

$$x \mapsto mx + n,$$

where $m \in \mathbf{R}^-, n \in \mathbf{R}$ (these functions correspond to g_1, g_3 and g_6);

$$x \mapsto mx,$$

where $m \in \mathbf{R}^+$ (these functions correspond to g_5);

$$x \mapsto \begin{cases} 2mx + n & \text{for } x \leq 0, \\ mx + n & \text{for } 0 < x \leq -nm^{-1}, \\ 2(mx + n) & \text{for } x > -nm^{-1}, \end{cases}$$

where $m \in \mathbf{R}^+, n \in \mathbf{R}^-$ (these functions correspond to g_2);

$$x \mapsto \begin{cases} 2(mx + n) & \text{for } x \leq -nm^{-1}, \\ mx + n & \text{for } -nm^{-1} < x \leq 0, \\ 2mx + n & \text{for } x > 0, \end{cases}$$

where $m, n \in \mathbf{R}^+$ (these functions correspond to g_4).

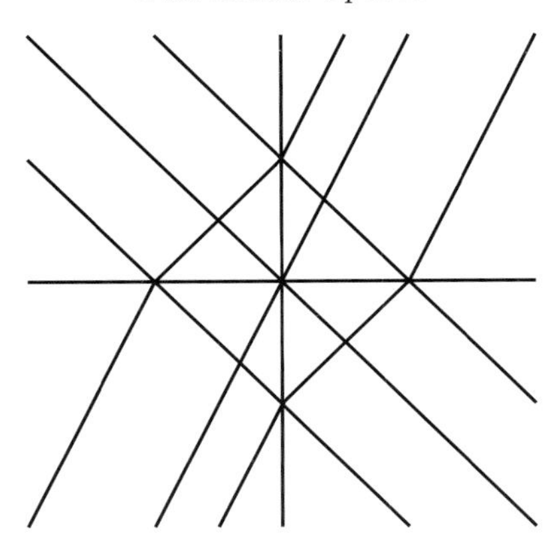

Fig. 2.20. Two parallel classes in an example of an affine triangle plane

2.7.12 Semioval Planes

The *semioval planes* are flat projective planes that were classified in Lippert [1986]. These planes are the flat projective planes whose automorphism groups contain a connected 2-dimensional subgroup isomorphic to L_2 that fixes a line, a topological semioval, and a number of points on the fixed line. Here a *topological semioval* S in a flat projective plane $\mathcal{P}$ is a topological arc homeomorphic to $\mathbf{R}$ satisfying the following conditions.

- Through every point of the arc passes exactly one supporting line.
- The topological closure of the arc S in the point set of $\mathcal{P}$ contains one or two (boundary) points that are not contained in S.

In this subsection we describe the two essentially different constructions of semioval planes corresponding to the semiovals having one and two boundary points.

Let Δ be the group isomorphic to L_2 consisting of the maps

$$\mathbf{R}^2 \to \mathbf{R}^2 : (x, y) \to (ax, ay + b),$$

where $a, b \in \mathbf{R}, a > 0$. In both constructions we first construct a flat affine plane with point set $\mathbf{R}^2$ from a number of generating curves using this group. The projective extension of this flat affine plane is in both cases a semioval plane.

2.7.12.1 Planes in Which an Oval and One of Its Tangents Are Fixed

We start by constructing a flat affine plane on $\mathbf{R}^2$. Choose three curves as in Figure 2.21. Then the lines of the flat affine plane are the horizontals plus the images of the three curves under Δ. In the projective extension of the resulting flat affine plane, the y-axis plus its boundary point on the line at infinity is a topological oval. Both this topological oval and the line at infinity which is a tangent of the oval are fixed by Δ. Conversely, Lippert [1986] Theorem 5.4 (p. 99) proves that a flat projective plane of group dimension 2 admits a group of automorphisms isomorphic to L_2 that fixes exactly a topological oval, and one of its tangents is isomorphic to one of the above planes.

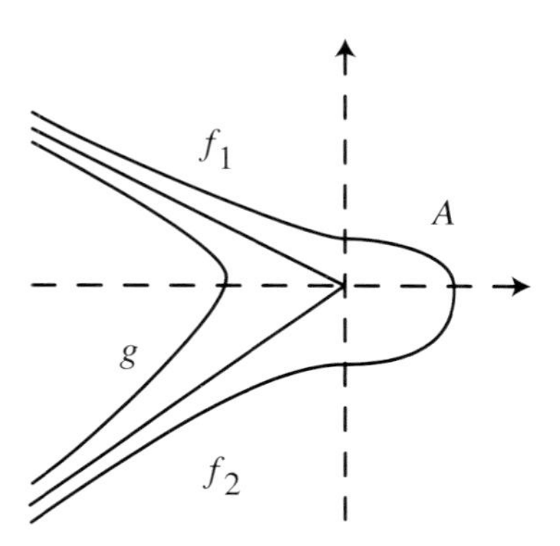

Fig. 2.21. Three curves generating a semioval plane

Here is a precise description of the possible curves that we can choose. Let $\alpha, \beta \in \mathbf{R}$, $\beta < 0 < \alpha$. Let f_1, f_2, A, g be continuous functions with the following properties.

(i) • $f_1 : \mathbf{R}^- \to \mathbf{R}^+$ is strictly convex,
- $\lim_{x\to 0} f_1(x) = a$, $a > 0$,
- $\lim_{x\to -\infty} |f_1(x) + \alpha x| = 0$,
- slp $f_1 = (-\alpha, 0)$.

(ii) • $f_2 : \mathbf{R}^- \to \mathbf{R}^-$ is strictly concave,
- $\lim_{x\to 0} f_2(x) = b$, $b < 0$,
- $\lim_{x\to -\infty} |f_2(x) + \beta x| = 0$,
- slp $f_2 = (0, -\beta)$.

(iii) • $g : \mathbf{R} \to \mathbf{R}^-$ is strictly concave,
- $\lim_{x\to \infty} |g(x) + (1/\alpha x)| = \lim_{x\to -\infty} |g(x) + (1/\beta x)| = 0$,
- slp $g = (-1/\alpha, -1/\beta)$.

(iv) • $A : (b, a) \to \mathbf{R}^+$ is strictly concave,
- $\lim_{x\to a} A(x) = \lim_{x\to b} A(x) = 0$,
- slp $A = \mathbf{R}$.

The first curve is the union of the graphs of f_1 and f_2, the two points $(0, a)$ and $(0, b)$, and the image of the graph of A under the homeomorphism

$$\sigma : \mathbf{R}^2 \to \mathbf{R}^2 : (x, y) \mapsto (y, x).$$

The second curve is the set

$$\{(x, -\alpha x) \in \mathbf{R}^2 \mid x < 0\} \cup \{(0, 0)\} \cup \{(x, -\beta x) \in \mathbf{R}^2 \mid x < 0\}.$$

The third curve is the image of the graph of the function g under the homeomorphism σ.

By Lippert [1986] Appendix D, the Euclidean plane is generated by the functions

$$\begin{aligned} f_1 :&\quad \mathbf{R}^- \to \mathbf{R}^+ : x \mapsto \sqrt{x^2 + 1}, \\ f_2 :&\quad \mathbf{R}^- \to \mathbf{R}^+ : x \mapsto -\sqrt{x^2 + 1}, \\ A :&\quad (-1, 1) \to \mathbf{R}^+ : x \mapsto \sqrt{1 - x^2}, \\ g :&\quad \mathbf{R} \to \mathbf{R} : x \mapsto -\sqrt{x^2 + 1/3}. \end{aligned}$$

Note that if we restrict this model of the Euclidean plane to the right half-plane, we arrive at the Poincaré model of the hyperbolic plane.

The affine plane that we arrive at by removing a tangent of the unit circle from one of the skew hyperbolic planes can also be represented in this way so that the restriction of the resulting model to the left half-plane is the Poincaré model of the hyperbolic plane.

2.7.12.2 Planes in Which a Proper Semioval and a Line Are Fixed

We start by constructing a flat affine plane on $\mathbf{R}^2$. Choose three curves as in Figure 2.22, a set of 'suitable bent Euclidean lines', and a countable family of 'suitable' further curves. Then the lines of the flat affine plane are the horizontals and the images of all these curves under Δ. In the projective extension of the resulting flat affine plane, the y-axis forms a topological semioval which has two boundary points on the line at infinity. Both this topological semioval and the line at infinity are fixed by Δ. Furthermore, Δ also fixes the two boundary points plus the infinite points of the countably many distinguished curves. Conversely, Lippert [1986] Theorem 5.5 (p. 101) proves that a flat projective plane of group dimension 2 that admits a group of automorphisms isomorphic to L_2 that fixes exactly a topological semioval with two boundary points, the connecting line of the two boundary points and countably many further points on this line, is isomorphic to one of these planes.

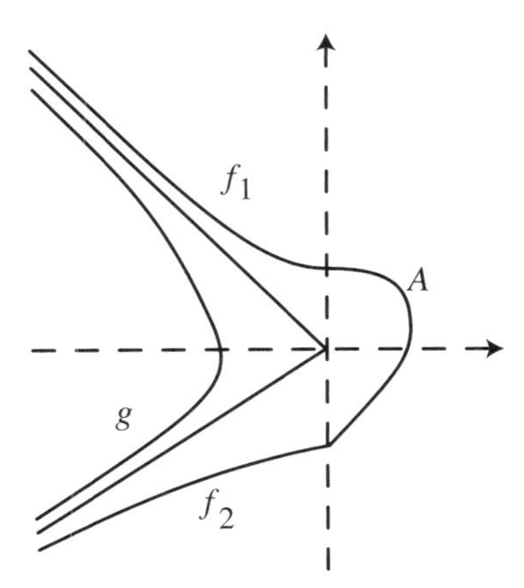

Fig. 2.22. Three curves generating a semioval plane

Here is a precise description of the possible curves that we can choose. Let $\beta, K \in \mathbf{R}$, $\beta < K < 0$. Let f_1, f_2, A, g be functions with the following properties.

(i) • $f_1 : \mathbf{R}^- \to \mathbf{R}^+$ is strictly convex,
- $\lim_{x\to 0} f_1(x) = a$, $a > 0$,
- $\lim_{x\to -\infty} |f_1(x) + x| = 0$,
- slp $f_1 = (-1, 0)$.

(ii) • $f_2 : \mathbf{R}^- \to \mathbf{R}^-$ is strictly concave,
- $\lim_{x\to 0} f_2(x) = b$, $b < -1$,
- $\lim_{x\to -\infty} |f_2(x) + \beta x| = 0$,
- slp $f_2 = (-K, -\beta)$.

(iii) • $g : \mathbf{R} \to \mathbf{R}^+$ is strictly concave,
- $\lim_{x\to \infty} |g(x) + x| = \lim_{x\to -\infty} |g(x) + (1/\beta x)| = 0$,
- slp $g = (-1, -1/\beta)$.

(iv) • $A : (b, a) \to \mathbf{R}^+$ is strictly concave,
- $\lim_{x\to a} A(x) = \lim_{x\to b} A(x) = 0$,
- slp $A = (-\infty, 1)$,
- $\lim_{x\to a} A'(x) = -\infty$,
- $\lim_{x\to b} A'(x) = 1$.

We define the three distinguished curves as in the previous subsubsection. The first curve is the union of the graphs of f_1 and f_2, the two points $(0, a)$ and $(0, b)$, and the image of the graph of A under the homeomorphism

$$\sigma : \mathbf{R}^2 \to \mathbf{R}^2 : (x, y) \mapsto (y, x).$$

The second curve is the set

$$\{(x, -x) \in \mathbf{R}^2 \mid x < 0\} \cup \{(0, 0)\} \cup \{(x, -\beta x) \in \mathbf{R}^2 \mid x < 0\}.$$

The third curve is the image of the graph of the function g under the homeomorphism σ.

Let $F_i = \{f_v^i \mid v \in I\}$, $i = 1, 2$, be two countable families of continuous functions $\mathbf{R}^+ \to \mathbf{R}$ and $\mathbf{R}^- \to \mathbf{R}$, respectively, that are indexed by the same subset I of the integers. Furthermore, these two sets have the following properties:

- $\lim_{x\to 0} f_v^i(x) = 0$ for $i = 1, 2$ and all $v \in I$;
- the graphs of the functions in F_1 generate an arc plane of type L_2 on the right half-plane and the union of the slopes of all these functions is a subset of the interval $[0, 1]$;
- the graphs of the functions in F_1 generate an arc plane of type L_2 on the left half-plane and the union of the slopes of all these functions is a subset of the interval $[0, K]$;
- given that $v < w$, $v, w \in I$ implies that the slopes of f_v^i are smaller than the slopes of f_w^i;
- all functions in F_i, $i = 1, 2$, are either strictly convex, or strictly concave; furthermore, f_v^1 convex implies that f_v^2 is concave, and vice versa, for all $v \in I$.

Choose a homeomorphism $\xi : [0, 1] \to [0, K]$ such that $\xi(0) = 0$ and such that for all $v \in I$, $\xi(\mathrm{slp}\ f_v^1) = f_v^2$.

Let S be the set of all numbers in $[0, 1]$ that are not slopes of any among the functions f_v^1, $v \in I$. For every $\lambda \in S$ define a 'bent Euclidean line' L_λ as the graph of the function

$$\mathbf{R} \to \mathbf{R} : x \mapsto \begin{cases} \lambda x & \text{for} \quad x \geq 0, \\ 0 & \text{for} \quad x = 0, \\ \xi(\lambda) x & \text{for} \quad x \leq 0. \end{cases}$$

For every $v \in I$ define a curve g_v as the graph of the function

$$\mathbf{R} \to \mathbf{R} : x \mapsto \begin{cases} f_v^1(x) & \text{for} \quad x \geq 0, \\ 0 & \text{for} \quad x = 0, \\ f_v^2(x) & \text{for} \quad x \leq 0. \end{cases}$$

These curves form the countable family of further 'suitable' curves that we mentioned at the beginning of this subsubsection.

2.7.13 The Modified Real Dual Cylinder Plane

Consider the hyperbolic arc plane (see p. 72) generated by the graph of the function

$$\mathbf{R}^+ \to \mathbf{R} : x \mapsto x^{-1}.$$

Extend parallel Euclidean lines of positive slope by a common point at infinity and assemble all infinite points into a line at infinity. The resulting geometry is isomorphic to a disk Möbius strip plane. It has, just like the $\mathbf{R}^2$-plane it extends, group dimension 3. It is called the *modified real dual cylinder plane*, and is denoted by MDC($\mathbf{R}$). See Löwen [1981c] for details about this construction.

2.8 Planes with Special Properties

In the following subsections we construct flat projective planes whose groups of automorphisms are compact, differentiable projective planes, and maximal flat linear spaces.

2.8.1 Compact Groups of Automorphisms

Schellhammer [1981] 8.1, 8.2 states that every compact group of automorphisms of a flat projective plane is isomorphic to a subgroup of $\mathrm{SO}_3(\mathbf{R})$. Among the flat projective planes only the classical plane admits the full group $\mathrm{SO}_3(\mathbf{R})$ as a group of automorphisms. The proper subgroups of $\mathrm{SO}_3(\mathbf{R})$ are the trivial group, the rotation groups of the tetrahedron, the octahedron/cube, and the dodecahedron/icosahedron, the cyclic groups, the dihedral groups, $\mathrm{SO}_2(\mathbf{R})$, and $\mathrm{SO}_2(\mathbf{R})$ 'with a flip'. Given such a subgroup, there are flat projective planes that have exactly this subgroup as their full group of automorphisms. We construct examples of such planes by deforming the classical flat projective plane in sets of disjoint circles. We follow the exposition in Schellhammer [1981] Section 8.

2.8.1.1 $SO_2(\mathbf{R})$ *'with a Flip'*

We modify the classical plane as follows; see Figure 2.23. All lines that do not intersect the open unit disk stay untouched. Consider a line that intersects the unit circle in the two points p and q. If the length of the line segment $[p,q]$ is greater than or equal to $\sqrt{2}$, we modify the line by replacing $[p,q]$ by the circle segment that intersects the unit circle at right angles in both p and q. If the length of the line segment $[p,q]$ is less than or equal to $\sqrt{2}$, we modify the line by replacing $[p,q]$ by the circle segment of radius 1 that intersects the unit circle in both p and q. Figure 2.23 shows how the verticals that meet the unit circle get modified.

Note that the restriction of the resulting flat projective plane to the interior of the smaller circle of radius $\sqrt{2}-1$ in the diagram coincides with the restriction to this set of the Poincaré model of the real hyperbolic plane on the unit disk. This implies that the plane is locally classical in this smaller circle and outside the unit circle. On the other hand, this is not the case in the annulus bounded by the two circles. Hence every automorphism of the plane leaves this ring fixed. Clearly, the rotations around the origin are automorphisms of the geometry. So is the reflection through any of the lines through the origin.

We show that the full group of automorphisms is $SO_2(\mathbf{R})$ 'with a flip', the group generated by these automorphisms. We do this by showing that every automorphism γ of the plane is the identity modulo this group. Let L be the line that connects the points $(0,1)$ and $(1,0)$; see the diagram on the right in Figure 2.23. Since L is a tangent of the smaller circle, every image of L under γ is also a tangent of this circle. Modulo the rotations and a reflection we may assume that L and its two points of intersection with K are fixed by γ. The line M that connects the two points $(0,-1)$ and $(1,0)$ is also a tangent of the smaller circle. Since it intersects L in a point of the unit circle, it is fixed under γ. The two points in which L and M touch the smaller circle and the points $(0,1)$ and $(0,-1)$ form the vertices of a nondegenerate quadrangle. Since γ fixes every point of this quadrangle, it is the identity by Corollary 2.4.4.

2.8.1.2 D_n

To construct a plane that has the dihedral group D_n as its full group of automorphisms, we modify the Euclidean plane as above in n nonoverlapping unit circles whose centres form the vertices of a regular n-gon.

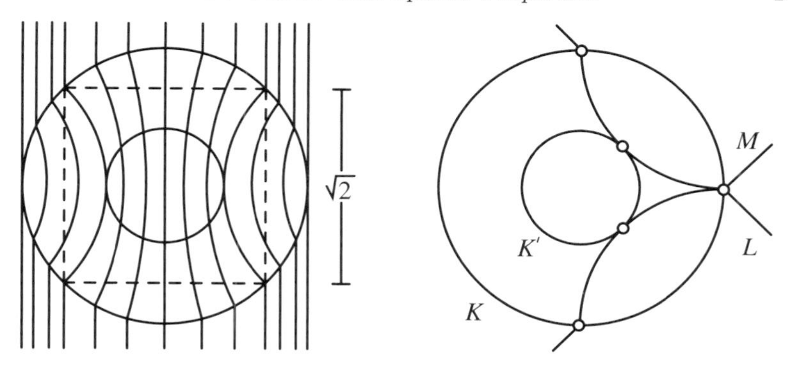

Fig. 2.23.

2.8.1.3 $SO_2(\mathbf{R})$

We modify the $SO_2(\mathbf{R})$-with-a-flip construction as follows. Consider a line that intersects the unit circle in two points p and q. If the length of the line segment $[p, q]$ is greater than or equal to $\sqrt{2}$, we replace $[p, q]$ by a certain curve that is not symmetric with respect to any line through the origin. Suitable curves are the intersections with the unit disk of the ellipses defined by the equation

$$\frac{(x-a)^2}{a^2} + (y-1)^2 = 1,$$

where $a > 1$, and all images of these intersections under rotations around the origin. Note that all these ellipses contain the point $(0, 1)$ and that for $a = 1$ the ellipse is a circle of radius 1 that also contains the point $(1, 0)$. Apart from this modification, the above construction stays unchanged. The resulting plane admits $SO_2(\mathbf{R})$ as its full group of automorphisms.

2.8.1.4 Z_n

To construct planes that have the cyclic group Z_n as their full groups of automorphisms, we modify the Euclidean plane as above in n nonoverlapping unit circles whose centres form the vertices of a regular n-gon.

2.8.1.5 The Groups of the Platonic Solids

Let P be a Platonic solid inscribed in a sphere. Identify the sphere via the antipodal map γ with the real projective plane. If P is either the octahedron/cube or the dodecahedron/icosahedron, pairs of opposite vertices of these solids get identified under γ. By modifying the classical flat projective plane in (nonoverlapping) unit circles around the images of the vertices of P as in the $\mathrm{SO}_2(\mathbf{R})$-with-a-flip case, we arrive at the planes we are looking for. In the case of the tetrahedron we modify as in the $\mathrm{SO}_2(\mathbf{R})$ case.

2.8.1.6 Identity—Rigid Planes

A flat projective plane with no nonidentity automorphisms is called *rigid*. To construct a rigid plane just modify the classical plane in a number of nonoverlapping wildly distributed unit circles as above.

2.8.2 More Rigid Planes

Choosing the homeomorphisms g and h suitably in the semi-classical flat projective planes defined by g and h (see Subsection 2.7.2) one obtains rigid flat projective planes. For example, let $m \geq 2$ be an integer and let $0 < r < 3$. Then

$$f_{m,r} : \mathbf{R} \to \mathbf{R} : x \mapsto (1 - r/3)x^{2m+1} + r(x^3/3 - x^2 + x)$$

is a strictly increasing homeomorphism. Let $\mathcal{P}(m, r; n, s)$ be the semi-classical flat projective plane with $g = f_{m,r}$ and $h = f_{n,s}$. Then the plane $\mathcal{P}(m, r; n, s)$ is rigid for $m \neq n$ or $r \neq s$. These planes are mutually nonisomorphic for $2 \leq m, n$, $0 < r, s < 3$, $n < m$ or $m = n$, $s < r$; see Steinke [1985a].

Since a projective plane and its dual have isomorphic automorphism groups, one also obtains rigid planes in the family of dual semi-classical flat projective planes for suitably chosen homeomorphisms f and g.

The restriction of the Euclidean plane to the interior of a topological oval without symmetries is a rigid $\mathbf{R}^2$-plane. Such an $\mathbf{R}^2$-plane is never a flat affine plane. The restriction of a rigid projective plane to the complement of a point is a rigid point Möbius strip plane.

2.8.3 Differentiable Planes

The classical flat affine planes and flat projective planes are *differentiable* in the sense that both their point and line sets are differentiable manifolds in a natural way and the geometric operations of joining points, intersecting lines, and drawing parallels are differentiable in their domains of definition.

Otte [1993] constructs nonclassical differentiable flat affine planes the projective extensions of which are nonclassical differentiable flat projective planes.

Let $\delta \in \mathbf{R}$, $0 < \delta < \sqrt{2} - 1$, and let $\psi : \mathbf{R} \to \mathbf{R}$ be differentiable such that the following conditions are satisfied:

(i) $\psi(r) = 0$ for every r not contained in $(0, \delta/2)$;
(ii) $|\psi'(r)| \leq 1$ for all r;
(iii) ψ is not identically 0.

Then the geometry $(\mathbf{R}^2, \mathcal{L})$ whose lines are the verticals in $\mathcal{L}$ and the graphs of the functions

$$\tau_{s,t} : \mathbf{R} \to \mathbf{R} : t \mapsto sx + t + \psi(|s|^2)\psi(|x|^2)\psi(|t|^2),$$

where $s, t \in \mathbf{R}$, is an example of a differentiable flat affine plane whose extension is a differentiable flat projective plane.

Of course, by restricting the corresponding differentiable projective planes to suitable subsets of their point sets, we can construct examples of the other types of flat linear spaces that are differentiable.

For further results about differentiable and analytic projective planes, flat linear spaces, and stable planes see Breitsprecher [1967a], [1967b], Bödi [1997], [1998a], [1998b], [1998c], and Bödi–Immervoll–Löwe [2000]. We only note that there has also been some interest in constructing flat projective planes that are 'differentiable' in a manner different from the one considered above. For example, Kuiper [1957] constructed a nonclassical flat projective plane all of whose lines are algebraic curves.

2.8.4 Maximal Flat Stable Planes and the First Nonclassical Flat Linear Space

We call a flat linear space *maximal* if it does not arise as the restriction of some other flat linear space $(P, \mathcal{L})$ to a nontrivial open subset of its point set P. Clearly, flat projective planes are maximal and flat affine planes and point Möbius strip planes are not.

Stroppel [1998] showed that the first example of a non-Desarguesian flat linear space mentioned in the literature is an $\mathbf{R}^2$-plane that is maximal. This special $\mathbf{R}^2$-plane $\mathcal{R}_{\text{Hilbert}} = (P, \mathcal{L})$ was constructed by Hilbert. He used it in his 1898/99 lecture at Göttingen entitled 'Grundlagen der Euklidischen Geometrie' (Foundations of Euclidean Geometry).

The construction of $\mathcal{R}_{\text{Hilbert}}$ runs as follows. Let N be the negative part of the x-axis, let U and D be the upper and lower open half-planes, and let $\overline{U} = U \cup N$ and $\overline{D} = D \cup N$. Then the point set P coincides

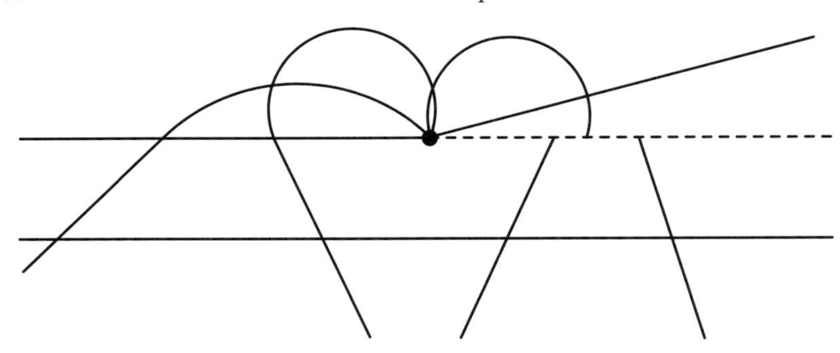

Fig. 2.24. Hilbert's $\mathbf{R}^2$-plane

with $U \cup D \cup N$ and the lines are constructed as follows. Let L be a Euclidean line that is partly contained in $\overline{D}$. If L does not intersect N, or contains N, its restriction to $\overline{D}$ is a line in $\mathcal{L}$. If it intersects N in exactly the point p, let K be the Euclidean circle through the origin that touches L at p. Then the union of the restrictions of L and K to $\overline{D}$ and $\overline{U}$, respectively, is a line in $\mathcal{L}$. If K is a Euclidean circle or line through the origin that is partly contained in U but does not intersect N, then its restriction to U is also a line in $\mathcal{L}$. See Figure 2.24 for examples of the different kinds of lines in Hilbert's plane.

The restrictions of $\mathcal{R}_{\text{Hilbert}}$ to U and D are both isomorphic to the restriction of the Euclidean plane to a half-plane, that is, every point in the interiors of U and D has a classical neighbourhood. In the case of D this is obvious. For U we observe that the inversion

$$U \to U : (x, y) \mapsto \frac{(x, y)}{x^2 + y^2}$$

is an isomorphism between the restriction of $\mathcal{R}_{\text{Hilbert}}$ to U and the restriction of the Euclidean plane to U. On the other hand, all neighbourhoods of points on N are nonclassical.

Further analysis shows that $\mathcal{R}_{\text{Hilbert}}$ is isomorphic to one of the pasted planes made up of exponential arc planes that we described in Subsubsection 2.7.11.1. This means that the automorphism group of the geometry is 3-dimensional; see Stroppel [1998] for more details. In this paper Stroppel also shows that $\mathcal{R}_{\text{Hilbert}}$ cannot be embedded as a 'proper open subplane in a flat linear space'. This implies the result mentioned at the beginning of this subsection.

PROPOSITION 2.8.1 (Maximality of Hilbert's Example) *The proper $\mathbf{R}^2$-plane $\mathcal{R}_{\text{Hilbert}}$ is a maximal flat linear space.*

For general criteria that can be used in deciding whether or not a flat linear space is maximal see Löwen [1981c] Section 5, Stroppel [1998] Section 6, Stroppel [1994] Sections 2 and 3. Here is an example of one such criterion.

PROPOSITION 2.8.2 (Embeddings in Flat Projective Planes) *Let $\mathcal{R}$ be a flat linear space, let L be one of its lines and let p and q be two of its points off L such that there are a unique parallel of L through p and more than one parallel of L containing the point q. Then $\mathcal{R}$ does not arise as the restriction of a flat projective plane to some open subset of its point set.*

This criterion is proved in Stroppel [1998] as a warm-up exercise for the proof of Proposition 2.8.1.

Proof of proposition. Assume that there is a flat projective plane $\mathcal{P}$ that extends $\mathcal{R}$. Since there is a unique parallel to L through p, the line L is extended by a single point of $\mathcal{P}$ to a line of $\mathcal{P}$. This implies that in $\mathcal{R}$ there is also a unique parallel to L through the point q, which is a contradiction. □

In his famous book *Grundlagen der Geometrie* (Foundations of Geometry) Hilbert replaced the above example by a non-Desarguesian flat affine plane. It is constructed by modifying the Euclidean plane in the interior of an ellipse; see Hilbert [1899]. This flat affine plane has an automorphism group that is believed to be isomorphic to $\mathbf{Z}_2$; see Anisov [1992] for an incomplete proof. In Stoppel [1993a] a generalization of Hilbert's construction is investigated.

Other early papers featuring constructions of flat projective planes with special properties include Moulton [1902], Mohrmann [1922], Tschetweruchin [1927], Levenberg [1950], Naumann [1954] Sections 7 and 8, Lombardo-Radice [1955], Pickert [1956], Kuiper [1957], Sitaram [1962].

2.9 Other Invariants and Characterizations

The group of automorphisms of a flat projective plane is one of its invariants. Other invariants of a flat projective plane include its Lenz–Barlotti type, its group of projectivities, and its semigroup of continuous lineations. We consider these invariants in the following subsections.

2.9.1 The Lenz–Barlotti Types

Given a point p and a line L in a projective plane, let Σ be the full automorphism group and let $\Sigma_{p,[L]}$ be the group made up of all automorphisms that fix a point p linewise and fix a line L pointwise that is, every nontrivial automorphism in $\Sigma_{p,[L]}$ fixes precisely the points on L and the point p and also fixes precisely every line through p and the line L globally. An automorphism in $\Sigma_{p,[L]}$ is called a *central* or *axial collineation with centre p and axis L* if $p \notin L$ or $p \in L$, respectively. Central and axial collineations are also referred to as *homologies* and *elations*, respectively. In an affine plane *homotheties* (also known as *dilatations*) and *translations* are automorphisms that extend to homologies and elations of the projective extension, respectively, where the line at infinity is the axis. A projective plane is called (p, L)*-transitive* if $\Sigma_{p,[L]}$ acts transitively on every set $M \setminus (\{p\} \cup L\})$, where M is a line through p.

Lenz [1954] classified projective planes with respect to the point–line pairs (p, L) with $p \in L$ for which the planes are (p, L)-transitive. This was later refined by Barlotti [1957] by admitting point–line pairs (p, L) with $p \notin L$. The *Lenz–Barlotti type of a projective plane* is determined by the set C of all point–line pairs (p, L) such that the plane is (p, L)-transitive. See Pickert [1975] Anhang 6, for a full list of possible Lenz–Barlotti types. In the following we give a list of Lenz–Barlotti types that can actually occur in flat projective planes.

- I.1. $C = \emptyset$.
- I.2. C consists of a single antiflag.
- I.4. $C = \{(a, L), (b, M), (c, N)\}$, where $a \in M, N$, $b \in L, N$, $c \in L, M$, $a \notin L$, $b \notin M$, $c \notin N$.
- II.1. C consists of a single flag.
- II.2. $C = \{(a, L), (b, M)\}$, where $a \in L, M$, $b \in L$, $a \neq b$, $L \neq M$.
- III.2. There is an antiflag (a, L) such that $C = \{(x, xa) \mid x \in L\} \cup \{(a, L)\}$.
- VII.2. $C = P \times \mathcal{L}$.

Examples for the respective Lenz–Barlotti types can be obtained as follows; see also Prieß-Crampe [1983] Section V.5.

- I.1. A rigid plane; see Subsection 2.8.2.
- I.2. A semi-classical plane $\mathcal{P}_{h,g}$ with $h = g$ defined by $g(x) = x^3$; compare Steinke [1997] Theorem 5.5. See Subsection 2.7.2 for a definition of semi-classical planes.

I.4. Naumann planes; see Naumann [1954] or Prieß-Crampe [1983] p. 268.

II.1. A semi-classical plane $\mathcal{P}_{h,g}$ with the functions $g = id$ and h defined by $h(x) = x^3$; compare Steinke [1997] Theorem 5.5.

II.2. See Jónsson [1963] p. 290 for an example due to H. Salzmann.

III.2. A nonclassical projective Moulton plane $\overline{\mathcal{M}}(s)$, $s > 0$; see Subsection 2.7.1. Here p and L are the point and line fixed by all automorphisms of the projective plane.

VII.2. The classical flat projective plane.

There are seven Lenz types numbered I to VII. A projective plane is of Lenz type at least III if its Lenz type is III, IV, V, VI, or VII.

THEOREM 2.9.1 (Lenz Type at Least III) *A flat projective plane of Lenz type at least III is isomorphic to a projective Moulton plane (this includes the classical plane). In particular, the nonclassical Moulton planes are the only flat projective planes of Lenz–Barlotti type III.2.*

For a proof of this fact see Salzmann et al. [1995] Theorem 64.18.

THEOREM 2.9.2 (The Possible Lenz–Barlotti Types) *A flat projective plane is of Lenz–Barlotti type I.1, I.2, I.4, II.1, II.2, III.2, or VII.2.*

For a proof see Salzmann et al. [1995] Chapter 6 and, also, Prieß-Crampe [1983] Chapter 5.5.

2.9.2 Groups of Projectivities

Given a nonincident point–line pair (p, L) in a projective plane $\mathcal{P}$, denote by $\pi(L, p)$ the bijection between L and the line pencil through p that maps a point on L to the line that connects this point with p. This bijection is called the *perspectivity* with *centre* p and *axis* L. Let $\pi(p, L)$ be the inverse of $\pi(L, p)$. The products

$$\pi(p_{n-1}, L_n) \cdots \pi(L_2, p_2)\pi(p_1, L_2)\pi(L_1, p_1)$$

are called the *projectivities* of L_1 onto L_n where p_i are points and L_i are lines such that $p_i \notin L_i, L_{i+1}$. The projectivities of a line L onto itself forms a permutation group the isomorphism type of which does not depend on the choice of the line L. This group is called the *group of projectivities* of the projective plane $\mathcal{P}$.

The *group of affine projectivities* of an affine plane $\mathcal{A}$ is the group consisting of all projectivities of the projective extension $\overline{\mathcal{A}}$ generated by the perspectivities whose centers are contained in the line at infinity of $\mathcal{A}$. This means that *affine projectivities* are products of parallel projections.

2.9.2.1 The Group of Fractional Linear Maps $\mathrm{PGL}_2(F)$

If $\mathcal{P}$ is the classical projective plane over the field F, then its group of projectivities is isomorphic to $\mathrm{PGL}_2(F)$ with its natural action on the projective line $\mathrm{PG}(1, F)$ as the so-called *group of fractional linear maps.* In the following we give a definition of $\mathrm{PGL}_2(F)$ and list some of its most important properties.

Let F be a field. The elements of the general linear group $\mathrm{GL}_2(F)$ are the 2×2 matrices with entries in F and nonzero determinant. Then $\mathrm{PGL}_2(F)$ is the quotient group of this general linear group by its normal subgroup consisting of the nonzero scalar matrices.

If $(x_0 : x_1)$ denotes the homogeneous coordinates of a point in the projective line corresponding to the 1-dimensional subspace spanned by the nonzero vector $(x_0, x_1) \in F^2$, then $\{(x : 1) \mid x \in F\} \cup \{(1 : 0)\}$ represents all points of the projective line. Therefore $\mathrm{PG}(1, F)$ can be identified with $F \cup \{\infty\}$ via the map $(x : 1) \mapsto x$ and $(1 : 0) \mapsto \infty$.

An element of $\mathrm{PGL}_2(F)$ represented by the matrix

$$\begin{pmatrix} a & b \\ c & d \end{pmatrix} \in \mathrm{GL}_2(F)$$

acts on $\mathrm{PG}(1, F)$ by

$$(x_0 : x_1) \mapsto (ax_0 + bx_1 : cx_0 + dx_1).$$

With the above identification this element then acts on $F \cup \{\infty\}$ by

$$x \mapsto \frac{ax + b}{cx + d}$$

with the obvious definitions for $x = \infty$ and when the denominator becomes 0.

The stabilizer of the point ∞ is (isomorphic to) the *group of affine maps of F*

$$F \to F : x \mapsto ax + b,$$

where $a, b \in F$, $a \neq 0$. The group of affine maps together with its natural action on F is isomorphic to the group of affine projectivities of the classical affine plane over F.

If $F = \mathbf{R}$, then the projective line is homeomorphic to $\mathbf{S}^1 = \mathbf{R} \cup \{\infty\}$

and $\mathrm{PGL}_2(F)$ is a 3-dimensional Lie transformation group acting on the circle $\mathbf{S}^1$. The group $\mathrm{PSL}_2(\mathbf{R})$ is a normal subgroup of index 2 of the group $\mathrm{PGL}_2(\mathbf{R})$ consisting of all orientation-preserving fractional linear maps. If $F = \mathbf{C}$, then $\mathrm{PGL}_2(F)$ is a 6-dimensional Lie transformation group acting on $\mathbf{S}^2 = \mathbf{C} \cup \{\infty\}$. For further information about all these groups see Section A2.2.

It can be shown that the group of projectivities of a line in a projective plane acts 3-transitively (see Subsection 2.10.1 for a definition) on the line. In the case of the group $\mathrm{PGL}_2(F)$ the action on a projective line is even sharply 3-transitive. Similarly, the group of affine projectivities of a line in an affine plane acts 2-transitively on the line. In the case of the group of affine maps of F the action on F is sharply 2-transitive.

THEOREM 2.9.3 (Characterizations of the Classical Plane I) *Let G be the group of projectivities of a line L in a flat projective plane and let G be endowed with the compact-open topology. Then the following conditions are equivalent.*

(i) *The group G is locally compact.*

(ii) *The group G acts sharply 3-transitively on L.*

(iii) *The group G is isomorphic and acts equivalently to* $\mathrm{PGL}_2(\mathbf{R})$ *in its natural action on* $\mathbf{S}^1$.

(iv) *The group G acts ω-regularly on L, that is, there exists a finite set $F \subseteq L$ such that the subgroup in G fixing F elementwise is discrete with respect to the compact-open topology.*

(v) *The flat projective plane under consideration is the classical flat projective plane.*

THEOREM 2.9.4 (Characterizations of the Classical Plane II) *Let H be the group of affine projectivities of a line L in a flat affine plane and let H be endowed with the compact-open topology. Then the following conditions are equivalent.*

(i) *The group H is locally compact.*

(ii) *The group H acts sharply 2-transitively on L.*

(iii) *The group H is isomorphic and acts equivalently to the group of affine maps of* $\mathbf{R}$ *in its natural action on* $\mathbf{R}$.

(iv) *The group H acts ω-regularly on L.*

(v) *The flat affine plane under consideration is the classical flat affine plane.*

These two theorems summarize classical results and results by Strambach [1977] and Löwen [1977]. Compare Löwen [1981d] 5.1 and Salzmann et al. [1995] Theorems 66.1 and 66.2.

2.9.2.2 The Group of Piecewise Projective Homeomorphisms

The groups of (affine) projectivities are invariants of affine and projective planes. However, unlike the groups of automorphisms, the groups of projectivities are not very useful tools when it comes to classifying affine and projective planes. In general, they are 'too large' to be of any use. For example, the group of projectivities of a finite nonclassical projective plane of order n is isomorphic to either the alternating group A_{n+1}, the symmetric group S_{n+1} or, if $n = 23$, the Mathieu group M_{24}; see Grundhöfer [1988]. Also the groups of (affine) projectivities of nonclassical flat affine and projective planes are not locally compact (see Theorems 2.9.3 and 2.9.4) and thus no Lie groups.

Apart from the group of projectivities $\mathrm{PGL}_2(\mathbf{R})$ of the classical flat projective plane, for flat projective planes only the groups of projectivities of the nonclassical projective Moulton planes (see Subsection 2.7.1) have been calculated explicitly. All these groups are isomorphic to the so-called *group of piecewise projective homeomorphisms* Γ of $\mathbf{S}^1$ to itself. Here a homeomorphism g of $\mathbf{S}^1$ to itself is called *piecewise projective* if $\mathbf{S}^1$ is the union of finitely many closed intervals such that for each such interval I there is a projective homeomorphism $h \in \mathrm{PGL}_2(\mathbf{R})$ such that the restrictions of g and h to I coincide. Note that this group contains $\mathrm{PGL}_2(\mathbf{R})$ and is really a much larger group than this classical group of projectivities. As the main step in his proof that Γ is the group of projectivities of a nonclassical projective Moulton plane Betten [1979] proves the following result.

LEMMA 2.9.5 (Piecewise Projective Homeomorphisms) *The group of piecewise projective homeomorphisms is generated by its subgroup* $\mathrm{PGL}_2(\mathbf{R})$ *and the set* H *of dilatations in one direction*

$$\mathbf{R} \cup \{\infty\} \to \mathbf{R} \cup \{\infty\} : x \mapsto \begin{cases} kx & \text{for } x \geq 0, \\ x & \text{for } x < 0, \\ \infty & \text{for } x = \infty, \end{cases}$$

where $k > 0$. *The orientation-preserving elements of* Γ *form a normal subgroup of index* 2 *that is generated by* $\mathrm{PSL}_2(\mathbf{R})$ *and the set* H.

This result greatly facilitates identifying a group of piecewise projective homeomorphisms as the group Γ and has also been used to construct abstract ovals and flat Minkowski planes that have groups of projectivities isomorphic to Γ; see Subsections 2.10.2 and 4.4.6.

For more information about Γ see Betten [1979], Betten–Wagner [1982], Betten–Weigand [1985], Löwen [1981d], and Polster [1992], [1996b].

2.9.3 Semigroups of Continuous Lineations

We have been able to classify the flat linear spaces that have a 'large' group of automorphisms. On the other hand, we have seen that it is not difficult to construct many mutually nonisomorphic rigid examples of the different types of flat linear spaces. This means that, in general, it is impossible to reconstruct a flat linear space from its group of automorphisms.

It comes as a pleasant surprise that the semigroup of continuous lineations of a flat linear space, which contains the group of automorphisms of such a plane, determines the plane completely.

A *lineation* of a geometry is a map from its point set to itself such that the image of every line under the map is contained in another (not necessarily unique) line. Of course, every automorphism of a geometry is also a lineation. A lineation is called *collapsed* if the image of the whole point set is contained in a line. The set of all lineations of a geometry forms a semigroup. From now on let $\mathcal{R}$ be a flat linear space with point set P. Provided with the compact-open topology, the semigroup Σ of all continuous lineations of such a plane is a topological semigroup. Let Π be the set of all continuous lineations that map the whole point set onto a point, let Λ be the set of all collapsed continuous lineations not in Π, and let E be the set of all injective continuous lineations.

We define an equivalence relation $\sim$ on Λ. We first define

$$\lambda \leq \mu \longleftrightarrow \lambda(P) \subseteq \mu(P)$$

for $\lambda, \mu \in \Pi \cup \Lambda$. Now define the relation $*$ on Λ by

$$\lambda * \mu \longleftrightarrow \text{ there is a } \nu \in \Lambda \text{ such that } \lambda \leq \nu \text{ and } \mu \leq \nu.$$

Note that $*$ is reflexive and symmetric. Finally, define $\sim$ as the transitive closure of the relation $*$, that is, $\lambda \sim \mu$ if and only if there are an integer n and $\lambda = \beta_1, \beta_2, \cdots, \beta_n = \mu \in \Lambda$ such that $\beta_1 * \beta_2, \beta_2 * \beta_3, \ldots, \beta_{n-1} * \beta_n$.

Construct a geometry $\mathcal{R}'$ as follows. Its points are the elements of Π and its lines are the equivalence classes of the equivalence relation $\sim$. An element $p \in \Pi$ is incident with an equivalence class if there exists a lineation λ in this equivalence class such that $p \leq \lambda$.

THEOREM 2.9.6 (Semigroup Determines Plane) *Let $\mathcal{R}$ be a flat linear space. Then the semigroup Σ of all continuous lineations of $\mathcal{R}$ is the disjoint union of Π, Λ, and E. This decomposition can be read off from Σ alone, without referring to the given action on $\mathcal{R}$. Furthermore, the geometry $\mathcal{R}$ is isomorphic to $\mathcal{R}'$.*

This implies that every flat linear space can be reconstructed from its semigroup of lineations.

For more details about this result see Stroppel [1997].

2.10 Related Geometries

In the following subsections we describe some links between flat linear spaces and some other important incidence geometric structures such as sharply transitive sets of permutations, semibiplanes, and pseudoline arrangements.

2.10.1 Sharply Transitive Sets

A set G of permutations acting on the set S containing more than n elements is called *n-transitive* if and only if, given two ordered n-tuples each of distinct elements of S, there is at least one element in G that maps the first n-tuple elementwise onto the second n-tuple. The set G is called *sharply n-transitive* if and only if, given two ordered n-tuples each of distinct elements of S, there is exactly one element in G that maps the first n-tuple elementwise onto the second n-tuple. Two n-transitive sets G and H of permutations of the set S are *isomorphic* if there are permutations h_1 and h_2 of S such that $h_1 H h_2 = G$.

Note that a sharply 1-transitive set is also '1-unisolvent' and that a sharply 2-transitive set can be made into a '2-unisolvent set' by adding the constant functions to it. Unisolvent sets of functions play an important role in the theory of higher-rank flat circle planes; see Chapter 7.

In this book we are primarily interested in sharply n-transitive sets of homeomorphisms $\mathbf{R} \to \mathbf{R}$ and $\mathbf{S}^1 \to \mathbf{S}^1$ as some of these sets have interpretations as geometries on surfaces. It turns out that only sharply 1-transitive and 2-transitive sets of homeomorphisms $\mathbf{R} \to \mathbf{R}$ exist; see Theorem 2.10.2 below. Similarly, only sharply 1-transitive and sharply 3-transitive sets of homeomorphisms $\mathbf{S}^1 \to \mathbf{S}^1$ exist; see Theorem 2.10.2 below. We list the classical examples of such sets.

(i) The set of translations

$$\mathbf{R} \to \mathbf{R} : x \mapsto x + a,$$

where $a \in \mathbf{R}$, is sharply 1-transitive.

(ii) The set of all affine maps

$$\mathbf{R} \to \mathbf{R} : x \mapsto ax + b,$$

where $a, b \in \mathbf{R}$, $a \neq 0$ is a sharply 2-transitive set.

(iii) The set of all rotations of the circle is a sharply 1-transitive set.

(iv) The classical sharply 3-transitive set on $\mathbf{S}^1$ is the group $\mathrm{PGL}_2(\mathbf{R})$ acting on $\mathbf{S}^1$ as the group of fractional linear maps; see Subsubsection 2.9.2.1.

Associated with every pair of lines K and L in a projective plane is a sharply 2-transitive set of permutations of $K \setminus \{p\}$, where p is the point of intersection of L and K. It is constructed as follows. Consider the set H of all projectivities of the form $\pi(q, L)\pi(K, q)$, where q is a point that is contained neither in L nor in K; see Figure 2.25 (in this diagram the point p is the common point at infinity of the two parallel lines).

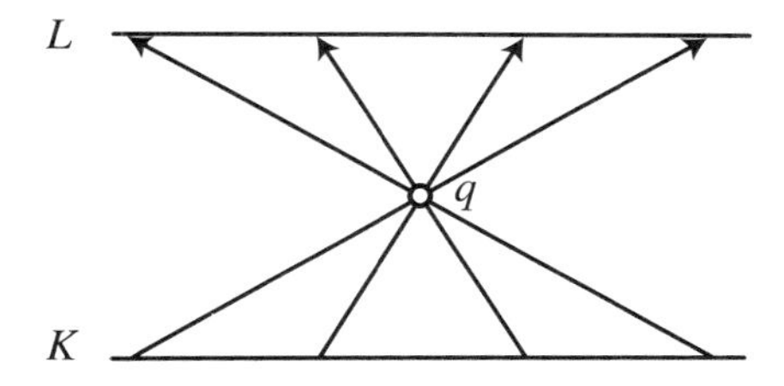

Fig. 2.25. The projectivity associated with q

Let φ be an element of this set. Then $\varphi^{-1}H$ is a sharply 2-transitive set when restricted to $K \setminus \{p\}$. The sharply 2-transitive sets associated with two lines in the classical projective plane over the field F are isomorphic to the group of affine bijections $F \to F$.

The sharply 2-transitive sets associated with flat projective planes are special sets of homeomorphisms $\mathbf{R} \to \mathbf{R}$. Conversely, given a sharply 2-transitive set of homeomorphisms $\mathbf{R} \to \mathbf{R}$, it is possible to construct an $\mathbf{R}^2$-plane whose point set is $\mathbf{R}^2$ and whose lines are the graphs of the functions in the set together with all horizontal and vertical Euclidean lines. Of course, this $\mathbf{R}^2$-plane is not necessarily a flat affine plane.

At this point we only remark that every sharply 3-transitive set of homeomorphisms $\mathbf{S}^1 \to \mathbf{S}^1$ corresponds to a toroidal circle plane; see Subsection 4.2.3. Since the results about sharply n-transitive sets of homeomorphisms $\mathbf{R} \to \mathbf{R}$ and $\mathbf{S}^1 \to \mathbf{S}^1$ are very similar, we list all relevant results simultaneously in this section.

The first part of the following result is also a corollary of Proposition 2.7.17.

THEOREM 2.10.1 (Cut and Paste) *Let G_i, $i = 1, 2$, be two sharply 2-transitive sets of homeomorphisms $\mathbf{R} \to \mathbf{R}$, or two sharply 3-transitive sets of homeomorphisms $\mathbf{S}^1 \to \mathbf{S}^1$. Let G_i^- and G_i^+ be the sets of orientation-reversing and orientation-preserving homeomorphisms in G_i. Then $G_1^+ \cup G_2^-$ is a sharply 2-transitive or sharply 3-transitive set, respectively.*

Proof. Let G_i, $i = 1, 2$, be two sharply 2-transitive sets of homeomorphisms $\mathbf{R} \to \mathbf{R}$. We show that $G_1^+ \cup G_2^-$ is sharply 2-transitive as well.

Let $(x_1, y_1), (x_2, y_2) \in \mathbf{R} \times \mathbf{R}$, where $x_i \neq y_i$, $i = 1, 2$, and let f and g be the unique homeomorphisms $\mathbf{R} \to \mathbf{R}$ in G_1 and G_2 that map x_1 to x_2 and y_1 to y_2. Then Proposition A1.4.1(i) guarantees that f and g are either both orientation-preserving or both orientation-reversing. This implies that there is a unique element in $G_1^+ \cup G_2^-$ that maps x_1 to x_2 and y_1 to y_2. Hence this set is sharply 2-transitive as well.

The second part of the theorem can be proved in a similar fashion using Proposition A1.4.1(iii). □

THEOREM 2.10.2 (Existence) *Sharply n-transitive sets of homeomorphisms of $\mathbf{R} \to \mathbf{R}$ exist if and only if $n = 1$ or $n = 2$.*

Sharply n-transitive sets of homeomorphisms of $\mathbf{S}^1 \to \mathbf{S}^1$ exist if and only if $n = 1$ or $n = 3$.

The following proof has been adapted from Polster [1998a] Propositions 2.1.2 and 2.1.3.

Proof of theorem. Let G be a set of homeomorphisms $\mathbf{R} \to \mathbf{R}$. Bearing in mind the examples given at the beginning of this section, we know that G could be sharply 1- or 2-transitive.

Proposition A1.4.1(i) implies that for $n \geq 3$ there exists no homeomorphism of $\mathbf{R} \to \mathbf{R}$ that fixes $2, 3, \ldots, n-1$ and exchanges 1 and -1. This implies that G cannot be a sharply n-transitive set.

Let G be a set of homeomorphisms $\mathbf{S}^1 \to \mathbf{S}^1$. Bearing in mind the examples given at the beginning of this section, we know that G could be sharply 1- or 3-transitive.

Let a, b, c, and d be four distinct points on the circle. Then Proposition A1.4.1(iii) implies that that there exists no homeomorphism h that fixes a and b and such that c and $h(c)$ are contained in the same connected component of $\mathbf{S}^1 \setminus \{a, b\}$ and such that d and $h(d)$ are contained in different connected components of the set $\mathbf{S}^1 \setminus \{a, b\}$. Hence there are no sharply n-transitive sets of homeomorphisms of $\mathbf{S}^1$ to itself for $n \geq 4$.

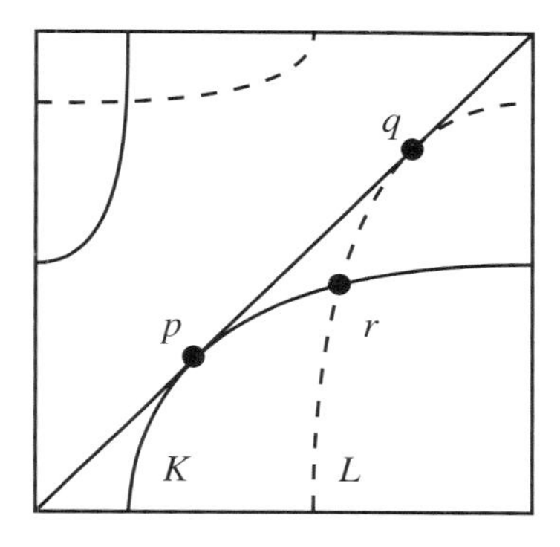

Fig. 2.26.

Assume that G is a sharply 2-transitive set of homeomorphisms of the circle to itself. Consider the geometry $\mathcal{T}$ whose point set coincides with the torus $\mathbf{S}^1 \times \mathbf{S}^1$ and whose lines are the horizontals and verticals on the torus, and the graphs of the functions in G. In this geometry every two points are contained in exactly one line. If f is an element of G, denote by $L(f)$ its associated line in $\mathcal{T}$. We identify $\mathbf{S}^1 \times \mathbf{S}^1$ with the unit square in the usual manner, that is, verticals turn into verticals, horizontals turn into horizontals and boundary points get identified in the obvious manner; see Figure 2.26.

Let $g \in G$. Then Gg^{-1} is a sharply 2-transitive set that contains the identity id. We may therefore assume that id is contained in G. By Proposition A1.4.1(iv), every orientation-reversing homeomorphism has exactly two fixed points, hence the graph of such a homeomorphism would intersect $L(id)$ in two points. Hence these two points would have two connecting lines in $\mathcal{T}$, which is a contradiction to what we deduced before. Hence G does not contain any orientation-reversing homeomorphisms. If $g, h \in G$ and $g(a) = h(a) = b$ for some $a, b \in \mathbf{S}^1$, then $L(g)$ and $L(h)$ are two topological circles that touch topologically at the point (a, b). They cannot intersect transversally since this would imply that they intersect in a second point, which is again impossible.

Consider two points p and q on $L(id)$ and one more point r that is not contained in any of the horizontals and verticals through p and q. Consider the lines K and L in the geometry $\mathcal{T}$ that connect p with r and q with r, respectively. Figure 2.26 shows that if these lines both touch $L(id)$, they necessarily intersect transversally in the point r. This is a contradiction. □

A sharply n-transitive set G is called *invertible* if G contains the identity and $G = G^{-1}$. The sets of involutions in the invertible sharply n-transitive sets that we are interested in turn out to be very interesting themselves.

THEOREM 2.10.3 (Sets of Involutions) *Let G be an invertible sharply 2-transitive set of homeomorphisms of $\mathbf{R}$ to itself and let H be the set of involutions in G. Then H is a sharply 1-transitive set of orientation-reversing homeomorphisms. If the $\mathbf{R}^2$-plane associated with G is a flat affine plane, then H corresponds to a parallel class of lines in this affine plane.*

For a proof of this result see Polster [1998a] Proposition 2.1.5. In the next subsection we will consider the set of involutions in sharply 3-transitive sets of homeomorphisms $\mathbf{S}^1 \to \mathbf{S}^1$.

For more information about sharply transitive sets of permutations see Dembowski [1968] and Polster [1998a].

2.10.2 Quasi-Sharply-2-Transitive Sets and Abstract Ovals

A set G of permutations acting on the set S is called *quasi-sharply-2-transitive* if each permutation $g \in G$ has order at most 2 and for any two pairs $(a_1, a_2), (b_1, b_2) \in S \times S$, with $a_i \neq b_j$, $i, j = 1, 2$, there is a unique permutation $h \in G$ such that $h(a_1) = a_2$, $h(b_1) = b_2$.

Quasi-sharply-2-transitive sets are also referred to as *Buekenhout ovals* or *abstract ovals.* An abstract oval that consists of homeomorphisms of the circle to itself is called a *real abstract oval.*

Given an oval $\mathcal{O}$ in a projective plane $\mathcal{P} = (P, \mathcal{L})$ and a point p not contained in the oval, the *bundle involution* of the oval associated with the point is the involution $\mathcal{O} \to \mathcal{O}$ that fixes a point of the oval if the tangent at this point contains p, and exchanges two distinct points of the oval if the line through the two points contains p. It is easy to check that the set of bundle involutions associated with the points of $P \setminus \mathcal{O}$ is an abstract oval on $\mathcal{O}$. Every abstract oval that arises like this from an oval in a projective plane is called *projective*. For examples of nonprojective abstract ovals see Krier [1977] and Faina [1984].

Associated with every abstract oval G on S is the *ambient of* G. This is a geometry whose points are the elements of $S \cup G$. If $a, b \in S$ (a and b not necessarily distinct), then the line defined by a and b is the set consisting of a and b and all elements $g \in G$ such that $g(a) = b$. If G comes from an oval $\mathcal{O}$ in a projective plane $\mathcal{P}$, then this ambient clearly corresponds to the geometry of all lines in $\mathcal{P}$ that intersect $\mathcal{O}$.

Two abstract ovals G and H of permutations acting on S are *isomorphic* if and only if there is a permutation h of S such that $h^{-1}Hh = G$. If G and H are isomorphic, then their ambients are isomorphic.

The classical abstract oval over the field F is the abstract oval associated with the nondegenerate conic sections in the projective plane over F. If F is of even characteristic, this abstract oval is isomorphic to the set of all elements of order at most 2 in $\mathrm{PGL}_2(F)$ (see Subsection 2.9.2.1), that is, the set consisting of the identity and the involutions in this permutation group. If F is of odd characteristic or characteristic 0, then the classical abstract oval associated with it consists of the involutions in $\mathrm{PGL}_2(F)$. Identify the conic section with the projective line $L = F \cup \{\infty\}$ via the usual stereographic projection. Then the elements of the classical abstract oval are the (fractional linear) involutions

$$L \to L : x \mapsto -x + b,$$
$$L \to L : x \mapsto \frac{ax - b}{x - a},$$

where $a, b \in F$ and $a^2 \neq b$.

Let G be a finite abstract oval on S. If S has an odd number of elements, then id_S is contained in G and all elements different from the identity fix precisely one point of S. An involution of a set S is called *elliptic* if it has no fixed points. It is called *hyperbolic* if it has exactly two fixed points. If S is finite and has an even number of elements, or if G is a real abstract oval, then each element of G is either an elliptic or a hyperbolic involution. This implies that for $|S| > 2$ id_S is not contained in G. Both elliptic and hyperbolic involutions are contained in G. The set of all elliptic (hyperbolic) involutions in G is called the *elliptic (hyperbolic) part* of G.

We call a set H of hyperbolic involutory homeomorphisms of the circle to itself a *real hyperbolic part* if it has the following property.

Let $(a_1, a_2), (b_1, b_2) \in \mathbf{S}^1 \times \mathbf{S}^1$, where $a_i \neq b_j$, $i, j = 1, 2$. If $a_1 = a_2$, or $b_1 = b_2$, or a_1, a_2, b_1, and b_2 are distinct and both a_1 and a_2 are contained in the same connected component of $\mathbf{S}^1 \setminus \{b_1, b_2\}$, then there is a unique $g \in H$ such that $g(a_1) = a_2$, $g(b_1) = b_2$; compare Proposition A1.4.2(iii).

We call a set E of elliptic involutory homeomorphisms of the circle to itself a *real elliptic part* if it has the following property.

Let $(a_1, a_2), (b_1, b_2) \in \mathbf{S}^1 \times \mathbf{S}^1$, with a_1, a_2, b_1, and b_2 distinct and a_1 and a_2 contained in different components of $\mathbf{S}^1 \setminus \{b_1, b_2\}$; see Proposition A1.4.2(iii). Then there is a unique element $g \in E$ having the property that $g(a_1) = a_2$, $g(b_1) = b_2$.

Of course, every hyperbolic part or elliptic part of a real abstract oval is a real hyperbolic part or a real elliptic part, respectively.

THEOREM 2.10.4 (Cut and Paste) *Let E be a real elliptic part and let H be a real hyperbolic part. Then $E \cup H$ is a real abstract oval.*

Use Proposition A1.4.2(iii) to arrive at a simple proof of this result similar to the proof of the corresponding cut-and-paste result Theorem 2.10.1 for sharply n-transitive sets of homeomorphisms.

This result also implies that every real elliptic (hyperbolic) part is the elliptic (hyperbolic) part of a real abstract oval.

It can be shown that every real abstract oval is projective and arises from a topological oval in a flat projective plane. In the following we only describe how, given a real abstract oval, such a projective plane and topological oval can be constructed.

We first need to recall some facts about topological ovals in flat projective planes; see Theorem 2.2.4. Let $\mathcal{O}$ be a topological oval in a flat projective plane $\mathcal{P}$. Then the set of tangents of $\mathcal{O}$ is a topological oval in the dual $\mathcal{P}^{dual}$ of $\mathcal{P}$. We call it $\mathcal{O}^{dual}$. The restriction of $\mathcal{P}$ to the exterior of a topological oval is a disk Möbius strip plane.

Let G be a real abstract oval. Let

$$M = \mathbf{S}^1 * \mathbf{S}^1 = \{\{x, y\} \mid x, y \in \mathbf{S}^1\}$$

denote the set of all unordered pairs of points in $\mathbf{S}^1$. We think of it as being provided with the natural topology that it inherits from $\mathbf{S}^1$. As a topological space M is homeomorphic to a Möbius strip with boundary $B = \{\{x, x\} | x \in \mathbf{S}^1\}$; see Salzmann et al. [1995] remark following

Corollary 55.12. For every involution $i \in G$ let $l_i = \{\{x, y\} \mid y = i(x)\}$. Clearly, if i is an elliptic involution, then l_i is a subset of $\mathbf{S}^1 * \mathbf{S}^1$ homeomorphic to the circle and if i is a hyperbolic involution, then l_i is a closed Jordan arc (homeomorphic to a closed interval) that intersects B in its two boundary points. Furthermore, B is homeomorphic to $\mathbf{S}^1$. Let t_a, $a \in \mathbf{S}^1$, be the set $\{\{a, y\} | y \in \mathbf{S}^1\}$. Let E and S be the sets of all l_i associated with elliptic and hyperbolic involutions in G, respectively, and let T be the set of all t_a, $a \in \mathbf{S}^1$. We define a geometry $\mathcal{R} = (M, \mathcal{L})$ whose line set $\mathcal{L}$ is $E \cup T \cup S$.

If G has been constructed from a topological oval $\mathcal{O}$ in a flat projective plane $\mathcal{P}$, then it is easy to see that this geometry is isomorphic to the geometry that one arrives at by removing the set of interior points of $\mathcal{O}^{dual}$ from the dual flat projective plane $\mathcal{P}^{dual}$. The set B corresponds to $\mathcal{O}^{dual}$. The elements of E, T, and S correspond to the exterior lines, the tangents, and secants of $\mathcal{O}^{dual}$ in $\mathcal{P}^{dual}$, respectively. In this special case it is also clear that any two points in $\mathcal{R}$ are connected by a unique line in $\mathcal{R}$. Since abstract ovals are quasi-sharply-2-transitive, it follows that this last statement is true for an $\mathcal{R}$ constructed from a general real abstract ovals. Consequently, the restriction of $\mathcal{R}$ to $M \setminus B$ is a disk Möbius strip plane. Now, it is easy to prove that we can construct a flat projective plane $\mathcal{P}_G$ as in Subsection 2.7.13 by gluing together $\mathcal{R}$ and the restriction of the Euclidean plane to the closed unit disk; see Polster [1998a] Section 2.4 for the details of this construction.

THEOREM 2.10.5 (Real Abstract Ovals Are Projective) *Let G be a real abstract oval. Then the following hold.*

(i) *The incidence structure $\mathcal{P}_G$ is a flat projective plane.*

(ii) *The set B is a topological oval in $\mathcal{P}_G$ whose sets of tangents, exterior lines, and secant lines coincide with the sets E, T, and S, respectively.*

(iii) *The real abstract oval G is projective and isomorphic to the real abstract oval associated with the dual topological oval B^{dual}.*

Remember that a sharply n-transitive set G is called invertible if G contains the identity and $G = G^{-1}$. Theorem 2.10.3 guarantees that the set of involutions in an invertible sharply 2-transitive set of homeomorphisms of $\mathbf{R}$ to itself forms a sharply transitive set. A similar result is true for invertible sharply 3-transitive sets of homeomorphisms of $\mathbf{S}^1$ to itself.

THEOREM 2.10.6 (Invertible Sharply 3-Transitive) *Let G be an invertible sharply 3-transitive set of homeomorphisms of $\mathbf{S}^1$ to itself. Let H be the set of hyperbolic involutions in G and let E be the set of elliptic involutions. Then the following hold.*

- (i) *The set H is a real hyperbolic part.*
- (ii) *The set of all involutions in G, that is, $E \cup H$ is a real abstract oval if and only if, given any two distinct points $a, b \in \mathbf{S}^1$, every element in G that exchanges a and b is an involution.*

For a proof see Polster [1998a] Theorem 2.2.1. Keeping in mind that every sharply 3-transitive set as in this theorem corresponds to a toroidal circle plane, this result together with Theorem 2.10.5 establishes a nontrivial connection between certain toroidal circle planes and flat projective planes via a real hyperbolic part/real abstract oval.

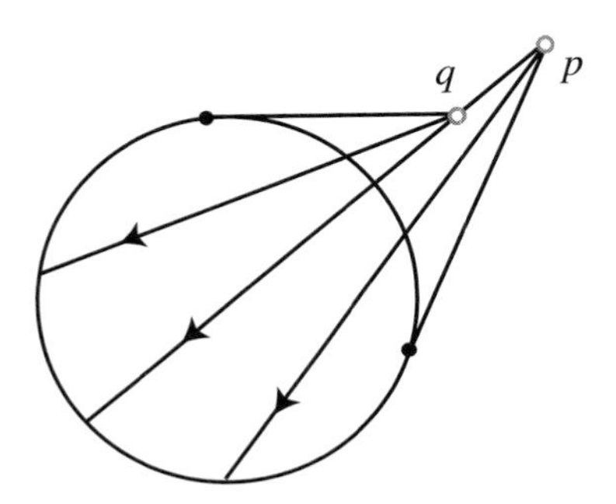

Fig. 2.27. A 'bent' bundle involution

The *group of projectivities* of an abstract oval G on S is the permutation group acting on S generated by G. The group of projectivities is 3-transitive and is isomorphic to $\mathrm{PGL}_2(F)$ if G is the classical abstract oval associated with the field F. Note that this group is also the group of projectivities of the classical projective plane; see Subsection 2.9.2.

Just as the group of projectivities of a nonclassical projective plane tends to be very large, so does the group of projectivities of a nonclassical abstract oval. For example, if G is a finite nonclassical abstract oval on a set containing n elements, then its group of projectivities is either the alternating or the symmetric group on the same n elements; see Polster [1992] Theorem 3.1.

Here is a construction of a real abstract oval around the unit circle in the Euclidean plane whose group of projectivities is isomorphic to the group of piecewise projective homeomorphisms of the circle to itself; see p. 114 for a definition of this group. We associate with every point of the projective plane over the reals off the unit circle an involutory homeomorphism of the unit circle. The set of all these involutions is the real abstract oval we are looking for; see Polster [1992] Lemma 8.8.2. The involutions associated with the interior points of the unit circle and with the points on the line at infinity of the Euclidean plane are the bundle involutions associated with these points. Let p be a point of the Euclidean plane that is an exterior point of the unit circle. Draw the line through p and the origin of the plane and let q be the point on this line that is at equal distance between p and the unit circle. Then the 'bent' involution associated with p is defined as in Figure 2.27.

The standard reference for abstract ovals is Buekenhout [1966]. For real abstract ovals see Polster [1992], [1998a], and Valette [1965]. For more information about groups of projectivities of projective real abstract ovals see Löwen [1977], [1981d].

2.10.3 Semibiplanes

A *semibiplane* is a point–block geometry that satisfies three axioms.

Axioms for Semibiplanes

(S1) Two distinct points are contained in no or exactly two distinct blocks.

(S2) Two distinct blocks intersect in no or exactly two distinct points.

(S3) The geometry is connected.

Here *connected* means that the point set of the semibiplane is not the disjoint union of two proper subsets such that every block is contained in either one of these subsets.

Two points in a semibiplane are *parallel* if they coincide of if there is no block that contains both of them. Similarly, two blocks are parallel if they coincide or do not intersect in a point. A semibiplane is called *divisible* if being parallel defines equivalence relations on both the point and block sets of the geometry. A divisible semibiplane is called *tactical* if the following hold.

(i) Given a parallel class of points, the set of all blocks that do not intersect this parallel class is a parallel class of blocks.
(ii) Given a parallel class of blocks, the set of all points that are not contained in any of the blocks in this parallel class is a parallel class of points.

Here is an example of a tactical semibiplane. Take as its points the vertices of the icosahedron and associate with every vertex v a block of the geometry that consists of the five vertices that are connected by an edge to the vertex v; see Figure 2.28. The parallel classes of points are the pairs of opposite vertices and the parallel classes of blocks are the pairs of blocks associated with opposite vertices.

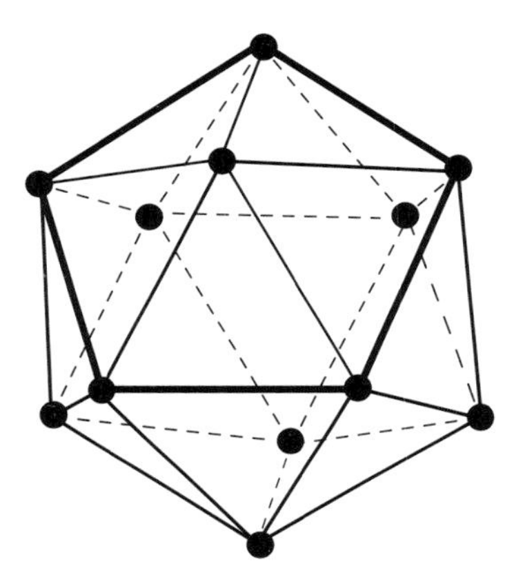

Fig. 2.28. A block of a semibiplane on the icosahedron

Associated with every involutory automorphism γ of a projective plane is a semibiplane that is constructed as follows.

(i) Points are the unordered pairs $\{p, \gamma(p)\}$, where p is a point of the projective plane with $p \neq \gamma(p)$.
(ii) Blocks are the unordered pairs $\{L, \gamma(L)\}$, where L is a line of the projective plane with $L \neq \gamma(L)$.

A *homology semibiplane* is a semibiplane associated with an involutory homology. It is easy to check that homology semibiplanes are tactical, that the parallel classes of points in the semibiplane correspond to the lines in the projective plane through the center of the homology, and the parallel classes of blocks to the pencils of lines through points on the axis of the homology. In fact, it can be shown that the above example on the icosahedron is isomorphic to a homology semibiplane associated with the projective plane of order 5.

By Salzmann et al. [1995] Corollary 55.29, every involutory automorphism of a flat projective plane is a homology and we call a homology semibiplane *flat* if it arises as described above from a flat projective plane.

Let us first have a look at the flat homology semibiplane associated with a reflection of the classical plane with center c and axis A (remember that $c \notin A$). We use the standard representation of the classical plane on the Möbius strip M, or equivalently the strip $[0, \pi] \times \mathbf{R}$ whose left and right boundaries have been identified as usual; see Subsection 2.1.3. Remember that the nonvertical lines are the graphs of the half-periodic functions

$$f_{a,b} : [0, \pi] \to \mathbf{R} : x \mapsto a \sin x + b \cos x,$$

where $a, b \in \mathbf{R}$. Then a reflection is given by the map

$$M \to M : (x, y) \mapsto (x, -y).$$

The center of this reflection is the infinite point of the verticals and its axis is the graph of the function $f_{0,0}$. Let C be the strip $[0, \pi] \times [0, +\infty)$ whose boundaries have been identified via the function $(0, y) \mapsto (\pi, y)$. This means that C is a cylinder with boundary c at the bottom of the cylinder.

The continuous map

$$M \to C : (x, y) \mapsto (x, |y|)$$

identifies pairs of points on M that get exchanged by our homology with one point on the cylinder. The axis is mapped to the circular boundary c at the bottom of this cylinder. This gives the following representation of the corresponding homology semibiplane on the cylinder. The point set is the cylinder $C \setminus c$, the blocks are the graphs of the functions

$$h_{a,b} : [0, \pi] \setminus \{b\} \to \mathbf{R}^+ : x \mapsto a \sin((x - b) \bmod \pi),$$

where $a \in \mathbf{R}^+$ and $b \in [0, \pi]$. The parallel classes of points are the verticals on the cylinder, and the parallel classes of blocks correspond to the sets of functions $\{h_{a,b} \mid a \in \mathbf{R}^+\}$ for $b \in [0, \pi)$.

A *cylinder semibiplane* is a tactical semibiplane that looks like this classical homology semibiplane on the cylinder, that is, has a representation as follows. The point set is the cylinder $\mathbf{S}^1 \times \mathbf{R}^+$, the verticals on the cylinder are the parallel classes of points. Every block is the restriction to this cylinder of the graph of a continuous function $\mathbf{S}^1 \to [0, +\infty)$ that has exactly one zero. To every $b \in \mathbf{S}^1$ corresponds exactly one parallel class of blocks consisting of all those blocks that correspond to functions having a zero at b.

Here is one of the simplest geometrical constructions of a cylinder semibiplane; see Figure 2.29. Consider the upper half of the vertical cylinder in $\mathbf{R}^3$ containing the unit circle. This upper half is itself a cylinder. Then the set of all intersections of this upper half with non-vertical planes that touch the unit circle forms the block set of a cylinder semibiplane. Figure 2.29 shows a parallel class of blocks in this semibiplane.

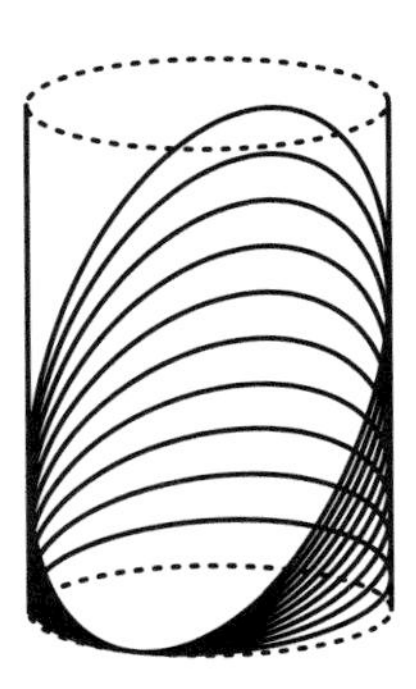

Fig. 2.29. A parallel class of blocks in a cylinder semibiplane

We note that this cylinder semibiplane is a subgeometry of the classical flat Laguerre plane. It is isomorphic to the classical homology semibiplane that we constructed above. We summarize the main results about flat homology semibiplanes in the following theorem.

THEOREM 2.10.7 (Homology = Cylinder Semibiplane) *Every flat homology semibiplane has a representation as a cylinder semibiplane. Every cylinder semibiplane is topologically isomorphic to a flat homology semibiplane.*

It is not hard to prove the first part of this result by generalizing the above construction of the classical cylinder semibiplane from the classical flat projective plane.

Since cylinder semibiplanes occur as geometries associated with many of the other geometries on surfaces, the above theorem constitutes an important link between flat projective planes and these other geometries; see Chapter 6 for more details.

For more details and proofs of the results mentioned in this subsection and more information about topological homology semibiplanes in general see Polster [1995d] and Polster–Schroth [20XX].

2.10.4 Pseudoline Arrangements, Universal Planes, Spreads

Any topological circle in the real projective plane whose complement is homeomorphic to $\mathbf{R}^2$ is also called a *pseudoline*. Note that every line in a flat projective plane is a pseudoline. Note also that two pseudolines that have exactly one point in common intersect each other transversally in this point, just like two lines in a flat projective plane.

A *pseudoline arrangement* is a finite set of pseudolines every two of which intersect in exactly one point. This means that every finite subset of lines in a flat projective plane is a pseudoline arrangement. Pseudoline arrangements are discrete counterparts of flat projective planes. They have interpretations in terms of various equivalent important combinatorial objects such as oriented matroids and switching sequences; see Grünbaum [1972], Knuth [1992], and Björner et al. [1993]. In this subsection we summarize some of the results about pseudoline arrangements that are important for us.

Two pseudoline arrangements are *isomorphic* if there is a homeomorphism of the real projective plane to itself that maps the first to the second arrangement. Up to isomorphism there is only a finite number of pseudoline arrangements having a fixed number of elements, and it is possible to compute these numbers.

A pseudoline arrangement is *embeddable* in a flat projective plane if it is isomorphic to a finite subset of lines of this flat projective plane. A pseudoline arrangement is *stretchable* if it can be embedded in the classical flat projective plane. It turns out that any pseudoline arrangement with less than nine lines is stretchable. On the other hand, there are nonstretchable pseudoline arrangements containing nine lines. An example of such an arrangement can be constructed as follows. Start with Pappus' configuration in the classical flat projective plane and modify it as in Figure 2.30. Note that we only draw that part of the arrangement in which something is happening, that is, lines are crossing.

A pseudoline arrangement is *simple* if no point in the real projective plane is contained in more than two pseudolines in the arrangement. The arrangement in Figure 2.30 is not simple.

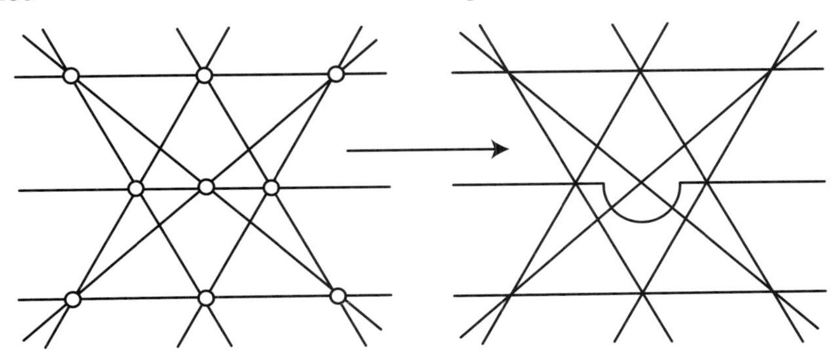

Fig. 2.30. A nonstretchable pseudoline arrangement

Figure 2.31 shows a nonstretchable *simple* arrangement having nine lines. For a nonstretchability proof based on Pappus' theorem see Grünbaum [1972] Theorem 3.2.

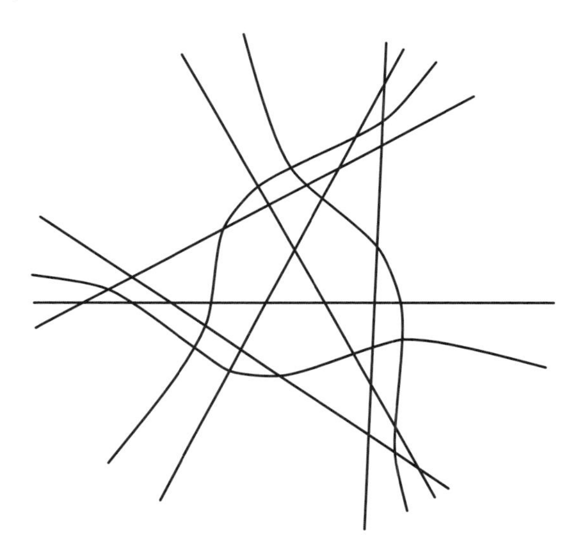

Fig. 2.31. A nonstretchable simple pseudoline arrangement

Of course, all this means that pseudoline arrangements are invariants of flat projective planes, and that it is possible to distinguish between certain such planes by the presence or absence of certain arrangements embedded in them. A natural question to ask is whether every pseudoline arrangement can be embedded in a flat projective plane. The following result states that there are no maximal pseudoline arrangements. For a proof see Grünbaum [1972] Theorem 3.4.

THEOREM 2.10.8 (Levi's Enlargement Theorem) *Let $\mathcal{P}$ be a pseudoline arrangement consisting of n pseudolines, and let p and q be points that are not both contained in any of these pseudolines. Then there exists a pseudoline through p and q which extends $\mathcal{P}$ to a pseudoline arrangement containing $n+1$ pseudolines.*

This means that, at least to start with, there is no reason to believe that we should not be able to extend any given pseudoline arrangement to a flat projective plane. In fact, while no proof along these lines is known, this result suggests one way to construct such a projective plane. Just extend the given arrangement a countably infinite number of times in such a way that the union of all pseudolines constructed contains a set of points that is dense in the real projective plane. Then form some kind of limit to arrive at the flat projective plane we are after.

Goodman et al. [1994b] prove the result we are interested in using a cut-and-paste technique.

THEOREM 2.10.9 (Embedding) *Any pseudoline arrangement can be embedded in a flat projective plane.*

Sketch of a proof. First, observe that a pseudoline arrangement induces a decomposition of the real projective plane into a cell complex consisting of faces, edges, and vertices. An isomorphism between two arrangements induces a one-to-one correspondence between their cell complexes such that neighbouring vertices, edges, and faces in one arrangement are mapped to neighbouring vertices, edges, and faces in the other. The converse of this statement is also true.

Second, given a pseudoline arrangement consisting of n pseudolines and having one face that is bounded by at least $n-1$ pseudolines, this pseudoline arrangement is stretchable. This intuitive result is one of the lemmas proved in Goodman et al. [1994b].

Let $\mathcal{P}$ be a pseudoline arrangement having at least three elements (arrangements with less than three lines are stretchable anyway). Fix one of the pseudolines L in $\mathcal{P}$. For every face F of $\mathcal{P}$ let h be a homeomorphism of the real projective plane to itself that 'stretches' the arrangement consisting of L and the pseudolines in $\mathcal{P}$ bounding F. The image $h(F)$ of the face F under the homeomorphism h is a convex disk bounded by straight line segments; see Figure 2.32.

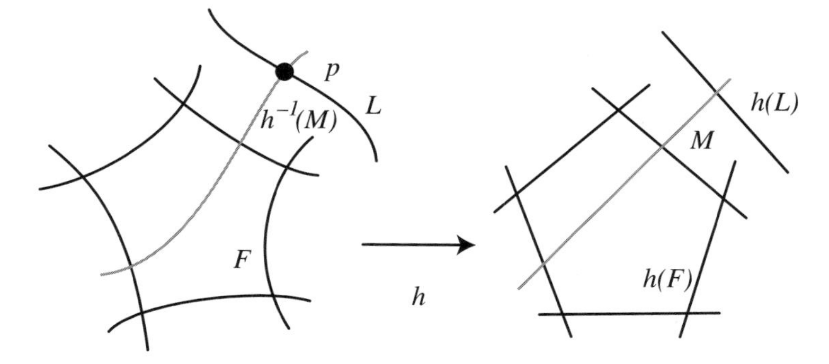

Fig. 2.32.

Let M be a straight line that intersects the boundary of the disk $h(F)$ in two points. Then $h^{-1}(M) \cap F$ is an arc in F with 'slope' p, where p is the point $h^{-1}(M)$ and L intersect in; see Figure 2.32. Consider the set of all these arcs in all faces. These arcs are now strung together to form pseudolines. Starting at a point q of L, travel along one of the arcs ending in this point. Once you hit the other boundary point on one of the original pseudolines continue on the other side of this pseudoline along the unique arc with slope q. Continue along that arc until you hit the next original pseudoline. Continue on the other side of this pseudoline along the unique arc with slope q. Continue like this until you arrive back at the point you started with. Now it is not hard to prove that the original pseudolines together with all the possible paths you can construct using the above rule form the line set of a flat projective plane. □

Goodman et al. [1994b] proved even more.

THEOREM 2.10.10 (Universal Planes) *Universal planes, that is, flat projective planes in which every pseudoline arrangement is embeddable, exist.*

Proof. We first observe that up to isomorphism there is only a countably infinite number of pseudoline arrangements. Given any pseudoline arrangement, we first embed it into a flat projective plane; by Theorem 2.10.9 this is possible. In this flat projective plane we can find three points such that all points of intersection of this arrangement are contained inside a triangle bounded by the lines through the three points.

Let the restriction of the flat projective plane to this triangle be the 'triangular patch' associated with the pseudoline arrangement. We can now cut countably many disjoint triangular holes in the classical flat projective plane, and fill these holes with the triangular patches every single one of which has been suitably deformed to fit its hole; see Figure 2.33. If we continue pieces of lines in the classical plane across the holes in the only possible way using the line segments in the triangular patches, we arrive at a universal flat projective plane. □

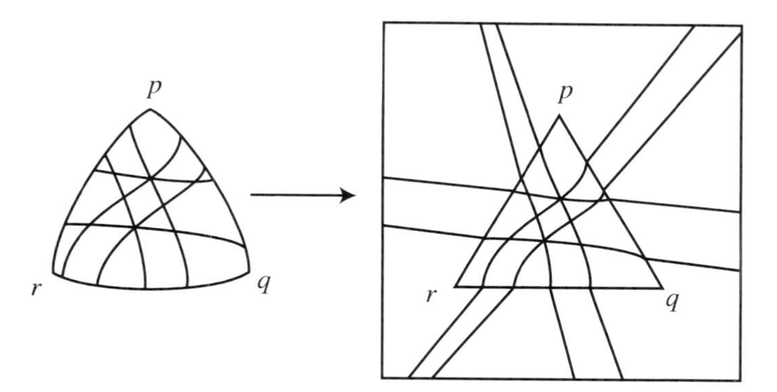

Fig. 2.33. Fitting a triangular patch into a triangular hole in the classical plane in the proof of Theorem 2.10.10

Note that there are many different ways to build the universal projective plane constructed in this proof. In fact, it is not difficult to construct infinitely many nonisomorphic universal planes.

An *isotopy* between two isomorphic pseudoline arrangements $\mathcal{P}$ and $\mathcal{P}'$ is an isotopy, or equivalently a continuous deformation, of the real projective plane that induces an isomorphism between $\mathcal{P}$ and $\mathcal{P}'$. The following result is due to Ringel [1956].

THEOREM 2.10.11 (Isotopy) *There exists an isotopy between any two isomorphic simple arrangements of pseudolines.*

A *classical* isotopy between two straight pseudoline arrangements is an isotopy in which every intermediate arrangement is straight, too. Pairs of isomorphic straight arrangements that are not classically isotopic have been constructed; see Björner et al. [1993]. Finally, we mention one more geometrical object that fits somewhere in between pseudoline arrangements and flat projective planes.

A *spread of pseudolines* $\mathcal{S}$ with respect to a pseudoline L consists of infinitely many pseudolines such that the following conditions are satisfied.

(i) The pseudoline L and an element of $\mathcal{S}$ intersect in exactly one point and every point of L is contained in one element of $\mathcal{S}$, called $\mathcal{S}(p)$.
(ii) The pseudoline $\mathcal{S}(p)$ depends continuously on p.
(iii) Two elements of $\mathcal{S}$ intersect in exactly one point.

One of the simplest ways to construct a spread is to start with a point p and a line L in a flat projective plane such that p is not contained in L. Then the pencil of lines through p is a spread of pseudolines with respect to L. The investigation of spreads was motivated by results in the theory of convex sets; see Grünbaum [1972] Chapter 4.

Before Goodman et al. succeeded in proving that all pseudoline arrangements can be embedded in a flat projective plane, they proved that they can be embedded in spreads; see Goodman et al. [1994a]. It is not known whether every spread can be embedded in a flat projective plane. For more information about spreads see Grünbaum [1972] Chapter 4.

2.11 Open Problems

PROBLEM 2.11.1 *Are there linear spaces on the Möbius strip in which the connecting line of two distinct points does not depend continuously on the two points? Is every linear space on the Möbius strip automatically a flat linear space, that is, a Möbius strip plane? See Subsection 1.2.1 for a definition of linear spaces on surfaces.*

A negative answer to this problem would allow us to develop the theory of flat linear spaces along the same lines as we developed the theory of ideal flat linear spaces.

PROBLEM 2.11.2 *Are there linear spaces on surfaces other than the ones considered in this chapter? See Subsection 1.2.1 for a definition of linear spaces on surfaces.*

As a first step it would be very interesting to classify the linear spaces on compact surfaces whose lines are topological circles.

It is easy to prove that no linear space $(\mathbf{S}^2, \mathcal{L})$ exists in which all lines are topological circles. Skornjakov [1954], [1957] investigates linear spaces whose point sets are either $\mathbf{R}^2$ or $\mathbf{S}^2$ and whose lines are fairly arbitrary curves in the point sets. Among other things, he proved that no such linear space can exist on $\mathbf{S}^2$. If further types of geometries on surfaces exist, then the results about flat linear spaces suggest that these geometries are not very attractive from a topological point of view.

Note that we did not consider linear spaces whose point sets are surfaces and whose lines may also be homeomorphic to closed or half-open intervals. It is easy to construct examples for such geometries that are extensions of flat linear spaces. Let C be the unit circle in the classical flat projective plane, C' a proper nonempty subset of C, I the set of interior points of C, and O the set of exterior points of C. Then the restrictions of the projective plane to $I \cup C'$ and $O \cup C'$ are examples of linear spaces in which lines are homeomorphic to open, closed, or half-open intervals, or circles. In these and similar examples the 'endpoints' of 'interval lines' are boundary points of the surfaces the geometries live on.

PROBLEM 2.11.3 *Develop all topological and geometrical aspects of a general convexity theory for* $\mathbf{R}^2$*-planes.*

We already summarized a few results in Section 2.2. Many more are scattered throughout the literature. However, a comprehensive transfer of results from the classical to the general case is still missing.

PROBLEM 2.11.4 *Given an arbitrary* $\mathbf{R}^2$*-plane on* $\mathbf{R}^2$*, is it always possible to find a family of closed topological arcs such that every three points that are not contained in a line are contained in exactly one of the arcs?*

A positive solution for this and similar problems is equivalent to embedding the $\mathbf{R}^2$-plane as a derived plane into a spherical circle plane; see the next chapter for definitions of these terms. For many more embedding problems involving flat linear spaces see the problem sections in the following chapters.

PROBLEM 2.11.5 *Are any two flat linear spaces of the same type isotopic (in some sense)?*

The idea is to start with two flat linear spaces on the same point set and then continuously deform the lines of the first plane into the lines of the second plane. See Rosehr [20XX] for a proof that all $\mathbf{R}^2$-planes are isotopic.

PROBLEM 2.11.6 *Is it possible to distinguish between the different families of flat projective planes with large group dimension by some simple pseudoline arrangements contained in their members?*

For example, start by considering the planes of group dimension at least 3 and try to find (small) pseudoline arrangements in one of the nonclassical projective Moulton planes that cannot be embedded in the classical flat projective plane. This is not a problem since it contains Desargues' configurations that do not close. What about a pseudoline arrangement that cannot be embedded in any nonclassical projective skew parabola plane but can be embedded in every projective Moulton plane?

Note that since universal flat projective planes exist, it is impossible to distinguish any two flat projective planes by just looking at the pseudoline arrangements contained in them.

PROBLEM 2.11.7 *Can every spread of pseudolines be embedded in a flat projective plane?*

This is one step up from the embedding of pseudoline arrangements in a flat projective plane. Even if the embedding of an arbitrary spread into a flat projective plane can be accomplished, it seems highly unlikely that there is a flat projective plane in which every spread can be embedded, that is, universal flat projective planes probably do not exist when it comes to spreads.

3
Spherical Circle Planes

Flat Möbius planes were first investigated by Wölk [1966] and Strambach [1967c]. Later, Strambach [1970d], [1972], [1973], [1974a], [1974b] studied the more general spherical circle planes. For more information about Möbius planes and, in particular, finite Möbius planes, we refer to Dembowski [1968], Delandtsheer [1995], Hering [1965], Mäurer [1967], Wilker [1981] and the references given there.

A spherical circle plane is a point–circle geometry whose point set is (homeomorphic to) $\mathbf{S}^2$ and whose circles are topological circles on $\mathbf{S}^2$. Furthermore, the Axiom of Joining B1 (see p. 7) is satisfied, that is, any three distinct points are contained in exactly one of the circles. A spherical circle plane is a flat Möbius plane if, in addition, the Axiom of Touching B2 is satisfied, that is, for each circle C and any two distinct points p, q with $p \in C$ there is precisely one circle through p and q that touches C at p geometrically, that is, intersects C only at the point p or coincides with C.

3.1 Models of the Classical Flat Möbius Plane

In this first section we describe a number of models of the classical flat Möbius plane. For detailed information about most of these models see Benz [1973].

3.1.1 The Geometry of Plane Sections

We start with the standard unit sphere $\mathbf{S}^2$ in 3-dimensional Euclidean space $\mathbf{R}^3$, that is,

$$\mathbf{S}^2 = \{(x, y, z) \in \mathbf{R}^3 \mid x^2 + y^2 + z^2 = 1\}.$$

Points of the classical flat Möbius plane are the points of $\mathbf{S}^2$, circles are the intersections of $\mathbf{S}^2$ with planes in $\mathbf{R}^3$ provided they intersect in more than one point. A plane E of $\mathbf{R}^3$ given by

$$E = \{(x, y, z) \in \mathbf{R}^3 \mid ax + by + cz + d = 0\}$$

for some $a, b, c, d \in \mathbf{R}$, $a^2 + b^2 + c^2 = 1$, intersects $\mathbf{S}^2$ in more than one point if and only if E has distance less than 1 from the origin $(0, 0, 0)$, that is, if and only if $|d| < 1$.

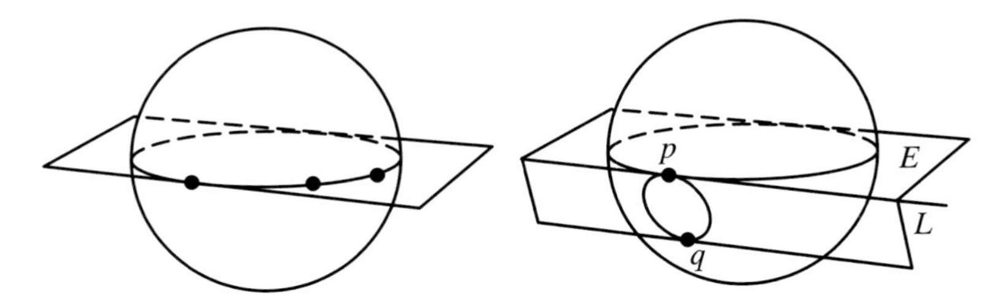

Fig. 3.1. The Axioms of Joining and Touching in the classical plane

In this model Axioms B1 and B2 are easily verified; see Figure 3.1. As for Axiom B1, three points of $\mathbf{S}^2$ determine a unique plane in $\mathbf{R}^3$ and its trace on $\mathbf{S}^2$ is the unique circle containing the three points. In Axiom B2 we have two points p and q and a circle. This circle is contained in a plane E in $\mathbf{R}^3$ and has a unique tangent L in E at the given point p. Then L and the other given point q determine a unique plane in $\mathbf{R}^3$. Its trace on $\mathbf{S}^2$ is the unique circle through the two points touching the given circle at the point p.

More generally, we can start off with any nondegenerate elliptic quadric in 3-dimensional real projective space and obtain the classical flat Möbius plane as the geometry of nontrivial plane sections of this quadric.

3.1.2 The Geometry of Euclidean Lines and Circles

We fix a point p on the unit sphere and consider the derived plane of the classical flat Möbius plane at the point p. Remember that the points of this derived plane are the points different from p and its lines are the circles passing through p that have been punctured at p. This derived plane is isomorphic to the Euclidean plane. We use stereographic projection from p onto a plane that does not contain p to convince ourselves of this fact. Without loss of generality we may assume that p

is the north pole $(0, 0, 1)$ of the sphere and that the plane we project on is the xy-plane; see Figure 3.2.

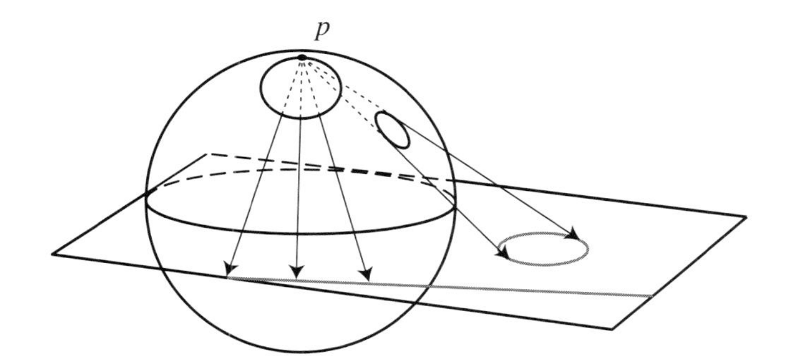

Fig. 3.2. Stereographic projection

This stereographic projection from the north pole $(0, 0, 1)$ onto the xy-plane is given by the formula

$$(x, y, z) \mapsto \left(\frac{x}{1-z}, \frac{y}{1-z} \right)$$

where $(x, y, z) \in \mathbf{S}^2$, $(x, y, z) \neq (0, 0, 1)$. A plane E of $\mathbf{R}^3$ given by

$$E = \{(x, y, z) \in \mathbf{R}^3 \mid ax + by + cz + d = 0\}$$

where $a, b, c, d \in \mathbf{R}$, $a^2 + b^2 + c^2 = 1$, intersects $\mathbf{S}^2$ in more than one point if and only if $|d| < 1$; see Subsection 3.1.1. Furthermore, this plane passes through the north pole if and only if $c + d = 0$. Under the stereographic projection $E \cap \mathbf{S}^2$ is taken to

$$\left\{(x, y) \in \mathbf{R}^2 \mid (c + d)(x^2 + y^2) + 2ax + 2by + d - c = 0\right\}.$$

This yields a Euclidean line for $c+d = 0$ and a Euclidean circle otherwise. Clearly, every Euclidean line or circle is the image of one of the circles on the sphere. This also means that a circle on the sphere not passing through the point p induces a conic in the derived plane.

Conversely, we can start off with the geometry on $\mathbf{R}^2$ whose circles are the Euclidean lines and circles. This geometry can be extended to another model of the classical flat Möbius plane. All we have to do is to one-point-compactify the point space $\mathbf{R}^2$ by a point ∞. We further one-point-compactify every Euclidean line by the point ∞.

In this model of the classical flat Möbius plane Axioms B1 and B2 are easily verified but the process is a bit cumbersome as one has to distinguish whether or not one of the points involved is the point ∞. As for Axiom B1, if ∞ is one of the points, then the circle through these points must be a Euclidean line which is uniquely determined by two of its points. If ∞ is not among the points, then one solves the homogeneous system consisting of the three linear equations in the four coefficients a, b, c, and d that we arrive at by substituting the coordinates of the three points into the general equation describing the Euclidean lines and circles

$$a(x^2 + y^2) + bx + cy + d = 0,$$

where $a, b, c, d \in \mathbf{R}$ are not all equal to zero. This system has a 1-dimensional space of solutions. (Note that the vector $(a, b, c, d) \in \mathbf{R}^4$ representing the above circle can be replaced by any nonzero scalar multiple, that is, circles correspond to points of a 3-dimensional projective space.) As for Axiom B2, if ∞ is the point in which the circles touch, we must have two parallel Euclidean lines; of course a parallel to a Euclidean line is uniquely determined by one of its points. If the circles touch at a point of $\mathbf{R}^2$, then their tangents at this point must be the same and we obtain again a homogeneous system of three linear equations in the four coefficients a, b, c, and d which has a 1-dimensional space of solutions.

3.1.3 Pentacyclic Coordinates

In this subsection we introduce a certain higher-dimensional space such that points and circles of the classical flat Möbius plane are both represented by points of this space. These ideas go back to Coolidge [1916].

Let $\tilde{\mathcal{Q}}$ be the nondegenerate quadric in the projective space $\mathrm{PG}(4, \mathbf{R})$ defined by

$$\tilde{\mathcal{Q}} = \{(x_0 : x_1 : x_2 : x_3 : x_4) \in \mathrm{PG}(4, \mathbf{R}) \mid x_0^2 + x_1^2 + x_2^2 - x_3^2 - x_4^2 = 0\},$$

where $(x_0 : x_1 : x_2 : x_3 : x_4)$ denotes the homogeneous coordinates of the point corresponding to the 1-dimensional subspace spanned by the nonzero vector $(x_0, x_1, x_2, x_3, x_4) \in \mathbf{R}^5$. We intersect $\tilde{\mathcal{Q}}$ with hyperplanes

$$E_{\mathbf{a}} = \{(x_0 : x_1 : x_2 : x_3 : x_4) \mid a_0x_0 + a_1x_1 + a_2x_2 + a_3x_3 + a_4x_4 = 0\},$$

where $\mathbf{a}$ is the point $\mathbf{a} = (a_0 : a_1 : a_2 : a_3 : a_4) \in \mathrm{PG}(4, \mathbf{R})$. Starting with the point $\mathbf{e} = (0 : 0 : 0 : 0 : 1)$ the intersection of $E_{\mathbf{e}}$ and $\tilde{\mathcal{Q}}$ is

$$\mathcal{Q} = E_{\mathbf{e}} \cap \tilde{\mathcal{Q}} = \{(x_0 : x_1 : x_2 : x_3 : 0) \mid x_0^2 + x_1^2 + x_2^2 - x_3^2 = 0\},$$

that is, an elliptic quadric. Introducing affine coordinates in such a way that $x_3 = 0$ describes the plane at infinity of the projective space $E_{\mathbf{e}}$ (this plane does not meet $\mathcal{Q}$) we just obtain the standard 2-sphere $\mathbf{S}^2$. We thus have the point space of the classical flat Möbius plane.

We further intersect $\mathcal{Q} = E_{\mathbf{e}} \cap \tilde{\mathcal{Q}}$ with hyperplanes $E_{\mathbf{a}}$ where $\mathbf{a} \in \tilde{\mathcal{Q}} \setminus E_{\mathbf{e}}$ to obtain the circles. Clearly, $E_{\mathbf{e}} \cap \tilde{\mathcal{Q}} \cap E_{\mathbf{a}}$ is the intersection of the elliptic quadric $\mathcal{Q}$ in the projective space $E_{\mathbf{e}} \simeq \mathrm{PG}(3, \mathbf{R})$ with the plane $E_{\mathbf{e}} \cap E_{\mathbf{a}}$ of $E_{\mathbf{e}}$. Note that

$$\begin{aligned} \tilde{\mathcal{Q}} \setminus E_{\mathbf{e}} &= \{(x_0 : x_1 : x_2 : x_3 : 1) \in \mathrm{PG}(4, \mathbf{R}) \mid x_0^2 + x_1^2 + x_2^2 - x_3^2 = 1\} \\ &\simeq \{(x_0, x_1, x_2, x_3) \in \mathbf{R}^4) \mid x_0^2 + x_1^2 + x_2^2 = x_3^2 + 1\}. \end{aligned}$$

If $\mathbf{a} = (a_0 : a_1 : a_2 : a_3 : 1) \in \tilde{\mathcal{Q}} \setminus E_{\mathbf{e}}$, then

$$\begin{aligned} E_{\mathbf{a}} \cap \mathcal{Q} &= \{(x_0 : x_1 : x_2 : x_3 : 0) \in \mathrm{PG}(4, \mathbf{R}) \mid x_0^2 + x_1^2 + x_2^2 = x_3^2, \\ &\qquad a_0 x_0 + a_1 x_1 + a_2 x_2 + a_3 x_3 = 0\} \\ &= \{(x_0 : x_1 : x_2 : 1 : 0) \in \mathrm{PG}(4, \mathbf{R}) \mid x_0^2 + x_1^2 + x_2^2 = 1, \\ &\qquad a_0 x_0 + a_1 x_1 + a_2 x_2 + a_3 = 0\} \\ &\simeq \{(x_0, x_1, x_2) \in \mathbf{S}^2 \mid a_0 x_0 + a_1 x_1 + a_2 x_2 + a_3 = 0\}. \end{aligned}$$

Here $a_0^2 + a_1^2 + a_2^2 = a_3^2 + 1 \geq 1$. Furthermore, if $b_i = a_i / \sqrt{a_0^2 + a_1^2 + a_2^2}$, then $b_0^2 + b_1^2 + b_2^2 = 1$ and $|b_3| < 1$ so that we obtain the intersection of $\mathbf{S}^2$ with a plane of $\mathbf{R}^3$. Moreover, all planes of $E_{\mathbf{e}}$ that intersect $\mathbf{S}^2$ nontrivially are covered.

More generally, one can use any point $\mathbf{p} = (p_0 : p_1 : p_2 : p_3 : p_4)$ such that $p_0^2 + p_1^2 + p_2^2 - p_3^2 - p_4^2 < 0$ instead of $\mathbf{e}$. Thus the point space is $E_{\mathbf{p}} \cap \tilde{\mathcal{Q}}$ and circles are of the form $E_{\mathbf{p}} \cap \tilde{\mathcal{Q}} \cap E_{\mathbf{q}}$ for $\mathbf{q} \in \tilde{\mathcal{Q}} \setminus E_{\mathbf{p}}$. Hence the quadric $\tilde{\mathcal{Q}}$ contains many copies of the classical flat Möbius plane. Points and circles of the Möbius plane $\mathcal{M}$ are both represented by points of $\tilde{\mathcal{Q}}$. Points of $\mathcal{M}$ are the points of $\tilde{\mathcal{Q}}$ that are orthogonal to a fixed point $\mathbf{p} \in \tilde{\mathcal{Q}}$. Circles of $\mathcal{M}$ correspond precisely to the remaining points of $\tilde{\mathcal{Q}}$. A point $\mathbf{u} \in \tilde{\mathcal{Q}}$ is incident with a circle $\mathbf{v} \in \tilde{\mathcal{Q}}$ if and only if $\mathbf{u}$ and $\mathbf{v}$ are orthogonal.

3.1.4 The Geometry of Chains

The *chain geometry* $\Sigma(\mathbf{R}, \mathbf{C})$ is the geometry whose point set is the complex projective line $\mathrm{PG}(1, \mathbf{C})$ and whose *chains* are the images under the group $\Gamma = \mathrm{PGL}_2(\mathbf{C})$ of the real projective line $\mathrm{PG}(1, \mathbf{R})$ that has been canonically embedded into $\mathrm{PG}(1, \mathbf{C})$; see Herzer [1995]. Incidence between points and chains is inclusion. We show that this chain geometry is yet another model of the classical flat Möbius plane. See Subsubsection 2.9.2.1 for basic information about the group Γ and its standard action on $\mathbf{S}^2$.

The 6-dimensional connected Lie group Γ acts in the standard way as a sharply 3-transitive group on $\mathrm{PG}(1,\mathbf{C})$. This fact readily implies Axiom B1.

Clearly, Γ is transitive on the set of chains. Moreover the stabilizer of the standard real projective line $\mathrm{PG}(1,\mathbf{R})$ is the canonical subgroup $\mathrm{PGL}_2(\mathbf{R})$. This subgroup in turn is transitive on $\mathrm{PG}(1,\mathbf{R})$. Thus the stabilizer of a chain is transitive on that chain. Remember that we call a pair (p, C), where p is a point and C is a chain through p, a *flag*. The above argument then shows that Γ is transitive on the set of all flags, that is, Γ is *flag-transitive.*

THEOREM 3.1.1 (Flag-Transitive Automorphism Group) *The group* $\mathrm{PGL}_2(\mathbf{C})$ *is a flag-transitive group of automorphisms of* $\Sigma(\mathbf{R},\mathbf{C})$.

To verify Axiom B2 we can, by Theorem 3.1.1, restrict ourselves to considering the tangent circles to the circle $C_0 = \mathrm{PG}(1,\mathbf{R}) = \mathbf{R}\cup\{\infty\}$ at the point ∞. The subgroup

$$T = \left\{ \begin{pmatrix} 1 & b \\ 0 & 1 \end{pmatrix} \;\middle|\; b \in \mathbf{C} \right\}$$

of Γ fixes ∞. Indeed, the point ∞ is the only fixed point of elements in T where $b \neq 0$ and C_0 is taken to a tangent circle to C_0 at ∞. Since T is transitive on $\mathbf{C}$, we find such a tangent circle through any given point. In fact, the tangent circle is unique since the stabilizer of ∞, C_0, and any further point not on C_0 only consists of the identity.

The chains are the sets

$$\left\{ \frac{ax+b}{cx+d} \in \mathbf{C}\cup\{\infty\} \;\middle|\; x \in \mathbf{R}\cup\{\infty\} \right\},$$

where $a,b,c,d \in \mathbf{C}$, $ad-bc \neq 0$. Futhermore, if $z = (ax+b)/(cx+d)$, then $x = (dz-b)/(-cz+a)$. The condition that $x \in \mathbf{R}\cup\{\infty\}$ yields

$$(dz-b)\overline{(a-cz)} = \overline{(dz-b)}(a-cz),$$

where $\bar{z}$ denotes the complex conjugate of the complex number z, or

$$(c\bar{d}-\bar{c}d)|z|^2 + (\bar{a}d-\bar{b}c)z + (b\bar{c}-a\bar{d})\bar{z} + a\bar{b} - \bar{a}b = 0.$$

Therefore

$$2i\ \mathrm{Im}(c\bar{d})|z|^2 + 2i\ \mathrm{Im}((\bar{a}d-\bar{b}c)z) + 2i\ \mathrm{Im}(a\bar{b}) = 0,$$

where Im denotes the imaginary part of the complex number. Dividing by $2i$ and writing $z = x + iy$ we obtain an equation of the form

$$r(x^2 + y^2) + \tilde{a}x + \tilde{b}y + \tilde{c} = 0,$$

where $r = \operatorname{Im}(c\bar{d})$, $\tilde{a} = \operatorname{Im}(\bar{a}d - \bar{b}c)$, $\tilde{b} = \operatorname{Re}(\bar{a}d - \bar{b}c)$, and $\tilde{c} = \operatorname{Im}(a\bar{b})$. Note that all these coefficients are real numbers and

$$\tilde{a}^2 + \tilde{b}^2 - 4r\tilde{c} = |ad - bc|^2 = 1.$$

This is the equation of a Euclidean line or Euclidean circle and we see that the chain geometry can be embedded into the geometry of Euclidean lines and circles described in Subsection 3.1.2. However, three points uniquely determine a chain and a Euclidean line or circle. Hence every Euclidean line and circle must occur. This shows that the geometry of chains is just another model of the classical flat Möbius plane.

3.1.5 The Geometry of the Group of Fractional Linear Maps

The group $\Gamma = \mathrm{PGL}_2(\mathbf{C}) = \mathrm{PSL}_2(\mathbf{C})$ from the previous subsection in fact determines the geometry. The stabilizer Γ_∞ of the point ∞ is

$$\Gamma_\infty = \left\{ \begin{pmatrix} a & b \\ 0 & d \end{pmatrix} \;\middle|\; a, b, d \in \mathbf{C}, ad = 1 \right\}$$

and the stabilizer $\Gamma_{\mathrm{PG}(1,\mathbf{R})}$ of the circle $\mathrm{PG}(1, \mathbf{R}) = \mathbf{R} \cup \{\infty\}$ is

$$\Gamma_{\mathrm{PG}(1,\mathbf{R})} = \left\{ \begin{pmatrix} a & b \\ c & d \end{pmatrix} \;\middle|\; a, b, c, d \in \mathbf{R}, ad - bc = 1 \right\}.$$

Since Γ is flag-transitive, we can identify the points with the cosets of Γ_∞ in Γ and the circles with the cosets of $\Gamma_{\mathrm{PG}(1,\mathbf{R})}$ in Γ. Furthermore, a point $\alpha\Gamma_\infty$ is on the circle $\beta\Gamma_{\mathrm{PG}(1,\mathbf{R})}$ if and only if the intersection of the two cosets is nonempty; see Higman–McLaughlin [1961].

We assign an element in Γ represented by the matrix

$$\begin{pmatrix} a & b \\ c & d \end{pmatrix} \in \mathrm{SL}_2(\mathbf{C})$$

the point $(a : c) \in \mathrm{PG}(1, \mathbf{C})$, that is, the point $a/c \in \mathbf{C} \cup \{\infty\}$. Elements in the same coset of Γ_∞ are assigned the same point. Likewise, an element in Γ represented by the above matrix is assigned the vector

$$(\operatorname{Im}(c\bar{d}), \operatorname{Im}(a\bar{d} - b\bar{c}), \operatorname{Re}(a\bar{d} - b\bar{c}), \operatorname{Im}(a\bar{b})) \in \mathbf{R}^4.$$

An easy computation shows that if v_i, $i = 1, 2, 3, 4$, are the components of this vector, then $v_2^2 + v_3^2 - 4v_1v_4 = |ad - bc|^2 = 1$ so that the vector cannot be the zero vector. Furthermore, elements in the same coset of $\Gamma_{PG(1,\mathbf{R})}$ are assigned a scalar multiple of the same vector, that is, the same point in projective 3-space. Note that both assignments depend only on the element of Γ and not on the particular representing matrix.

Let $\alpha, \beta \in \Gamma$ be represented by the matrices

$$\begin{pmatrix} a & b \\ c & d \end{pmatrix}, \begin{pmatrix} a' & b' \\ c' & d' \end{pmatrix} \in \mathrm{SL}_2(\mathbf{C}),$$

respectively. Then the point $\alpha\Gamma_\infty$ is on the circle $\beta\Gamma_{PG(1,\mathbf{R})}$ if and only if $\beta^{-1}\alpha\Gamma_\infty$ contains an element of $\Gamma_{PG(1,\mathbf{R})}$. But

$$\begin{pmatrix} a' & b' \\ c' & d' \end{pmatrix}^{-1} \begin{pmatrix} a & b \\ c & d \end{pmatrix} = \begin{pmatrix} d'a - b'c & d'b - b'd \\ a'c - c'a & a'd - c'b \end{pmatrix}$$

represents $\beta^{-1}\alpha$ so that $\frac{d'a-b'c}{a'c-c'a} \in \mathbf{R} \cup \{\infty\}$ in order for $\alpha\Gamma_\infty$ to be on $\beta\Gamma_{PG(1,\mathbf{R})}$. Thus

$$\mathrm{Im}(d'a - b'c)\overline{(a'c - c'a)} = 0.$$

Expanding and dividing by $|c|^2$ we obtain

$$\mathrm{Im}(c'\bar{d}')\left|\frac{a}{c}\right|^2 + \mathrm{Im}\left((\bar{a}'d' - \bar{b}'c')\frac{a}{c}\right) + \mathrm{Im}(a'\bar{b}') = 0,$$

that is, $\frac{a}{c}$ is on the Euclidean circle

$$\mathrm{Im}(c'\bar{d}')(x^2 + y^2) + \mathrm{Im}(\bar{a}'d' - \bar{b}'c')x + \mathrm{Re}(\bar{a}'d' - \bar{b}'c')y + \mathrm{Im}(a'\bar{b}') = 0;$$

compare Subsection 3.1.4.

The above two subgroups of Γ are essentially unique. Any 4-dimensional connected subgroup of Γ is conjugate to Γ_∞, by Theorem A2.3.16. A connected 3-dimensional subgroup of Γ is conjugate to one of the three groups $\mathrm{SO}_3(\mathbf{R})$, $\mathrm{PSL}_2(\mathbf{R})$, and $\{z \mapsto e^{ict}z + b \mid b \in \mathbf{C}, t \in \mathbf{R}\}$ for some $c \in \mathbf{R}$, $c \neq 0$; see also Theorem A2.3.16. A group of the last kind cannot occur as the stabilizer of a circle since such a group fixes a point. Similarly, $\mathrm{SO}_3(\mathbf{R})$ cannot occur either because this group can only act transitively or trivially on $\mathbf{S}^2$; see Montgomery–Zippin [1955] Theorem 6.7.1. This shows that the stabilizer in Γ of a circle must be isomorphic to $\mathrm{PGL}_2(\mathbf{R})$. In fact, up to conjugacy in Γ_∞, we can assume that this stabilizer is the standard subgroup $\Gamma_{PG(1,\mathbf{R})}$.

3.2 Derived Planes and Topological Properties

In this section we describe the main link between spherical circle planes and $\mathbf{R}^2$-planes. Later this relation will be an important tool in the classification of spherical circle planes with respect to their automorphism groups. We also determine the topological structure of the circle space and investigate the topological behaviour of the geometric operations.

3.2.1 Derived $\mathbb{R}^2$*-Planes*

The idea of representing the classical flat Möbius plane in the Euclidean plane can be extended as follows.

Remember that the *derived plane* at a point p of a spherical circle plane consists of all points different from p and, as lines, all circles passing through p but punctured at p. This is an $\mathbf{R}^2$-plane. In the case of a flat Möbius plane we even obtain a flat affine plane, the *derived affine plane at* p; Axiom B2 implies the parallel axiom of an affine plane. Even stronger we obtain the following theorem.

THEOREM 3.2.1 (Characterization via Derived Planes) *A geometry* $\mathcal{M}$ *of topological circles on* $\mathbf{S}^2$ *is a spherical circle plane if and only if the derived geometry at every point of* $\mathbf{S}^2$ *is an* $\mathbf{R}^2$*-plane. It is a flat Möbius plane if and only if the derived plane at every point is a flat affine plane.*

Each derived affine plane of the classical flat Möbius plane is the Euclidean plane; see Subsection 3.1.2.

The derived affine plane at a point p of a flat Möbius plane extends to a flat projective plane. We call this plane the *derived projective plane at* p. Derived planes link spherical circle planes to $\mathbf{R}^2$-planes or flat affine and projective planes. One can consider a spherical circle plane as many intersecting $\mathbf{R}^2$-planes glued together. We may therefore use many of the results on these flat linear spaces in proving results about spherical circle planes. This, in fact, is one of the main techniques in studying spherical circle planes since many results on $\mathbf{R}^2$-planes or flat affine and projective planes carry over from derived planes to spherical circle planes. For example, the proof of the following result shows how to use the fact that lines in an $\mathbf{R}^2$-plane intersect transversally to prove the corresponding result for spherical circle planes.

THEOREM 3.2.2 (Intersection of Circles) *Two distinct circles in a spherical circle plane intersect in at most two points. If they intersect in exactly one point, then they touch topologically. If they intersect in two points, then they intersect transversally at these points.*

Proof. Axiom B1 implies that two circles in a spherical circle plane can intersect in no more than two points. Intersecting circles either intersect transversally in a point of intersection or they touch topologically at this point. By the Jordan–Schoenflies Separation Theorem A1.2.4 and the connectedness of circles, two circles that intersect transversally in a point certainly have a second point in common. If two circles intersect in two points p and q, consider the two circles as lines of the derived plane at p. Then it follows from Theorem 2.2.1 that the lines/circles have to intersect transversally in q and therefore also in p. □

This result implies that we can decide locally in the neighbourhood of a point of intersection of two circles whether or not these circles have a second point in common.

3.2.2 Affine Parts

A circle C not passing through the distinguished point p induces a closed topological arc in the derived plane at p. In the case of a flat Möbius plane we even obtain a topological oval in the derived affine plane at p. A spherical circle plane can be described in one of its derived planes $\mathcal{A}$ by the lines of $\mathcal{A}$ and a collection of closed topological arcs. We call the induced geometry the *affine part*. The affine part of the classical flat Möbius plane is just the geometry of Euclidean lines and circles on $\mathbf{R}^2$; see Subsection 3.1.2.

Conversely, let $\mathcal{A}$ be a geometry with point set $\mathbf{R}^2$ whose line set contains both topological lines and circles. Let $\mathcal{L}$ be the set of all lines that are topological lines and let $\mathcal{C}$ be the set of all lines that are topological circles. Define a geometry $\mathcal{M}$ whose point set is the one-point compactification of $\mathbf{R}^2$ by a point ∞ and whose circles are all elements of $\mathcal{C}$ and all elements of $\mathcal{L}$ to which the extra point ∞ has been adjoined. This means that $\mathcal{M}$ is a geometry of circles on a sphere and $\mathcal{A}$ is the affine part of a spherical circle plane or flat Möbius plane plane if and only if $\mathcal{M}$ is such a plane. If $\mathcal{M}'$ is a second spherical circle plane that has $\mathcal{A}$ as one of its affine parts, then $\mathcal{M}$ is topologically isomorphic to $\mathcal{M}'$; see also Polster–Steinke [1994] Proposition 1.

3.2.3 Continuity of Geometric Operations

In a spherical circle plane $\mathcal{M} = (\mathbf{S}^2, \mathcal{C})$ there are two geometric operations; joining three points by a circle and intersecting two circles. A flat

Möbius plane further admits the operation of forming tangent circles. In order to be more precise, let P^{3*} denote the set of all triples of pairwise distinct points of $\mathbf{S}^2$. Furthermore, let $\mathcal{C}^{2*}$ be the collection of all pairs of distinct circles that have nonempty intersection and let $P^{1,2}$ be the quotient space of $\mathbf{S}^2 \times \mathbf{S}^2$ under the equivalence relation $(x, y) \sim (u, v)$ if and only if $x = u$, $y = v$ or $x = v$, $y = u$. Then $\alpha : P^{3*} \to \mathcal{C}$ is defined by $\alpha(x, y, z)$ being the unique circle through the points x, y, and z and $\gamma : \mathcal{C}^{2*} \to P^{1,2}$ is defined by $\gamma(C, D) = C \cap D$. In the case that $\mathcal{M}$ is a flat Möbius plane we also have the map $\beta : \mathcal{B} \to \mathcal{C}$ from the set $\mathcal{B}$ of all triples (p, C, q), where p and q are distinct points and C is a circle through p, to the set of circles which assigns (p, C, q) the unique circle through p and q that touches C at p.

The sets P^{3*} and $P^{1,2}$ carry natural topologies. As for the line set of a flat linear space, it is possible to provide the circle set of $\mathcal{M}$ with a number of equivalent 'natural' topologies. For example, we can define convergence of circles in the sense of Hausdorff and the H-topology as in Section 2.3. Alternatively, we may topologize $\mathcal{C}$ with respect to the Hausdorff metric, that is,

$$d(C, D) = \max \left\{ \sup_{y \in D} \inf_{x \in C} |x - y|, \sup_{x \in C} \inf_{y \in D} |x - y| \right\}$$

for two circles C and D, where $|x-y|$ is the Euclidean distance between x and y; compare p. 430. This metric yields topologies on $\mathcal{C}^{2*}$ and the domain $\mathcal{B}$ of β.

With respect to these topologies the geometric operations are continuous; see Wölk [1966] and Strambach [1970d].

THEOREM 3.2.3 (Continuity) *Let $\mathcal{M}$ be a spherical circle plane with circle set $\mathcal{C}$. Then the following hold:*

(i) *α is continuous and open;*

(ii) *the collection of all pairs of circles that intersect in precisely two points is an open subset of $\mathcal{C} \times \mathcal{C}$ and γ is continuous.*

In the case of a flat Möbius plane we further have

(iii) *β is continuous.*

The Euclidean topology of $\mathbf{S}^2$ and the continuity of the geometric operations in fact uniquely determine the topology on $\mathcal{C}$; see Wölk [1966] Satz 5.2 and Strambach [1970d] 1.2 and 2.8.

THEOREM 3.2.4 (Topology of $\mathbf{S}^2 \to$ Topology of Circle Space) *The topology of the circle space $\mathcal{C}$ of a spherical circle plane is uniquely determined by the topology of $\mathbf{S}^2$; it is the finest topology with respect to which the map α of joining three pairwise distinct points by a circle is continuous. In particular, the sets $\alpha(O_1, O_2, O_3)$, where O_1, O_2, O_3 are disjoint nonempty open subsets of the point space $\mathbf{S}^2$, form a basis for the topology of the circle space.*

We can obtain neighbourhoods of a circle of the form $\alpha(O_1, O_2, O_3)$ for much smaller sets O_1, O_2, and O_3; see Wölk [1966] Lemma 5.3 and Strambach [1970d] 1.3.

THEOREM 3.2.5 (Neighbourhood Bases for Circles) *Let C be a circle of a spherical circle plane $\mathcal{M}$. For three distinct points p_1, p_2, p_3 on C let C_1, C_2, C_3 be three circles such that C_i intersects C in precisely two points including the point p_i, $i = 1, 2, 3$, and let U_i be a neighbourhood of p_i on C_i such that U_1, U_2, U_3 are pairwise disjoint. Then the set $\alpha(U_1, U_2, U_3)$ is a neighbourhood of C and the sets of this type form a neighbourhood basis of C.*

As an immediate consequence of Theorem 3.2.5 and the fact that all circles are homeomorphic to $\mathbf{S}^1$ we derive the following result; see Strambach [1970d] 2.9.

COROLLARY 3.2.6 (Circle Space Locally $\mathbf{R}^3$) *The circle space of a spherical circle plane is locally homeomorphic to $\mathbf{R}^3$.*

Conversely, we can obtain the topology of $\mathbf{S}^2$ from the topology of the circle set.

LEMMA 3.2.7 (Topology of Circle Space $\to$ Topology of $\mathbf{S}^2$) *Let $\mathcal{K}$ be an open subset of the circle space of a spherical circle plane. Then the set $\tilde{\mathcal{K}} = \bigcup_{C \in \mathcal{K}} C$ of all points on circles in $\mathcal{K}$ is open in $\mathbf{S}^2$. Furthermore the sets of the form $\tilde{\mathcal{K}}$ form a subbasis for the topology of $\mathbf{S}^2$.*

Since the topologies are uniquely determined, we do not have to worry about them. So, in this chapter, we do not refer to the topologies involved and we always assume that these are the topologies as above.

The topology of the point space is, of course, the usual Euclidean topology of $\mathbf{S}^2$ but one can also describe it geometrically; compare also Lemma 3.2.13. Let C_1, C_2, C_3 be three circles that pairwise intersect in two points and such that $C_1 \cap C_2 \cap C_3$ consists of a single point q. Let the *triode* $\mathcal{D}(C_1, C_2, C_3; q)$ be the collection of all points in the interior of the triangle formed by the lines induced by C_1, C_2, C_3 in the derived plane at the point q. The *vertices* of the triode $\mathcal{D}(C_1, C_2, C_3; q)$ are the points $(C_i \cap C_j) \setminus \{q\}$, $i, j \in \{1, 2, 3\}$, $i \neq j$. With this definition Strambach [1970d] 3.1 shows the following.

LEMMA 3.2.8 (Triodes Form Neighbourhood Basis) *For a fixed point q the triodes that contain the point $p \neq q$ form a basis for the neighbourhoods of p.*

3.2.4 Topological Möbius Planes

A topological Möbius plane is a Möbius plane whose sets of points and circles carry nonindiscrete topologies such that the geometric operations of joining, touching, and intersecting distinct circles are continuous on their domains of definition. The results of the previous sections imply that the flat Möbius planes are topological Möbius planes.

We refer to a topological geometry as an X geometry if its point space is a topological space of type X. The following result characterizes the flat Möbius planes among the topological Möbius planes; see Wölk [1966] Satz 6.1.

THEOREM 3.2.9 (Topological Characterization) *The flat Möbius planes are precisely the 2-dimensional compact connected Möbius planes.*

The dimension of a topological locally compact finite-dimensional Möbius plane is either 0 or 2, there are no '4-dimensional Möbius planes'; see Buchanan–Hähl–Löwen [1980] and Löwen [1981d].

From our definitions of the topologies on the circle and line spaces of spherical circle planes and $\mathbf{R}^2$-planes, it is clear that the natural topology on the line space of a derived $\mathbf{R}^2$-plane coincides with the topology the line set inherits from the topology on the circle space of the spherical circle plane.

A fundamental question for general topological Möbius planes is to decide whether the continuity of the geometric operations in such planes gives rise to continuous geometric operations in the derived affine planes, that is, whether the derived planes are topological, and whether the ovals induced by circles in derived affine planes are 'topological', or equivalently, have the secants-tend-to-tangents property specified in Theorem 2.4.13. By the preceding remark, we know that the first question can be answered in the affirmative for flat Möbius planes.

THEOREM 3.2.10 (Derived Planes, Arcs and Ovals) *Each derived plane of a spherical circle plane at a point p is an $\mathbf{R}^2$-plane. The natural topology on the line set of the $\mathbf{R}^2$-plane is the restriction of the natural topology on the circle set to the set of circles through p.*

Circles not passing through the point p are closed topological arcs in the derived plane at p. They are topological ovals if the spherical circle plane is a flat Möbius plane.

The second part of this result is an immediate consequence of Axioms B1 and B2.

Wölk [1966] for flat Möbius planes and later Strambach [1970d] for spherical circle planes considered the so-called *coherence axiom*. In fact Wölk made this coherence condition part of his definition of flat Möbius planes. Strambach then showed that the coherence axiom is automatically satisfied in flat Möbius planes. The coherence axiom as introduced by Wölk as Axiom K1 reflects the fact that ovals induced in derived projective planes must be topological. In flat Möbius planes this translates into the fact that touching is the limit of proper intersection. More precisely, we say a flat Möbius plane is *coherent* if for each circle C and points $x_n, p \in C$, $x_n \neq p$ for $n \in \mathbf{N}$, and $q \notin C$ the circles $\alpha(x_n, p, q)$ joining x_n, p, and q converge to the tangent circle $\beta(p, C, q)$ through p and q that touches C at p if the x_n converge to p on C.

THEOREM 3.2.11 (Coherence) *Every flat Möbius plane is coherent.*

To prove this just note that in the derived flat affine plane at q the circle C becomes a topological oval. Now the above result stated in the language of topological ovals in flat linear spaces is Theorem 2.4.13.

The following lemma shows that spherical circle planes are, in a way, not too far away from flat Möbius planes. We always find a tangent circle as in Axiom B2 of a flat Möbius plane. However, there may be more than one.

LEMMA 3.2.12 (Existence of Touching Circles) *Let C be a circle of a spherical circle plane $\mathcal{M}$ and let $p \in C$ and $q \notin C$ be points. Then there is at least one circle through q that touches C at p.*

The above lemma follows from the fact that in the derived $\mathbf{R}^2$-plane at p there is at least one line L through q that does not meet the line induced by C; see Theorem 2.3.6(i). The circle corresponding to L then touches C at p and passes through q.

The circles that touch the circle C at p induce in the derived plane $\mathcal{A}_p$ at p a family of asymptotes to the line that comes from C. This family contains a sequence of lines that converge to infinity. In particular, given a compact subset K of $\mathcal{A}_p$, there is such a line that does not meet K. Since the complement of an open neighbourhood U of p is compact, we thus can find a circle completely contained in U. By the continuity of the geometric operations, we can change that circle a little bit so that the resulting circle C' is still contained in U and such that the connected component of $\mathbf{S}^2 \setminus C'$ entirely contained in U contains the point p; compare Strambach [1970d] 2.5.

LEMMA 3.2.13 (Small Circles) *Let $\mathcal{M}$ be a spherical circle plane. Then the connected components of the complements of circles that contain a point p form a neighbourhood basis of p. In particular, for every nonempty open subset U of $\mathbf{S}^2$ there is a circle of $\mathcal{M}$ entirely contained in U, that is, $\mathcal{M}$ contains 'small circles'.*

For more information about topological circle planes see Heise [1969], Löwen [1981d], and Förtsch [1982]. Also, Petkantschin [1940], [1941], Kroll [1970], [1971], [1977a], and Meyer [1987] are concerned with ordered circle planes.

3.2.5 Circle Space and Flag Space

From Corollary 3.2.6 and Theorems 2.3.6(i) and (iii), which describe the line space of an $\mathbf{R}^2$-plane and the shape of a pencil of lines, we readily obtain the following topological determination of the circle set and certain bundles of circles.

PROPOSITION 3.2.14 (The Space of Circles I) *The circle space of a spherical circle plane is a 3-dimensional manifold. With respect to the induced topology the bundle of circles through a point is homeomorphic to a Möbius strip and the bundle of circles through two points is homeomorphic to the 1-sphere $\mathbf{S}^1$.*

For flat Möbius planes we can determine explicitly what the circle space looks like.

THEOREM 3.2.15 (The Space of Circles II) *The circle space of a flat Möbius plane is homeomorphic to the real 3-dimensional projective space with one point deleted.*

For the proof see Strambach [1974b]. In the following we only sketch this proof. The first step is to introduce coordinates for the circle space of a flat Möbius plane. To this end it is shown that $\mathcal{M}$ admits for any two points n and s a *flock with carriers n and s*, that is, a collection of circles of $\mathcal{M}$ that partition $\mathbf{S}^2 \setminus \{n, s\}$; see Subsection 3.8.4. For each circle C in such a flock the points n and s are in different connected components of $\mathbf{S}^2 \setminus C$. Furthermore, the bundle $\mathcal{B}$ of circles through n and s is used. The points n and s represent the north and south poles, respectively, circles in a flock $\mathcal{F}$ with carriers n and s represent the parallels of latitude and circles in $\mathcal{B}$ represent meridians. We parametrize $\mathcal{F}$ by the open interval $(-1, 1)$ and $\mathcal{B}$ by the half-open interval $[0, 2\pi)$. Each circle C of $\mathcal{M}$ is then given a quintuple of coordinates $(c_1, \ldots, c_5)$. The first two entries c_1 and c_2 are the northern and southern latitude, respectively, that is, the parameter values of the unique circles C_1 and C_2 in $\mathcal{F}$ that touch C, or $+1$ if C passes through n and -1 if it passes through s. This means that $c_1 \in (-1, 1]$ and $c_2 \in [-1, 1)$ and $c_2 \leq c_1$. The third coordinate c_3 equals $\max\{|c_1|, |c_2|\} - 1$ or $1 - \max\{|c_1|, |c_2|\}$ depending on whether or not C separates the points n and s; if C passes through n or s, then $c_3 = 0$. The coordinate c_4 is determined as follows. Let m_1 and m_2 be the parameter values of the meridians of the points of intersection of C with C_1 and C_2, respectively. (The circle C touches C_1 and C_2 at these points.) If $|c_1| \neq |c_2|$, then $c_4 = m_1$ as long as $1 - c_1 > 1 + c_2$; otherwise $c_4 = m_2$. If $|c_1| = |c_2|$, then $c_4 = \infty$. Finally, the last entry c_5 is determined by $c_5 = \frac{1}{2}(m_2 + f_{c_2,c_1}(m_1) - f_{c_2,c_1}(0))$ if C is not in $\mathcal{F}$ and does not pass through n or s. Here $f_{c_2,c_1}(m)$ is the parameter value of the meridian of the point of intersection of C_1 with the unique circle through the point on C_2 of meridian m that touches both circles C_1 and C_2 and does not separate the points n and s. (Such a circle exists by Groh [1972] Theorem 1.) If C is in $\mathcal{B}$, then c_5 is the parameter value of C, and if $C \in \mathcal{F}$ or C passes through either n or s, we set $c_5 = \infty$. Each circle uniquely determines its coordinates and the set of all coordinates forms a well-defined subset of

$$(-1, 1] \times [-1, 1) \times [-1, 1) \times ([0, 2\pi) \cup \{\infty\}) \times ([0, 2\pi) \cup \{\infty\})$$

which is the same for any flat Möbius plane.

Furthermore, the map $C \mapsto (c_1, \ldots, c_5)$ is continuous. If the symbol ∞ does not occur in the coordinates, continuity is a simple consequence of the continuity of the geometric operations in $\mathcal{M}$ and coherence. If ∞ does occur, appropriate neighbourhoods are defined so that the above map indeed becomes continuous. It then follows that the map that takes a circle in one flat Möbius plane to the circle in a second Möbius plane that has the same coordinates is a homeomorphism between the circle spaces. In particular, the circle space of any flat Möbius plane must be homeomorphic to the circle space of the classical flat Möbius plane.

In the classical flat Möbius plane the topological structure of the circle space can be determined as follows. We have a nondegenerate elliptic quadric $\mathcal{Q}$ in 3-dimensional real projective space $\mathrm{PG}(3, \mathbf{R})$. This quadric defines a polarity π on $\mathrm{PG}(3, \mathbf{R})$. A point on $\mathcal{Q}$ is assigned the tangent plane to $\mathcal{Q}$ at this point. A point p not on $\mathcal{Q}$ is an exterior point if the plane $\pi(p)$ intersects $\mathcal{Q}$ in at least three points; p is called an inner point if $\pi(p)$ does not intersect $\mathcal{Q}$. If $\mathcal{Q}$ is given in homogeneous coordinates by the symmetric bilinear form β, then a point with homogeneous coordinates x is an exterior point if and only if $\beta(x, x) > 0$, a point on $\mathcal{Q}$ if and only if $\beta(x, x) = 0$, and an inner point if and only if $\beta(x, x) < 0$. The circle space of the classical flat Möbius plane is therefore homeomorphic to the space of exterior points, that is, $\mathrm{PG}(3, \mathbf{R}) \setminus (\mathcal{Q} \cup I(\mathcal{Q}))$, where $I(\mathcal{Q})$ denotes the set of inner points of $\mathcal{Q}$. However, $\mathcal{Q} \cup I(\mathcal{Q})$ can be contracted to an inner point. Hence $\mathcal{C}$ is homeomorphic to $\mathrm{PG}(3, \mathbf{R})$ with one point deleted.

We include one more proof of the fact that the circle space of the classical flat Möbius plane is homeomorphic to $\mathrm{PG}(3, \mathbf{R})$ with one point deleted. Recall that the affine part of this plane consists of the curves

$$\{(x, y) \in \mathbf{R}^2 \mid a(x^2 + y^2) + bx + cy + d = 0\}$$

where $a, b, c, d \in \mathbf{R}$, $b^2 + c^2 - 4ad > 0$; see Subsection 3.1.2. Of course, each Euclidean circle or line is uniquely determined by the vector (a, b, c, d) up to real scalar multiples. This shows that Euclidean circles and Euclidean lines correspond to elements of real 3-dimensional projective space in the outside of the quadric given by $b^2 + c^2 - 4ad = 0$. This argument generalizes to ovoidal flat Möbius planes; see Subsection 3.3.1 for ovoidal planes. The affine part of an ovoidal flat Möbius plane has circles

$$\{(x, y) \in \mathbf{R}^2 \mid af(x, y) + bx + cy + d = 0\}$$

where $a, b, c, d \in \mathbf{R}$, not all zero, provided that set contains at least three points. We again identify circles with certain points in projective 3-space. Depending on the explicit form of f a condition on the four coefficients a, b, c, and d is obtained.

Strambach's proof does not carry over to spherical circle planes that are not Möbius planes as tangent circles are used in an essential way at several stages in the process of introducing coordinates for circles.

We just mention one more result about circle sets in spherical circle planes; see Strambach [1970d] 2.6.

LEMMA 3.2.16 (Large Circles) *Let $\mathcal{M}$ be a spherical circle plane. Then for every $\varepsilon \in \mathbf{R}^+$, the set $\mathcal{C}_\varepsilon = \{C \in \mathcal{C} \mid \sup_{x,y \in C}|x - y| \geq \varepsilon\}$ of 'large circles' is compact.*

Recall that a flag is an incident point–circle pair (p, C), where p is a point and C is a circle through p.

PROPOSITION 3.2.17 (The Space of Flags) *The flag space $\mathcal{F}$ of a spherical circle plane is a 4-dimensional connected manifold.*

3.3 Constructions

In this section we present some of the most important construction methods for nonclassical flat Möbius planes and spherical circle planes. These methods allow us to construct most of the most homogeneous spherical circle planes and flat Möbius planes and, on the other end of the spectrum, rigid planes.

3.3.1 Ovoidal Planes

Let $\mathcal{O}$ be an ovoid in 3-dimensional real projective space $\mathrm{PG}(3, \mathbf{R})$, that is, a set of points such that each line intersects $\mathcal{O}$ in at most two points and such that, given any point of the set, the collection of all *tangent lines* at that point fills a plane, called the *tangent plane* at that point. Examples for ovoids are the elliptic quadrics, or take a hemisphere and half of an ellipsoid and paste the two together along a common equator, for example,

$$\begin{aligned} & \{(x, y, z) \in \mathbf{R}^3 \mid x^2 + y^2 + z^2 = 1, x \geq 0\} \\ \cup\ & \{(x, y, z) \in \mathbf{R}^3 \mid (x/a)^2 + y^2 + z^2 = 1, x \leq 0\}, \end{aligned}$$

where $a \in \mathbf{R}^+$. One obtains a quadric for $a = 1$ and one obtains proper ovoids, that is, not quadrics, for $a \neq 1$.

Points of the *ovoidal Möbius plane* $\mathcal{M}(\mathcal{O})$ are the points of $\mathcal{O}$; circles are the intersections of $\mathcal{O}$ with planes in $\mathrm{PG}(3, \mathbf{R})$ provided they intersect in at least three points. Axioms B1 and B2 are verified in the same way as for the classical flat Möbius plane. Of course, for flat Möbius planes one requires ovoids that are homeomorphic to $\mathbf{S}^2$. These ovoids are exactly the *topological ovoids*, that is, the differentiable strictly convex topological spheres. We only remark that it is possible to construct ovoids in $\mathrm{PG}(3, \mathbf{R})$ that are not of this form using transfinite induction; see Heise [1971] Satz 3.4.

The affine part can be found as follows. Let $f : \mathbf{R}^2 \to \mathbf{R}$ be a function such that $\{(x, y, f(x, y)) \mid x, y \in \mathbf{R}\}$ extended by the infinite point of the third coordinate axis describes an ovoid $\mathcal{O}(f)$ in 3-dimensional real projective space. We obtain a Möbius plane as follows. Stereographic projection from that infinite point onto the xy-plane, that is, the projection $(x, y, z) \mapsto (x, y)$ of points of $\mathbf{R}^3$ onto the first and second coordinates, yields the following planar description of the ovoidal Möbius plane $\mathcal{M}(f)$ over $\mathcal{O}(f)$. The point set of $\mathcal{M}(f)$ is $\mathbf{R}^2 \cup \{\infty\}$ and the circles of $\mathcal{M}(f)$ are the Euclidean lines

$$\{(x, y) \in \mathbf{R}^2 \mid ax + by + c = 0\}$$

extended by ∞ and circles of the form

$$\{(x, y) \in \mathbf{R}^2 \mid f(x, y) + ax + by + c = 0\}$$

provided this set contains at least three points. As for the classical flat Möbius plane we can combine both descriptions of the affine parts of circles into

$$\{(x, y) \in \mathbf{R}^2 \mid af(x, y) + bx + cy + d = 0\},$$

where $a, b, c, d \in \mathbf{R}$, a, b, c not all zero, provided that set contains at least three points. In particular, we see that the derived affine plane at the distinguished point from which we project is the Euclidean plane. This remains true for any point of an ovoidal Möbius plane.

Note that every ovoid in $\mathrm{PG}(3, \mathbf{R})$ can be obtained in the form $\mathcal{O}(f)$. One just has to choose one point p of the ovoid. One then introduces coordinates $(x_1 : x_2 : x_3 : x_4)$ in such a way that the tangent plane to the ovoid at p becomes the plane $x_4 = 0$ at infinity and p becomes the point $(0 : 0 : 1 : 0)$. If the ovoid we started with is topological, then the function f is clearly a strictly convex or concave differentiable function.

PROPOSITION 3.3.1 (Maps Describing Topological Ovoids) *Let $f : \mathbf{R}^2 \to \mathbf{R}$ be a continuous strictly convex function and define $\mathcal{O}(f)$ as above. Then $\mathcal{O}(f)$ is a topological ovoid in 3-dimensional real projective space if and only if f satisfies the following conditions.*

- *The function f is differentiable.*
- *The function grad f is 'strictly increasing', that is,*

$$(grad\ f(x) - grad\ f(y)) \cdot (x - y) > 0$$

  for all $x, y \in \mathbf{R}^2$, $x \neq y$.
- $\lim_{t\to\infty}(grad\ f(x + ty) \cdot y) = \infty$ *for all $x, y \in \mathbf{R}^2$, $y \neq 0$.*

For a proof see Hartmann [1984] Satz 2. Note that a strictly concave continuous function $\mathbf{R}^2 \to \mathbf{R}$ describes an ovoid if and only the strictly convex function $-f$ does.

We obtain a spherical circle plane if we replace the ovoid by a strictly convex topological sphere 'with corners'. For example,

$$\begin{aligned} &\{(x, y, z) \in \mathbf{R}^3 \mid x^2 + y^2 + z^2 = 1, x \geq 0\} \\ \cup\ &\{(x, y, z) \in \mathbf{R}^3 \mid x^2 - 2ax + y^2 + z^2 = 1, x \leq 0\}, \end{aligned}$$

where $a \in \mathbf{R}$, $a \geq 0$, is such a convex set. It is made up of a hemisphere and a spherical cap from a bigger sphere. One obtains an ovoid only for $a = 0$. For $a > 0$ the resulting set is not an ovoid since at the points $(0, y, z)$, $y^2 + z^2 = 1$, there are too many tangents and therefore Axiom B2 is no longer satisfied.

Finally, we consider isomorphisms between ovoidal flat Möbius planes, and their automorphisms. Let $\mathcal{O}$ be some topological ovoid in 3-dimensional projective space $\mathrm{PG}(3, \mathbf{R})$ and let $\mathcal{M}(\mathcal{O})$ be the associated flat Möbius plane. It is obvious that every collineation of $\mathrm{PG}(3, \mathbf{R})$ produces another copy of the ovoidal plane and each collineation that leaves $\mathcal{O}$ invariant induces an automorphism of $\mathcal{M}(\mathcal{O})$. Conversely, Mäurer [1967] showed that every isomorphism between ovoidal flat Möbius planes and every automorphism of $\mathcal{M}(\mathcal{O})$ can be obtained in this way.

THEOREM 3.3.2 (Isomorphisms of Ovoidal Planes) *Two ovoidal flat Möbius planes $\mathcal{M}(\mathcal{O})$ and $\mathcal{M}(\mathcal{O}')$ are isomorphic if and only if the ovoids $\mathcal{O}$ and $\mathcal{O}'$ are projectively equivalent in* $\mathrm{PG}(3, \mathbf{R})$, *that is, there is a collineation of* $\mathrm{PG}(3, \mathbf{R})$ *that maps $\mathcal{O}$ to $\mathcal{O}'$. In particular, the automorphisms of an ovoidal flat Möbius plane $\mathcal{M}(\mathcal{O})$ are precisely the collineations of the ambient projective space* $\mathrm{PG}(3, \mathbf{R})$ *that leave the ovoid $\mathcal{O}$ invariant.*

For example, if $\mathcal{O}$ is an elliptic quadric, we can represent points of the projective space not on $\mathcal{O}$ by certain involutions (compare Subsection 3.3.4), and the associated collection of fixed circles. An automorphism $\mathcal{M}(\mathcal{O})$ must preserve these configurations and therefore extends to a collineation of the ambient projective space.

3.3.2 Ewald's Planes

The first examples of nonovoidal flat Möbius planes were constructed by Ewald [1960]. In Ewald [1967] seven constructions of Möbius planes are presented to account for some Hering types; see Section 3.5. Several of these planes are ovoidal and the other constructions are in terms of affine parts. In any case, when representing these planes by their affine parts with respect to suitable points, they all use Euclidean lines. The Euclidean circles, as used in the classical flat Möbius plane, become modified and replaced by topological circles. Besides the representations of ovoidal flat Möbius planes by ovoids in 3-dimensional projective space there are three basic principles involved in Ewald's constructions. In the following we describe the affine parts at the point ∞ of the flat Möbius planes in these constructions. In all three cases all Euclidean lines (extended by ∞) are included as circles.

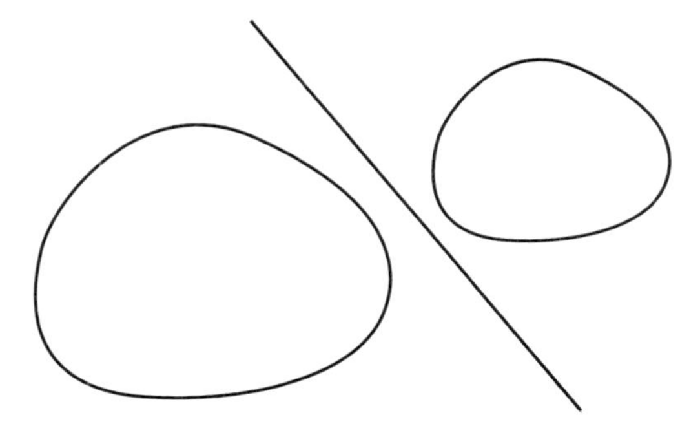

Fig. 3.3. Circles are the Euclidean lines and the images of a topological oval under all possible translations and positive dilatations

Type 1: Let O be an arbitrary topological oval in the Euclidean plane, that is, a differentiable strictly convex simply closed curve. We consider all Euclidean lines and all possible images of O under the group

$$\{(x, y) \mapsto (rx + a, ry + b) \mid a, b, r \in \mathbf{R}, r > 0\}$$

as circles, that is, we first stretch and shrink O from one point with all possible positive factors and then translate the resulting ovals in all possible directions. The nonclassical among these planes are precisely the flat Möbius planes of Hering type III; see Subsection 3.5.2. Note that if we start with a Euclidean circle, we generate all Euclidean circles in this manner and, consequently, we obtain the affine part of the classical flat Möbius plane.

Type 2: Again we start off with an arbitrary topological oval O in the Euclidean plane. We choose $r \geq 0$ and construct the *outer envelope at distance r to O*, that is, the boundary curve of the convex hull formed by all Euclidean circles of radius r with centres on O; see Figure 3.4. This outer envelope is again an oval in the Euclidean plane.

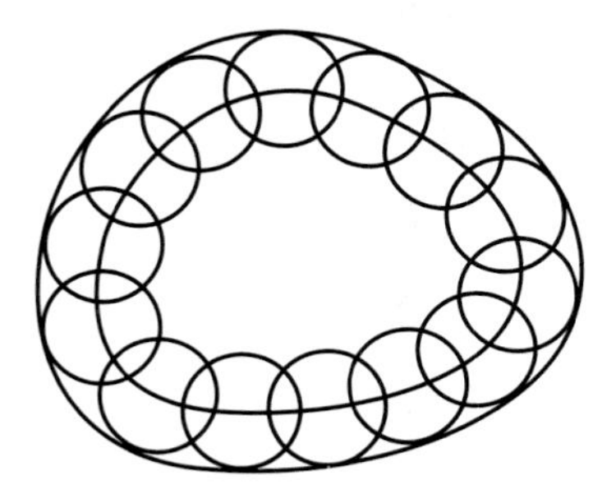

Fig. 3.4. The outer envelope at distance r of a topological oval

Now, the circles of the affine part of a flat Möbius plane are all Euclidean lines, Euclidean circles of radius at most r, and all outer envelopes at distance r to images of O under all possible translations and positive dilatations.

Type 3: This construction is a special cut-and-paste method; see Subsection 3.3.4 for more information about cut and paste in flat Möbius planes. We start with two ovals O_1 and O_2 and the group G of translations and positive dilatations. (In Ewald's constructions O_2 is a Euclidean circle or an outer envelope of O_1.) Then circles of an affine part of a flat Möbius plane are the Euclidean lines and the images of O_1 under G entirely contained in the left half-plane $H_L = \{(x, y) \in \mathbf{R}^2 \mid x \leq 0\}$, the images of the oval O_2 under G entirely contained in the right half-plane $H_R = \{(x, y) \in \mathbf{R}^2 \mid x \geq 0\}$, and suitable unions of traces in H_L of images of O_1 under G that intersect the y-axis in two points with traces in H_R of images of O_2 under G that intersect the y-axis in the same two points.

3.3.3 Semi-classical Flat Möbius Planes

The construction of the so-called semi-classical flat Möbius planes is a special case of the cut-and-paste constructions described in Subsection 3.3.4. Furthermore, some of these planes can be found among Ewald's type 3 planes described above.

The construction of semi-classical flat Möbius planes can be imagined as two halves of a classical flat Möbius plane being pasted together along a circle. One can abstractly define a *semi-classical flat Möbius plane* to be a flat Möbius plane whose point set is the union of two closed connected subsets P_1 and P_2 intersecting in a circle such that the induced geometry on each P_i is isomorphic to the geometry induced from the classical flat Möbius plane on a hemisphere. This construction extends a corresponding method for flat projective planes; see Subsection 2.7.2.

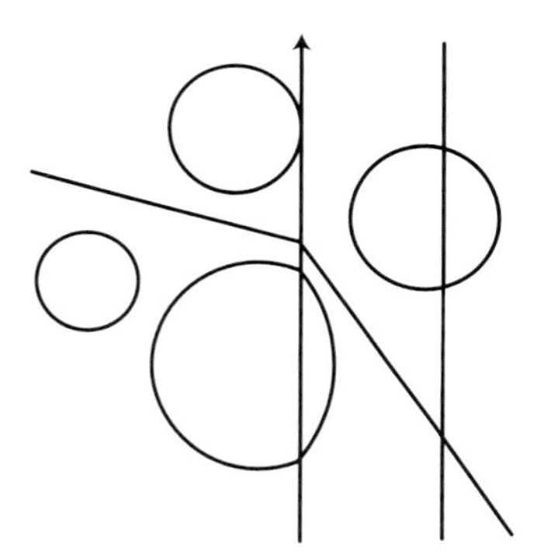

Fig. 3.5. The different kinds of circles in the affine part of a semi-classical flat Möbius plane

Semi-classical flat Möbius planes were classified by Steinke [1986a]. Both closed connected subsets are hemispheres and their common boundary consists of precisely one circle. Explicitly, we use the right hemisphere

$$\{(x, y) \in \mathbf{R}^2 \mid x \geq 0\} \cup \{\infty\}$$

and the left hemisphere

$$\{(x, y) \in \mathbf{R}^2 \mid x \leq 0\} \cup \{\infty\}$$

such that the common boundary consists of the circle

$$\{(0, y) \mid y \in \mathbf{R}\} \cup \{\infty\},$$

that is, the extended y-axis. Now let g and h be orientation-preserving homeomorphisms of $\mathbf{R}$ fixing 0, and further suppose that the graph of h is symmetric about the origin, that is, $h(-x) = -h(x)$ for all $x \in \mathbf{R}$. The point set of a semi-classical flat Möbius plane $\mathcal{M}(g,h)$ is $\mathbf{R}^2 \cup \{\infty\}$. Circles of $\mathcal{M}(g,h)$ are of the following form; see also Figure 3.5.

- The vertical Euclidean lines extended by ∞
$$\{(c,y) \mid y \in \mathbf{R}\} \cup \{\infty\},$$
where $c \in \mathbf{R}$.
- Halves of nonvertical lines pasted together at the y-axis and extended by ∞
$$\begin{array}{ll} & \{(x, mx+t) \mid x \in \mathbf{R}, x \geq 0\} \\ \cup & \{(x, g^{-1}(h(m)x + g(t))) \mid x \in \mathbf{R}, x \leq 0\} \cup \{\infty\}, \end{array}$$
where $m, t \in \mathbf{R}$.
- The Euclidean circles in the right half-plane intersecting the y-axis in at most one point
$$\{(x,y) \in \mathbf{R}^2 \mid (x-a)^2 + (y-b)^2 = r^2\},$$
where $a \geq r > 0$.
- The modified Euclidean circles in the left half-plane intersecting the y-axis in at most one point
$$\{(x,y) \in \mathbf{R}^2 \mid (x-a)^2 + (g(y)-b)^2 = r^2\},$$
where $-a \geq r > 0$.
- Circles intersecting the y-axis in two points
$$\begin{array}{ll} & \{(x,y) \in \mathbf{R}^2 \mid x \geq 0, (x-a)^2 + (y-b)^2 = r^2\} \\ \cup & \{(x,y) \in \mathbf{R}^2 \mid x \leq 0, (x-a')^2 + (g(y)-b')^2 = (r')^2\}, \end{array}$$
where
$$\begin{array}{rcl} a' &=& \frac{1}{2}(g(b+\sqrt{r^2-a^2}) - g(b-\sqrt{r^2-a^2}))h(a/\sqrt{r^2-a^2}), \\ b' &=& \frac{1}{2}(g(b+\sqrt{r^2-a^2}) + g(b-\sqrt{r^2-a^2})), \\ r' &=& \frac{1}{2}(g(b+\sqrt{r^2-a^2}) - g(b-\sqrt{r^2-a^2}))\sqrt{1+h(a/\sqrt{r^2-a^2})^2}, \end{array}$$
and $a, b, r \in \mathbf{R}$, $|a| < r$.

Clearly, the geometries induced by $\mathcal{M}(g,h)$ on the left and right half-planes of the y-axis are isomorphic to the geometries induced by the classical flat Möbius plane on the same sets; an isomorphism for the left hemisphere is given by the coordinate transformation $(x,y) \mapsto (x, g(y))$. Circles that intersect the y-axis in precisely two finite points p and q are pasted together in such a way that the two halves have the same tangent line at p and q.

THEOREM 3.3.3 (Characterization of Semi-classical Planes) *A flat Möbius plane $\mathcal{M}$ is semi-classical, that is, $\mathcal{M}$ is isomorphic to a Möbius plane $\mathcal{M}(g,h)$ for suitable orientation-preserving homeomorphisms g and h, if and only if there is a circle C in $\mathcal{M}$ such that the induced geometry on each connected component of $\mathbf{S}^2 \setminus C$ is isomorphic to the geometry induced from the classical flat Möbius plane on a hemisphere. Furthermore, each semi-classical flat Möbius plane is isomorphic to a plane $\mathcal{M}(g,h)$, where $g(1) = 1$.*

Each derived affine plane at a point of C_0, the circle at which both halves of a semi-classical flat Möbius plane are glued together, is a semi-classical flat affine plane; see Subsection 2.7.2 for more information about these special kinds of affine planes. Since each derived affine plane of an ovoidal flat Möbius plane is the Euclidean plane, we obtain from Proposition 2.7.1 the following characterization of the ovoidal planes among the semi-classical flat Möbius planes.

PROPOSITION 3.3.4 (Ovoidal Semi-classical Planes) *Let $\mathcal{H}_a$ be the collection of all orientation-preserving additive homeomorphisms of $\mathbf{R}$ to itself, that is, $\mathcal{H}_a$ comprises precisely the homeomorphisms of the form $\mathbf{R} \to \mathbf{R} : x \mapsto rx$, where $r \in \mathbf{R}^+$. With this notation we have the following characterizations.*

(i) *A semi-classical flat Möbius plane $\mathcal{M}(g,h)$ is ovoidal if and only if $g, h \in \mathcal{H}_a$.*

(ii) *A plane $\mathcal{M}(g,h)$ is classical if and only if $h = id$ and $g \in \mathcal{H}_a$.*

An ovoidal semi-classical flat Möbius plane can be obtained from a hemisphere and half of an ellipsoid pasted together along a common equator, for example,

$$\begin{array}{ll} & \{(x,y,z) \in \mathbf{R}^3 \mid x^2 + y^2 + z^2 = 1, x \geq 0\} \\ \cup & \{(x,y,z) \in \mathbf{R}^3 \mid (x/a)^2 + y^2 + z^2 = 1, x \leq 0\}, \end{array}$$

where $a \in \mathbf{R}^+$.

Of course, the circle $C_0 = \{(0,y) \mid y \in \mathbf{R}\} \cup \{\infty\}$ at which the two halves of a semi-classical flat Möbius plane are glued together plays a special role and is distinguished among all circles of a semi-classical flat Möbius plane, at least in the nonclassical case. It therefore must be taken to itself under any automorphism or isomorphism.

PROPOSITION 3.3.5 (Semi-classical Isomorphisms I) *An isomorphism between nonclassical semi-classical flat Möbius planes $\mathcal{M}(g,h)$ and $\mathcal{M}(g',h')$ takes the distinguished circle C_0 to itself. In particular, every automorphism of a nonclassical plane $\mathcal{M}(g,h)$ fixes C_0.*

Since both classical subgeometries of a semi-classical flat Möbius plane can be uniquely embedded, up to automorphisms of the classical flat Möbius plane, into the classical flat Möbius plane (see Steinke [1983a]), automorphisms of and isomorphisms between semi-classical planes can be explicitly determined. This was carried out in Steinke [1986a] and the semi-classical flat Möbius planes were classified according to their group dimensions and their Hering types.

We say that two homeomorphisms f and f' of $\mathbf{S}^1$ are *projectively equivalent* if and only if there are fractional linear functions $\alpha, \beta \in \mathrm{PGL}_2(\mathbf{R})$ such that $f' = \alpha \circ f \circ \beta$. Obviously this defines an equivalence relation on the set of all homeomorphisms of $\mathbf{S}^1$ and, consequently, also on the set of all homeomorphisms of $\mathbf{S}^1$ that fix ∞ and 0.

THEOREM 3.3.6 (Semi-classical Isomorphisms II) *Two semi-classical flat Möbius planes $\mathcal{M}(g,h)$ and $\mathcal{M}(g',h')$ are isomorphic if and only if g' is projectively equivalent to g and $h' = h$ or g' is projectively equivalent to g^{-1} and $h' = h^{-1}$.*

For example, the transformation $(x,y) \mapsto (-x, g(y))$ and $\infty \mapsto \infty$ is an isomorphism from $\mathcal{M}(g,h)$ onto $\mathcal{M}(g^{-1},h^{-1})$. Under this isomorphism the right half of $\mathcal{M}(g,h)$ is taken onto the left half of $\mathcal{M}(g^{-1},h^{-1})$ and the left half of $\mathcal{M}(g,h)$ is taken onto the right half of the other plane. This phenomenon essentially accounts for the two different conditions in the above theorem. Note also that Theorem 3.3.6 implies that two semi-classical flat Möbius planes $\mathcal{M}(g,h)$ and $\mathcal{M}(g',h')$ are never isomorphic if $h' \neq h, h^{-1}$, irrespective of g and g'. The family of semi-classical flat Möbius planes therefore contains many nonisomorphic planes.

We conclude this subsection with the group-dimension classification of the semi-classical flat Möbius planes. These planes account for most of the possible group dimensions (= dimension of the full automorphism group; see Subsection 3.4.1). In order to be able to state this result, we define a *semi-multiplicative homeomorphism* $\mathbf{R} \to \mathbf{R}$ to be one of the maps $h_{d,r}$, where $d, r \in \mathbf{R}^+$ given by

$$h_{d,r}(x) = \begin{cases} x^d & \text{for } x \geq 0, \\ -r|x|^d & \text{for } x < 0; \end{cases}$$

compare the end of Section A1.4.

THEOREM 3.3.7 (Classification of Semi-classical Planes) *A semi-classical flat Möbius plane $\mathcal{M}(g, h)$ has group dimension*

6 *if and only if $h = id$ and $g \in \mathcal{H}_a$;*
3 *if and only if $h \neq id$ and $g \in \mathcal{H}_a$;*
1 *if and only if $g \notin \mathcal{H}_a$ is projectively equivalent to a semi-multiplicative homeomorphism;*
0 *if and only if g is not projectively equivalent to a semi-multiplicative homeomorphism.*

3.3.4 Different Ways to Cut and Paste

There are essentially two different kinds of cut-and-paste techniques to construct new spherical circle planes or flat Möbius planes from old ones.

One kind uses a separating set of the point set $\mathbf{S}^2$. Examples for planes that have been constructed using this kind of cut-and-paste technique are the semi-classical flat Möbius planes discussed in Subsection 3.3.3 and Ewald's type 3 planes described in Subsection 3.3.2. More generally, one can start with two spherical circle planes or flat Möbius planes $\mathcal{M}_1$ and $\mathcal{M}_2$ that have a circle C in common. The circle C separates $\mathbf{S}^2$ into two connected components C^+ and C^- by the Jordan–Schoenflies Separation Theorem A1.2.4. The circle set of a new circle plane consists of three types of circles: (1) all circles of $\mathcal{M}_1$ that are contained in the topological closure of C^+; (2) all circles of $\mathcal{M}_2$ that are contained in the topological closure of C^-; (3) here one first has to specify a rule that matches circles of $\mathcal{M}_1$ that intersect C in two points with circles of $\mathcal{M}_2$ that intersect C in the same two points; circles of the pasted spherical circle plane that intersect C in two points are of the form $(D \cap C^+) \cup (D' \cap C^-) \cup \{p, q\}$, where D is a circle of $\mathcal{M}_1$ that intersect C in the two points p and q, and D' is its corresponding circle in $\mathcal{M}_2$ intersecting C in the same two points p and q; if one matches circles carefully enough one obtains a spherical circle plane or even a flat Möbius plane; a general rule for a matching such as this is difficult to specify, may even be impossible in general, and has only been carried out in full generality for the case that both $\mathcal{M}_1$ and $\mathcal{M}_2$ are isomorphic to the classical flat Möbius plane and the new plane is a flat Möbius plane; see Subsection 3.3.3. We have already encountered a corresponding general kind of cut-and-paste construction in the case of flat linear spaces; see Subsections 2.7.2, 2.7.10, and 2.7.11.

In the following we focus on a second kind of cut-and-paste method, namely cutting and pasting along *(X-embedded) separating sets* in the circle space. Here X stands for some geometric property that describes a collection of circles. More precisely, an X-embedded separating set is a set S of topological circles on $\mathbf{S}^2$ that satisfies the following conditions.

(i) If S is X-embedded in the circle set of a spherical circle plane, that is, is embedded in a certain prescribed way X, then it separates the circle set into a finite number $n \geq 2$ of connected components.
(ii) The number n is the same for all X-embeddings of S.
(iii) Let $\mathcal{M}_i$, $i = 1, 2$, be two spherical circle planes containing S. Then there are labellings of the connected components of the circle sets of both planes with numbers from 1 to n such that replacing component number m in the circle set of $\mathcal{M}_1$ by the respective component in the circle set of $\mathcal{M}_2$ gives a circle set of a new spherical circle plane.

If there is no special requirement on how S is to be embedded in circle sets, we omit the 'X-embedded' and just refer to S as a *separating set.* Finally, *strong* (X-embedded) separating sets have the additional property that the spherical circle plane we arrive at by combining two flat Möbius planes as in (iii) is also a flat Möbius plane.

Let $\mathcal{M} = (\mathbf{S}^2, \mathcal{C})$ be a spherical circle plane. We fix one circle C and consider the set $\mathcal{C}^1$ consisting of C and all circles that intersect C in exactly one point. The circle C separates $\mathbf{S}^2$ into two connected components C^+ and C^-. We define $\mathcal{C}^\pm$ to be the collection of all circles that are completely contained in $C^\pm$. Finally, let $\mathcal{C}^2$ be the set of all circles that intersect C in precisely two points. The following result shows that $\mathcal{C}^1$ is a strong X-embedded separating set. Here X stands for the requirement that there is a circle C as above that defines the set $\mathcal{C}^1$. Clearly, the number of connected components n as in the above definition is 3 and the three connected components are $\mathcal{C}^2$, $\mathcal{C}^+$, and $\mathcal{C}^-$.

PROPOSITION 3.3.8 (Cut and Paste I) *Let $\mathcal{M}_i = (\mathbf{S}^2, \mathcal{C}_i)$, $i = 1, 2, 3$, be three spherical circle planes. Suppose that $\mathcal{C}_1 \cap \mathcal{C}_2 \cap \mathcal{C}_3 \supseteq \mathcal{C}_1^1$ for some common circle C. Let $\mathcal{C} = \mathcal{C}_1^1 \cup \mathcal{C}_1^2 \cup \mathcal{C}_2^+ \cup \mathcal{C}_3^-$. Then $\mathcal{M} = (\mathbf{S}^2, \mathcal{C})$ is a spherical circle plane. If the three planes we started with are flat Möbius planes, then so is $\mathcal{M}$.*

Proof. We show that each derived geometry $\mathcal{M}_p = (\mathbf{S}^2 \setminus \{p\}, \mathcal{C}_p)$ of $\mathcal{M}$

at a point $p \in \mathbf{S}^2$ is an $\mathbf{R}^2$-plane. Then $\mathcal{M}$ is a spherical circle plane by Theorem 3.2.1.

If $p \in C$, then obviously $\mathcal{C}_p = (\mathcal{C}_1)_p$. Hence $\mathcal{M}_p = (\mathcal{M}_1)_p$ is an $\mathbf{R}^2$-plane. If $p \notin C$, say $p \in C^+$, then the derived planes $(\mathcal{M}_1)_p$ and $(\mathcal{M}_2)_p$ are two $\mathbf{R}^2$-planes that have a simply closed topological arc $\mathcal{O}$ induced by C in common and whose line sets have all tangents to $\mathcal{O}$ in common. The set $\mathcal{M}_p$ is obtained from $(\mathcal{M}_1)_p$ and $(\mathcal{M}_2)_p$ by replacing the exterior lines of $\mathcal{O}$ in $(\mathcal{M}_1)_p$ by the exterior lines of $\mathcal{O}$ in $(\mathcal{M}_2)_p$. By Proposition 2.7.18 this yields an $\mathbf{R}^2$-plane. If $p \in C^-$, then $\mathcal{M}_p$ is obtained from the $\mathbf{R}^2$-planes $(\mathcal{M}_1)_p$ and $(\mathcal{M}_3)_p$ in a similar fashion. This shows that each derived geometry $\mathcal{M}_p$ is an $\mathbf{R}^2$-plane.

If the three planes we started with are flat Möbius planes, then the same arguments show that all derived planes of the new geometry are flat affine planes. This in turn implies that the new geometry is a flat Möbius plane as well. □

Further X-embedded separating sets are generalizations of special sets of circles in the ovoidal spherical circle plane $\mathcal{M}(\mathcal{O})$ over an ovoid $\mathcal{O}$. Circles of the Möbius plane are the nontrivial intersections of planes in projective 3-dimensional space $\mathrm{PG}(3, \mathbf{R})$ with the ovoid $\mathcal{O}$. We now choose two points p and q of $\mathrm{PG}(3, \mathbf{R})$ and let L be the line through them. Since the line L is homeomorphic to $\mathbf{S}^1$, the complement $L \setminus \{p, q\}$ has precisely two connected components C_1 and C_2. Every plane of $\mathrm{PG}(3, \mathbf{R})$ that does not contain L intersects L in precisely one point. It readily follows that the collection of all planes of $\mathrm{PG}(3, \mathbf{R})$ that pass through the points p and/or q is a 'separating set' in the space of all planes of $\mathrm{PG}(3, \mathbf{R})$. The complement has two connected components, namely all planes that intersect L in precisely one point of the component C_1 and all planes that intersect L in precisely one point of the component C_2. Since we can always find a plane in each of the two components that intersects $\mathcal{O}$ in a circle, we correspondingly obtain a 'separating set' $\mathcal{C}^{p,q}$ in the circle set $\mathcal{C}$ of the ovoidal spherical circle plane $\mathcal{M}(\mathcal{O})$ by taking all circles that come from planes through the point p or q. Moreover, each of these components is path-connected. A point in $\mathrm{PG}(3, \mathbf{R})$ is contained in $\mathcal{O}$, it is an inner point, or it is an outer point of the ovoid. In Polster–Steinke [20XXa] all possible configurations of the two points p and q and their connecting line L with respect to the ovoid $\mathcal{O}$ were considered and the corresponding 'separating sets' in the ovoidal plane $\mathcal{M}(\mathcal{O})$ were generalized to X-embedded separating sets in spherical circle planes.

If p and q are two points of $\mathbf{S}^2$ this is no problem. Let $\mathcal{M} = (\mathbf{S}^2, \mathcal{C})$

be a spherical circle plane and let $\mathcal{C}^{p,q}$ be the set of all circles in $\mathcal{C}$ that contain p and/or q. Then $\mathcal{C}^{p,q}$ separates $\mathcal{C}$ into two connected components $\mathcal{C}^+$ and $\mathcal{C}^-$. Here $\mathcal{C}^+$ and $\mathcal{C}^-$ are the sets of all circles C such that the points p and q are contained in the same or different components of $\mathbf{S}^2 \setminus C$, respectively. Then the following result shows that $\mathcal{C}^{p,q}$ is a strong separating set; see Polster–Steinke [20XXa].

PROPOSITION 3.3.9 (Cut and Paste II) *Let $\mathcal{M}_i = (\mathbf{S}^2, \mathcal{C}_i)$, $i = 1, 2$, be two spherical circle planes. Suppose that $\mathcal{C}_1 \cap \mathcal{C}_2 \supseteq \mathcal{C}_1^{p,q}$ for two points p and q. Let $\mathcal{C} = \mathcal{C}_1^{p,q} \cup \mathcal{C}_1^+ \cup \mathcal{C}_2^-$. Then $\mathcal{M} = (\mathbf{S}^2, \mathcal{C})$ is a spherical circle plane. If the two planes we started with are flat Möbius planes, then so is $\mathcal{M}$.*

3.3.4.1 Inner and Outer Involutions

It does not make sense to speak of inner and outer points of general spherical circle planes, but it turns out that certain involutory homeomorphisms of $\mathbf{S}^2$ to itself can play the role of such points.

Consider the ovoidal flat Möbius plane $\mathcal{M} = \mathcal{M}(\mathcal{O})$ associated with an ovoid $\mathcal{O}$ in $\mathrm{PG}(3, \mathbf{R})$ and let p be a point of this space not contained in $\mathcal{O}$. The *bundle involution* γ of $\mathcal{O}$ associated with p is defined just as the bundle involution of a topological oval in a flat projective plane associated with a point not on the oval; see p. 120. If a line through the point p intersects the ovoid in one point, this point is fixed by γ. If it intersects in two points, these two points get exchanged by γ. The bundle involution γ has the following properties.

(i) If p is an inner point of $\mathcal{O}$, then γ is fixed-point-free.
(ii) If p is an exterior point of $\mathcal{O}$, then the fixed-point set of γ is a topological circle F on $\mathcal{O}$ and γ exchanges the two connected components of $\mathcal{O} \setminus F$.
(iii) A circle is contained in the set $Fix(\gamma)$ of all circles in $\mathcal{M}$ that are globally but not pointwise fixed if and only if it contains a point q and its image $\gamma(q)$, where $q \neq \gamma(q)$.

Note that γ is not necessarily an automorphism of $\mathcal{M}$ and that the set F is not necessarily a circle of $\mathcal{M}$. Bundle involutions associated with points of $\mathrm{PG}(3, \mathbf{R}) \setminus \mathcal{O}$ should not be confused with bundle involutions associated with two points of a spherical circle plane; see Subsection 3.4.2.

Now, let $\mathcal{M} = (\mathbf{S}^2, \mathcal{C})$ be a spherical circle plane and let γ be an involutory homeomorphism of the point set $\mathbf{S}^2$ to itself. We call γ an *inner involution* if it satisfies the properties listed in (i) and (iii) above.

We call it an *outer involution* if it satisfies the properties listed in (ii) and (iii). Note that both inner and outer involutions are orientation-reversing homeomorphisms. The following result is an immediate consequence of the above definition of inner and outer involutions.

LEMMA 3.3.10 (Inner and Outer Involutions) *Let $\mathcal{M}_i = (\mathbf{S}^2, \mathcal{C}_i)$, $i = 1, 2$, be two spherical circle planes. Suppose that γ is an inner or outer involution of $\mathcal{M}_1$ and that $Fix(\gamma)$ is a subset of $\mathcal{C}_2$. Then γ is also an inner or outer involution of $\mathcal{M}_2$, respectively.*

Inner and outer involutions are 'virtual' inner and outer points that can play the same role in the construction of separating sets as inner and outer points.

Let γ be an inner involution of a spherical circle plane $\mathcal{M} = (\mathbf{S}^2, \mathcal{C})$ with fixed-circle set $Fix(\gamma)$. A circle is contained in $Fix(\gamma)$ if and only if it contains some point p and its image $\gamma(p)$. This property implies that the complete bundle of circles through a point and its image under γ is contained in $Fix(\gamma)$. Hence any two points of $\mathbf{S}^2$ that are not exchanged by γ are contained in exactly one circle in $Fix(\gamma)$.

Here are some other useful properties of inner involutions.

LEMMA 3.3.11 (Properties of Inner Involutions) *Let γ be an inner involution of a spherical circle plane $\mathcal{M} = (\mathbf{S}^2, \mathcal{C})$. Then the following hold.*

(i) *Let p be a point of $\mathcal{M}$. Then every circle in $\mathcal{M}$ through p and $\gamma(p)$ is contained in $Fix(\gamma)$.*

(ii) *Two distinct circles in $Fix(\gamma)$ have exactly two points in common. These two points get exchanged by γ.*

(iii) *Let C be a circle not in $Fix(\gamma)$. Then C is disjoint from its image under the involution.*

(iv) *Let r, s, $\gamma(r)$, and $\gamma(s)$ be four distinct points on a circle C in $Fix(\gamma)$. Then r and $\gamma(r)$ are contained in different connected components of $C \setminus \{s, \gamma(s)\}$.*

For a proof of this lemma see Polster–Steinke [20XXa].

We give two examples of cut-and-paste construction techniques based on inner involutions. Let $\mathcal{M} = (\mathbf{S}^2, \mathcal{C})$ be a spherical circle plane that admits the inner involution γ, let $p \in \mathbf{S}^2$, and let $\mathcal{C}^{p,\gamma}$ be the set of all circles in $\mathcal{C}$ that contain p and/or are contained in $Fix(\gamma)$. If a circle is not fixed under γ, then it is disjoint from its image under γ; see Lemma 3.3.11. Consequently, $\mathcal{C}^{p,\gamma}$ separates $\mathcal{C}$ into two connected components $\mathcal{C}^+$ and $\mathcal{C}^-$. Here $\mathcal{C}^+$ consists of all circles $C \notin \mathcal{C}^{p,\gamma}$ such that the point p and the topological circle $\gamma(C)$ are in different components of $\mathbf{S}^2 \setminus C$ and $\mathcal{C}^- = \mathcal{C} \setminus (\mathcal{C}^+ \cup \mathcal{C}^{p,\gamma})$. The following result together with Lemma 3.3.10 shows that $\mathcal{C}^{p,\gamma}$ is a strong separating set.

PROPOSITION 3.3.12 (Cut and Paste III) *Let $\mathcal{M}_i = (\mathbf{S}^2, \mathcal{C}_i)$, $i = 1, 2$, be two spherical circle planes and let γ be an inner involution of $\mathcal{M}_1$. Let p be a point and suppose that $\mathcal{C}_1 \cap \mathcal{C}_2 \supseteq \mathcal{C}_1^{p,\gamma}$. Furthermore, let $\mathcal{C} = \mathcal{C}_1^{p,\gamma} \cup \mathcal{C}_1^+ \cup \mathcal{C}_2^-$. Then $\mathcal{M} = (\mathbf{S}^2, \mathcal{C})$ is a spherical circle plane. If the two planes we started with are flat Möbius planes, then so is $\mathcal{M}$.*

Let $\mathcal{M} = (\mathbf{S}^2, \mathcal{C})$ be a spherical circle plane with two inner involutions γ_1 and γ_2 such that $\gamma_1(p) = \gamma_2(p)$ is satisfied for exactly two points p_1 and p_2. If a circle is not fixed under γ_i, then it is disjoint from its image under γ_i; see Lemma 3.3.11. Then $\mathcal{C}^{\gamma_1,\gamma_2} = Fix(\gamma_1) \cup Fix(\gamma_2)$ separates the circle set $\mathcal{C}$ into two connected components $\mathcal{C}^+$ and $\mathcal{C}^-$. Here $\mathcal{C}^+$ consists of all circles $C \in \mathcal{C} \setminus \mathcal{C}^{\gamma_1,\gamma_2}$ such that $\gamma_1(C)$ and $\gamma_2(C)$ are in different connected components of $\mathbf{S}^2 \setminus C$, and $\mathcal{C}^- = \mathcal{C} \setminus (\mathcal{C}^+ \cup \mathcal{C}^{p,\gamma})$. The following result together with Lemma 3.3.10 shows that $\mathcal{C}^{\gamma_1,\gamma_2}$ is a strong separating set.

PROPOSITION 3.3.13 (Cut and Paste IV) *Let $\mathcal{M}_i = (\mathbf{S}^2, \mathcal{C}_i)$, $i = 1, 2$, be two spherical circle planes and let γ and δ be two inner involutions of $\mathcal{M}_1$ that agree at exactly two points. Furthermore, suppose that $\mathcal{C}_1 \cap \mathcal{C}_2 \supseteq \mathcal{C}_1^{\gamma,\delta}$ and let $\mathcal{C} = \mathcal{C}_1^{\gamma,\delta} \cup \mathcal{C}_1^+ \cup \mathcal{C}_2^-$. Then $\mathcal{M} = (\mathbf{S}^2, \mathcal{C})$ is a spherical circle plane. If the two planes we started with are flat Möbius planes, then so is $\mathcal{M}$.*

Certain X-embedded separating sets involving outer involutions have also been constructed. In all known cases the condition X is rather complicated and we will not give any examples. For the above and further cut-and-paste constructions involving separating sets associated with points and virtual points of spherical circle planes; see Polster–Steinke [20XXa] and, for some preliminary results, Polster–Steinke [1995].

In Subsection 3.8.2 we investigate the fix-geometries of inner and outer involutions.

3.3.5 Integrals of $\mathbf{R}^2$-Planes

In Subsubsection 5.3.4.1 we describe a way to construct flat Laguerre planes by 'integrating' flat affine planes. It is also possible to integrate

$\mathbf{R}^2$-planes in such a way that the resulting geometries can be turned into spherical circle planes. Not much is known about these construction techniques. The known results are summarized in Subsubsection 5.3.4.1.

3.4 Groups of Automorphisms and Groups of Projectivities

In this section we investigate automorphisms and projectivities in spherical circle planes. These transformations are homeomorphisms of $\mathbf{S}^2$ and $\mathbf{S}^1$, respectively. With respect to composition we obtain groups, and, with respect to automorphisms, we classify the most homogeneous spherical circle planes. Most of the results presented in this section on groups of automorphisms are due to Strambach [1967c], [1970d], [1972], [1973], and [1974a].

3.4.1 Automorphisms and Automorphism Groups

An automorphism of a spherical circle plane that fixes a point p induces an automorphism of the derived plane at p. This allows us to prove results about the automorphisms and automorphism groups of spherical circle planes based on the respective results for flat linear spaces.

LEMMA 3.4.1 (Induced Automorphism Groups) *Let Γ be a group of automorphisms of a spherical plane $\mathcal{M}$. If Γ fixes a point p of $\mathcal{M}$, then Γ induces a group of collineations of the derived $\mathbf{R}^2$-plane at p.*

As a first application of essentially this technique we find that, just as for $\mathbf{R}^2$-planes, each isomorphism between spherical circle planes is continuous; see Strambach [1970d] Satz 3.4.

THEOREM 3.4.2 (Isomorphisms Are Continuous) *Every isomorphism between spherical circle planes is continuous. In particular, every automorphism of a spherical circle plane is continuous.*

One can model a proof on the corresponding result for $\mathbf{R}^2$-planes; see Theorem 2.4.2. Alternatively, one uses that result and derived planes. If γ is an isomorphism from a spherical circle plane $\mathcal{M}_1$ to a spherical circle plane $\mathcal{M}_2$ and p is a point of $\mathcal{M}_1$, then the restriction γ_p of γ to the derived planes at p and $\gamma(p)$ is an isomorphism of $\mathbf{R}^2$-planes. But such a map is continuous by Theorem 2.4.2. Hence γ is continuous at all points except p. Passing over to a different point yields that γ is continuous everywhere.

Remember that any tripod and its image under an automorphism of a flat linear space determine the automorphism; see Corollary 2.4.3. Similarly, we want to show that four points in 'general position' and their images under an automorphism of a spherical circle plane determine the automorphism. As a first step in this direction we consider the subplane generated by four points; compare Strambach [1970d] 3.3.

LEMMA 3.4.3 (Triodes Generate Dense Subplanes) *The vertices of a triode $\mathcal{D}(C_1, C_2, C_3; q)$ (as defined on p. 149) together with a point in the interior of the triode generate a dense subplane.*

Proof. Let p_1, p_2, and p_3 be the three vertices of the triode and let p_0 be a point in its interior. In the derived plane at q the four points generate a dense (linear) subplane by Theorem 2.4.1. Hence the points of the circular subgeometry $\mathcal{S}$ generated by p_0, p_1, p_2, and p_3 also form a dense subset of $\mathbf{S}^2$. Consequently, the circles of $\mathcal{S}$ are dense in $\mathcal{C}$, too, because the sets of the form $\alpha(O_1, O_2, O_3)$, where O_1, O_2, O_3 are disjoint open subsets of the point space $\mathbf{S}^2$, form a basis for the topology of the circle space; see Theorem 3.2.4. □

Given four nonconcircular points, we can always find a triode that has three of the points as vertices and the fourth point in its interior. Since an isomorphism between spherical circle planes takes nonconcircular points to nonconcircular points, and by continuity, we readily obtain that an isomorphism is uniquely determined by four nonconcircular points and their images under the isomorphism; see Strambach [1970d] Satz 3.4.

LEMMA 3.4.4 (Triodes Determine Isomorphisms) *Each isomorphism between two spherical circle planes is uniquely determined by four nonconcircular points and their images. In particular, the identity is the only automorphism of a spherical circle plane that fixes four points not on a circle.*

An immediate consequence of the above lemma is the following result.

COROLLARY 3.4.5 (Fixed Points Are Concircular) *Fixed points of an automorphism $\gamma \neq id$ of a spherical circle plane are concircular.*

We want to show that the automorphism group Γ of a spherical circle plane is a Lie group. Clearly, Γ acts effectively on $\mathbf{S}^2$. Hence by Theorem A2.3.5 all we have to show is that Γ is locally compact. When endowed with the compact-open topology, a sequence of automorphisms (γ_n) in Γ converges to an automorphism γ if and only if $\gamma_n(p_n)$ converges to $\gamma(p)$ whenever the sequence (p_n) of points converges to p. We fix four nonconcircular points p_1, p_2, p_3, p_4. The first three points form the vertices of a triode $t = D(C_1, C_2, C_3; p_4)$. We now identify an automorphism γ with the quadruple $(\gamma(p_1), \gamma(p_2), \gamma(p_3), \gamma(p_4))$ and then carry over the topology on the set of quadruples of nonconcircular points onto Γ. This topology on Γ is called the *topology of convergence on t*.

THEOREM 3.4.6 (Automorphism Group) *Let $\mathcal{M}$ be a spherical circle plane and Γ its automorphism group. Then the compact-open topology on Γ coincides with the topology of convergence on any of the triodes of $\mathcal{M}$.*

Let $\mathcal{Q}$ be the collection of all quadruples of nonconcircular points of a spherical circle plane $\mathcal{M}$ and let $q = (p_1, p_2, p_3, p_4) \in \mathcal{Q}$. Let Γ be the automorphism group of $\mathcal{M}$. Then $\mathcal{Q}$ is an 8-dimensional manifold and the map given by $\gamma \mapsto (\gamma(p_1), \gamma(p_2), \gamma(p_3), \gamma(p_4))$ is a homeomorphism from Γ onto a closed subset of $\mathcal{Q}$. In particular, Γ is locally compact, separable and of dimension at most 8.

For the proof see Strambach [1970d] 3.6–3.8. Theorem 2.4.5 is the corresponding results for ideal flat linear spaces.

In Subsection 3.1.4 we have seen that the automorphism group of the classical flat Möbius plane contains the group $\mathrm{PSL}_2(\mathbf{C})$. From the description of the classical flat Möbius plane in Subsection 3.1.2 we readily see that complex conjugation $\kappa : z \mapsto \bar{z}$, that is, the transformation $(x, y) \mapsto (x, -y)$, $\infty \mapsto \infty$ is also an automorphism. In fact, the full automorphism group Γ of the classical flat Möbius plane is generated by $\mathrm{PSL}_2(\mathbf{C})$ and κ, that is, $\Gamma = \mathrm{PSL}_2(\mathbf{C}) \cup \mathrm{PSL}_2(\mathbf{C})\kappa$. Hence Γ is 6-dimensional. Since the automorphism group of the classical model of a geometry usually has the highest dimension, we expect that the upper bound given in Theorem 3.4.6 for the dimension of the automorphism group of a spherical circle plane $\mathcal{M}$, that is, the group dimension of $\mathcal{M}$, can be improved to 6. We say that a spherical circle plane $\mathcal{M}$ has *group dimension* n if the (full) automorphism group of $\mathcal{M}$ has dimension n.

To this end we study inversions of a spherical circle plane. Let C be a circle of a spherical circle plane $\mathcal{M}$. An *inversion at C* is an automorphism of $\mathcal{M}$ that fixes precisely the points of C. For example, in the classical flat Möbius plane complex conjugation $\kappa : z \mapsto \bar{z}$ is an inversion. If $\mathcal{O}$ is a topological ovoid in 3-dimensional projective space $\mathrm{PG}(3, \mathbf{R})$ and $\mathcal{M}(\mathcal{O})$ is the associated flat Möbius plane, then every automorphism of $\mathcal{M}(\mathcal{O})$ is induced by a collineation of the ambient projective space $\mathrm{PG}(3, \mathbf{R})$ that leaves the ovoid $\mathcal{O}$ invariant; see Theorem 3.3.2. Hence, in this case, an inversion at a circle C comes from a reflection in $\mathrm{PG}(3, \mathbf{R})$ about the plane that determines C such that $\mathcal{O}$ is invariant.

In the derived $\mathbf{R}^2$-plane at a point p of C an involution at C induces a collineation that fixes precisely the points of the line $C \setminus \{p\}$. In $\mathbf{R}^2$-planes there can be many such collineations. For spherical circle planes, however, the possibilities for inversions are severely restricted; see Strambach [1970d] Satz 3.12.

PROPOSITION 3.4.7 (Inversions Are Unique) *Each inversion at a circle C in a spherical circle plane $\mathcal{M}$ is involutory and exchanges both connected components of $\mathbf{S}^2 \setminus C$. Moreover, $\mathcal{M}$ admits at most one inversion at C.*

If γ is an inversion at the circle C of the spherical circle plane $\mathcal{M}$, let q be a point not on C. We can then find a circle D through q and $\gamma(q)$ that intersects C in two points p_1 and p_2. This, of course, is no problem, if q and $\gamma(q)$ are separated by C. If q and $\gamma(q)$ are in the same connected component of $\mathbf{S}^2 \setminus C$, then Strambach [1970d] even shows the following; see also Theorem 6.8.1.

LEMMA 3.4.8 *Let C be a circle in a spherical circle plane $\mathcal{M}$ and let p and q be two distinct points in the same connected component of $\mathbf{S}^2 \setminus C$. Then the pencil of circles through p and q contains precisely two circles that touch C.*

By the continuity of intersection there is another circle $D' \neq D$ through q and $\gamma(q)$ that intersects C in two points p_1' and p_2'. It readily follows that both D and D' are fixed under γ so that $\{q, \gamma(q)\}$ is invariant as well. Hence γ^2 fixes both q and $\gamma(q)$ and all points on C and $\gamma^2 = id$ by Lemma 3.4.4. This shows that γ is involutory. In the derived plane at a point of C the inversion γ then induces a reflection at the line L induced by C. But such a collineation must exchange the two sides of L. This further implies that an inversion at a circle C is unique (if one exists). If γ_1 and γ_2 are inversions at C, then $\gamma = \gamma_1\gamma_2$ fixes all points of C and both connected components of $\mathbf{S}^2 \setminus C$ so that γ

cannot be an inversion. Hence γ fixes at least one further point $p \notin C$. But then $\gamma = id$ by Lemma 3.4.4.

We are now ready to show that the upper bound given in Theorem 3.4.6 for the group dimension of a spherical circle plane can be reduced to 6, the group dimension of the classical flat Möbius plane; see Strambach [1970d] Satz 3.4.

THEOREM 3.4.9 (Maximal Group Dimension) *The automorphism group Γ of a spherical circle plane $\mathcal{M}$ is a Lie group with respect to the compact-open topology (or, equivalently, the topology of uniform convergence on $\mathbf{S}^2$) of dimension at most* 6.

Proof. The automorphism group Γ of a spherical circle plane $\mathcal{M}$ is a locally compact finite-dimensional transformation group acting effectively on $\mathbf{S}^2$ by Theorem 3.4.6. Therefore the connected component Γ^1 of Γ is a Lie group by Theorem A2.3.5 and Γ is a Lie group, too.

From Theorem 3.4.6 we know that Γ is at most 8-dimensional. Suppose that the dimension of Γ is at least 7. Then the stabilizer Γ_p of a point p of $\mathcal{M}$ has dimension at least 5 by the dimension formula A2.3.6. The derived plane of $\mathcal{M}$ at p must be the Euclidean plane as a consequence of Theorems 2.6.3 and 2.6.4. Furthermore, the group of collineations induced by Γ_p contains the group of orientation-preserving collineations of the Euclidean plane fixing some point at infinity or the group of area-preserving affinities; see Salzmann [1967b] 4.3 and Theorem 4.4. In any case, we have noninvolutory axial collineations. However, this contradicts Proposition 3.4.7. □

In fact, if the group dimension of a spherical circle plane is at least 4, then the plane is classical; see Theorem 3.4.17. Examples in all other possible dimensions can be found as follows. We define $f(x) = x^3$ and $g(x) = \sinh(x)$ for $x \in \mathbf{R}$ extended onto $\mathbf{S}^1 = \mathbf{R} \cup \{\infty\}$ by setting $f(\infty) = g(\infty) = \infty$. Then we obtain the following flat Möbius planes; compare Theorem 3.3.7.

(i) The classical flat Möbius plane has group dimension 6.
(ii) The semi-classical plane $\mathcal{M}(id, f)$ has group dimension 3. This plane admits the transformations

$$(x, y) \mapsto \left(\frac{x}{c^2x^2 + (cy + d)^2}, \frac{acx^2 + (ay + b)(cy + d)}{c^2x^2 + (cy + d)^2} \right),$$

where $a, b, c, d \in \mathbf{R}$, $ad - bc = 1$, as automorphisms. After identifying the point (x, y) with the complex number $z = x + iy$ these maps turn into

$$z \mapsto \frac{az + ib}{-icz + d}.$$

(The set of all these transformations is just the stabilizer of the circle $C = \{iy \mid y \in \mathbf{R}\} \cup \{\infty\}$ in the classical flat Möbius plane.)

(iii) Let $\mathcal{M}$ be the flat Möbius plane whose circles are of the following form.

- All Euclidean lines extended by ∞.
- All Euclidean circles

$$\{(x, y) \in \mathbf{R}^2 \mid (x - a)^2 + (y - b)^2 = r^2\},$$

where $a, b, r \in \mathbf{R}$, $r > 0$, $a \leq -r$, that are entirely contained in the left half-plane $H_L = \{(x, y) \in \mathbf{R}^2 \mid x \leq 0\}$.

- All curves

$$\{(x, y) \in \mathbf{R}^2 \mid (x - a)^4 + (y - b)^4 = r^4\},$$

where $a, b, r \in \mathbf{R}$, $r > 0$, $a \geq r$, that are entirely contained in the right half-plane $H_R = \{(x, y) \in \mathbf{R}^2 \mid x \geq 0\}$.

- All curves that are made up of $O' \cap H_R$, where O' is a curve

$$\{(x, y) \in \mathbf{R}^2 \mid (x - a)^4 + (y - b)^4 = r^4\},$$

where $a, b, r \in \mathbf{R}$, $r > 0$, $|a| < r$, that intersects the y-axis in precisely two points p and q, and of $C' \cap H_L$, where C' is a Euclidean circle through p and q that has the same tangent at p as O', that is,

$$\begin{aligned} & \{(x, y) \in \mathbf{R}^2 \mid (x - a)^4 + (y - b)^4 = r^4, x \geq 0\} \\ \cup \quad & \{(x, y) \in \mathbf{R}^2 \mid (x - \tfrac{a}{\sqrt{r^4 - a^4}})^2 + (y - b)^2 \\ & \qquad = \sqrt{r^4 - a^4} + \tfrac{a^6}{r^4 - a^4}\}, \end{aligned}$$

where $a, b, r \in \mathbf{R}$, $r > 0$, $|a| < r$.

Circles are pasted together along the circle $C_0 = (\{0\} \times \mathbf{R}) \cup \{\infty\}$. On the left half, circles come from the classical flat Möbius plane, and on the right half, circles come from an Ewald type 1 plane and we have a flat Möbius plane; see Ewald [1967]. Moreover, $\mathcal{M}$ has group dimension 2. The derived plane at ∞ is the only derived affine plane that is Desarguesian so that every automorphism of $\mathcal{M}$ fixes ∞. Also each automorphism must fix the circle C_0.

Hence the automorphisms of $\mathcal{M}$ are extensions of collineations of the Euclidean plane that fix the y-axis, that is, on the affine part we have the transformations $(x,y) \mapsto (rx, sy + tx + u)$ for $r,s,t,u \in \mathbf{R}$, $r,s \neq 0$. It readily follows that this plane admits the transformations

$$\infty \mapsto \infty, (x,y) \mapsto (rx, ry + t),$$

where $r,t \in \mathbf{R}$, $r > 0$, as automorphisms and that these are the only automorphisms that correspond to elements of the above group of automorphisms of the Euclidean plane.

(iv) The semi-classical plane $\mathcal{M}(f, id)$ has group dimension 1. This plane admits the transformations

$$\infty \mapsto \infty, (x,y) \mapsto \begin{cases} (rx, ry) & \text{for } x \geq 0, \\ (f(r)x, ry) & \text{for } x \leq 0, \end{cases}$$

where $r \in \mathbf{R}^+$, as automorphisms.

(v) The semi-classical plane $\mathcal{M}(g, id)$ has group dimension 0.

3.4.2 Compact Groups of Automorphisms

Compact groups play a crucial role in the theory of locally compact groups and Lie groups. They can often act only in certain ways on certain manifolds; see for example Theorem A2.3.15. Compact groups furthermore contain involutions which again have a prominent role; see for example Subsection 3.6.3.

We begin with a classification of involutory automorphisms of spherical circle planes. We already encountered inversions. A second kind of involutory automorphism is a *bundle involution*, that is, an involutory automorphism that fixes two points and every circle through both points. It readily follows that the two fixed points are the only points fixed by a bundle involution; compare the beginning of Subsection 3.5.2. The two fixed points are the *centres* of the bundle involution. By Theorem A1.4.5, every involutory homeomorphism of $\mathbf{S}^2$ is topologically equivalent to a reflection of $\mathbf{S}^2$ about its centre, a line through the centre, or an equatorial plane. We thus have the following result.

LEMMA 3.4.10 (Involutory Automorphisms) *An involutory automorphism of a spherical circle plane is either an inversion, a bundle involution, or fixed-point-free on $\mathbf{S}^2$. Furthermore, in the first and third cases, the automorphism is orientation-reversing.*

Note that bundle involutions, as defined above, should not be confused with bundle involutions of an ovoid $\mathcal{O}$ associated with points of $\mathrm{PG}(3, \mathbf{R}) \setminus \mathcal{O}$; see Subsubsection 3.3.4.1.

Since a bundle involution induces an involutory automorphism in the derived plane at one of its centres, we see from Salzmann et al. [1995] Lemma 32.12 that a bundle involution is determined by its fixed points.

LEMMA 3.4.11 (Centers Determine Bundle Involution) *For two distinct points in a spherical circle plane there is at most one bundle involution with centres these two points.*

A compact connected Lie group Σ acting effectively on $\mathbf{S}^2$ has dimension at most 3 by Theorem A2.3.13. If Σ is a 1-dimensional compact connected Lie group, then $\Sigma \simeq \mathrm{SO}_2(\mathbf{R})$. If Σ is 2-dimensional, then $\Sigma \simeq \mathrm{SO}_2(\mathbf{R}) \times \mathrm{SO}_2(\mathbf{R})$. Finally, if Σ is 3-dimensional, then Σ contains a torus subgroup $\mathrm{SO}_2(\mathbf{R}) \times \mathrm{SO}_2(\mathbf{R})$ or Σ is isomorphic to a covering group of $\mathrm{SO}_3(\mathbf{R})$, that is, $\Sigma \simeq \mathrm{SO}_3(\mathbf{R})$ or $\Sigma \simeq \mathrm{SU}_2(\mathbf{C})$. The torus group $\mathrm{SO}_2(\mathbf{R}) \times \mathrm{SO}_2(\mathbf{R})$ and the group $\mathrm{SU}_2(\mathbf{C})$ cannot act effectively on $\mathbf{S}^2$. Hence, we only have to consider the groups $\mathrm{SO}_2(\mathbf{R})$ and $\mathrm{SO}_3(\mathbf{R})$.

PROPOSITION 3.4.12 (Actions of $\mathrm{SO}_2(\mathbf{R})$) *Let $\Phi \cong \mathrm{SO}_2(\mathbf{R})$ be a group of automorphisms of a spherical circle plane $\mathcal{M}$. Then Φ fixes precisely two points a and b of $\mathcal{M}$ and operates effectively and sharply transitively on every orbit $\Phi(c)$, $c \neq a, b$. Furthermore, the orbits $\Phi(c)$ are circles of $\mathcal{M}$ and every arc from a to b along a circle of $\mathcal{M}$ meets every orbit of Φ in precisely one point.*

Note that in the notation of Subsection 3.8.4 the orbits of Φ form a flock of $\mathcal{M}$ with carriers a and b.

For the proof see Strambach [1970d] Lemma 4.1. The existence of the two fixed points a and b and the partition of $\mathbf{S}^2 \setminus \{a, b\}$ into orbits homeomorphic to $\mathbf{S}^1$ and the statement on arcs from a to b follow from the fact that any action of $\mathrm{SO}_2(\mathbf{R})$ on $\mathbf{S}^2$ is equivalent to the standard action of $\mathrm{SO}_2(\mathbf{R})$ on $\mathbf{S}^2$ as a group of Euclidean rotations about a fixed axis; see Theorem A2.3.15.

In order to verify that nontrivial orbits of Φ are circles Strambach uses the involution σ contained in Φ. This automorphism, as an element in the connected group Φ, is orientation-preserving and must be a bundle involution by Lemma 3.4.10. It fixes the two points a and b and every circle through both of them. Using order properties of $\mathbf{R}$ Strambach then determines the structure of all circles fixed by σ.

PROPOSITION 3.4.13 (Structure of Bundle Involutions) *Let σ be a bundle involution of a spherical circle plane $\mathcal{M}$ with centres a and b. For each point $p \neq a, b$ of $\mathcal{M}$ there are precisely two circles through p that are invariant under σ. One is the circle through a, b, and p and the other circle separates the centres a and b.*

Proof. The circles fixed by σ and not passing through a and b partition the set $\mathbf{S}^2 \setminus \{a, b\}$ and the collection $\mathcal{F}$ of these circles is homeomorphic to $\mathbf{R}$. Since Φ is abelian, $\mathcal{F}$ is invariant under Φ. Therefore $\Phi \simeq \mathrm{SO}_2(\mathbf{R})$ acts on $\mathcal{F} \simeq \mathbf{R}$ as a transformation group. By Corollary A2.3.10, this action must be trivial so that Φ fixes every circle in $\mathcal{F}$. If $p \in \mathbf{S}^2 \setminus \{a, b\}$ and C is the circle in $\mathcal{F}$ that passes through p, then the orbit $\Phi(p)$ of p must be contained in C. In fact, $\Phi(p) = C$, because $\Phi(p)$ is closed and open in C and C is connected. □

The existence of a subgroup isomorphic to $\mathrm{SO}_3(\mathbf{R})$ leads to the classical flat Möbius plane.

PROPOSITION 3.4.14 ($\mathrm{SO}_3(\mathbf{R})$ Equals Classical) *If a spherical circle plane $\mathcal{M}$ admits a group isomorphic to* $\mathrm{SO}_3(\mathbf{R})$ *as a group of automorphisms, then $\mathcal{M}$ is isomorphic to the classical flat Möbius plane.*

Proof. Proposition 3.4.12 is the crucial step here. To begin with, let the group $\Sigma \simeq \mathrm{SO}_3(\mathbf{R})$ be a group of automorphisms of the spherical circle plane $\mathcal{M}$. By Theorem A2.3.11 the action of Σ on $\mathbf{S}^2$ is equivalent to the standard action of $\mathrm{SO}_3(\mathbf{R})$ on $\mathbf{S}^2$ as group of Euclidean rotations, that is, there is a homeomorphism ψ of $\mathbf{S}^2$ such that $\psi\sigma\psi^{-1}$ is a rotation for each $\sigma \in \Sigma$. It then follows that ψ is even an isomorphism from $\mathcal{M}$ onto the classical flat Möbius plane. To see this one only has to verify that $\psi(C)$ is a circle of the classical flat Möbius plane for every circle C of $\mathcal{M}$. The stabilizer Σ_C of C contains a subgroup Φ isomorphic to $\mathrm{SO}_2(\mathbf{R})$. By Proposition 3.4.12, the circle C then is an orbit under the group Φ. Thus $\psi\Phi\psi^{-1} \leq \mathrm{SO}_3(\mathbf{R})$ is a group of rotations. But the nontrivial orbits of this group are all circles of the classical flat Möbius plane. In particular, $\psi(C)$ is, as claimed, a circle of the classical flat Möbius plane. □

A direct consequence of Proposition 3.4.14 is the following characterization of the classical flat Möbius plane in terms of transitivity of its automorphism group on $\mathbf{S}^2$; see Strambach [1970d] Korollar 4.3.

THEOREM 3.4.15 (Transitive Group Equals Classical) *A spherical circle plane is isomorphic to the classical flat Möbius plane if and only if its automorphism group is transitive on the set of points.*

Proof. As seen in Subsection 3.1.4, the automorphism group of the classical flat Möbius plane is flag-transitive and thus transitive on the point set.

Conversely, suppose that the automorphism group Γ of a spherical circle plane $\mathcal{M}$ is point-transitive. Then Γ^1, the connected component of the identity, is an effective transitive transformation group of $\mathbf{S}^2$. By Theorem A2.3.11, such a group is isomorphic and acts equivalently to either $SO_3(\mathbf{R})$, $PSL_2(\mathbf{C})$, or $PSL_3(\mathbf{R})$. The latter group is not possible, since the group Γ can be at most 6-dimensional. In the first two cases the spherical circle plane admits a group of automorphisms isomorphic to $SO_3(\mathbf{R})$ so that $\mathcal{M}$ must be isomorphic to the classical flat Möbius plane by Proposition 3.4.14. □

As outlined at the beginning of this subsection, Proposition 3.4.14 implies the following characterization of the classical flat Möbius plane; compare Strambach [1970d] Satz 4.2.

THEOREM 3.4.16 (Compact 2-Dimensional Equals Classical) *A spherical circle plane is isomorphic to the classical flat Möbius plane if and only if it admits a compact group of automorphisms of dimension at least* 2.

For further characterizations of the classical flat Möbius plane in terms of transitive groups of automorphisms see Theorem 3.6.7.

3.4.3 Classification with Respect to Group Dimension

Strambach classified spherical circle planes admitting an automorphism group of dimension at least 3 in a series of papers; see Strambach [1972], [1973], and [1974a]. In most cases Möbius planes are obtained. The results are rather involved and the description of the planes obtained takes several pages in Strambach's papers. We only summarize his results.

THEOREM 3.4.17 (4-Dimensional Equals Classical) *A spherical circle plane is isomorphic to the classical flat Möbius plane if and only if it admits a group of automorphisms of dimension at least* 4.

To prove the above theorem Strambach assumes that the spherical circle plane $\mathcal{M}$ admits a closed connected group Σ of automorphisms and considers the cases dim $\Sigma = 6, 5, 4$ separately. Some arguments in the different cases are very similar, so we outline a different route.

By Halder's Theorem A2.3.14, the group Σ has a fixed point, or a 1-dimensional orbit homeomorphic to $\mathbf{S}^1$, or is transitive on $\mathbf{S}^2$. In our situation, Σ cannot have an orbit homeomorphic to $\mathbf{S}^1$. Since Σ cannot be effective on such an orbit B by Corollary A2.3.9, the kernel $\Sigma_{[B]}$ of the action of Σ on B is at least 1-dimensional. By Corollary 3.4.5, the orbit B must be contained in a circle C so that $B = C$. But then $\Sigma_{[B]}$ consists of inversions at C, which is impossible by Proposition 3.4.7.

If Σ is point-transitive, then $\mathcal{M}$ is classical by Proposition 3.4.15. This case can only occur if Σ is 6-dimensional; see Theorem A2.3.11.

If Σ fixes a point p and dim $\Sigma \geq 5$, then the derived plane at p is the Euclidean plane by Theorems 2.6.3 and 2.6.4. The induced group of collineations contains noninvolutory axial collineations with finite axis that comes from a circle of $\mathcal{M}$, which is a contradiction to Proposition 3.4.7.

In the remaining case where Σ fixes a point p and is 4-dimensional, one uses the classification of $\mathbf{R}^2$-planes of group dimension at least 4 and Proposition 3.4.7 to show that the derived plane at p is the Euclidean plane. Furthermore, the induced group of collineations consists of the group of similarities of the Euclidean plane. One can then use the following proposition by Buckel [1953] and Van Heemert [1955]. They essentially show that a spherical circle plane that admits every orientation-preserving Euclidean motion in one of its derived planes as an automorphism must be the classical flat Möbius plane.

PROPOSITION 3.4.18 (Special Group Implies Classical) *Let* $\mathcal{M}$ *be a spherical circle plane that admits the automorphisms*

$$(x, y) \mapsto (x \cos t - y \sin t + a, x \sin t + y \cos t + b), \infty \mapsto \infty,$$

where $a, b, t \in \mathbf{R}$. *Then* $\mathcal{M}$ *is the classical flat Möbius plane.*

As seen at the end of Subsection 3.4.1, there are nonclassical flat Möbius planes of group dimension 3. A 3-dimensional connected Lie group is either almost simple or solvable. More precisely, if such a group is not simple, it contains a nontrivial abelian closed normal subgroup. As for the actions on $\mathbf{S}^2$ of such 3-dimensional groups, we have the following result; see Strambach [1970d] Lemma 4.10.

LEMMA 3.4.19 (Common Fixed Point of Special Groups) *Let Σ be a connected group of automorphisms of a spherical circle plane $\mathcal{M}$. Suppose that Σ has a nontrivial abelian normal subgroup. Then Σ fixes a point of $\mathcal{M}$.*

By the above lemma, a connected 3-dimensional group Σ of automorphisms of a spherical circle plane either fixes a point or is simple. In the latter case, Σ is isomorphic to $\mathrm{SO}_3(\mathbf{R})$ (and the plane is classical by Proposition 3.4.14) or to $\mathrm{PSL}_2(\mathbf{R})$. In the former case, the resulting planes are classified in Strambach [1972] Hauptsatz, pp. 292–295.

THEOREM 3.4.20 (Nonsimple 3-Dimensional) *Let Σ be a connected 3-dimensional group of automorphisms of a spherical circle plane. If, in addition, Σ is nonsimple, then Σ fixes precisely one point ∞, the derived plane $\mathcal{A}_\infty$ at ∞ is an affine plane, and Σ operates transitively on the points of $\mathcal{A}_\infty$. Furthermore, Σ induces the full translation group of $\mathcal{A}_\infty$ and $\mathcal{A}_\infty$ is the Euclidean plane. More precisely, Σ is the semidirect product of the translation group with one of the one-parameter groups*

(i) $\Delta_d = \{(x, y) \mapsto (xd^t \cos t - yd^t \sin t, xd^t \sin t + yd^t \cos t) \mid t \in \mathbf{R}\}$, *where* $d \in \mathbf{R}$, $d \geq 1$,

(ii) $\Xi_c = \{(x, y) \mapsto (tx, t^c y) \mid t \in \mathbf{R}^+\}$, *where* $c \in \mathbf{R}$, $c \geq 1$,

(iii) $\Pi = \{(x, y) \mapsto (e^t x + te^t y, e^t y) \mid t \in \mathbf{R}\}$.

The group Σ has precisely two orbits on the set of circles. One orbit consists of the circles through ∞; these are the Euclidean lines extended by ∞. The other orbit consist of all images under Σ of a single simply closed, strictly convex curve C.

Two spherical circle planes $\mathcal{M}_1$ and $\mathcal{M}_2$ obtained from C_1 and C_2, respectively, are isomorphic if and only if the same group Σ operates on them and the curve C_1 can be taken to C_2 by an affinity of the Euclidean plane that normalizes Σ.

Depending on the form of a 1-dimensional complement of the translation group, four different (uncountable) families of planes are obtained.

If $\Sigma = \Delta_1 \cdot \mathbf{R}^2$, one obtains the classical flat Möbius plane by Proposition 3.4.18.

If $\Sigma = \Xi_1 \cdot \mathbf{R}^2$, one obtains a flat Möbius plane, more precisely, an Ewald plane of type 1.

If $\Sigma = \Delta_d \cdot \mathbf{R}^2$, where $d > 1$, then the resulting spherical circle plane is a Möbius plane. The plane is generated by a circle C which does not pass through ∞ but contains the point $(0, 0)$ and has the x-axis as tangent at this point. Furthermore, C is entirely contained in the upper half-plane. Then C can be described by a function $h : \mathbf{C} \setminus \{0\} \to \mathbf{R}$ as follows. The value $h(z)$ is nonnegative and equals $|z|$ times the distance of the origin to the oriented tangent to C that is parallel to iz. If $z = x + iy$, then

$$C = \left\{ \left(\frac{\partial h}{\partial x}(z), \frac{\partial h}{\partial y}(z) \right) \;\middle|\; z \in \mathbf{C} \setminus \{0\} \right\}.$$

Note that $h(rz) = rh(z)$ for $r > 0$. This implies that $\frac{\partial h}{\partial x}(rz) = \frac{\partial h}{\partial x}(z)$ and $\frac{\partial h}{\partial y}(rz) = \frac{\partial h}{\partial y}(z)$; therefore the circle

$$C = \left\{ \left(\frac{\partial h}{\partial x}(e^{it}), \frac{\partial h}{\partial y}(e^{it}) \right) \;\middle|\; t \in \mathbf{R} \right\}$$

is homeomorphic to $\mathbf{S}^1$. The function h further satisfies various other conditions; see Strambach [1972]. Planes for different values of d are never isomorphic and there are uncountably many nonisomorphic spherical circle planes for given $d > 1$.

If $\Sigma = \Xi_c \cdot \mathbf{R}^2$, where $c > 1$, then the resulting spherical circle plane is generated by a circle C that does not pass through ∞ but contains the points $(0, 0)$ and $(a_0, 1)$ for some $a_0 \in \mathbf{R}$ and has the horizontal lines defined by $y = 0$ and $y = 1$, respectively, as tangents at these points. Let $a_1 < 0 < a_2$ such that C is completely contained between the vertical lines $x = a_1$ and $x = a_2$ and has precisely one point on each of these lines. The four arcs on C from $(0, 0)$ to (a_1, b_1), from (a_1, b_1) to $(a_0, 1)$, from $(0, 1)$ to (a_2, b_2) and from (a_2, b_2) to $(0, 0)$ (for suitable $b_i \in [0, 1]$) are described by four continuously differentiable functions g_j that satisfy various conditions; see Strambach [1972]. The resulting spherical circle plane $\mathcal{M}$ is a Möbius plane if and only if the curve C is continuously differentiable, that is, C has unique tangents at the four exceptional points $(0, 0)$, (a_1, b_1), $(a_0, 1)$, and (a_2, b_2). If $\mathcal{M}$ is not a Möbius plane, then ∞ is the only point from which there is more than one tangent to circles in $\mathcal{M}$. Planes for different values of c are never isomorphic and there are uncountably many nonisomorphic spherical circle planes for given $c > 1$.

If $\Sigma = \Pi \cdot \mathbf{R}^2$, then the resulting spherical circle plane is generated by a circle C that does not pass through ∞ but contains the points $(0, 0)$

and $(0,1)$ and has the horizontal lines defined by $y = 0$ and $y = 1$, respectively, as tangents at these points. The two arcs on C from $(0,0)$ to $(0,1)$ can be described in the form $\{(h_j(y), y) \mid y \in [0,1]\}$, where the two functions $h_j : [0,1] \to \mathbf{R}$, $j = 1, 2$, are continuously differentiable. These functions satisfy various conditions; see Strambach [1972]. The resulting spherical circle plane $\mathcal{M}$ is a Möbius plane if and only if the curve C is continuously differentiable, that is, C has unique tangents at the two exceptional points $(0,0)$ and $(0,1)$. If $\mathcal{M}$ is not a Möbius plane, then ∞ is the only point from which there is more than one tangent to circles in $\mathcal{M}$.

Strambach further determines the structure of the full automorphism groups and Hering types for each of these planes; compare also Proposition 3.5.5.

In Strambach [1973] spherical circle planes $\mathcal{M}$ that admit 3-dimensional simple groups of automorphisms were determined and classified. As seen before, only the case $\Gamma \cong \mathrm{PSL}_2(\mathbf{R})$ needs to be examined. The classification of such spherical circle planes is achieved in a series of steps which make strong use of Haupt's theory of geometric orders; see Haupt–Künneth [1967]. The results are rather involved.

THEOREM 3.4.21 (Simple 3-Dimensional) *Let Σ be a connected 3-dimensional simple group of automorphisms of a spherical circle plane. If Σ is isomorphic to $\mathrm{PSL}_2(\mathbf{R})$, then Σ acts in the standard way as a subgroup of the group $\mathrm{PSL}_2(\mathbf{C})$ of fractional linear maps*

$$\left\{ z \mapsto \frac{az+b}{cz+d} \;\middle|\; a, b, c, d \in \mathbf{R}, ad - bc = 1 \right\}$$

on $\mathbf{S}^2 \cong \mathbf{C} \cup \{\infty\} \cong \mathbf{R}^2 \cup \{\infty\}$; see Subsubsection 2.9.2.1.

All such planes are flat Möbius planes and are of Hering type IV.1 or VII.2. The group Σ has precisely three orbits on $\mathbf{S}^2$. One of these orbits is the circle $C = \{(x,0) \mid x \in \mathbf{R}\} \cup \{\infty\}$ and the other two orbits are the two open connected components of $\mathbf{S}^2 \setminus C$.

The subgeometry of all circles intersecting C in at most one point is as in the classical flat Möbius plane (that is, these circles are described by Euclidean circles or horizontal lines). The collection of all circles that intersect C in two points is an orbit of the pencil $\mathcal{C}_{\infty,(0,0)}$ of circles through $\infty, (0,0) \in C$ under Σ.

The resulting Möbius planes split into three distinct uncountable families according to the action on the set of circles through two points of C.

The three families of flat Möbius planes can be distinguished according to the action on the set $\mathcal{C}_{\infty,(0,0)}$ of the stabilizer $\Sigma_{\infty,(0,0)}$ and its normalizer $N = N_\Sigma(\Sigma_{\infty,(0,0)})$ in Σ.

In the first family, $\Sigma_{\infty,(0,0)}$ acts trivially on $\mathcal{C}_{\infty,(0,0)}$. The Möbius planes obtained are isomorphic to the semi-classical planes $\mathcal{M}(id, h)$ (see Subsection 3.3.3) under the isomorphism $(x, y) \mapsto (y, x), \infty \mapsto \infty$. These planes are nonclassical for $h \neq id$ and classical for $h = id$. In the former case, the full automorphism group is isomorphic to $\mathrm{PGL}_2(\mathbf{R})$ and is generated by Σ and the involution $z \mapsto -\bar{z}$.

In the other two families $\Sigma_{\infty,(0,0)}$ acts effectively on $\mathcal{C}_{\infty,(0,0)}$. In the second family, N fixes no circle in $\mathcal{C}_{\infty,(0,0)} \setminus \{C\}$ whereas in the third family of flat Möbius planes N fixes a circle in $\mathcal{C}_{\infty,(0,0)} \setminus \{C\}$. Let E_1 and E_2 be the two open connected components of $\mathbf{S}^2 \setminus C$. Furthermore, let $\bar{E}_j = E_j \cup C$, $j = 1, 2$. Then $K \cap \bar{E}_j$, $j = 1, 2$, for $K \in \mathcal{C}_{\infty,(0,0)}$ can be described by one or two continuously differentiable functions from $[0, \infty)$ to $\mathbf{R}$ whose derivatives are strictly monotonic. They depend on various parameters. The final step is to paste together suitable halves in $\bar{E}_1$ and $\bar{E}_2$. Describing functions of the circles that intersect C in two points are not easily written down (it takes several pages in Strambach's paper) and we only refer to Strambach [1973].

Strambach [1974a] further investigated what groups Σ can possibly occur as closed subgroups of the automorphism group of a spherical circle plane. This is mainly a problem of extensions of the connected component Σ^1 of Σ. Under fairly mild restrictions, e.g. Σ is of dimension at least 2 or the connected component Σ^1 that contains the identity contains nontrivial compact subgroups, it is shown that Σ *splits* over Σ^1, that is, Σ is the semi-direct product of Σ^1 by a subgroup $H \cong \Sigma/\Sigma^1$. For more detailed information, see Strambach [1974a] Satz 1.

3.4.4 Von Staudt's Point of View—Groups of Projectivities

Following von Staudt's point of view, the group of all projectivities of a circle was investigated by Strambach [1977] and Löwen [1977], [1981d]. A *projectivity* of a fixed circle C is a finite composition of perspectivities between circles, the first and last circle being C. A *perspectivity* between two distinct circles B and C is defined via a pencil of circles as follows. Let $b \in B \setminus C$ and $c \in C \setminus B$. Then $\pi = \pi(B, b, c, C) : B \to C$ is given by the condition that x, $\pi(x)$, b, and c are concircular and different with the obvious modifications for points of tangency and for $x = b$ and $x = c$.

(The circle B is projected onto C via the bundle of circles through the points b and c.) These maps are continuous by coherence.

For example, in the classical flat Möbius plane let C, C_1, C_2, and C_3 be the circles

$$\begin{aligned} C &= (\mathbf{R} \times \{0\}) \cup \{\infty\}, \\ C_1 &= (\mathbf{R} \times \{2\}) \cup \{\infty\}, \\ C_2 &= (\{0\} \times \mathbf{R}) \cup \{\infty\}, \\ C_3 &= \{(x, y) \in \mathbf{R}^2 \mid x^2 + y^2 = 1\}. \end{aligned}$$

Easy computations show that the projectivities

$$\begin{aligned} &\pi(C_1, (t,2), (0,0), C) \circ \pi(C, (0,0), (0,2), C_1) \text{ for } t \in \mathbf{R}, \\ &\pi(C_2, (0,1), (r,0), C) \circ \pi(C, (1,0), (0,1), C_2) \text{ for } r \in \mathbf{R} \setminus \{0\}, \\ &\pi(C_3, (0,-1), \infty, C) \circ \pi(C, \infty, (0,1), C_3) \end{aligned}$$

are the homeomorphisms of C given by

$$\begin{aligned} &\infty \mapsto \infty \text{ and } (x,0) \mapsto (x+t, 0), \\ &\infty \mapsto \infty \text{ and } (x,0) \mapsto (rx, 0), \\ &\infty \mapsto 0, 0 \mapsto \infty \text{ and } (x,0) \mapsto (\tfrac{1}{x}, 0) \text{ for } x \neq 0, \end{aligned}$$

respectively. The perspectivity $\pi(C_3, (0,-1), \infty, C)$ is just stereographic projection from the south pole of the unit circle onto the x-axis in $\mathbf{R}^2$ and $\pi(C, \infty, (0,1), C_3)$ is the inverse of the stereographic projection from the north pole of the unit circle onto the x-axis in $\mathbf{R}^2$.

The collection of all projectivities of a circle C forms a group Π_C. Clearly, this group is independent of the circle C, that is, for two circles C and C' the corresponding groups Π_C and $\Pi_{C'}$ are isomorphic. One therefore simply speaks of the group of projectivities of a spherical circle plane. For example, the projectivities of the classical flat Möbius plane constructed above generate the group $\mathrm{PGL}_2(\mathbf{R})$ of fractional linear transformations of $\mathbf{S}^1$; see Subsubsection 2.9.2.1 for the definition of this group. In fact, the group of projectivites of the classical flat Möbius plane is isomorphic to $\mathrm{PGL}_2(\mathbf{R})$.

We endow the group Π_C with the compact-open topology τ. In general, Π_C is not locally compact and rather large. Note, however, that for the classical flat Möbius plane this group is locally compact. Strambach [1977] characterizes the classical flat Möbius plane by this property.

THEOREM 3.4.22 (Locally Compact Group Equals Classical) *Let $\mathcal{M}$ be a flat Möbius plane and let Π_C be the group of projectivities of a circle C. Then $\mathcal{M}$ is classical if the closure of Π_C in the group of* *all homeomorphisms of C with respect to the compact-open topology τ is locally compact with respect to τ. In this case, Π_C itself is closed and locally compact and isomorphic to* $\mathrm{PGL}_2(\mathbf{R})$ *as a transformation group of the circle C.*

Note that the group $\mathrm{PGL}_2(\mathbf{R})$ is sharply 3-transitive on $\mathbf{S}^1$ in its standard action. Hence the group of projectivities Π_C of the classical flat Möbius plane is sharply 3-transitive on C. Löwen [1977] investigated Π_C with respect to transitivity properties.

PROPOSITION 3.4.23 (High Degree of Transitivity) *Let $\mathcal{M}$ be a flat Möbius plane and let Σ be the pathwise connected component containing the identity in the group of projectivities of a circle C. Then the group Σ is 2-transitive on C.*

Clearly, if Σ is 2-transitive on C, then the full group Π_C of projectivities of C is 2-transitive on C, too. In general, Π_C is highly transitive on C and the classical flat Möbius plane is the one with lowest degree of transitivity; see Freudenthal–Strambach [1975].

PROPOSITION 3.4.24 (Sharply 3-Transitive Equals Classical) *A flat Möbius plane is classical if and only if the group of projectivities of a circle C is sharply 3-transitive on C.*

The above proposition is true for every Möbius plane, finite or infinite, where in the finite case one has to exclude the Möbius planes of order 9.

Pursuing this line of investigation, the following characterization of the classical flat Möbius plane was obtained by Löwen [1977] and [1981d].

THEOREM 3.4.25 (ω-Regular Equals Classical) *Let $\mathcal{M}$ be a flat Möbius plane and let Π_C be the group of projectivities of a circle C. Then $\mathcal{M}$ is classical, if Π_C acts ω-regularly on C, that is, there exists a finite set $F \subseteq C$ such that the subgroup in Π fixing F elementwise is discrete with respect to the compact-open topology τ.*

3.5 The Hering Types

Similarly to the Lenz-Barlotti classification of projective planes with respect to central collineations, Möbius planes can be classified with respect to *central automorphisms*, that is, automorphisms that fix at least one point and induce a central collineation in the derived projective plane at that fixed point. This was carried out by Hering [1965]. More precisely, he considered subgroups of central automorphisms that are linearly transitive, that is, the induced groups of central collineations are transitive on each central line except for the obvious fixed points, the centre and the point of intersection with the axis.

Hering studied two types of central automorphisms in Möbius planes. These are automorphisms that fix precisely one or two points (except the identity) and that induce a translation or homothety in the derived projective plane at each of these fixed points. In fact, in his classification Hering considered groups of automorphisms and determined their types according to transitive subgroups of central automorphisms contained in them. In the following we only deal with the full automorphism group. We then say that a Möbius plane $\mathcal{M}$ is of Hering type X if the full automorphism group of $\mathcal{M}$ is of Hering type X.

Strambach [1970e] determined the possible Hering types in flat Möbius planes. Whereas Hering obtained 18 different types, only 8 can be realized in flat Möbius planes. Strambach [1970d] investigated central automorphisms as above in spherical circle planes and obtained topological results that can serve as a starting point for an analogous classification of spherical circle planes with respect to central automorphisms.

3.5.1 q-Translations

The first kind of central automorphisms investigated by Hering are the q-translations. They have exactly one fixed point. More precisely, a *q-translation* or *translation with centre* q of a Möbius plane $\mathcal{M} = (P, \mathcal{C})$ is an automorphism of $\mathcal{M}$ that either is the identity, or fixes precisely the point q and induces a translation of the derived affine plane $\mathcal{A}_q$ at q. In order to specify the translation direction, let C be a circle passing through q and let $B(q, C)$ denote the *touching pencil with support* q, that is, $B(q, C)$ consists of all circles that touch the circle C at the point q. In the derived affine plane at q the touching pencil represents a parallel class of lines and we can look at translations in this direction. Then a $(q, B(q, C))$-*translation* of $\mathcal{M}$ is a q-translation that fixes C (and thus each circle in $B(p, C)$) globally.

A group of $(q, B(q, C))$-translations of $\mathcal{M}$ is called $(q, B(q, C))$-*transitive*, if it acts transitively on $C \setminus \{q\}$; a group of q-translations is called *q-transitive*, if it acts transitively on $P \setminus \{q\}$. We say that the automorphism group Γ of $\mathcal{M}$ is $(q, B(q, C))$-transitive if the group Γ contains a $(q, B(q, C))$-transitive subgroup of $(q, B(q, C))$-translations. Let $\mathcal{H}$ be the collection of all pairs $(p, B(p, C))$, $p \in C$ for which the Möbius plane is $(p, B(p, C))$-transitive. Possible types with respect to transitive sets of $(p, B(p, C))$-translations are denoted by the Roman numerals I to VII. Hering [1965] shows that exactly one of the following statements is valid.

I. $\mathcal{H}$ is empty.
II. $\mathcal{H}$ consists of a single pair.
III. Γ fixes precisely one point p and $\mathcal{H} = \{(p, B(p, C)) \mid p \in C\}$.
IV. Γ fixes precisely one circle C and $\mathcal{H} = \{(p, B(p, C)) \mid p \in C\}$.
V. Each point of $\mathcal{M}$ occurs in at most one pair in $\mathcal{H}$ and the incidence structure consisting of the points that occur in $\mathcal{H}$ and the circles all of whose points occur as centres of translations is a linear space with at least two circles.
VI. For every point p of $\mathcal{M}$ there is exactly one touching pencil $B(p, C)$ such that $(p, B(p, C))$ is in $\mathcal{H}$.
VII. $\mathcal{H} = \{(p, B(p, C)) \mid p \in P, C \in \mathcal{C}, p \in C\}$, that is, $\mathcal{H}$ contains all possible pairs.

Planes of Hering type III have a distinguished point p such that the derived affine plane at p is a translation plane and such that all translations are induced by automorphisms of the Möbius plane. In flat Möbius planes of Hering type III the derived affine plane at p is the Euclidean plane; see Proposition 2.4.9. The affine part at p of such a plane contains all Euclidean lines and all translates of circles in a touching pencil $B(q, C)$ for some point $q \neq p$.

Planes of Hering type IV have a distinguished circle C and the automorphism group acts 2-transitively on C. In flat Möbius planes of Hering type IV the automorphism group must be at least 3-dimensional by Brouwer's Theorem A2.3.8, and must contain a subgroup locally isomorphic to $\mathrm{PSL}_2(\mathbf{R})$. By Theorem 3.4.20, a proper covering group of the group $\mathrm{PSL}_2(\mathbf{R})$ cannot occur. We thus must have the classical flat Möbius plane or a plane listed in Theorem 3.4.21. This gives us the following characterization of planes of Hering type IV; see Strambach [1974a].

THEOREM 3.5.1 (Hering Type IV) *A flat Möbius plane is of Hering type IV if and only if the connected component of the identity in the full automorphism group is isomorphic to* $\mathrm{PSL}_2(\mathbf{R})$.

Hering types V and VI cannot occur in flat Möbius planes. Given a point p in such a Möbius plane $\mathcal{M}$, we can find four nonconcircular points p_1, p_2, p_3, and p_4 that occur as translation centres in $\mathcal{H}$ such that $p \neq p_i$, $i = 1, 2, 3, 4$. Then p has a 2-dimensional orbit under the group generated by all p_i-translations, $i = 1, 2, 3, 4$. But $\mathbf{S}^2$ is connected and 2-dimensional so that the connected component of the automorphism group of $\mathcal{M}$ must be transitive on $\mathbf{S}^2$. Hence $\mathcal{M}$ is classical by Theorem 3.4.15 and thus of Hering type VII.

Strambach uses weaker assumptions to exclude Hering types V and VI. He proves the following result.

THEOREM 3.5.2 (Hering Types V and VI Are Not Possible) *Let $\mathcal{M}$ be a flat Möbius plane and let T be a collection of translations of $\mathcal{M}$ such that the centres of translations in T are uncountable and not all on a circle. Then $\mathcal{M}$ is the classical flat Möbius plane.*

The classical flat Möbius plane is the flat Möbius plane of highest Hering type VII.

PROPOSITION 3.5.3 (Possible Types) *A flat Möbius plane is of Hering type I, II, III, IV, or VII.*

For examples for types I, II, III, IV, and VII see Subsection 3.5.3.

3.5.2 The Classification

Hering further considered a second kind of central automorphisms. These are automorphisms that fix precisely two points. Let p and q be two distinct points of a Möbius plane $\mathcal{M} = (P, \mathcal{C})$. A *$(p, q)$-homothety* of $\mathcal{M}$ is an automorphism of $\mathcal{M}$ that either is the identity, or fixes precisely the points p and q and induces a homothety with centre q in the derived affine plane $\mathcal{A}_p$ at p. (Then we also obtain a homothety with centre p in the derived affine plane $\mathcal{A}_q$ at q.) A group of (p, q)-homotheties is called *(p, q)-transitive* if it acts transitively on each circle through p and q minus the two points p and q. We say that the automorphism group Γ of $\mathcal{M}$ is (p, q)-transitive if Γ contains a (p, q)-transitive subgroup of (p, q)-homotheties.

Let $\mathcal{K}$ be the collection of all unordered pairs $\{p, q\}$ of points for which the Möbius plane is (p, q)-transitive. If the Möbius plane under consideration is not the finite Möbius plane with five points, all possible configurations of $\mathcal{K}$ given the type with respect to $(p, B(p, C))$-translations were determined. The possible types with respect to $(p, B(p, C))$-translations are further distinguished by Arabic numerals which indicate types with respect to (p, q)-homotheties.

This leads to 18 different types of Möbius planes, some of which are known to be empty or to occur only in finite Möbius planes; see Hering [1965], Krier [1973], and Yaqub [1978]. For flat Möbius planes the number of Hering types reduces to 8; see Strambach [1970e]. In the following theorem we give a list of those Hering types that can actually occur in flat Möbius planes.

THEOREM 3.5.4 (The Possible Hering Types) *A flat Möbius plane $\mathcal{M} = (\mathbf{S}^2, \mathcal{C})$ with automorphism group Γ is of Hering type I.1, I.2, II.1, II.2, III.1, III.2, IV.1, or VII.2. For these possible types the sets $\mathcal{H}$ and $\mathcal{K}$ are as follows.*

I.1. $\mathcal{H} = \emptyset$ *and* $\mathcal{K} = \emptyset$.

I.2. $\mathcal{H} = \emptyset$ *and* $\mathcal{K}$ *consists of a single unordered pair of points.*

II.1. $\mathcal{H}$ *consists of a single pair and* $\mathcal{K} = \emptyset$.

II.2. There is a distinguished flag (p, C) *such that* $\mathcal{H} = \{(p, B(p, C))\}$ *and* $\mathcal{K} = \{\{p, q\} \mid q \in C \setminus \{p\}\}$.

III.1. Γ *fixes precisely one point* p *and* $\mathcal{H} = \{(p, B(p, C)) \mid p \in C\}$ *and* $\mathcal{K} = \emptyset$.

III.2. Γ *fixes precisely one point* p *and* $\mathcal{H} = \{(p, B(p, C)) \mid p \in C\}$ *and* $\mathcal{K} = \{\{p, q\} \mid q \in P, q \neq p\}$.

IV.1. Γ *fixes precisely one circle* C *and* $\mathcal{H} = \{(p, B(p, C)) \mid p \in C\}$ *and* $\mathcal{K} = \emptyset$.

VII.2. In this case $\mathcal{H} = \{(p, B(p, C)) \mid p \in P, C \in \mathcal{C}, p \in C\}$ *and* $\mathcal{K} = \{\{p, q\} \in P \times P \mid p \neq q\}$.

No other Hering types can occur in flat Möbius planes.

For examples for all the types listed in the above theorem see the following subsection.

The stabilizer of the distinguished flag (p, C) in a flat Möbius plane of Hering type II.2 is 2-transitive on $C \setminus \{p\}$. Thus such a plane has group dimension at least 2.

The q-translations and (p, q)-homotheties of a flat Möbius plane of Hering type III.2 generate a group of dimension at least 3 and, by Theorem 3.4.17, a flat Möbius plane of Hering type III.2 must therefore have group dimension 3. The affine part of such a plane at the distinguished point p contains all Euclidean lines and images of a single circle under the group of (positive) dilatations. Thus we have a flat Möbius plane as in Ewald's type 1 construction; see Subsection 3.3.2. If we identify p with the point ∞ of $\mathbf{S}^2$, then the automorphism group contains the transformations $(x, y) \mapsto (rx + b, ry + c)$, where $b, c, r \in \mathbf{R}$, $r \neq 0$. Since these transformations form a nonsimple 3-dimensional group, these planes occur in Theorem 3.4.20.

PROPOSITION 3.5.5 (Hering Types III.1, III.2, and VII.2) *A flat Möbius plane that admits a nonsimple connected 3-dimensional group of automorphisms is of Hering type III.1, III.2, or VII.2.*

As seen in Subsection 3.5.1, planes of Hering type IV must occur among the planes described in Theorem 3.4.21. They can be characterized as those flat Möbius planes having the property that the connected components of the identity in their full automorphism groups are isomorphic to $\mathrm{PSL}_2(\mathbf{R})$; see Theorem 3.5.1.

THEOREM 3.5.6 (Hering Type IV Equals Hering Type IV.1) *A flat Möbius plane of Hering type IV is of Hering type IV.1. Thus the flat Möbius planes of Hering type IV are precisely those planes having the property that the connected components of the identity in their full automorphism groups are isomorphic to the group* $\mathrm{PSL}_2(\mathbf{R})$.

Combining Proposition 3.5.5 and Theorem 3.5.6, we obtain the following result.

COROLLARY 3.5.7 (Hering Types III.1, III.2, and IV.1) *A flat Möbius plane of group dimension* 3 *is of Hering type III.1, III.2, or IV.1.*

For spherical circle planes Strambach [1970d], more generally, defines a (p, q)*-homothety* as an automorphism that fixes the points p and q and every circle through both points. He shows that p and q are the only fixed points of a nonidentical (p, q)-homothety and that the group of all (p, q)-homotheties with fixed centres p and q is commutative and at most 1-dimensional.

3.5.3 Examples

Before we proceed to give examples of flat Möbius planes of the different Hering types listed in Theorem 3.5.4, we determine the Hering types of the semi-classical flat Möbius planes; see Steinke [1986a] Proposition 6.4.

Since the circle $C_0 = (\{0\} \times \mathbf{R}) \cup \{\infty\}$ along which a semi-classical flat Möbius plane is pasted together is fixed by any automorphism of $\mathcal{M}(g,h)$ in case we are dealing with a nonclassical plane, each centre q of a q-translation and both points p and q of a (p,q)-homothety must be on C_0 and C_0 must be in the touching pencil of a q-translation. This restriction immediately excludes Hering type III. Recall from Subsection 3.3.3 that two homeomorphisms f and f' of $\mathbf{S}^1$ to itself are projectively equivalent if and only if there are fractional linear functions $\alpha, \beta \in \mathrm{PGL}_2(\mathbf{R})$ such that $f' = \alpha \circ f \circ \beta$. We call a homeomorphism $g : \mathbf{R} \to \mathbf{R}$ *semi-linear* if it is of the form

$$g(x) = \begin{cases} px & \text{for } x \geq 0, \\ qx & \text{for } x \leq 0, \end{cases}$$

where $p, q \in \mathbf{R}^+$. For $p = q$ we obtain a *multiplication*.

THEOREM 3.5.8 (Hering Types of the Semi-classical Planes) *A semi-classical flat Möbius plane is of Hering type VII.2, IV.1, I.2, or I.1. More precisely, a semi-classical flat Möbius plane $\mathcal{M}(g,h)$ is of Hering type*

VII.2 if and only if $h = id$ and g is a multiplication;

IV.1 if and only if $h \neq id$ and g is a multiplication;

I.2 if and only if $h = id$ and g is not a multiplication but is projectively equivalent to a semi-linear homeomorphism;

I.1 if and only if $h \neq id$ and g is not a multiplication or $h = id$ and g is not projectively equivalent to a semi-linear homeomorphism.

We now continue by giving examples for flat Möbius planes of the Hering types listed in Theorem 3.5.4. In the following let DT (DT^+) be the group generated by all Euclidean translations and (positive) dilatations of the Euclidean plane.

I.1

A semi-classical flat Möbius plane $\mathcal{M}(g,h)$ with the functions $g = h$ defined by $g(x) = h(x) = x^3$; compare Theorem 3.5.8.

Of course, any rigid plane is of this type; see Subsection 3.7.1. For rigid ovoidal flat Möbius planes see Ewald [1967].

I.2

A semi-classical flat Möbius plane $\mathcal{M}(g,id)$ with g being semi-linear but not a multiplication; compare Theorem 3.5.8. Type I.2 is the only Hering type not accounted for in Strambach's [1970e] classification.

In this case $\mathcal{K}$ consists of $\{(0,0),\infty\}$ only and the automorphisms

$$\infty \mapsto \infty, (x,y) \mapsto (rx, ry),$$

where $r \in \mathbf{R}^+$, and

$$\infty \mapsto \infty, (x,y) \mapsto (g(rx), rg(y)),$$

where $r \in \mathbf{R}^-$ are $((0,0),\infty)$-homotheties so that the Möbius plane becomes $((0,0),\infty)$-transitive.

II.1

Let O be an oval in the Euclidean plane which is centrally symmetric about the origin and also symmetric about the x-axis. Circles of the Möbius plane consist of the following curves; see Ewald [1967].

- All Euclidean lines extended by ∞.
- All Euclidean circles that are entirely contained in the left half-plane $H_L = \{(x,y) \in \mathbf{R}^2 \mid x \leq 0\}$.
- All images of the oval O under the group DT that are entirely contained in the right half-plane $H_R = \{(x,y) \in \mathbf{R}^2 \mid x \geq 0\}$.
- All curves that are made up of $O' \cap H_R$, where O' is an image of O under an element of DT that intersects the y-axis in precisely two points p and q, and of $C' \cap H_L$, where C' is a Euclidean circle through p and q that has the same tangent at p as O'.

In this flat Möbius plane circles are pasted together along the distinguished circle $C_0 = (\{0\} \times \mathbf{R}) \cup \{\infty\}$. On the left half circles come from the classical flat Möbius plane, and on the right half circles come from an Ewald type 1 plane. Furthermore, the resulting plane is of Hering type II.1 for suitable O. In particular, O cannot be a conic (this just yields the classical flat Möbius plane). An example for O is the set

$$\{(x,y) \in \mathbf{R}^2 \mid x^4 + y^4 = 1\}.$$

In this case the derived plane at ∞ is the only derived affine plane that is Desarguesian. Hence ∞ must occur as the centre of any translation or as one of the two centres of any homothety. One has $\mathcal{H} = \{(\infty, B(\infty, C_0))\}$ and

$$\infty \mapsto \infty, (x,y) \mapsto (x, y+t),$$

where $t \in \mathbf{R}$, are $(\infty, B(\infty, C_0))$-translations so that the Möbius plane becomes $(\infty, B(\infty, C_0))$-transitive. Furthermore, all maps

$$\infty \mapsto \infty, (x,y) \mapsto (rx, r(y-b)+b),$$

where $b, r \in \mathbf{R}$, $r > 0$, are $(\infty, (0, b))$-homotheties. However, we do not obtain automorphisms as above for $r < 0$. Hence, the group of all $(\infty, (0, b))$-homotheties is not $(\infty, (0, b))$-transitive. Note that the group of $(\infty, (0, b))$-homotheties is relatively large. For a Möbius plane of Hering type II.1 that admits no nontrivial homotheties one pastes circles together at a second line. More precisely, the circles are as described below.

- All Euclidean lines extended by ∞.
- All Euclidean circles that are entirely contained in the right half-plane $H'_R = \{(x, y) \in \mathbf{R}^2 \mid x \geq 1\}$.
- All circles of the above Möbius plane that are entirely contained in the left half-plane $H'_L = \{(x, y) \in \mathbf{R}^2 \mid x \leq 1\}$. (These circles are possibly pasted together along the y-axis from Euclidean circles.)
- All curves that are made up of $C \cap H'_L$, where C is a circle in the above Möbius plane that intersects the line $x = 1$ in precisely two points p and q, and of $C' \cap H'_R$, where C' is a Euclidean circle through p and q that has the same tangent at p as C. (These circles are possibly pasted together along the y-axis and the vertical line $x = 1$.)

II.2

Let O be an oval in the Euclidean plane that is centrally symmetric about the origin and also symmetric about the x-axis such that the points of intersection of O with the x-axis are the boundaries of a smallest diameter of O. Circles of the Möbius plane consist of the following curves; see Ewald [1967].

- All Euclidean lines extended by ∞.
- All images of O under the group DT that intersect the y-axis in at most one point. (This means that these images are entirely contained in the left half-plane $H_L = \{(x, y) \in \mathbf{R}^2 \mid x \leq 0\}$ or the right half-plane $H_R = \{(x, y) \in \mathbf{R}^2 \mid x \geq 0\}$.)
- All curves that are made up as follows. Let O' be an image of O under an element of the group DT that intersects the y-axis in precisely two points p and q and let $r \geq 0$ be the distance of the centre of O' from the y-axis. Among all outer envelopes at distance r to images of O under DT one finds exactly one such envelope O''_r that passes through p and q and has the same tangent at p as O'. Then $(O' \cap H_L) \cup (O''_r \cap H_R)$ or $(O' \cap H_R) \cup (O''_r \cap H_L)$ is a circle depending on whether the centre of O' is in H_L or in H_R.

The resulting plane is of Hering type II.2 for suitable O. In particular, O cannot be a conic (this just yields the classical flat Möbius plane). An example for O is again $\{(x,y) \in \mathbf{R}^2 \mid x^4 + y^4 = 1\}$. In this case the derived affine plane at ∞ is the only derived affine plane that is Desarguesian. Hence ∞ must occur as the centre of any translation or as one of the two centres of any homothety. Again let $C_0 = (\{0\} \times \mathbf{R}) \cup \{\infty\}$. In this case $\mathcal{H} = \{(\infty, B(\infty, C_0))\}$ and $\mathcal{K} = \{\{\infty, (0,b)\} \mid b \in \mathbf{R}\}$. The maps

$$\infty \mapsto \infty, (x,y) \mapsto (x, y+t),$$

where $t \in \mathbf{R}$, are $(\infty, B(\infty, C_0))$-translations so that the Möbius plane becomes $(\infty, B(\infty, C_0))$-transitive. Furthermore, all maps

$$\infty \mapsto \infty, (x,y) \mapsto (rx, r(y-b)+b),$$

where $b, r \in \mathbf{R}$, $r \neq 0$, are $(\infty, (0,b))$-homotheties. This means that the plane is $(\infty, (0,b))$-transitive.

III.1

Let O be an oval in the Euclidean plane that is not symmetric about any point. Then Ewald's type 1 construction yields a Möbius plane of type III.1. The distinguished point is ∞ and all translations of the Euclidean plane extend to ∞-translations of the Möbius plane. Furthermore, all maps of the form

$$\infty \mapsto \infty, (x,y) \mapsto (r(x-a)+a, r(y-b)+b),$$

where $r \in \mathbf{R}^+$, turn out to be $(\infty, (a,b))$-homotheties. However, since O is not symmetric about any point, we do not obtain automorphisms as above for $r < 0$. Of course, this means that the group of all $(\infty, (a,b))$-homotheties is not $(\infty, (a,b))$-transitive. Note, however, that the group of $(\infty, (a,b))$-homotheties is relatively large. For a Möbius plane of Hering type III.1 that admits no nontrivial dilatations one employs Ewald's type 2 construction.

III.2

Let O be a nonconic oval in the Euclidean plane that is symmetric about some point. This means that there is a point $(a,b) \in \mathbf{R}^2$ such that the function $(x,y) \mapsto (2a-x, 2b-y)$ takes O to itself. Our standard example

$$\{(x,y) \in \mathbf{R}^2 \mid x^4 + y^4 = 1\}$$

has this property with respect to the origin. Then Ewald's type 1 construction yields a Möbius plane of type III.2. The distinguished point is ∞ and all translations of the Euclidean plane extend to ∞-translations of the Möbius plane. Furthermore, because O is symmetric about some point, all transformations

$$\infty \mapsto \infty, (x, y) \mapsto (r(x-a)+a, r(y-b)+b),$$

where $r \in \mathbf{R}$, $r \neq 0$, are $(\infty, (a, b))$-homotheties.

IV.1

A semi-classical flat Möbius plane $\mathcal{M}(id, h)$ with $h \neq id$; compare Theorem 3.5.8. The distinguished circle C is the circle along which the semi-classical flat Möbius plane is pasted together; see Subsection 3.3.3.

Clearly, the maps

$$\infty \mapsto \infty, (x, y) \mapsto (x, y+t),$$

where $t \in \mathbf{R}$, are $(\infty, B(\infty, C))$-translations. This implies that the Möbius plane is $(\infty, B(\infty, C))$-transitive. Furthermore, the transformations

$$(x, y) \mapsto \left(\frac{x}{t^2x^2 + (ty - st - 1)^2}, \frac{(st-1)tx^2 + ((st-1)y - s^2t)(ty - st - 1)}{t^2x^2 + (ty - st - 1)^2} \right),$$

where $s, t \in \mathbf{R}$, are $((0, s), B((0, s), C))$-translations and the Möbius plane is $((0, s), B((0, s), C))$-transitive.

VII.2

The classical flat Möbius plane. Here all admissible subgroups of central automorphisms are linearly transitive.

3.6 Characterizations of the Classical Plane

In this section we characterize the classical flat Möbius plane with respect to various geometrical and topological properties.

3.6.1 The Locally Classical Plane

In Steinke [1983b] it was shown that being classical is a local property of a flat Möbius plane. That is, if a flat Möbius plane looks like the classical flat Möbius plane around each point, then the plane is classical.

THEOREM 3.6.1 (Locally Classical Equals Classical) *A locally classical flat Möbius plane is classical, that is, isomorphic to the classical Möbius plane.*

To prove the above theorem one shows in a first step that if there are two nonempty open sets U_1 and U_2 such that $U_1 \cap U_2$ is nonempty and isomorphisms φ_i from the geometry induced on U_i into the geometry induced on some open subset of the classical flat Möbius plane such that the two restrictions to $U_1 \cap U_2$ agree, then, using transitivity properties of the automorphism group of the classical flat Möbius plane, we can always find an extension of φ_1 onto U_2. From this property we then can construct extensions along paths of a local isomorphism defined in a neighbourhood of a point p. Since $\mathbf{S}^2$ is simply connected, a monodromy argument yields a global isomorphism.

3.6.2 The Miquelian Plane

The classical flat Möbius plane is geometrically characterized by *Miquel's configuration* and can algebraically be represented as a chain geometry.

Miquel's configuration in its generic form involves eight points and six circles as in Figure 3.6. Think of this configuration as the geometry whose points are the vertices of a cube and whose circles are the faces of the cube.

Remember that the classical flat projective plane is characterized among its relatives by the fact that Desargues' configuration closes in it; see, for example, Subsection 1.1.1. A similar result holds for the classical flat Möbius plane.

THEOREM 3.6.2 (Miquelian Equals Classical) *Miquel's configuration closes in a flat Möbius plane if and only if it is classical.*

There is even a local version of this result. We call a flat Möbius plane *locally Miquelian* if each point possesses a neighbourhood in which Miquel's configuration closes.

THEOREM 3.6.3 (Locally Miquelian Equals Classical) *A locally Miquelian flat Möbius plane is classical.*

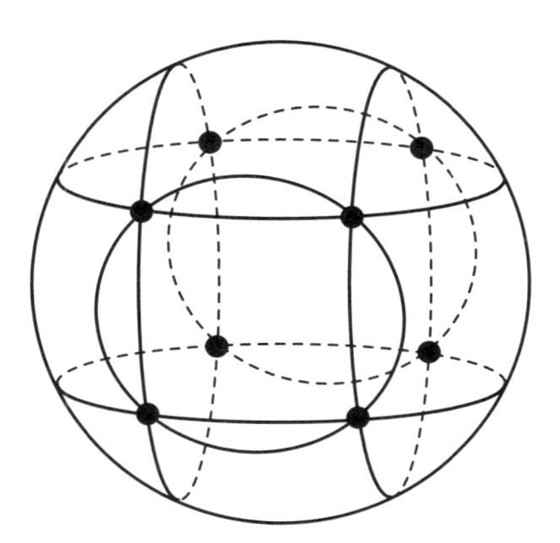

Fig. 3.6. A Miquel's configuration in the classical flat Möbius plane

The proof of this result, as given in Torrechante [1980] for flat Möbius planes and Steinke [1984a] for all Benz planes, is a local version of van der Waerden–Smid [1935] and heavily relies on order properties of $\mathbf{R}$ for the coordinatization of sufficiently small neighbourhoods of points to show that a locally Miquelian flat Möbius plane is locally classical; see Steinke [1984a] and the previous subsection.

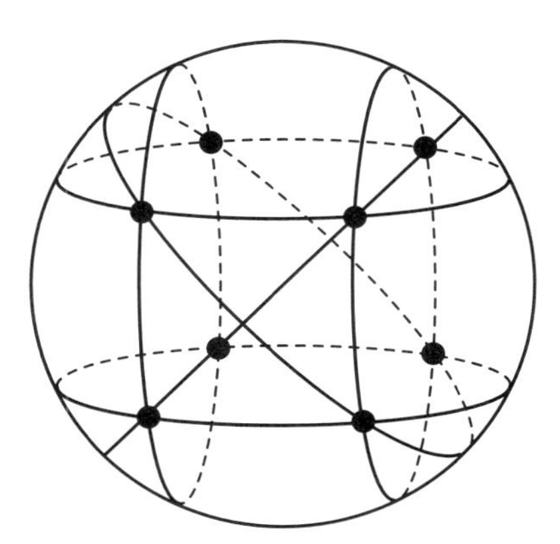

Fig. 3.7. A bundle configuration in the classical flat Möbius plane

The ovoidal Möbius planes are geometrically characterized by the bundle configuration; see Kahn [1980]. Just like Miquel's configuration, the bundle configuration in its generic form involves eight points and six circles. Think of this geometry as the geometry whose points are the eight points of a cube and whose circles are the six planes determined by two each of the four parallel edges along a given perimeter of the cube; see Figure 3.7.

THEOREM 3.6.4 (Bundle Equals Ovoidal) *The bundle configuration closes in a flat Möbius plane if and only if it is ovoidal.*

3.6.3 The Symmetric Plane

Recall that an inversion at a circle C of a spherical circle plane is an involutory automorphism that fixes precisely the points on the circle C; see Section 3.4. Such an automorphism must exchange the two hemispheres determined by C. For the collection of all inversions of a spherical circle plane Strambach [1970d] Folgerung 3.24 shows the following.

PROPOSITION 3.6.5 (The Set of Inversions) *The collection of all inversions of a spherical circle plane is closed in the full automorphism group of the spherical circle plane.*

We say that a spherical circle plane is *symmetric* if it admits an inversion at every circle. In fact, the only symmetric spherical circle plane is the classical flat Möbius plane; see Strambach [1970d] Korollar 4.4.

THEOREM 3.6.6 (Symmetric Equals Classical) *A spherical circle plane is isomorphic to the classical flat Möbius plane if and only if each circle is the axis of an inversion. In this case, the closed subgroup generated by all inversions is the full automorphism group of the classical flat Möbius plane.*

3.6.4 The Plane with Transitive Group

From Theorem 3.4.15 we know that the classical flat Möbius plane is the only spherical circle plane that admits a point-transitive automorphism group. In Subsection 3.1.4 we have seen that the classical flat Möbius plane is even flag-transitive. More generally, we say that a spherical circle plane is *flexible* if its automorphism group has an open orbit in the flag space of the plane. Clearly, flag-transitive implies flexible, point-transitive, and circle-transitive so that the classical flat Möbius plane has all four properties. Indeed, each of these properties implies classical.

THEOREM 3.6.7 (Transitive Equals Classical) *A spherical circle plane $\mathcal{M}$ is isomorphic to the classical flat Möbius plane if and only if any one of the following holds.*

(i) *The automorphism group Γ of $\mathcal{M}$ is point-transitive.*
(ii) *Γ is circle-transitive.*
(iii) *Γ is flag-transitive.*
(iv) *$\mathcal{M}$ is flexible.*

Proof. From what we said at the beginning of this subsection, and by Theorem 3.4.15, we only have to show that circle-transitive or flexible results in the classical flat Möbius plane. If $\mathcal{M}$ is flexible, then Γ must be at least 4-dimensional (see Proposition 3.2.17) and $\mathcal{M}$ is classical by Proposition 3.4.17.

Suppose that Γ is circle-transitive. Then the connected component Γ^1 must also be transitive on the circle set. By Theorem A2.3.14 the group Γ^1 has a closed orbit in $\mathbf{S}^2$, that is, a fixed point, an orbit homeomorphic to $\mathbf{S}^1$ or all of $\mathbf{S}^2$. Clearly, Γ^1 cannot be circle-transitive in the first two cases because the complement of a point or an embedded $\mathbf{S}^1$ contains a circle by Lemma 3.2.13. Hence the group Γ^1, and thus Γ, must be point-transitive. □

We can also use the method applied in Subsection 3.1.5 to obtain a different proof of the fact that flag-transitive implies classical. The connected component Γ^1 of the identity is an effective transitive transformation group of the the 2-sphere $\mathbf{S}^2$ and thus isomorphic and acting equivalently to either $\mathrm{SO}_3(\mathbf{R})$, $\mathrm{PSL}_2(\mathbf{C})$, or $\mathrm{PSL}_3(\mathbf{R})$. The last group is not possible and, because the flag space of a spherical circle plane is 4-dimensional, the first group cannot occur either. Hence Γ^1 is isomorphic and acts equivalently to $\mathrm{PSL}_2(\mathbf{C})$. The stabilizer of a point is 4-dimensional and the stabilizer of a circle is 3-dimensional. As we have seen in Subsection 3.1.5, the stabilizer of a point must be conjugate to the group $\mathrm{L}_2(\mathbf{C}) = \{z \mapsto az + b \mid a, b \in \mathbf{C}, a \neq 0\}$ and the stabilizer of a circle to $\mathrm{PSL}_2(\mathbf{R})$. Since Γ is flag-transitive, we now can reconstruct the classical plane as in Subsection 3.1.5.

3.6.5 The Plane of Hering Type at Least V

We say that a flat Möbius plane $\mathcal{M}$ is of Hering type at least V if the automorphism group of $\mathcal{M}$ is of type V, VI, or VII; see Subsection 3.5.1.

From the results in Subsection 3.5.2 we readily obtain the following characterization of the classical flat Möbius plane.

PROPOSITION 3.6.8 (Large Hering Type Equals Classical) *A flat Möbius plane $\mathcal{M}$ is of Hering type at least V if and only if $\mathcal{M}$ is the classical flat Möbius plane.*

3.6.6 Summary

THEOREM 3.6.9 (Characterizations of the Classical Plane) *A spherical circle plane $\mathcal{M}$ is isomorphic to the classical flat Möbius plane if and only if any one of the following holds.*

(i) *The automorphism group Γ contains a compact subgroup of dimension > 1.*
(ii) Γ *is point-transitive.*
(iii) Γ *is circle-transitive.*
(iv) Γ *is flag-transitive.*
(v) $\mathcal{M}$ *is flexible.*
(vi) *Each circle is the axis of an inversion.*
(vii) *Each pair of distinct points occur as centres of a nontrivial central automorphism.*
(viii) Γ *is at least 4-dimensional.*

Furthermore, a flat Möbius plane $\mathcal{M}$ is isomorphic to the classical flat Möbius plane if and only if any one of the following holds.

(ix) $\mathcal{M}$ *is Miquelian.*
(x) $\mathcal{M}$ *is locally Miquelian.*
(xi) $\mathcal{M}$ *is locally classical.*
(xii) $\mathcal{M}$ *is of Hering type at least V.*
(xiii) $\mathcal{M}$ *is symmetric.*

3.7 Planes with Special Properties

In the following we consider flat Möbius planes that do not admit any nontrivial automorphisms, and differentiable flat Möbius planes.

3.7.1 Rigid Planes

We can choose homeomorphisms g and h such that the semi-classical flat Möbius plane defined by g and h (see Subsection 3.3.3) is a *rigid* flat Möbius plane, that is, a plane whose only automorphism is the identity. For example, let $m \geq 2$ be an integer and let $0 < r < 1$. Then

$$f_{m,r} : \mathbf{R} \to \mathbf{R} : x \mapsto (1 - r)x^{2m+1} + r(x^3 - 3x^2 + 3x)$$

is a strictly increasing homeomorphism of $\mathbf{R}$. One can use this kind of homeomorphism to obtain rigid semi-classical flat Möbius planes. Let $\mathcal{M}(m, r; h)$ be the semi-classical flat Möbius plane with $g = f_{m,r}$ and h an orientation-preserving symmetric homeomorphism of $\mathbf{R}$ fixing 0. Then $\mathcal{M}(m, r; h)$ is rigid. The planes $\mathcal{M}(m, r; h)$ and $\mathcal{M}(m', r'; h')$ are nonisomorphic if $h' \neq h, h^{-1}$; see Steinke [1986a] 6.6.

In Ewald [1967] it is noted that one can construct rigid ovoidal flat Möbius planes from suitable ovoids.

3.7.2 Differentiable Planes

The point set $\mathbf{S}^2$ and the circle set $\mathcal{C}$ of a spherical circle plane naturally carry smooth differentiable structures which make them into smooth manifolds. In the classical flat Möbius plane each circle is then a submanifold of $\mathbf{S}^2$ and the geometric operations are not only continuous (see Theorem 3.2.3) but even smooth differentiable maps with respect to the appropriate smooth manifold structures. We can now replace the topological conditions in the definition of a spherical circle plane or flat Möbius plane by differentiability assumptions and ask what planes can occur. More precisely, a *smooth (differentiable) spherical circle plane or flat Möbius plane* is a spherical circle plane or flat Möbius plane such that circles are submanifolds of the point space and such that the geometric operations are smooth (differentiable) on their respective domains of definition.

THEOREM 3.7.1 (Smooth Ovoidal Planes) *An ovoidal flat Möbius plane over an ovoid $\mathcal{O}$ in $\mathbf{R}^3$ is smooth if and only if $\mathcal{O}$ is a smooth submanifold of $\mathbf{R}^3$.*

As for projective planes (see Subsection 2.8.3), being smooth is rather restrictive, but so far no classification of smooth flat Möbius planes has been carried out.

3.8 Subgeometries and Lie Geometries

Apart from the constructions in this section, a large number of constructions of other types of geometries from flat Möbius planes are described in detail in Chapter 6. Also included in Chapter 6 is the solution of the Apollonius problem for flat Möbius planes.

3.8.1 Recycled Flat Projective Planes

In the following we describe a generalization of the disk-model construction of the classical flat projective plane that we introduced in Subsection 2.1.2.

Let D be a closed topological disk with boundary C and let $\gamma : C \to C$ be a fixed-point-free continuous involution. Furthermore, let $\mathcal{L}$ be a set of Jordan curves on D such that every $L \in \mathcal{L}$ has the following properties.

(i) The curve L is homeomorphic to a closed interval.
(ii) The curve L intersects C only in its two boundary points.
(iii) The two boundary points of L get exchanged by γ.

By identifying points of D that get exchanged by γ, we arrive at a topological space $\overline{D}$ homeomorphic to the real projective plane. Furthermore, every element of $\mathcal{L}$ turns into a topological circle on $\overline{D}$. Let $\overline{\mathcal{L}}$ consist of all such topological circles plus the quotient space C/γ (this set is also a topological circle in $\overline{D}$). Then $(D, \gamma, \mathcal{L})$ is a *disk model* of a flat projective plane if and only if $(\overline{D}, \overline{\mathcal{L}})$ is a flat projective plane.

We proceed to describe a simple construction of disk models of flat projective planes from spherical circle planes.

THEOREM 3.8.1 (Disk Model) *Let $\mathcal{M}$ be a spherical circle plane, let D be a closed topological disk bounded by a circle C, let $\gamma : C \to C$ be a fixed-point-free continuous involution, and let $\mathcal{L}$ be the set consisting of all restrictions to D of circles of $\mathcal{M}$ through points that get exchanged by γ. Then $(D, \gamma, \mathcal{L})$ is a disk model of a flat projective plane.*

Let I be the set of all circles completely contained in the interior of D, let T be the set of all circles completely contained in D and touching C at exactly one point, and let S be the restrictions to D of all circles that intersect C in two points that do not get exchanged by the involution γ. Then the elements of I, T, and S are topological arcs in the flat projective plane $\mathcal{P} = (\overline{D}, \overline{\mathcal{L}})$. If $\mathcal{M}$ is a flat Möbius plane, then the elements of I and T are topological ovals in $\mathcal{P}$.

Proof. Let $p \in D \setminus C$ and $q \in C$. By Axiom B1 there is exactly one circle $C(q)$ in $\mathcal{M}$ that contains the points p, q, and $\gamma(q)$. This circle depends continuously on q. Let $r, s, \gamma(r), \gamma(s) \in C$ be distinct. Then, because γ is fixed-point-free, the points r and $\gamma(r)$ are contained in different connected components of $C \setminus \{s, \gamma(s)\}$; compare Proposition A1.4.2. Hence the two circles $C(r)$ and $C(s)$ intersect in two points; the point p and one point not contained in D. Hence every point $\neq p$ in D is contained in exactly one of the circles $C(q)$ for some q. This translates to the fact that two points in $\overline{D}$ are contained in exactly one line in $\overline{\mathcal{L}}$. By Theorem 2.2.7 this implies that $\mathcal{P}$ is a flat projective plane.

As an immediate consequence of Axiom B1, we conclude that all elements of I, T, and S are topological arcs in $\mathcal{P}$.

Assume that $\mathcal{M}$ is a flat Möbius plane and let O be an element of I or T. It remains to show that every point p of O is contained in exactly one tangent line in $\mathcal{P}$. If $p \in C$, then the line that corresponds to C is this tangent. Let p be an interior point of D. The set $\mathcal{K}$ of circles in $\mathcal{M}$ that touch O at the point p define a continuous involution γ' of C as follows. Let q be a point on C and let $K(q)$ be the unique circle in $\mathcal{K}$ that contains q (by Axiom B2). Then either $K(q)$ touches C at q or it intersects C in a second point. In the first case let $\gamma'(q) = q$. In the second case let $\gamma'(q)$ be the second point of intersection. This involution has necessarily two fixed points corresponding to the two circles in $\mathcal{K}$ that touch C; see Theorem 6.8.1. Since γ does not fix any point, there is exactly one pair of points q, $\gamma(q)$ on C that is also exchanged by γ'. The line in $\mathcal{P}$ that corresponds to the circle in $\mathcal{K}$ through these two points is a tangent line of the topological arc O at this point. This tangent is unique since all lines through p that do not arise from circles in $\mathcal{K}$ intersect O in a second point. □

In Subsection 2.1.2 we described the special case where C is a Euclidean unit circle in the affine part of the classical flat Möbius plane and γ is the antipodal map. In this case $(D, \gamma, \mathcal{L})$ is the disk model of the classical flat projective plane. The same is true if we leave C unchanged and replace γ by any bundle involution corresponding to an inner point of the circle C.

Let us return to the general case considered in the theorem. Let D' be the set of interior points of D and let $\mathcal{L}'$ be the set of restrictions of elements of $\mathcal{L}$ to D'. Then $\mathcal{A} = (D', \mathcal{L}')$ is the flat affine plane that arises from the projective plane $(\overline{D}, \overline{\mathcal{L}})$ by removing the line C/γ.

All the elements of I, T, and S are topological arcs in this flat affine plane and the geometry $(D', \mathcal{L}' \cup I)$ almost looks like the planar representation of a flat spherical circle plane. Of course, it is not, because there are triples of points of this geometry that get connected by elements of T in the original circle plane $\mathcal{M}$. Still, it seems worthwhile to ask the question, whether this partial geometry can be made into a planar representation of a flat spherical plane by adding some further simply closed curves to its circle set. The answer to this question is 'No'. We can convince ourselves of this fact as follows. We can easily find a sequence of triples of distinct points that correspond to connecting circles in I that converge to a triple of distinct points that corresponds to a connecting circle E in T. Because the spherical circle plane we are looking for would be topological, its planar representation would have to contain the intersection of E with D'. The resulting curve would be a Jordan arc, but all Jordan arcs have already been accounted for by the lines in the flat affine plane $\mathcal{A}$.

For more information about disk models see Polster [1996a].

3.8.2 Double Covers of $\mathbb{R}^2$*-Planes and Flat Projective Planes*

In Subsection 3.3.4 we introduced inner and outer involutions of spherical circle planes and used them to construct separating sets in such planes. In the following we consider the fixed-circle sets of these involutions.

3.8.2.1 Inner Involutions

Let γ be an inner involution of a spherical circle plane $\mathcal{M} = (\mathbf{S}^2, \mathcal{C})$ with fixed-circle set $Fix(\gamma)$. Recall from Subsubsection 3.3.4.1 that γ is a fixed-point-free homeomorphism $\mathbf{S}^2 \to \mathbf{S}^2$ such that a circle is contained in $Fix(\gamma)$ if and only if it contains some point p and its image $\gamma(p)$ and that any two points of $\mathbf{S}^2$ that are not exchanged by γ are contained in exactly one circle in $Fix(\gamma)$.

Let P be the quotient space $\mathbf{S}^2/\gamma$ and let $\mathcal{L}$ be the set of all C/γ, where $C \in Fix(\gamma)$. We conclude that $\mathcal{P} = (P, \mathcal{L})$ is a geometry of topological circles on a surface that is homeomorphic to the real projective plane and that Axiom P1 (see p. 2) is satisfied. By Theorem 2.2.7, this geometry is a flat projective plane. We express this by saying that $(\mathbf{S}^2, Fix(\gamma))$ is a *double cover* of the flat projective plane $\mathcal{P}$.

Let D be a closed disk bounded by a circle $C \in Fix(\gamma)$, let γ_C be the restriction of γ to C, and let $Fix_D(\gamma)$ be the set of restrictions of elements of $Fix(\gamma)$ to D. Then $(D, \gamma_C, Fix_D(\gamma))$ is a disk model of the flat projective plane $\mathcal{P}$; see the previous subsection for details about disk models of flat projective planes.

If $\mathcal{M}$ is the classical flat Möbius plane on $\mathbf{S}^2$ and γ is the antipodal map of $\mathbf{S}^2$, that is, the bundle involution associated with the origin, then $Fix(\gamma)$ consists of all great circles on the sphere.

We summarize our considerations in the following theorem.

THEOREM 3.8.2 (Double Covers of Flat Projective Planes) *Let γ be an inner involution of a spherical circle plane $\mathcal{M} = (\mathbf{S}^2, \mathcal{C})$. Then $(\mathbf{S}^2, Fix(\gamma))$ is the double cover of a flat projective plane $\mathcal{P}$.*

Let D be a closed disk bounded by a circle $C \in Fix(\gamma)$. Then the triple $(D, \gamma_C, Fix_D(\gamma))$ is a disk model of $\mathcal{P}$.

3.8.2.2 Outer Involutions

Let γ be an outer involution of a spherical circle plane $\mathcal{M} = (\mathbf{S}^2, \mathcal{C})$. Recall from Subsection 3.3.4.1 that $Fix(\gamma)$ consists of all circles that are globally but not pointwise fixed by γ, that the fixed-point set F of γ is a topological circle on $\mathbf{S}^2$, and that the two connected components of $\mathbf{S}^2 \setminus F$ get exchanged by γ. Furthermore, a circle is contained in $Fix(\gamma)$ if and only if it contains some point p and its image $\gamma(p)$, where $p \neq \gamma(p)$. These properties imply that the complete bundle of circles through a point and its image under γ is contained in $Fix(\gamma)$ provided that $p \neq \gamma(p)$. Hence any two points of $\mathbf{S}^2$ that are not exchanged by γ are contained in exactly one circle in $Fix(\gamma)$.

Here is a list of useful properties of outer involutions.

LEMMA 3.8.3 (Properties of Outer Involutions) *Let γ be an outer involution of a spherical circle plane $\mathcal{M} = (\mathbf{S}^2, \mathcal{C})$ with fixed-point set F. Then the following hold.*

(i) *Let p be a point that is not fixed by γ. Then every circle through p and $\gamma(p)$ is contained in $Fix(\gamma)$.*
(ii) *Let $C \in \mathcal{C} \setminus \{F\}$. Then $C \cap \gamma(C) = C \cap F$.*
(iii) *Let r, s, $\gamma(r)$, and $\gamma(s)$ be four distinct points on a circle C in $Fix(\gamma)$. Then r and $\gamma(r)$ are contained in the same connected component of $C \setminus \{s, \gamma(s)\}$.*

For a proof of this lemma see Polster–Steinke [20XXa].

Let D_i, $i = 1, 2$, be the two connected components of $\mathbf{S}^2 \setminus F$ and let $\mathcal{L}_i$ be the set of restrictions of circles in $Fix(\gamma)$ to D_i. Then the geometry $\mathcal{R}_i = (D_i, \mathcal{L}_i)$ is an $\mathbf{R}^2$-plane. This $\mathbf{R}^2$-plane cannot be a flat affine plane. Furthermore, γ defines an isomorphism between $\mathcal{R}_1$ and $\mathcal{R}_2$. We express all this by saying that $(\mathbf{S}^2, Fix(\gamma))$ is a *double cover* of an $\mathbf{R}^2$-plane.

If $\mathcal{M}$ is the classical flat Möbius plane on $\mathbf{S}^2$ and γ is the bundle involution of $\mathbf{S}^2$ associated with an exterior point, then this $\mathbf{R}^2$-plane is isomorphic to the real hyperbolic plane.

We summarize our considerations in the following theorem.

THEOREM 3.8.4 (Double Covers of $\mathbf{R}^2$-Planes) *Let γ be an outer involution of a spherical circle plane $\mathcal{M} = (\mathbf{S}^2, \mathcal{C})$. Then $(\mathbf{S}^2, Fix(\gamma))$ is the double cover of an $\mathbf{R}^2$-plane.*

3.8.3 3-*Ovals*

Recall that an arc in a flat projective plane is a nonempty set of points such that every line has at most two points in common with it. An oval then is an arc every point of which is contained in exactly one tangent line; see Subsection 2.2.1. Starting with these definitions Groh–Heise [1973] investigated the analogous setting in Möbius planes.

To begin with, a 3-*arc* in a spherical circle plane $\mathcal{M}$ is a nonempty set of points such that every circle has at most three points in common with it. A 3-*oval* of $\mathcal{M}$ is a 3-arc $\mathcal{O}$ with at least two points such that for any two distinct points $p, q \in \mathcal{O}$ there is precisely one circle in $\mathcal{M}$ that intersects $\mathcal{O}$ only in p and q. In analogy to arcs in flat linear spaces, 3-ovals in spherical circle planes are convex with respect to all circles.

Using transfinite induction, Groh and Heise show that 3-ovals exist in every spherical circle plane. However, there is no such thing as a 'topological' 3-oval.

PROPOSITION 3.8.5 (3-Ovals Are Wild) *A* 3-*oval in a flat Möbius plane is pathwise totally disconnected, that is, it contains no homeomorphic image of the closed real interval* $[0, 1]$.

3.8.4 Flocks and Resolutions

Let p and q be two distinct points of a flat Möbius plane $\mathcal{M}$. A *flock of $\mathcal{M}$ with carriers p and q* is a partition of $\mathbf{S}^2 \setminus \{p, q\}$ into circles of $\mathcal{M}$.

Examples of flocks for ovoidal flat Möbius planes can be obtained as follows. Let $\mathcal{O}$ be an ovoid in 3-dimensional projective space $\mathrm{PG}(3, \mathbf{R})$ and let L be a line of $\mathrm{PG}(3, \mathbf{R})$ that has no points in common with $\mathcal{O}$.

There are exactly two planes through L that are tangent to $\mathcal{O}$ at points p and q, respectively. All other planes of $\mathrm{PG}(3, \mathbf{R})$ either intersect $\mathcal{O}$ in a circle of $\mathcal{M}(\mathcal{O})$ or have no point in common with $\mathcal{O}$. The collection of circles obtained by the former planes yields a flock of $\mathcal{M}(\mathcal{O})$ with carriers p and q; see Figure 3.8.

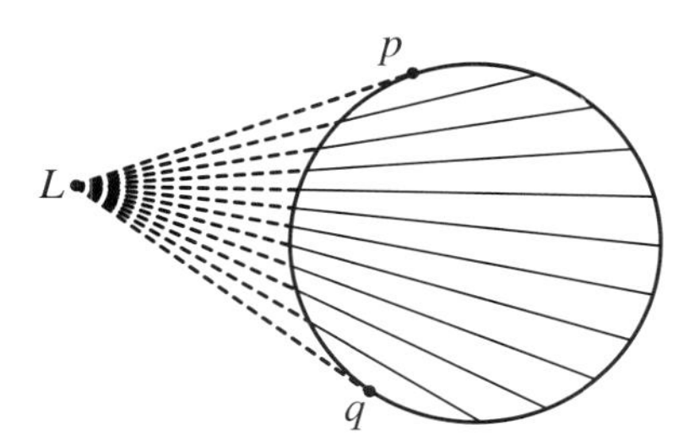

Fig. 3.8. Construction of a linear flock (the line L is perpendicular to the picture plane)

This kind of flock is called *linear*. Of course, for any two points p and q of $\mathcal{O}$ there is a linear flock with these two points as carriers. One just takes the intersection of the tangent planes to $\mathcal{O}$ at p and q as the exterior line to construct the linear flock from. We denote this flock by $\mathcal{F}_{\mathcal{M}(\mathcal{O})}(p, q)$.

Note, however, that there are nonlinear flocks of $\mathcal{M}(\mathcal{O})$ as well. Consider a plane E that intersects $\mathcal{M}(\mathcal{O})$ in a circle and in this plane two distinct exterior lines of $\mathcal{O}$. The plane E defines two hemispheres; let us call them the upper and a lower hemispheres. Our nonlinear flock consists of all circles in the upper hemisphere that come from planes through the first line and all circles in the lower hemisphere that come from planes through the second line. This kind of flock is called *bilinear*.

Let $\mathcal{F}$ be a flock with carriers p and q of a flat Möbius plane $\mathcal{M}$. The two carrier points are in different connected components of $\mathbf{S}^2 \setminus C$ for every circle C in $\mathcal{F}$. Let A be an arc from p to q on a circle of $\mathcal{M}$ through these two points. Every point on A except the two endpoints p and q determines a unique circle in $\mathcal{F}$ and, vice versa, every circle in $\mathcal{F}$ intersects A in exactly one point. This shows that $\mathcal{F}$, as a subset of the circle space, is homeomorphic to $\mathbf{R}$; see Figure 3.8.

PROPOSITION 3.8.6 (Flocks) *Every flock of a flat Möbius plane is homeomorphic to* $\mathbf{R}$.

Strambach [1974b] pp. 158–159 showed that given two points of a flat

Möbius plane $\mathcal{M}$ one can always find a flock of $\mathcal{M}$ with carriers these two points; see also Rosehr [1998] Theorem 1.9. Moreover, any partial closed flock can be extended to a flock. This generalizes Strambach's construction of flocks.

A *resolution* of a Möbius plane $\mathcal{M}$ is a partition of the circle space of $\mathcal{M}$ into flocks of $\mathcal{M}$. If $\mathcal{M}$ admits a resolution, the Möbius plane is called *resolvable*. For example, the ovoidal flat Möbius plane $\mathcal{M}(\mathcal{O})$ admits a resolution into linear flocks by taking a plane E of $\mathrm{PG}(3,\mathbf{R})$ that has no point in common with $\mathcal{O}$. Each line of E determines a linear flock and each circle of $\mathcal{M}(\mathcal{O})$ belongs to the unique flock determined by the line of intersection of E with the plane that induces the given circle. In this example we see that the flocks in a resolution correspond to the lines of a projective plane.

THEOREM 3.8.7 (Ovoidal Planes Are Resolvable) *Every ovoidal flat Möbius plane is resolvable.*

3.9 Open Problems

PROBLEM 3.9.1 *Let $(P,\mathcal{C})$ be a geometry and let $n > 1$ be a fixed integer such that P is a surface, every element of $\mathcal{C}$ is a topological circle on this surface, and any n distinct points are contained in exactly one element of $\mathcal{C}$. Does it follow that $n \in \{2,3\}$ and that $(P,\mathcal{C})$ is either a flat projective plane or a spherical circle plane?*

See Problem 2.11.1.

PROBLEM 3.9.2 *Assume that all derived affine planes of a flat Möbius plane are Desarguesian. Does this imply that the flat Möbius plane is ovoidal?*

The answer to the corresponding question for finite Möbius planes of odd order is affirmative. In fact, these planes are Miquelian. Furthermore, finite Möbius planes of even order are ovoidal. The corresponding question for flat Minkowski planes also has a positive answer; see Theorem 4.6.3.

PROBLEM 3.9.3 *What $\mathbf{R}^2$-planes and flat affine planes can occur as derived planes of spherical circle planes and flat Möbius planes? What flat projective planes have disk models associated with circles in spherical* *circle planes? Classify the flat projective planes that have disk models associated with the classical flat Möbius plane. Double covers of what flat projective planes and* $\mathbf{R}^2$*-planes occur as subgeometries of spherical circle planes?*

Some more constructions of flat linear spaces from flat Möbius planes via generalized quadrangles (see Chapter 6) correspond to some further similar questions.

Note that embedding an $\mathbf{R}^2$-plane as a derived plane in a spherical circle plane involves constructing many topological ovals in the $\mathbf{R}^2$-plane. Theorem 2.2.4(i) guarantees that every $\mathbf{R}^2$-plane does contain topological ovals.

PROBLEM 3.9.4 *Develop the theory of flat Möbius planes that are integrals of* $\mathbf{R}^2$*-planes; see Subsubsection 5.3.4.1.*

PROBLEM 3.9.5 *Are there any further essentially new ways to construct flat linear spaces from spherical circle planes apart from the ones that come about as combinations of the links between the different classes of geometries described in this book?*

PROBLEM 3.9.6 *Are there spherical circle planes all of whose derived planes are* $\mathbf{R}^2$*-planes that are not flat affine planes?*

None of the known spherical circle planes has this property. In particular, note that an ovoidal spherical circle plane corresponding to a strictly convex topological sphere O has derived planes that are not flat affine planes at exactly those points at which O is not differentiable.

PROBLEM 3.9.7 *Extend the classification of the semi-classical flat Möbius planes to a classification of the semi-classical spherical circle planes. Following this classify the semi-ovoidal spherical circle planes and flat Möbius planes, that is, spherical circle planes and flat Möbius planes that are pasted together along a common circle from two halves of ovoidal spherical circle planes.*

The corresponding question has been investigated for flat Laguerre planes. The semi-ovoidal flat Laguerre planes that are made up of two ovoidal halves glued together along two parallel classes have been determined in Polster–Rosehr–Steinke [1998].

PROBLEM 3.9.8 *Are there halves taken from two different spherical circle planes that cannot be glued together along a circle into a new spherical circle plane?*

PROBLEM 3.9.9 *Are all (X-embedded) separating sets strong? Are there separating sets that only allow one to combine flat Möbius planes but not proper spherical circle planes?*

PROBLEM 3.9.10 *Are there purely topological criteria that determine whether a set is a (strong) separating set?*

Theorem 5.3.7 seems to indicate that the answer to the respective question for cylindrical circle planes may, surprisingly, not be very 'far' away from 'Yes'.

PROBLEM 3.9.11 *Are the circle and flag spaces of a spherical circle plane homeomorphic to the respective spaces of the classical flat Möbius plane?*

It is easy to answer the corresponding question for toroidal and cylindrical circle planes in the affirmative. So far it is only known that the circle space of any flat Möbius plane is homeomorphic to the circle space of the classical plane; see Theorem 3.2.15.

PROBLEM 3.9.12 *Develop a classification of spherical circle planes with respect to central automorphisms similar to the Hering classification for flat Möbius planes.*

PROBLEM 3.9.13 *Characterize locally ovoidal flat Möbius planes, that is, flat Möbius planes such that each point has an open neighbourhood whose induced geometry is isomorphic to the induced geometry on some open set of some ovoidal flat Möbius plane.*

PROBLEM 3.9.14 *Are all spherical circle planes isotopic (in some sense) to the classical plane?*

See the remarks after Problem 2.11.5 for some references.

PROBLEM 3.9.15 *Let S be a set of topological circles on the sphere. Assume that any two curves in S intersect in at most two points and if they do intersect in two points, then they intersect transversally. Is it possible to embed S in a spherical circle plane? Are there 'universal* *spherical circle planes', that is, spherical circle planes in which every set S as above can be embedded?*

Remember that the corresponding questions for flat projective planes do have positive answers; see Subsection 2.10.4. A lot of information about finite sets of topological circles on the sphere has been collected in Grünbaum [1972].

PROBLEM 3.9.16 *Develop a theory of continuous lineations (circulations?) for spherical circle planes.*

See Subsection 2.9.3 for a summary of the known results about continuous lineations of flat linear spaces.

PROBLEM 3.9.17 *Use the emerging classification of 3-dimensional generalized quadrangles (see Chapter 6) to construct and classify flat Möbius planes.*

4
Toroidal Circle Planes

Flat Minkowski planes were first investigated by Schenkel [1980] in her dissertation. Later, Polster [1998b] studied the more general toroidal circle planes. For more information about general Minkowski planes and, in particular, finite Minkowski planes, we refer to the papers by Hartmann [1982a], Klein–Kroll [1989], Delandtsheer [1995], and the references given there.

A toroidal circle plane is a point–circle geometry whose point set is (homeomorphic to) the torus $\mathbf{S}^1 \times \mathbf{S}^1$. The point set is equipped with two nontrivial parallelisms. The parallel classes of these parallelisms are the horizontals and verticals on the torus. The circles of the toroidal circle plane are graphs of homeomorphisms $\mathbf{S}^1 \to \mathbf{S}^1$ that form a system of topological circles on the torus such that the Axiom of Joining B1 (see p. 7) is satisfied, that is, any three pairwise nonparallel points determine exactly one curve in the system. A toroidal circle plane is a flat Minkowski plane if it also satisfies the Axiom of Touching B2, that is, for each circle C and any two nonparallel points p, q with $p \in C$ there is precisely one circle through p and q that touches C (geometrically) at p, that is, intersects C only at the point p or coincides with C.

As in the case of spherical circle planes and flat Möbius planes, derived planes are an important tool in the investigation of toroidal circle planes and flat Minkowski planes. Since we have two different parallelisms on the point set, each derived plane admits two distinguished pencils of parallel lines. This fact facilitates many of the arguments encountered in the application of this tool in the case of toroidal circle planes.

Since Schenkel's dissertation is not readily available we include more proofs than usual in this chapter. Also some results are proved in a new way.

4.1 Models of the Classical Flat Minkowski Plane

In this first section we describe a number of models of the classical flat Minkowski plane. For in depth information about most of these models see Benz [1973].

4.1.1 The Geometry of Plane Sections

The point set of the classical flat Minkowski plane is the standard nondegenerate ruled quadric $\mathcal{Q}$ in 3-dimensional projective space $\mathrm{PG}(3, \mathbf{R})$, that is,

$$\mathcal{Q} = \{(x_0 : x_1 : x_2 : x_3) \in \mathrm{PG}(3, \mathbf{R}) \mid x_0x_3 + x_1x_2 = 0\},$$

where $(x_0 : x_1 : x_2 : x_3)$ are the homogeneous coordinates of the point corresponding to the 1-dimensional subspace spanned by the nonzero vector $(x_0, x_1, x_2, x_3) \in \mathbf{R}^4$. A plane

$$\{(x_0 : x_1 : x_2 : x_3) \in \mathrm{PG}(3, \mathbf{R}) \mid a_0x_0 + a_1x_1 + a_2x_2 + a_3x_3 = 0\},$$

where $(a_0 : a_1 : a_2 : a_3) \in \mathrm{PG}(3, \mathbf{R})$, is tangent to $\mathcal{Q}$ if and only if $a_0a_3 + a_1a_2 = 0$, that is, if and only if the point $(a_0 : a_1 : a_2 : a_3)$ is on $\mathcal{Q}$. A tangent plane at a point p of $\mathcal{Q}$ intersects $\mathcal{Q}$ in precisely two distinct lines through p. Therefore, through every point of $\mathcal{Q}$ there are precisely two distinct lines entirely contained in $\mathcal{Q}$. For example, the lines

$$\{(s : sx : t : -tx) \in \mathrm{PG}(3, \mathbf{R}) \mid s, t \in \mathbf{R}, (s, t) \neq (0, 0)\}$$

and

$$\{(s : t : sy : -ty) \in \mathrm{PG}(3, \mathbf{R}) \mid s, t \in \mathbf{R}, (s, t) \neq (0, 0)\}$$

pass through the point $(1 : x : y : -xy)$ and are entirely contained in $\mathcal{Q}$. The lines entirely contained in $\mathcal{Q}$ are the *parallel classes* of the two parallelisms of the classical flat Minkowski plane. We call the two different types of parallel classes (+)- and (−)-parallel classes. Remember that a (+)-parallel class and a (−)-parallel class intersect in exactly one point.

Every plane in $\mathrm{PG}(3, \mathbf{R})$ that is not tangent to $\mathcal{Q}$ intersects $\mathcal{Q}$ in a nondegenerate conic. Figure 4.1 shows part of the hyperbolic quadric (a hyperboloid) and its two essentially different plane intersections in a point—a conic and a pair of intersecting lines. The circles of the classical flat Minkowski plane are intersections of $\mathcal{Q}$ with planes in $\mathrm{PG}(3, \mathbf{R})$ that are not tangent to $\mathcal{Q}$.

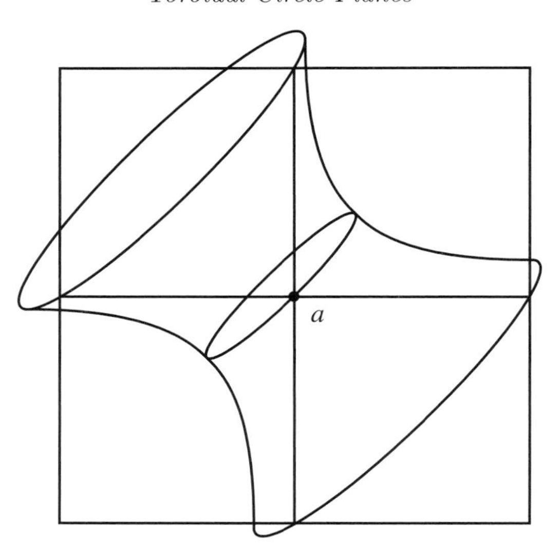

Fig. 4.1. The two essentially different plane intersections of a hyperbolic quadric—a conic and a pair of intersecting lines

In this model of the classical flat Minkowski plane the two Axioms B1 and B2 are readily verified. As for B1, three points of $\mathcal{Q}$, no two of which are on a parallel class, determine a unique plane in $\mathrm{PG}(3, \mathbf{R})$, and this plane cannot be tangent to $\mathcal{Q}$. Therefore its trace on $\mathcal{Q}$ is the unique circle containing the three points. In Axiom B2 we have two points p and q not on a common parallel class and a circle C through p, that is, a nontangent plane E in $\mathrm{PG}(3, \mathbf{R})$ that contains p. Now $C = E \cap \mathcal{Q}$ is a nondegenerate conic in $E \cong \mathrm{PG}(2, \mathbf{R})$. This conic has a unique tangent L in E at the given point p. Then L and the other given point q determine a unique plane in $\mathrm{PG}(3, \mathbf{R})$. Its trace on $\mathcal{Q}$ is the unique circle through the two points touching the given circle at p.

Note that there are two types of planes described by $(a_0 : a_1 : a_2 : a_3)$ that are not tangent to $\mathcal{Q}$ — those for which $a_0a_3 + a_1a_2 > 0$ and those for which $a_0a_3 + a_1a_2 < 0$. This means that the circle space of the classical flat Minkowski plane is not connected but has precisely two connected components.

4.1.2 The Geometry of Euclidean Lines and Hyperbolas

We fix a point p on the quadric $\mathcal{Q}$ and let G and H denote the two parallel classes through p. We now consider the derived plane at the point p. Remember that its points are all points not on one of the two parallel classes through p. Its lines are all circles passing through p that have been punctured at p, and all parallel classes different from G and H that have been punctured in their points of intersection with G and H. The derived plane can be identified with the Euclidean plane via a stereographic projection (see below) such that the lines corresponding to the parallel classes are the horizontals and verticals. The rest of the lines correspond to the nonvertical and nonhorizontal lines. Under this identification circles not passing through the point p induce conics in the affine plane. More precisely, these conics are Euclidean hyperbolas with asymptotes a horizontal and a vertical. In fact, any Euclidean hyperbola with asymptotes a horizontal and a vertical occurs. Hence we are dealing with a system of curves consisting of the Euclidean lines

$$\{(x, mx + t) \mid x \in \mathbf{R}\},$$

where $m, t \in \mathbf{R}$, $m \neq 0$, and the Euclidean hyperbolas

$$\{(x, y) \in \mathbf{R}^2 \mid (x - a)(y - b) = r\},$$

where $a, b, r \in \mathbf{R}$, $r \neq 0$. Two Euclidean points (x_1, y_1) and (x_2, y_2) are $(+)$-parallel if and only if $x_1 = x_2$, and they are $(-)$-parallel if and only if $y_1 = y_2$. Thus the vertical Euclidean lines form one type of parallel classes, the $(+)$-parallel classes, and the horizontal Euclidean lines form the other type of parallel classes, the $(-)$-parallel classes.

Conversely, we can start off with $\mathbf{R}^2$ together with all nonvertical and nonhorizontal Euclidean lines, and all Euclidean hyperbolas with asymptotes the horizontals and verticals. This essentially describes the classical flat Minkowski plane. All we have to do to reconstruct this geometry is to augment the point space $\mathbf{R}^2$ of the Euclidean plane by two parallel classes at infinity both homeomorphic to $\mathbf{S}^1 = \mathbf{R} \cup \{\infty\}$. This gives us a torus. Parallel classes are canonically extended by the point they intersect the correponding parallel class at infinity in to become the horizontals and verticals on the torus, that is, we obtain the sets

$$\{(x, y_0) \mid x \in \mathbf{S}^1\} \text{ and } \{(x_0, y) \mid y \in \mathbf{S}^1\}$$

for $x_0, y_0 \in \mathbf{S}^1$. We then one-point-compactify every one of the Euclidean lines with the point the parallel classes at infinity intersect in. Hyperbolas are extended by the same two infinite points as their asymptotes. Then all compactified Euclidean lines and all extended Euclidean hyperbolas are homeomorphic to the 1-sphere $\mathbf{S}^1$. In the above explicit description of the circles we extend the Euclidean lines by a point usually denoted by (∞, ∞). Thus

$$\{(x, mx + t) \mid x \in \mathbf{R}\} \cup \{(\infty, \infty)\}$$

describes a circle through the point (∞, ∞) and

$$\{(x, y) \in \mathbf{R}^2 \mid (x - a)(y - b) = r\} \cup \{(a, \infty), (\infty, b)\}$$

describes a circle arising from a hyperbola in this model of the classical flat Minkowski plane.

As mentioned above, the spatial description of the classical flat Minkowski plane as the geometry of nontrivial plane sections of a ruled quadric is related to the planar description in one derived plane by stereographic projection from one point of the quadric onto a plane not passing through the point of projection. Note that we cannot project those points parallel to the centre of projection.

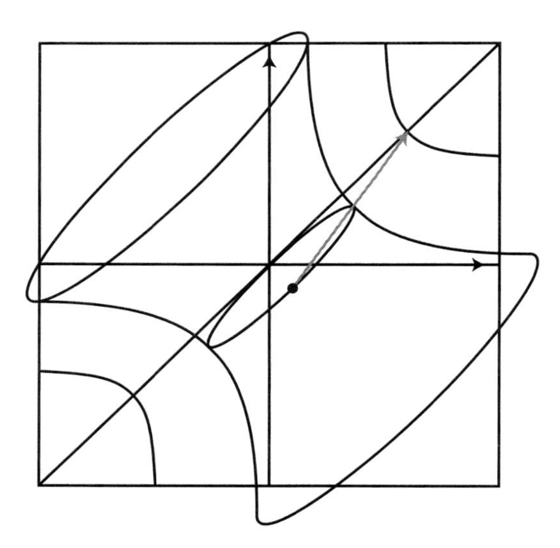

Fig. 4.2. A stereographic projection establishes the isomorphism between the geometry of plane sections and the geometry of Euclidean lines and hyperbolas

For example, if we use the quadric $\mathcal{Q}$ in Subsection 4.1.1, the stereographic projection from the point $(0 : 0 : 0 : 1)$ onto the plane $x_3 = 0$ cannot cover the points on the lines

$$\{(0 : 0 : s : t) \in \mathrm{PG}(3, \mathbf{R}) \mid s, t \in \mathbf{R}, (s, t) \neq (0, 0)\}$$

and

$$\{(0 : s : 0 : t) \in \mathrm{PG}(3, \mathbf{R}) \mid s, t \in \mathbf{R}, (s, t) \neq (0, 0)\}$$

through $(0 : 0 : 0 : 1)$. Then the stereographic projection is given by the map

$$(x_0 : x_1 : x_2 : x_3) \mapsto (x_0 : x_1 : x_2)$$

for

$$(x_0 : x_1 : x_2 : x_3) \in \mathcal{Q} \setminus \{(0 : r : s : t) \in \mathrm{PG}(3, \mathbf{R}) \mid r, s, t \in \mathbf{R}, rs = 0\}.$$

After introducing affine coordinates in the plane we project onto ($x_0 = 0$ defines the line at infinity) we can write the stereographic projection in the form

$$(1 : x : y : -xy) \mapsto (x, y)$$

for $x, y \in \mathbf{R}$. A plane E of $\mathrm{PG}(3, \mathbf{R})$ given by

$$E = \{(x_0 : x_1 : x_2 : x_3) \in \mathrm{PG}(3, \mathbf{R}) \mid ax_0 + bx_1 + cx_2 + dx_3 = 0\}$$

for some $(a : b : c : d) \in \mathrm{PG}(3, \mathbf{R})$, is tangent to the quadric $\mathcal{Q}$ if and only if $(a : b : c : d) \in \mathcal{Q}$, that is, $ad + bc = 0$. Furthermore, this plane passes through the above point of projection if and only if $d = 0$. The stereographic projection maps $E \cap \mathcal{Q}$ to

$$\{(u, v) \in \mathbf{R}^2 \mid a + bu + cv - duv = 0\}.$$

This yields a Euclidean line for $d = 0$ and a Euclidean hyperbola with horizontal and vertical asymptotes for $d \neq 0$.

Figure 4.2 illustrates what is happening. Note that the conic through the point we project from is indeed mapped onto a Euclidean line and that the contour of the hyperboloid, which is a conic not through this special point, is mapped onto a Euclidean hyperbola with one horizontal and one vertical asymptote.

4.1.3 The Pseudo-Euclidean Geometry

In this subsection we imitate the construction of the classical flat Möbius plane as the collection of all Euclidean lines and Euclidean circles. To this end, we replace the Euclidean distance between points by the so-called pseudo-Euclidean distance; see Benz [1973] I.4.1.

The *pseudo-Euclidean distance* between two points $(x_1, y_1), (x_2, y_2)$ in $\mathbf{R}^2$ is given by

$$d((x_1, y_1), (x_2, y_2)) = (x_2 - x_1)^2 - (y_2 - y_1)^2.$$

A *pseudo-Euclidean circle* with centre (a, b) and radius $r \neq 0$ is the set

$$\begin{aligned} & \{(x, y) \in \mathbf{R}^2 \mid d((x, y), (a, b)) = r\} \\ = \;& \{(x, y) \in \mathbf{R}^2 \mid x^2 - y^2 - 2ax + 2by + a^2 - b^2 - r = 0\}. \end{aligned}$$

From this we see that pseudo-Euclidean circles have the form

$$\{(x, y) \in \mathbf{R}^2 \mid x^2 - y^2 + \beta x + \gamma y + \delta = 0\},$$

where $\beta^2 - \gamma^2 \neq 4\delta$. Indeed, this set is the pseudo-Euclidean circle with centre $(-\frac{\beta}{2}, \frac{\gamma}{2})$ and radius $\frac{\beta^2 - \gamma^2}{4} - \delta$. In this way we obtain all Euclidean hyperbolas with asymptotes being Euclidean lines of slopes ± 1.

We now extend the notion of pseudo-Euclidean circles by also allowing all Euclidean lines of slopes $\neq \pm 1$. Clearly, a common description of both types of pseudo-Euclidean circles is given by

$$\{(x, y) \in \mathbf{R}^2 \mid a(x^2 - y^2) + bx + cy + d = 0\},$$

where $a, b, c, d \in \mathbf{R}$, $b^2 - c^2 \neq 4ad$. Note that the condition $b^2 - c^2 \neq 4ad$ excludes Euclidean lines of slopes ± 1. (For Euclidean lines we have $a = 0$ so that $|b| \neq |c|$.)

In order to obtain the full geometry we have to find the pseudo-Euclidean closure of the geometry of pseudo-Euclidean circles on $\mathbf{R}^2$. We can do this by augmenting the point set $\mathbf{R}^2$ by a symbol ∞ and all Euclidean lines of slopes ± 1. We then adjoin ∞ to all Euclidean lines and a hyperbola is extended by its two asymptotes.

Clearly, the above model and the model described in Subsection 4.1.2 are isomorphic. All we have to do is to rotate everything by $45°$. One can also give a spatial model analogous to the one described in Subsection 4.1.1. Instead of the quadric $\mathcal{Q}$ we use the quadric

$$\mathcal{Q}' = \{(x_0 : x_1 : x_2 : x_3) \in \mathrm{PG}(3, \mathbf{R}) \mid x_0^2 + x_1^2 - x_2^2 - x_3^2 = 0\}.$$

As in Subsection 4.1.2 both models are then related by stereographic projection from a point of $\mathcal{Q}'$ onto a plane not passing through this point.

4.1.4 Pentacyclic Coordinates

Let $\tilde{\mathcal{Q}}$ be the nondegenerate quadric in $\mathrm{PG}(4,\mathbf{R})$ defined by

$$\tilde{\mathcal{Q}} = \{(x_0 : x_1 : x_2 : x_3 : x_4) \in \mathrm{PG}(4,\mathbf{R}) \mid x_0x_3 + x_1x_2 + x_4^2 = 0\},$$

where again $(x_0 : x_1 : x_2 : x_3 : x_4)$ are the homogeneous coordinates of the point corresponding to the 1-dimensional subspace spanned by the nonzero vector $(x_0, x_1, x_2, x_3, x_4) \in \mathbf{R}^5$. We intersect $\tilde{\mathcal{Q}}$ with hyperplanes

$$\begin{aligned} E_{\mathbf{a}} &= \{(x_0 : x_1 : x_2 : x_3 : x_4) \in \mathrm{PG}(4,\mathbf{R}) \mid \\ &\qquad a_0x_0 + a_1x_1 + a_2x_2 + a_3x_3 + a_4x_4 = 0\}, \end{aligned}$$

where $\mathbf{a}$ is the point $\mathbf{a} = (a_0 : a_1 : a_2 : a_3 : a_4)$. For $\mathbf{e} = (0:0:0:0:1)$ the intersection of $E_{\mathbf{e}}$ and $\tilde{\mathcal{Q}}$ is

$$E_{\mathbf{e}} \cap \tilde{\mathcal{Q}} = \{(x_0 : x_1 : x_2 : x_3 : 0) \in \mathrm{PG}(4,\mathbf{R}) \mid x_0x_3 + x_1x_2 = 0\},$$

that is, the ruled quadric $\mathcal{Q}$ introduced in Subsection 4.1.1. We thus have the point space of the classical flat Minkowski plane.

We further intersect $E_{\mathbf{e}} \cap \tilde{\mathcal{Q}}$ with hyperplanes $E_{\mathbf{a}}$ for $\mathbf{a} \in \tilde{\mathcal{Q}} \setminus E_{\mathbf{e}}$ to obtain the circles. Clearly, $E_{\mathbf{e}} \cap \tilde{\mathcal{Q}} \cap E_{\mathbf{a}}$ is the intersection of the ruled quadric

$$\mathcal{Q} \simeq E_{\mathbf{e}} \cap \tilde{\mathcal{Q}} \subset E_{\mathbf{e}}$$

in 3-dimensional projective space $E_{\mathbf{e}} \simeq \mathrm{PG}(3,\mathbf{R})$ with the plane $E_{\mathbf{e}} \cap E_{\mathbf{a}}$ of $E_{\mathbf{e}}$. However,

$$\begin{aligned} \tilde{\mathcal{Q}} \setminus E_{\mathbf{e}} &= \{(x_0 : x_1 : x_2 : x_3 : 1) \in \mathrm{PG}(4,\mathbf{R}) \mid x_0x_3 + x_1x_2 = 1\} \\ &\simeq \{(x_0, x_1, x_2, x_3) \in \mathbf{R}^4) \mid x_0x_2 + x_1x_2 = 1\}. \end{aligned}$$

This set is connected so that, in this way, we obtain only half the circles of the classical flat Minkowski plane, that is, not every plane of $E_{\mathbf{e}}$ is of the form $E_{\mathbf{e}} \cap E_{\mathbf{a}}$ for some $\mathbf{a} \in \tilde{\mathcal{Q}} \setminus E_{\mathbf{e}}$. In order to get the missing other half we 'double up' the points of $\tilde{\mathcal{Q}} \setminus E_{\mathbf{e}}$ and add the images of the circles in the first half under the involutory collineation i of $\mathrm{PG}(4,\mathbf{R})$ given by

$$i : (x_0 : x_1 : x_2 : x_3 : x_4) \mapsto (x_0 : x_1 : -x_2 : -x_3 : x_4);$$

compare Schroth [1995a] Chapter 5. Note that i takes a point $\mathbf{a} \in \tilde{\mathcal{Q}} \setminus E_{\mathbf{e}}$ to a point $\mathbf{b}$ not belonging to $\tilde{\mathcal{Q}}$. If

$$\begin{aligned} \mathbf{a} &= (a_0 : a_1 : a_2 : a_3 : 1), \\ \mathbf{b} &= (b_0 : b_1 : b_2 : b_3 : 1), \end{aligned}$$

then

$$b_0b_3 + b_1b_2 = -(a_0a_3 + a_1a_2) = -1.$$

Hence all planes of $E_{\mathbf{e}}$ that are not tangent to $\mathcal{Q}$ are covered.

More generally, one can use any point $\mathbf{p} = (p_0 : p_1 : p_2 : p_3 : p_4)$ such that $p_0p_3 + p_1p_2 + p_4^2 > 0$ instead of $\mathbf{e}$ and an involutory collineation i of $\mathrm{PG}(4, \mathbf{R})$ that leaves the quadric $\tilde{\mathcal{Q}} \cap E_{\mathbf{p}}$ invariant and fixes pointwise two nonintersecting lines in $E_{\mathbf{p}}$. Thus the point space is $E_{\mathbf{p}} \cap \tilde{\mathcal{Q}}$ and circles are of the form $E_{\mathbf{p}} \cap \tilde{\mathcal{Q}} \cap E_{\mathbf{q}}$ or $i(E_{\mathbf{p}} \cap \tilde{\mathcal{Q}} \cap E_{\mathbf{q}})$ for $\mathbf{q} \in \tilde{\mathcal{Q}} \setminus E_{\mathbf{P}}$.

Points and circles of the Minkowski plane are both represented by points of $\tilde{\mathcal{Q}}$. Points of the Minkowski plane are the points of $\tilde{\mathcal{Q}}$ that are orthogonal to a fixed point $\mathbf{p} \in \tilde{\mathcal{Q}}$. Circles are in two-to-one correspondence to the remaining points of $\tilde{\mathcal{Q}}$. A point $\mathbf{u} \in \tilde{\mathcal{Q}}$ is incident with a circle $\mathbf{v} \in \tilde{\mathcal{Q}}$ if and only if $\mathbf{u}$ and $\mathbf{v}$ are orthogonal or $i(\mathbf{u})$ and $\mathbf{v}$ are orthogonal.

4.1.5 The Geometry of the Group of Fractional Linear Maps

Let Ξ be the 3-dimensional Lie group $\mathrm{PGL}_2(\mathbf{R})$ acting on $\mathbf{S}^1$ as the group of fractional linear maps; see p. 112 for the definition of (the group of) fractional linear maps. The group Ξ is sharply 3-transitive on $\mathbf{S}^1$, that is, given three pairwise distinct elements $x_1, x_2, x_3 \in \mathbf{S}^1$ and another three pairwise distinct elements $y_1, y_2, y_3 \in \mathbf{S}^1$, there is precisely one $\xi \in \Xi$ such that $\xi(x_j) = y_j$ for $j = 1, 2, 3$. This implies that the graphs $\{(x, \xi(x)) \mid x \in \mathbf{S}^1\}$ for all $\xi \in \Xi$ form a system of circles on the torus that satisfies Axiom B1. It can easily be verified that Axiom B2 is also satisfied, so that we have a flat Minkowski plane. Since

$$\frac{ax+b}{cx+d} = \frac{a}{d}x + \frac{b}{d}$$

describes a Euclidean line for $c = 0$ and

$$\frac{ax+b}{cx+d} = \frac{a}{c} - \frac{ad-bc}{c^2}\frac{1}{x+(d/c)}$$

describes a Euclidean hyperbola for $c \neq 0$, we see that we just obtain the classical flat Minkowski plane.

Obviously, $\Xi \times \Xi$ operates on $\mathbf{S}^1 \times \mathbf{S}^1$ as a group of automorphisms of the classical flat Minkowski plane by $(\xi_1, \xi_2)(x, y) = (\xi_1(x), \xi_2(y))$. In fact, $\Xi \times \Xi$ is flag-transitive. The group of automorphisms

$$\{(id, \xi) \mid \xi \in \Xi\}$$

is sharply transitive on the set of circles, and the stabilizer

$$\{(\xi, \xi) \mid \xi \in \Xi\}$$

of the diagonal circle $\{(x, x) \mid x \in \mathbf{R} \cup \{\infty\}\}$ is transitive on this circle. In fact, $\Xi \times \Xi$ and the automorphism $(x, y) \mapsto (y, x)$ generate the full automorphism group of the classical flat Minkowski plane.

THEOREM 4.1.1 (Flag-Transitive Automorphism Group) *The automorphism group of the classical flat Minkowski plane is flag-transitive, that is, transitive on the incident point–circle pairs of the classical flat Minkowski plane.*

Conversely, we can start with the group $\Gamma = \Xi \times \Xi$. The group Γ in fact determines the geometry. The stabilizer $\Gamma_{(\infty,\infty)}$ of the point (∞, ∞) is

$$\left\{ \left(\begin{pmatrix} a & b \\ 0 & d \end{pmatrix}, \begin{pmatrix} a' & b' \\ 0 & d' \end{pmatrix} \right) \,\middle|\, a, b, d, a', b', d' \in \mathbf{R}, ad = a'd' = 1 \right\}$$

and the stabilizer Γ_C of the diagonal circle $C = \{(x, x) \mid x \in \mathbf{R} \cup \{\infty\}\}$ is

$$\Gamma_C = \{(\xi, \xi) \mid \xi \in \Xi\}.$$

Since Γ is flag-transitive we can identify the points with the cosets of $\Gamma_{(\infty,\infty)}$ in Γ and the circles with the cosets of Γ_C in Γ. Furthermore, a point $\alpha\Gamma_{(\infty,\infty)}$ is on the circle $\beta\Gamma_C$ if and only if the intersection of the two cosets is nonempty; see Higman–McLaughlin [1961].

Note, however, that Γ is not connected. Indeed, Γ has four connected components, the cosets of $\Gamma^1 = \mathrm{PSL}_2(\mathbf{R}) \times \mathrm{PSL}_2(\mathbf{R})$. Furthermore, the stabilizers $\Gamma_{(\infty,\infty)}$ and Γ_C, as above, are not connected either. This makes it complicated to abstractly identify these subgroups in Γ. If we use $\Sigma = \Gamma^1$ instead, then $\Sigma_{(\infty,\infty)}$ and Σ_C are now both closed connected subgroups of respective dimensions 4 and 3. However, Σ is no longer flag-transitive. Nevertheless, Σ has only two orbits in the flag space. In fact, the two orbits geometrically are rather independent of each other; see Subsection 4.3.1.

We can also represent the classical flat Minkowski plane by using Ξ as its circle set. Of course, circles are the elements of Ξ. Points are the double cosets of the stabilizer

$$\Xi_\infty = \{x \mapsto ax + b \mid a, b \in \mathbf{R}, a \neq 0\}$$

of the point $\infty \in \mathbf{S}^1$. Note that the double coset $\xi_1 \Xi_\infty \xi_2^{-1}$ is the set of all transformations $\xi \in \Xi$ such that $\xi(\xi_2(\infty)) = \xi_1(\infty)$. Two points $\xi_1\Xi_\infty\xi_2$ and $\xi_1'\Xi_\infty\xi_2'$, where $\xi_1, \xi_2, \xi_1', \xi_2' \in \Xi$, are (+)-parallel if and only if the left cosets $\xi_1\Xi_\infty$ and $\xi_1'\Xi_\infty$ coincide. The points are (−)-parallel if and only if the right cosets $\Xi_\infty\xi_2$ and $\Xi_\infty\xi_2'$ coincide. The point $\xi_1\Xi_\infty\xi_2^{-1}$ is incident with the circle $\xi \in \Xi$ if and only if $\xi \in \xi_1\Xi_\infty\xi_2^{-1}$. With this definition Axiom B1 again is just a consequence of the fact that Ξ is sharply 3-transitive on $\mathbf{S}^1$.

4.1.6 The Geometry of Chains

Every 2-dimensional ring extension of $\mathbf{R}$ is isomorphic to one of the rings $R_p = \mathbf{R}[X]/p(X)$, where $p(X)$ is a quadratic polynomial over $\mathbf{R}$. We consider the projective line $\mathrm{PG}(1, R_p)$ over the extension ring R_p; see below for the construction of this projective line over the special ring of anormal-complex numbers. The associated chain geometry has as points the points of $\mathrm{PG}(1, R_p)$ and as blocks the images of the real projective line $\mathrm{PG}(1, \mathbf{R}) \subset \mathrm{PG}(1, R_p)$ under the projective group $\mathrm{PGL}_2(R_p)$; see Herzer [1995]. If $p(X)$ is irreducible, then R_p is a field, which leads to the classical flat Möbius plane; see Subsection 3.1.4. If $p(X)$ is the square of a linear polynomial, we obtain the classical flat Laguerre plane; see Subsection 5.1.6.

In this subsection we consider the case that $p(X)$ is the product of two distinct linear polynomials. Up to isomorphism, we may assume that

$$p(x) = (x-1)(x+1) = x^2 - 1.$$

Then

$$\mathbf{A} = R_p = \{a + bj \mid a, b \in \mathbf{R}\},$$

where $j^2 = 1$. The algebraic operations in $\mathbf{A}$ are given by

$$(a+bj) + (c+dj) = (a+c) + (b+d)j$$

and

$$(a+bj)(c+dj) = (ac+bd) + (ad+bc)j.$$

The ring $\mathbf{A}$ is called the ring of *anormal numbers over* $\mathbf{R}$ or *anormal-complex numbers*. Clearly, $\mathbf{A}$ is commutative and contains $\mathbf{R}$ as a subring. As for the usual complex numbers, $\mathbf{A}$ admits precisely two ring automorphisms that fix $\mathbf{R}$ globally. These are the identity and the involution

$$a + bj \mapsto \overline{a+bj} = a - bj.$$

Note that a ring automorphism that fixes $\mathbf{R}$ globally induces on $\mathbf{R}$ an automorphism of $\mathbf{R}$ and thus must be the identity on $\mathbf{R}$. Furthermore, apart from $\mathbf{A}$ and $\{0\}$, the ring $\mathbf{A}$ only has the two ideals

$$\begin{aligned} \mathcal{J}_+ &= \mathbf{R}(1+j) = \{r(1+j) \mid r \in \mathbf{R}\}, \\ \mathcal{J}_- &= \mathbf{R}(1-j) = \{r(1-j) \mid r \in \mathbf{R}\}. \end{aligned}$$

Let $\mathcal{R} = \mathbf{A} \setminus (\mathcal{J}_+ \cup \mathcal{J}_-)$. The elements of $\mathcal{R}$ are precisely those elements that are invertible in $\mathbf{A}$.

The invertible 2×2 matrices with entries in $\mathbf{A}$ form a group $\mathrm{GL}_2(\mathbf{A})$ with respect to the usual matrix multiplication. A 2×2 matrix

$$\begin{pmatrix} a & b \\ c & d \end{pmatrix}$$

is invertible if and only if its determinant

$$\begin{vmatrix} a & b \\ c & d \end{vmatrix} = ad - bc$$

is an element of $\mathcal{R}$.

We now consider the canonical free module $\mathbf{A}^2$ of rank 2 over $\mathbf{A}$. The group $\mathrm{GL}_2(\mathbf{A})$ acts on $\mathbf{A}^2$ by multiplication on the left. We further say that an element

$$\begin{pmatrix} x \\ y \end{pmatrix} \in \mathbf{A}^2 = \mathbf{A} \times \mathbf{A}$$

is *unimodular* if the only ideal of $\mathbf{A}$ containing both x and y equals $\mathbf{A}$. Note that for $r \in \mathcal{R}$

$$r \begin{pmatrix} x \\ y \end{pmatrix} = \begin{pmatrix} rx \\ ry \end{pmatrix}$$

is unimodular, too. It readily follows that the unimodular elements are precisely those that can occur as a column of a matrix in $\mathrm{GL}_2(\mathbf{A})$.

The projective line $\mathrm{PG}(1, \mathbf{A})$ over the anormal-complex numbers then consists of all free cyclic submodules of $\mathbf{A}^2$. Each such submodule is generated by a unimodular $v \in \mathbf{A}^2$. We therefore can identify $\mathrm{PG}(1, \mathbf{A})$ with the collection of all $\mathcal{R}v$, where $v \in \mathbf{A}^2$ is unimodular. Explicitly, the elements of $\mathrm{PG}(1, \mathbf{A})$ can be written as follows.

$$\mathcal{R}\begin{pmatrix} 1 \\ a \end{pmatrix} \quad \text{for} \quad a \in \mathbf{A},$$

$$\mathcal{R}\begin{pmatrix} 1+j \\ 1-j \end{pmatrix}, \mathcal{R}\begin{pmatrix} 1+j \\ r \end{pmatrix} \quad \text{for} \quad r \in \mathbf{R}, r \neq 0,$$

$$\mathcal{R}\begin{pmatrix} 1-j \\ 1+j \end{pmatrix}, \mathcal{R}\begin{pmatrix} 1-j \\ r \end{pmatrix} \quad \text{for} \quad r \in \mathbf{R}, r \neq 0,$$

$$\mathcal{R}\begin{pmatrix} 0 \\ 1 \end{pmatrix}.$$

Two points

$$\mathcal{R}\begin{pmatrix} u_1 \\ v_1 \end{pmatrix}, \mathcal{R}\begin{pmatrix} u_2 \\ v_2 \end{pmatrix}$$

of $\mathrm{PG}(1, \mathbf{A})$ are $(+)$-parallel if and only if

$$\begin{vmatrix} u_1 & u_2 \\ v_1 & v_2 \end{vmatrix} = u_1 v_2 - u_2 v_1$$

is an element of $\mathcal{J}_+$. Note that this condition is independent of the representatives of the points. The two points are $(-)$-parallel if and only if

$$\begin{vmatrix} u_1 & u_2 \\ v_1 & v_2 \end{vmatrix} \in \mathcal{J}_-.$$

For example,

$$\mathcal{R}\begin{pmatrix} 0 \\ 1 \end{pmatrix}$$

is $(+)$-parallel to all points

$$\mathcal{R}\begin{pmatrix} 1+j \\ a \end{pmatrix}$$

and $(-)$-parallel to all points

$$\mathcal{R}\begin{pmatrix} 1-j \\ a \end{pmatrix}.$$

Two points

$$\mathcal{R}\begin{pmatrix} 1 \\ a \end{pmatrix}, \mathcal{R}\begin{pmatrix} 1 \\ b \end{pmatrix}$$

for $a, b \in \mathbf{A}$ are $(+)$-parallel if and only if $b-a$ is a real multiple of $1+j$ so that the $(+)$-parallel classes on $\mathbf{A} \simeq \mathbf{R}^2$ correspond to the Euclidean lines of $\mathbf{R}^2$ of slope 1. Likewise, the $(-)$-parallel classes on $\mathbf{A} \simeq \mathbf{R}^2$ correspond to the Euclidean lines of $\mathbf{R}^2$ of slope -1.

Given a fixed basis $\{u, v\}$ of $\mathbf{A}^2$, we have the canonical embedding $\beta_{u,v}$ of the real projective line $\mathrm{PG}(1, \mathbf{R})$ into $\mathrm{PG}(1, \mathbf{A})$ given by

$$\mathbf{R}\begin{pmatrix} s \\ t \end{pmatrix} \mapsto \mathcal{R}(su + tv).$$

Note that $su + tv$ is again unimodular. The image of $\beta_{u,v}$, denoted by $C(u, v)$, is called an **R**-*chain of* PG$(1, \mathbf{A})$, that is,

$$C(u, v) = \left\{ \mathcal{R}(su + tv) \;\middle|\; \mathbf{R}\begin{pmatrix} s \\ t \end{pmatrix} \in \mathrm{PG}(1, \mathbf{A}) \right\}.$$

The *chain geometry* $\Sigma(\mathbf{R}, \mathbf{A})$ is the incidence structure whose point set is the projective line PG$(1, \mathbf{A})$ over **A** and whose blocks are the chains of PG$(1, \mathbf{A})$, where incidence between points and chains is defined by containment.

We can also use the group $\mathrm{GL}_2(\mathbf{A})$ to obtain the chains. The canonical embedding from **R** into **A** extends to a natural embedding of PG$(1, \mathbf{R})$ into PG$(1, \mathbf{A})$ and this embedded real projective line is moved around by the group $\mathrm{GL}_2(\mathbf{A})$. As in the case of the projective line over **R**, each 'scalar matrix'

$$\begin{pmatrix} r & 0 \\ 0 & r \end{pmatrix},$$

where $r \in \mathcal{R}$, fixes every point of PG$(1, \mathbf{A})$ so that, in effect, the projective group $\mathrm{PGL}_2(\mathbf{A})$, that is, the quotient group of $\mathrm{GL}_2(\mathbf{A})$ over its centre, acts on points and chains.

By construction, the group $\mathrm{PGL}_2(\mathbf{A})$ is transitive on the set of chains. The stabilizer of the chain PG$(1, \mathbf{R})$ is obtained from the matrices in $\mathrm{GL}_2(\mathbf{R})$ (canonically embedded into $\mathrm{GL}_2(\mathbf{A})$). This stabilizer is transitive on PG$(1, \mathbf{R})$.

THEOREM 4.1.2 (Flag-Transitive Automorphism Group) *The group* $\mathrm{PGL}_2(\mathbf{A})$ *is the automorphism group of* $\Sigma(\mathbf{R}, \mathbf{A})$. *Furthermore, the group* $\mathrm{PGL}_2(\mathbf{A})$ *is flag-transitive, that is, transitive on the incident point–chain pairs of* $\Sigma(\mathbf{R}, \mathbf{A})$.

To see that $\Sigma(\mathbf{R}, \mathbf{A})$ essentially is the classical flat Minkowski plane note that Euclidean hyperbolas can be described by anormal-complex numbers as follows; see Benz [1973] I.4.3.

Let $a, b, c, d \in \mathbf{R}$ such that $b^2 - c^2 \neq 4ad$. Then the set of points

$$\mathcal{R}\begin{pmatrix} u \\ v \end{pmatrix}$$

of PG$(1, \mathbf{A})$ that satisfy the equation

$$(u \; v)\begin{pmatrix} a & \frac{b+cj}{2} \\ \frac{b-cj}{2} & d \end{pmatrix}\begin{pmatrix} \bar{u} \\ \bar{v} \end{pmatrix} = 0$$

corresponds to the points of the Euclidean hyperbola

$$\{(x,y) \in \mathbf{R}^2 \mid a(x^2 - y^2) + bx + cy + d = 0\}$$

including its two asymptotes of slopes ± 1.

We therefore obtain just the geometry of pseudo-Euclidean circles described in Subsection 4.1.3.

4.1.7 The Beck Model

In this model points of the classical flat Minkowski plane are represented by *spears*, that is oriented Euclidean lines. Given two distinct points p and q on the unit circle $\mathbf{S}^1$, the oriented Euclidean line 'from p to q' determines the spear (p,q); see Figure 4.3. The unoriented Euclidean line tangent to $\mathbf{S}^1$ at p is also called a spear, denoted by (p,p). The collection of all spears forms the point set of the geometry. The point set becomes homeomorphic to $\mathbf{S}^1 \times \mathbf{S}^1$ when the spears that are tangent lines are fitted in appropriately. Two spears are (+)-parallel if and only if they pass through the same point of the unit circle and point away from this point. They are (−)-parallel if and only if they point to the same point on $\mathbf{S}^1$. Circles of the geometry are of the form $\{(p, \sigma(p)) \mid p \in \mathbf{S}^1\}$ for $\sigma \in \mathrm{PGL}_2(\mathbf{R})$.

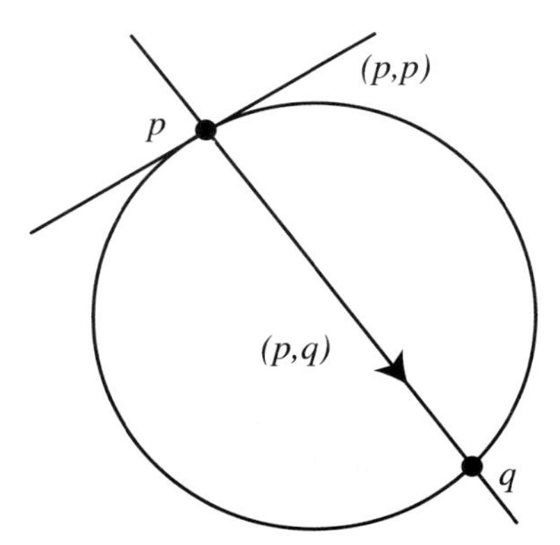

Fig. 4.3. The points of the Beck model are the tangents and oriented secant lines of the unit circle

Circles in this model can be geometrically described using points or certain oriented arcs in $\mathbf{R}^2$; see Benz [1973] I.4.9 for details. In the following we only give a number of examples that illustrate the essentially different types of circles; see Figure 4.4.

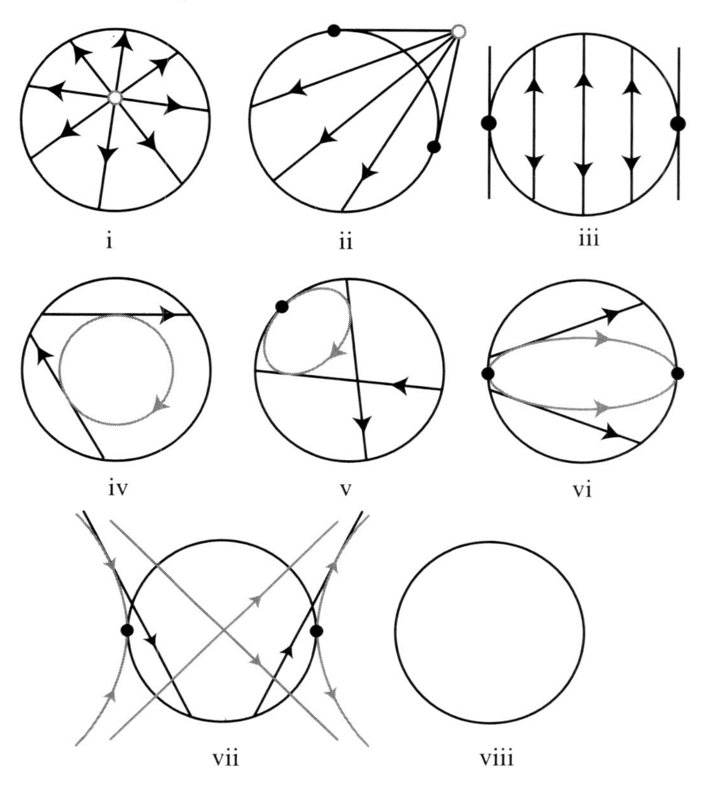

Fig. 4.4. Examples of circles in the Beck Model

(i) Each point of $\mathbf{R}^2$ inside the unit circle $\mathbf{S}^1$ determines a circle consisting of all the spears through it. This circle corresponds to a fixed-point-free involution in $\mathrm{PGL}_2(\mathbf{R})$.

(ii) Each point of $\mathbf{R}^2$ outside the unit circle determines a circle consisting of all the spears through this point that intersect $\mathbf{S}^1$. Note that this includes two tangent lines. This circle corresponds to an involution in $\mathrm{PGL}_2(\mathbf{R})$ with two fixed points. The same is true for the circles of the following type.

(iii) Each parallel class of Euclidean lines determines a circle consisting of all spears that intersect $\mathbf{S}^1$ and whose underlying Euclidean lines are lines of the parallel class.

This takes care of all involutions in $\mathrm{PGL}_2(\mathbf{R})$. In fact, it shows that the involutions in this group are precisely the bundle involutions of the unit circle in the classical flat projective plane; see Subsection 2.10.2.

(iv) Each oriented Euclidean circle E inside the unit circle, with centre at the origin, gives rise to a circle consisting of all the spears tangent to E whose orientation agrees with that of E. This circle corresponds to a fixed-point-free noninvolutory permutation in $\mathrm{PGL}_2(\mathbf{R})$.

(v) Here E is a certain oriented Euclidean ellipse inscribed in the unit circle and touching it at exactly one point p. It corresponds to the circle consisting of all spears tangent to E whose orientation agrees with that of E.

(vi) Here E is a certain Euclidean ellipse inscribed in the unit circle and touching it at the two points p_1 and p_2. The two arcs connecting these two points are oriented in the same way. The ellipse E corresponds to the circle consisting of the two tangents at the points p_1 and p_2 and all spears tangent to either one of the two arcs and sharing the orientation with the arcs they touch.

(vii) Here E is a Euclidean hyperbola H circumscribed to the unit circle and touching it at two points p_1 and p_2. The four arcs of the hyperbola determined by p_1 and p_2 are oriented consistent with an orientation of the two asymptotes. The corresponding circle consists of the two oriented asymptotes, the two tangents at the two points, and all spears touching one of the four arcs and sharing the orientation with the arc they touch.

(viii) All Euclidean tangents of the unit circle form another circle in this model. This circle corresponds to the identity in $\mathrm{PGL}_2(\mathbf{R})$.

4.2 Derived Planes and Topological Properties

In this section we define the two standard representations of toroidal circle planes. The first generalizes the representation of the classical plane as a geometry of Euclidean lines and hyperbolas, the second the representation as a 3-transitive set of homeomorphisms of the circle to itself. Following this we consider the continuity of the geometric operations, some questions of coherence, and the spaces of lines and flags of a toroidal circle plane.

4.2.1 Derived $\mathbb{R}^2$-Planes

The idea of representing the classical flat Minkowski plane in the Euclidean plane can be extended as follows.

Let p be a point of a toroidal circle plane and let G and H be the parallel classes containing p. Remember that the point set of the derived plane at the point p consists of all points not parallel to p. Its lines are the circles through p that have been punctured at p and all parallel classes, except G and H, that have been punctured at their points of intersection with G and H. This is an $\mathbf{R}^2$-plane. In the case of a flat Minkowski plane we even obtain a flat affine plane, the *derived affine plane at* p. This flat affine plane extends uniquely to a flat projective plane, which we call the *derived projective plane at* p.

THEOREM 4.2.1 (Characterization via Derived Planes) *Let $\mathcal{M}$ be a geometry whose point set the torus $\mathbf{S}^1 \times \mathbf{S}^1$ is equipped with two nontrivial parallelisms the parallel classes of which are the horizontals and verticals on the torus. Furthermore, circles of $\mathcal{M}$ are graphs of homeomorphism $\mathbf{S}^1 \to \mathbf{S}^1$. Then $\mathcal{M}$ is a toroidal circle plane or a flat Minkowski plane if and only if all its derived planes are $\mathbf{R}^2$-planes or flat affine planes, respectively.*

Each derived affine plane of the classical flat Minkowski plane is isomorphic to the Euclidean plane. Essentially, we already convinced ourselves of this fact in Subsection 4.1.2. In the notation at the end of that subsection the circle that corresponds to the plane

$$E = \{(x_0 : x_1 : x_2 : x_3) \in \mathrm{PG}(3, \mathbf{R}) \mid ax_0 + bx_1 + cx_2 + dx_3 = 0\}$$

for some $(a : b : c : 0) \in \mathrm{PG}(3, \mathbf{R})$, $bc \neq 0$, through $(0 : 0 : 0 : 1)$ is taken under the stereographic projection from $(0 : 0 : 0 : 1)$ to the Euclidean line $\{(x, y) \in \mathbf{R}^2 \mid a + bx + cy = 0\}$. Obviously, every nonhorizontal and nonvertical Euclidean line can be obtained in this way. Furthermore, the horizontal and vertical Euclidean lines are covered by the parallel classes not passing though $(0 : 0 : 0 : 1)$. Since the group of collineations of $\mathrm{PG}(3, \mathbf{R})$ that leave the quadric $\mathcal{Q}$ invariant is transitive on $\mathcal{Q}$, we obtain the Euclidean plane as derived affine plane at any point of $\mathcal{Q}$.

In fact, a flat Minkowski plane that has sufficiently many points at which the derived affine planes are isomorphic to the Euclidean plane must be classical; see Theorem 4.6.3.

Just as in the case of spherical circle planes, it is possible to use results about the derived planes of toroidal circle planes to prove similar results for toroidal circle planes. Here is an important example; compare the corresponding Theorem 3.2.2 for spherical circle planes.

THEOREM 4.2.2 (Intersection of Circles) *Two circles in a toroidal circle plane intersect in at most two points. If they intersect in exactly one point, then they touch topologically. If they intersect in two points, then they intersect transversally at these points.*

4.2.2 Affine Parts

A circle C not passing through the distinguished point p intersects each of the two parallel classes through p in one point. If we remove these two points, we are left with two connected components homeomorphic to $\mathbf{R}$. We call the restriction of the circle C to the point set of the derived plane at p a *hyperbolic curve*. Clearly, this hyperbolic curve is the union of two disjoint topological arcs homeomorphic to $\mathbf{R}$, and each of the two arcs separates the point set of the derived plane into two open components. In the case of a flat Minkowski plane, a closer look at the derived projective plane at p reveals that the hyperbolic curve can be extended to a topological oval by adding two distinguished points on the line at infinity. In fact, every topological oval in the derived projective plane that arises from a circle in the Minkowski plane in this way intersects the line at infinity in the same two points. These are the infinite points on lines that come from parallel classes.

A toroidal circle plane can thus be described in one derived plane $\mathcal{A}$ at a point p by the lines of $\mathcal{A}$ and a collection of hyperbolic curves. We call the induced geometry the *affine part at* p.

As demonstrated at the end of Subsection 4.1.2, the affine part (at any point) of the classical flat Minkowski plane can be identified with the Euclidean plane and all Euclidean hyperbolas that have horizontal and vertical asymptotes. There we looked at the affine part at the special point $(0 : 0 : 0 : 1)$. The hyperbolic curves obtained were the Euclidean hyperbolas

$$\left\{(x, y) \in \mathbf{R}^2 \;\middle|\; \left(x - \frac{c}{d}\right)\left(y - \frac{b}{d}\right) = \frac{ad + bc}{d}\right\}$$

in the notation of Subsection 4.1.2. Clearly, each Euclidean hyperbola with horizontal and vertical asymptotes occurs. Since the group of collineations of $\mathrm{PG}(3, \mathbf{R})$ that leave the quadric $\mathcal{Q}$ invariant is transitive on $\mathcal{Q}$, we obtain basically the same affine part at any other point of $\mathcal{Q}$.

In Polster–Steinke [1994] Proposition 3, it was shown that if one has a flat affine plane and a collection of hyperbolic curves such that the two axioms of a Minkowski plane are satisfied, we can topologically extend the affine plane by two parallel classes at infinity and obtain a flat Minkowski plane. Inspection of the proof of this result yields that the corresponding result is true for toroidal circle planes.

4.2.3 Standard Representation and Sharply 3-Transitive Sets

We say that a toroidal circle plane is in *standard representation* if its point set is the torus $T = \mathbf{S}^1 \times \mathbf{S}^1$, its parallel classes are the horizontals and verticals on the torus, its circles are graphs of homeomorphisms $\mathbf{S}^1 \to \mathbf{S}^1$, and the set $\{(x,x) \mid x \in \mathbf{S}^1\}$ is one of its circles. Starting with any toroidal circle plane $(T, \mathcal{C})$ we can construct a standard representation of this plane as follows.

We denote the two types of parallel classes of a point p in a toroidal circle plane by $|p|_+$ and $|p|_-$, respectively. The set of all $(+)$-parallel classes or all $(-)$-parallel classes is denoted by Π^+ or Π^-, respectively. We fix a point $p_\infty \in T$ and identify both the $(+)$-parallel class $|p_\infty|_+$ and the $(-)$-parallel class $|p_\infty|_-$ of p_∞ with the 1-sphere $\mathbf{S}^1$. Then two points $p_1 = (x_1, y_1)$, $p_2 = (x_2, y_2)$ are $(+)$-parallel or $(-)$-parallel if and only if $x_1 = x_2$ or $y_1 = y_2$, respectively. With this identification, each circle $C \in \mathcal{C}$ is uniquely determined by the map

$$f_C : |p_\infty|_+ \to |p_\infty|_- : z \mapsto |\, (|z|_- \cap C)\, |_+ \cap |p_\infty|_-.$$

More precisely,

$$C = \{(x, f_C(x)) \mid x \in \mathbf{S}^1\}$$

is just the graph of f_C.

Furthermore, if we choose a circle C_1 through p_∞ and identify the parallel classes $|p_\infty|_+$ and $|p_\infty|_-$ via C_1, then the C_1 describing homeomorphism is just the identity. Hence, without loss of generality, we may assume that the diagonal $\{(x,x) \mid x \in \mathbf{S}^1\}$ is a circle of the toroidal circle plane.

Remember that a set Σ of permutations of a set S is called sharply 3-transitive, if given two triples (x_1, x_2, x_3) and (y_1, y_2, y_3) of pairwise distinct elements of S, there is a unique permutation p in the set Σ such that $p(x_i) = y_i$ for $i = 1, 2, 3$; see Subsection 2.10.1.

THEOREM 4.2.3 (Characterization via Sharply 3-Transitive) *Let Σ be a set of homeomorphisms $\mathbf{S}^1 \to \mathbf{S}^1$ containing the identity and let $\mathcal{M}$ be a geometry whose point set is the torus $\mathbf{S}^1 \times \mathbf{S}^1$ and whose circles are the graphs of the elements of Σ. Then $\mathcal{M}$ is a toroidal circle plane if and only if Σ is a sharply 3-transitive set.*

Remember that the group $\mathrm{PGL}_2(\mathbf{R})$ with its standard action on $\mathbf{S}^1$ corresponds to the classical flat Minkowski plane. In fact, if Σ is a sharply 3-transitive group of homeomorphisms of the circle, then Σ contains the group $\mathrm{PSL}_2(\mathbf{R})$; this is a consequence of Brouwer's Theorem A2.3.8. It follows that the group Σ is isomorphic to $\mathrm{PGL}_2(\mathbf{R})$ and that the associated toroidal circle plane is classical.

THEOREM 4.2.4 (3-Transitive Group Equals Classical) *Let Σ be a set of homeomorphisms $\mathbf{S}^1 \to \mathbf{S}^1$ such that the graphs of its elements form the circle set of a toroidal circle plane $\mathcal{M}$ in standard form. Then the plane $\mathcal{M}$ is classical if and only if Σ is a group.*

A simple modification of the group $\mathrm{PGL}_2(\mathbf{R})$ leads to examples where the set Σ is not a group; see Subsection 4.3.1.

4.2.4 Continuity of the Geometric Operations

As in the case of flat linear spaces and spherical circle planes, the topology on the circle set of a toroidal circle plane is uniquely determined by the topology on its point set; compare Schenkel [1980] Satz 4.4. More precisely we have the following; see Schenkel [1980] 1.20.

THEOREM 4.2.5 (Topology Torus $\to$ Topology Circle Space) *The topology of the circle space of a toroidal circle plane is the finest topology with respect to which the map α of joining three pairwise nonparallel points by a circle is continuous. The sets of circles $\alpha(U_1, U_2, U_3)$, where U_1, U_2, U_3 are nonempty open subsets of the torus $\mathbf{S}^1 \times \mathbf{S}^1$ such that any three points $p_i \in U_i$, $i = 1, 2, 3$, are pairwise nonparallel, form a basis for the topology of the circle space.*

We can even obtain neighbourhoods of a circle of the form $\alpha(U_1, U_2, U_3)$ for much smaller sets U_1, U_2, and U_3.

THEOREM 4.2.6 (Neighbourhood Bases for Circles) *Let C be a circle of a toroidal circle plane $\mathcal{M}$. For three distinct points p_1, p_2, p_3 on C let U_i be a neighbourhood of p_i on the $(+)$-parallel class of p_i. Then the set $\alpha(U_1, U_2, U_3)$ is a neighbourhood of C and the sets of this type form a neighbourhood basis of C.*

Using the standard representation of toroidal circle planes and the description of circles as graphs of homeomorphisms of $\mathbf{S}^1$ to itself, the topology of the circle set can also be obtained from the topology of uniform convergence on $\mathbf{S}^1$; see Schenkel [1980] 3.4.

PROPOSITION 4.2.7 (Parallel Metric on Circle Space) *Let $\mathcal{M} = (\mathbf{S}^1 \times \mathbf{S}^1, \mathcal{C})$ be a toroidal circle plane in standard representation. Then*

$$d(C, D) = \sup\{|f_C(x) - f_D(x)| \mid x \in \mathbf{S}^1\},$$

where f_C and f_D are the homeomorphisms whose graphs are the circles C and D, respectively, and $|u-v|$ denotes the Euclidean distance between u and v, is a metric on $\mathcal{C}$ that induces the natural topology on $\mathcal{C}$.

Conversely, we can obtain the topology of the torus $\mathbf{S}^1 \times \mathbf{S}^1$ from the topology of the circle set.

LEMMA 4.2.8 (Topology Circle Space $\rightarrow$ Topology Torus) *Let $\mathcal{K}$ be a nonempty open subset of the circle space of a toroidal circle plane. Then the set $\tilde{\mathcal{K}} = \bigcup_{C \in \mathcal{K}} C$ of all points on circles in $\mathcal{K}$ is open in $\mathbf{S}^1 \times \mathbf{S}^1$. Furthermore the sets of the form $\tilde{\mathcal{K}}$ form a subbasis for the topology of $\mathbf{S}^1 \times \mathbf{S}^1$.*

In the following we always assume that both point and circle spaces of a toroidal circle plane carry their natural topologies.

The topology of the point space is, of course, the usual Euclidean topology of the torus but one can also describe it geometrically. To this end let C_1, C_2, C_3 be three circles that pairwise intersect in two points and such that $C_1 \cap C_2 \cap C_3$ consists of a single point q. Let the *triode* $\mathcal{D}(C_1, C_2, C_3; q)$ be the collection of all points in the interior of the triangle formed by the lines induced by C_1, C_2, C_3 in the derived plane at q.

PROPOSITION 4.2.9 (Triodes Form Neighbourhood Basis) *The triodes for a fixed q that contain a point p that is not parallel to q form a basis for the neighbourhoods of p.*

In fact, one can generalize the notion of a triode and replace one or two of the circles by parallel classes so that in the derived plane one or two sides of the triangle are vertical or horizontal lines (that is, they come from parallel classes of the toroidal circle plane).

It now follows that the other geometric operations are also continuous with respect to the topologies defined above.

THEOREM 4.2.10 (Continuity) *In a toroidal circle plane the operations of*

- *joining three pairwise nonparallel points by a circle,*
- *intersecting two parallel classes of different types in a point,*
- *intersecting a parallel class and a circle in a point,*
- *intersecting two distinct circles*

are continuous on their respective domains of definition.

In a flat Minkowski plane forming the tangent circle to a circle through two nonparallel points as in Axiom B2 is also a continuous operation.

4.2.5 Topological Minkowski Planes

A topological Minkowski plane is a Minkowski plane whose sets of points and circles carry nonindiscrete topologies such that the geometric operations of joining, touching, intersecting distinct circles, intersecting circles and parallel classes, and intersecting parallel classes of different types are continuous on their domains of definition. The results of the previous subsections imply that the flat Minkowski planes are topological Minkowski planes.

As usual, we call a topological Minkowski plane an X Minkowski plane if its point space is a topological space of type X. The following result characterizes the flat Minkowski planes among the topological Minkowski planes; see Schenkel [1980].

THEOREM 4.2.11 (Topological Characterization) *The flat Minkowski planes are precisely the 2-dimensional compact connected Minkowski planes.*

It can be shown that a topological locally compact finite-dimensional Minkowski plane has dimension 0, 2, or 4; see the Buchanan–Hähl–Löwen [1980] and Löwen [1981d]. For example, one obtains a 4-dimensional Minkowski plane as in Subsections 4.1.1 and 4.1.2 by replacing the real numbers **R** by the complex numbers **C**.

A fundamental question is whether the derived affine planes of a topological Minkowski plane are topological, too. Of course, in the case of the flat Minkowski planes the answer is 'Yes' since the derived affine planes are automatically flat affine planes.

To answer the more general topological question, Schenkel [1980] considered a number of *coherence axioms* and the problem was completely solved in Steinke [1989] for finite-dimensional, locally compact connected Minkowski planes. The seven coherence axioms introduced by Schenkel deal with arbitrary topological Minkowski planes and therefore use nets over directed sets. Since we are only concerned with flat Minkowski planes, we can use sequences to formulate the coherence axioms in flat Minkowski planes. One of these coherence axioms, Axiom K1, reflects the fact that ovals induced in derived projective planes must be 'topological', or equivalently, must have the secants-tend-to-tangents property specified in Theorem 2.4.13. In flat Minkowski planes this becomes that touching is the limit of proper intersection. Axiom K1 runs as follows.

(K1) Let (C_n) be a sequence of circles, and let (x_n), (y_n), (z_n) be three sequences of points such that the following conditions are satisfied.

- The points x_n, y_n, z_n are pairwise nonparallel, $x_n, y_n \in C_n$.
- The sequences (x_n), (y_n), (z_n) converge to x, y, z, respectively, with $x = y$ nonparallel to z.
- The sequence (C_n) converges to a circle C.

Then the sequence of circles joining x_n, y_n, z_n converges to the circle passing through z that touches C at x.

Specializing points and circles yields the continuity of forming parallel lines in derived affine planes. Other coherence axioms deal with the intersection of circles and parallel classes as the circles pass through points that converge to parallel points. For example, Axiom K2 can be expressed as follows.

(K2) Let (C_n) be a sequence of circles, and (x_n), (y_n), (z_n) be three sequences of points such that the following conditions are satisfied.

- The points x_n, y_n, z_n are pairwise nonparallel.
- The sequences (x_n), (y_n), (z_n) converge to x, y, z, respectively, with x, z nonparallel and $y \neq x$ parallel to x, say y $(+)$-parallel to x.

- The sequence (C_n) converges to a circle C.
- The circle C_n and the circle D_n joining the points x_n, y_n, z_n have nonempty intersection.

Then the sequence of intersections of $C_n \cap D_n$ converges to the intersection of C with $|x|_+$ and $|z|_-$.

One calls a flat Minkowski plane *Kn-coherent* if the coherence axiom Kn is satisfied and *coherent* if all seven coherence axioms are satisfied. In a number of steps Schenkel [1980] showed that every K1-coherent flat Minkowski plane (in fact, compact connected Minkowski plane) is coherent and finally that every flat Minkowski plane is K1-coherent.

THEOREM 4.2.12 (Coherence) *Flat Minkowski planes are coherent.*

Steinke [1989] extended this result by showing that derived planes of finite-dimensional locally compact connected Minkowski planes are topological. We emphasize again that for flat Minkowski planes this is a simple consequence of the fact that flat affine planes are topological geometries.

THEOREM 4.2.13 (Derived Planes, Arcs, and Ovals) *The derived plane of a toroidal circle plane at a point p is an $\mathbf{R}^2$-plane. The restriction of the natural topology on the circle set of the toroidal circle plane to the set of circles through p coincides with the restriction of the natural topology on the line set of the derived $\mathbf{R}^2$-plane at p to the set of lines that do not correspond to parallel classes of the circle plane.*

A circle not passing through the point p turns into the union of two topological arcs in the derived plane at p. In a flat Minkowski plane such a union of arcs extends to a topological oval in the derived flat projective plane at p that intersects the line at infinity of the derived affine plane at p in the infinite points of the horizontals and verticals.

For further information about topological Minkowski planes see the references given above and at the end of Subsection 3.2.4 and Castro–Wernicke [1990].

4.2.6 Circle Space and Flag Space

From the description of parallel classes as verticals and horizontals of the torus $\mathbf{S}^1 \times \mathbf{S}^1$ it is clear that each parallel class and also the set $\Pi^{\pm}$ of all $(\pm)$-parallel classes are homeomorphic to $\mathbf{S}^1$.

THEOREM 4.2.14 (The Space of Parallel Classes) *The set of* $(\pm)$-*parallel classes of a toroidal circle plane is homeomorphic to* $\mathbf{S}^1$.

Using the fact that each circle in a toroidal circle plane is uniquely determined by its intersection with three distinct parallel classes of the same type and that the projective group $\mathrm{PGL}_2(\mathbf{R})$ operates sharply 3-transitively on $\mathbf{S}^1$, the circle space of a toroidal circle plane can be explicitly described topologically.

THEOREM 4.2.15 (The Space of Circles) *The set of circles of a toroidal circle plane is homeomorphic to the real 3-dimensional projective group* $\mathrm{PGL}_2(\mathbf{R})$, *that is, homeomorphic to* $\mathbf{S}^1 \times \mathbf{R} \times (\mathbf{R} \setminus \{0\})$.

A *flag* in a toroidal circle plane $\mathcal{M}$ is an incident point–circle pair, that is, a pair (p, C), where p is a point and C a circle of $\mathcal{M}$ such that $p \in C$. The set of all flags $\mathcal{F}$ thus is a closed subset of the Cartesian product of the point set $\mathbf{S}^1 \times \mathbf{S}^1$ and the circle set $\mathcal{C} \simeq \mathrm{PGL}_2(\mathbf{R})$. More precisely, the map $(p, C) \mapsto (|p|_+, C)$ provides a homeomorphism from $\mathcal{F}$ onto the Cartesian product $\Pi^+ \times \mathcal{C}$ of the set Π^+ of all $(+)$-parallel classes and the circle set $\mathcal{C}$.

THEOREM 4.2.16 (The Space of Flags) *The flag space* $\mathcal{F}$ *of a toroidal circle plane is a 4-dimensional manifold that has exactly two connected components. More precisely,* $\mathcal{F}$ *is homeomorphic to the topological space* $\mathbf{S}^1 \times \mathrm{PGL}_2(\mathbf{R})$.

4.3 Constructions

In this section we present some of the most important construction methods for nonclassical flat Minkowski planes and toroidal circle planes. The planes obtained here comprise the most homogeneous flat Minkowski planes.

4.3.1 The Two Halves of a Toroidal Circle Plane

Let $\mathcal{M} = (\mathbf{S}^1 \times \mathbf{S}^1, \mathcal{C})$ be a toroidal circle plane in standard representation. Let $\mathcal{C}^+$ and $\mathcal{C}^-$ be the sets of all circles in $\mathcal{C}$ that are graphs of orientation-preserving and orientation-reversing homeomorphisms $\mathbf{S}^1 \to \mathbf{S}^1$, respectively. Clearly, $\mathcal{C} = \mathcal{C}^+ \cup \mathcal{C}^-$. It turns out that these two parts are completely independent of each other, that is, we can combine parts from different toroidal circle planes to obtain another toroidal circle plane.

THEOREM 4.3.1 (Swapping Halves) *Let $\mathcal{M}_i = (\mathbf{S}^1 \times \mathbf{S}^1, \mathcal{C}_i)$, $i = 1, 2$, be two toroidal circle planes and let $\mathcal{C} = \mathcal{C}_1^+ \cup \mathcal{C}_2^-$. Then the geometry $\mathcal{M} = (\mathbf{S}^1 \times \mathbf{S}^1, \mathcal{C})$ is a toroidal circle plane. If both $\mathcal{M}_1$ and $\mathcal{M}_2$ are flat Minkowski planes, so is $\mathcal{M}$.*

Proof. The first part is an immediate consequence of Theorem 2.10.1. We now assume that both planes $\mathcal{M}_i$ are Minkowski planes and verify that Axiom B2 is satisfied in $\mathcal{M}$. Let C and C^* be two circles in $\mathcal{C}$ that touch and let f and g be the corresponding homeomorphisms of $\mathbf{S}^1$, respectively. Then $f^{-1}g$ has precisely one fixed point. By Proposition A1.4.1(v), the homeomorphisms f and g are either both orientation-preserving or both orientation-reversing, that is, either both C and C^* are contained in $\mathcal{C}_1^+$ or both are contained in $\mathcal{C}_2^-$. Furthermore, given two circles $C \in \mathcal{C}_1^+$ and $C^* \in \mathcal{C}_2^-$ with corresponding homeomorphisms f and g, we know that fg^{-1} is orientation-reversing, that is, by Proposition A1.4.1(iv), fg^{-1} has precisely two fixed points. This means that C does not touch C^*. Hence the set of all circles in $\mathcal{C}$ that touch a circle $C \in \mathcal{C}_1^+$ ($C \in \mathcal{C}_2^-$) is the pencil of circles in $\mathcal{C}_1$ ($\mathcal{C}_2$) that touch C. All this implies that $\mathcal{M}$ is a flat Minkowski plane. □

Starting with the classical flat Minkowski plane as in Subsection 4.1.5, we can use this method to construct a variety of flat Minkowski planes as follows. Let Ξ denote the projective linear group PGL(2,**R**) as in Subsection 4.1.5. We further denote the subgroup PSL(2,**R**) by Λ. This is a normal subgroup of index 2 in Ξ. In the natural operation of Ξ on $\mathbf{S}^1$ as fractional linear transformations this subgroup consists of orientation-preserving homeomorphisms of $\mathbf{S}^1$ whereas $\Xi \setminus \Lambda$ consists of orientation-reversing homeomorphisms of $\mathbf{S}^1$. Recall that the circles of the classical flat Minkowski plane are the graphs of the homeomorphisms in Ξ. Furthermore, the two connected components of the circle set are obtained from Λ and $\Xi \setminus \Lambda$.

Let f and g be two orientation-preserving homeomorphisms of $\mathbf{S}^1$. We define

$$\mathcal{C}_{f,g} = \Lambda \cup g^{-1}(\Xi \setminus \Lambda)f.$$

Now $\mathcal{M}(f,g)$ is the following geometry. The point set of $\mathcal{M}(f,g)$ is the torus $\mathbf{S}^1 \times \mathbf{S}^1$, parallel classes are of the form $\{x_0\} \times \mathbf{S}^1$ and $\mathbf{S}^1 \times \{y_0\}$ for $x_0, y_0 \in \mathbf{S}^1$. The set of circles of $\mathcal{M}(f,g)$ 'is' $\mathcal{C}_{f,g}$, that is, circles are of the form

$$\{(x, \gamma(x)) \mid x \in \mathbf{S}^1\},$$

where $\gamma \in \Lambda \cup g^{-1}(\Xi \setminus \Lambda)f$; see also Subsection 4.2.3.

Choosing $f = g = id$ yields the classical flat Minkowski plane. We obtain another copy of this plane by taking the graphs of the homeomorphisms in $g^{-1}\Xi f$ as circle set. Indeed, the map $(x,y) \mapsto (f(x), g(y))$ is an isomorphism from this plane onto the standard model of the classical flat Minkowski plane. Then $\mathcal{M}(f,g)$ can be seen as being obtained from these two Minkowski planes by swapping corresponding halves of the circle sets as in Theorem 4.3.1. We therefore have the following.

THEOREM 4.3.2 (Nonclassical from Classical) *The point–circle geometry $\mathcal{M}(f,g)$ is a flat Minkowski plane.*

Replacing f and g by λf and $\lambda' g$, respectively, for $\lambda, \lambda' \in \Lambda$ does not alter the circle set, that is, $\mathcal{M}(f,g) = \mathcal{M}(\lambda f, \lambda' g)$. Since Λ is 2-transitive and because the stabilizer $\Lambda_{x,y}$ of two points $x, y \in \mathbf{S}^1$ is transitive on each connected component of $\mathbf{S}^1 \setminus \{x,y\}$, we may assume, if necessary, that f and g both fix ∞, 1, and 0.

Also note that the homeomorphism $\mathbf{S}^1 \times \mathbf{S}^1 \to \mathbf{S}^1 \times \mathbf{S}^1 : (x,y) \mapsto (y,x)$ defines an isomorphism from $\mathcal{M}(f,g)$ to $\mathcal{M}(g,f)$.

The planes for $g = id$ were first introduced by Schenkel [1980]. They occur in the classification of flat Minkowski planes of group dimension 4. The more general planes $\mathcal{M}(f,g)$ were investigated in Steinke [1994]. We summarize the group-dimension classification of these planes below.

Let $\mathcal{S}$ be the collection of all orientation-preserving homeomorphisms of $\mathbf{S}^1$ that are projectively equivalent to a semi-multiplicative homeomorphism, that is, $h \in \mathcal{S}$ if and only if $h = \delta \circ f$, where $\delta \in \mathrm{PSL}_2(\mathbf{R})$ and f is a semi-multiplicative homeomorphism; see the end of Section A1.4 for a definition of semi-multiplicative homeomorphisms.

THEOREM 4.3.3 (Group-Dimension Classification of $\mathcal{M}(f, id)$) *A flat Minkowski plane $\mathcal{M}(f, id)$, where f is an orientation-preserving homeomorphism of $\mathbf{S}^1$, has group dimension*

6 *if and only if $f \in \mathrm{PSL}_2(\mathbf{R})$;*

4 *if and only if f belongs to $\mathcal{S}$ but not to $\mathrm{PSL}_2(\mathbf{R})$;*

3 *if and only if f is an orientation-preserving homeomorphism of $\mathbf{S}^1$ that does not belong to $\mathcal{S}$.*

If Ξ and Λ denote the groups $\mathrm{PGL}_2(\mathbf{R})$ and $\mathrm{PSL}_2(\mathbf{R})$, respectively, and Σ_f is the connected component of the automorphism group of the plane $\mathcal{M}(f, id)$ that contains the identity, then an automorphism in Σ_f leaves each connected component of the circle space invariant and takes (+)-parallel classes to (+)-parallel classes and (−)-parallel classes to (−)-parallel classes. One readily concludes that such an automorphism has to be of the form $(x, y) \mapsto (\alpha(x), \beta(y))$, where α and β are homeomorphisms of $\mathbf{S}^1$ such that $\alpha \in \Lambda \cap f^{-1}\Lambda f$, $\beta \in \Lambda$. In particular, the group dimension d_f of $\mathcal{M}_f$ satisfies $d_f = 3 + \dim(\Lambda \cap f^{-1}\Lambda f)$.

If $\dim\,(\Lambda \cap f^{-1}\Lambda f) \geq 1$, then there must exist a closed connected 1-dimensional subgroup Φ of Λ such that $f\Phi f^{-1} \subset \Lambda$. We will see that such a group is conjugate to one of three possible groups, and f must be in $\mathcal{S}$.

Up to conjugacy in Ξ, the 1-dimensional closed connected subgroups Φ of Λ are determined as follows. A group Φ either acts transitively on $\mathbf{S}^1$ or has a fixed point. In the former case the subgroup must be isomorphic to $\mathrm{SO}_2(\mathbf{R})$ and so be a maximal compact subgroup of Λ. Furthermore, all maximal compact subgroups are conjugate to each other. In the latter case the subgroup is isomorphic to $\mathbf{R}$ and is, up to conjugacy, a subgroup of $\mathrm{L}_2 = \{x \mapsto rx + t \mid r, t \in \mathbf{R}, r > 0\}$ (with fixed point ∞). Furthermore, this subgroup can either act transitively on $\mathbf{R} = \mathbf{S}^1 \setminus \{\infty\}$ or have a second fixed point. Hence, there are three different types of closed connected 1-dimensional subgroups.

(i) Subgroups transitive on $\mathbf{S}^1$: each such subgroup is conjugate to

$$\Theta = \left\{ x \mapsto \frac{x\cos t - \sin t}{x \sin t + \cos t} \;\middle|\; t \in \mathbf{R}, 0 \leq t < 2\pi \right\} \simeq \mathrm{SO}_2(\mathbf{R}).$$

(ii) Subgroups fixing precisely one point: each such subgroup is conjugate to

$$\Omega = \{x \mapsto x + t \mid t \in \mathbf{R}\} \simeq \mathbf{R}.$$

(iii) Subgroups fixing precisely two points: each such subgroup is conjugate to

$$\Psi = \{x \mapsto rx \mid r \in \mathbf{R}^+\} \simeq \mathbf{R}.$$

For each of the three subgroups Θ, Ω, and Ψ one obtains functional equations for f from the condition that $f\Theta f^{-1} \subseteq \Lambda$, $f\Omega f^{-1} \subseteq \Lambda$ or $f\Psi f^{-1} \subseteq \Lambda$, respectively, if the respective subgroup occurs as a group of automorphisms of $\mathcal{M}(f, id)$. One finds that $f \in \Delta$ in the first two cases and $f \in \mathcal{S}$ in the case of the subgroup Ψ; see Steinke [1994] Lemmas 3.5, 3.2, and 3.6, respectively. Hence in any case one obtains that $f \in \mathcal{S}$. (Note that $\Delta \subset \mathcal{S}$.)

This proves the statement on group dimension 4.

If even dim $(\Delta \cap f^{-1}\Delta f) \geq 2$, then $\Delta = f^{-1}\Delta f$ or there must be a closed connected 2-dimensional subgroup Φ of Δ such that $f\Phi f^{-1} \subset \Delta$. However, Φ is conjugate to L_2 which contains the group Ω. Hence in any case we have, up to conjugacy in Ξ, that $f\Omega f^{-1} \subset \Delta$. But this implies that $f \in \Delta$. Therefore $\mathcal{M}(f, id)$ is classical in this case and thus has group dimension 6.

The group dimensions of the more general planes $\mathcal{M}(f, g)$ for orientation-preserving homeomorphisms f and g of $\mathbf{S}^1$ were determined in Steinke [1994] Theorem 3.9.

THEOREM 4.3.4 (Group-Dimension Classification of $\mathcal{M}(f, g)$) *Let $\mathcal{H}^+$ be the collection of all orientation-preserving homeomorphisms of $\mathbf{S}^1$ and let $\mathcal{S}$ be defined as in Theorem 4.3.3. Then a flat Minkowski plane $\mathcal{M}(f, g)$ has group dimension*

6 *if and only if $f, g \in \mathrm{PSL}_2(\mathbf{R})$;*
4 *if and only if $f \in \mathrm{PSL}_2(\mathbf{R})$, $g \in \mathcal{S} \setminus \mathrm{PSL}_2(\mathbf{R})$, or $g \in \mathrm{PSL}_2(\mathbf{R})$ and $f \in \mathcal{S} \setminus \mathrm{PSL}_2(\mathbf{R})$;*
3 *if and only if $f \in \mathrm{PSL}_2(\mathbf{R})$ and $g \in \mathcal{H}^+ \setminus \mathcal{S}$, or $g \in \mathrm{PSL}_2(\mathbf{R})$ and $f \in \mathcal{H}^+ \setminus \mathcal{S}$;*
2 *if and only if $f, g \in \mathcal{S} \setminus \mathrm{PSL}_2(\mathbf{R})$;*
1 *if and only if $f \in \mathcal{S} \setminus \mathrm{PSL}_2(\mathbf{R})$ and $g \in \mathcal{H}^+ \setminus \mathcal{S}$, or $g \in \mathcal{S} \setminus \mathrm{PSL}_2(\mathbf{R})$ and $f \in \mathcal{H}^+ \setminus \mathcal{S}$;*
0 *if and only if $f, g \in \mathcal{H}^+ \setminus \mathcal{S}$.*

4.3.2 Different Ways to Cut and Paste

In this subsection we look at different ways to construct new toroidal circle planes or flat Minkowski planes from old ones using a variety of 'cut-and-paste' techniques. As in the case of spherical and cylindrical circle planes, there are basically two types of techniques. Both techniques are much less developed than the respective techniques for the circle planes on the sphere and the cylinder.

The first type of cut-and-paste technique involves a separating set in the point set $\mathbf{S}^1 \times \mathbf{S}^1$. Examples for planes that can be constructed like this are the *semi-classical flat Minkowski planes.* Such a plane can be imagined as two halves of the classical Minkowski plane being pasted together along one or two disjoint circles or parallel classes. Since both classical subgeometries can be uniquely embedded, up to automorphisms of the classical flat Minkowski plane, into the classical flat Minkowski plane (see Steinke [1983a]), automorphisms of semi-classical Minkowski planes can be explicitly determined and a complete classification of all flat Minkowski planes of this type is possible.

The classification of semi-classical Minkowski planes is still unpublished. Although there are many combinations of circles and parallel classes possible whose removal will disconnect the point space of the classical flat Minkowski plane, there are only a few proper semi-classical Minkowski planes. These are obtained by removing a set consisting of two parallel classes of the same type. This set separates the torus into two open connected subsets that are both homeomorphic to the cylinder. Explicitly, let $k \in \mathbf{R}^+$, and let

$$f_k : \mathbf{R} \cup \{\infty\} \to \mathbf{R} \cup \{\infty\} : x \mapsto \begin{cases} x & \text{for } x \geq 0, \\ kx & \text{for } x \leq 0, \\ \infty & \text{for } x = \infty. \end{cases}$$

Then a semi-classical Minkowski plane is isomorphic to a plane $\mathcal{M}(f_k, id)$ in the notation of Subsection 4.3.1, where $k \geq 1$.

The second type of technique uses a separating set of the circle set. For example, in Subsection 4.3.1 we replaced one entire connected component of the circle space by a corresponding component from another plane. In this case the empty set forms a separating set in the circle space and we can regard this construction as a special case of the method of cutting and pasting along separating sets in the circle space that we used for the other types of flat circle planes.

Let $\mathcal{M} = (\mathbf{S}^1 \times \mathbf{S}^1, \mathcal{C}^+ \cup \mathcal{C}^-)$ be a toroidal circle plane such that $\mathcal{C}^+$ and $\mathcal{C}^-$ are the connected components of the circle space of $\mathcal{M}$ consisting of graphs of orientation-preserving and orientation-reversing homeomorphisms of the unit circle to itself. Then Theorem 4.3.1 implies that the *positive* half $(\mathbf{S}^1 \times \mathbf{S}^1, \mathcal{C}^+)$ and the *negative half* $(\mathbf{S}^1 \times \mathbf{S}^1, \mathcal{C}^-)$ of $\mathcal{M}$

are completely independent. Furthermore, any positive half is the image of a negative half (and vice versa) under a suitable homeomorphism of the torus $\mathbf{S}^1 \times \mathbf{S}^1$ to itself. Hence any further cutting and pasting can be restricted to cutting and pasting among positive halves of toroidal circle planes. A formal definition of (strong) (X-embedded) separating sets of toroidal circle planes or halves of such planes can be modelled after the definition of such sets in the case of spherical circle planes; see Subsection 3.3.4.

For example, if we fix a circle $C \in \mathcal{C}^+$ and consider the sets $\mathcal{C}^1$ consisting of C and all circles that intersect C in exactly one point and the sets $\mathcal{C}^i$, $i = 0, 2$, consisting of all circles in $\mathcal{C}^+$ that intersect C in precisely i points; then $\mathcal{C}^1$ is a separating set of $\mathcal{C}$. Indeed, $\mathcal{C} \setminus \mathcal{C}^1$ has three connected components; these are $\mathcal{C}^2$, $\mathcal{C}^0$, and $\mathcal{C}^-$. However, it is not clear whether or not $\mathcal{C}^1$ is a (strong) X-embedded separating set. Here X stands for the requirement that there is a circle C as above that defines the set $\mathcal{C}^1$.

Also, as in the case of spherical and cylindrical circle planes, it makes sense to try to generalize certain 'separating sets' associated with pairs of points p, q, and their connecting line L in the geometry of nontrivial plane sections of a hyperbolic quadric $\mathcal{Q}$ in $\mathrm{PG}(3, \mathbf{R})$. The quadric separates $\mathrm{PG}(3, \mathbf{R})$ into two connected components that can be exchanged by an automorphism of the space. Since it therefore does not make sense to speak of inner and outer points of the quadric, we only have to worry about points on the quadric and exterior points. We can define 'virtual' exterior points in terms of *exterior involutions* of a toroidal circle plane along the same lines as we defined inner and outer involutions of spherical and cylindrical circle planes. In fact, we do this in Subsection 4.8.2. Given two points, both either real or virtual, we consider the set of circles through at least one of the two points. Many of these sets turn out to be X-embedded separating sets. Usually, the condition X for X-embedded separating sets involving virtual exterior points are rather complicated and do not seem to be very useful when it comes to constructing new planes. We restrict ourselves to giving the simplest example of a construction technique.

Let $\mathcal{M} = (\mathbf{S}^1 \times \mathbf{S}^1, \mathcal{C})$ be a toroidal circle plane, let p and q be two distinct parallel points of the torus $\mathbf{S}^1 \times \mathbf{S}^1$, and let G denote the parallel class both points are contained in (note that in the classical setting described above G corresponds to the connecting line of the two points). Then $G \setminus \{p, q\}$ has two connected components G^1 and G^2. Let $\mathcal{C}^{p,q}$ be the set of all circles in $\mathcal{C}$ that contain p or q, let $\mathcal{C}^1$ and $\mathcal{C}^2$ be the circles that intersect G in G^1 and G^2, respectively.

PROPOSITION 4.3.5 (Cut and Paste) *Let $\mathcal{M}_i = (\mathbf{S}^1 \times \mathbf{S}^1, \mathcal{C}_i)$, $i = 1, 2$, be toroidal circle planes. Suppose that $\mathcal{C}_1 \cap \mathcal{C}_2 \supseteq \mathcal{C}_1^{p,q}$ for two distinct parallel points p and q. Let $\mathcal{C} = \mathcal{C}_1^{p,q} \cup \mathcal{C}_1^1 \cup \mathcal{C}_2^2$. Then the geometry $\mathcal{M} = (\mathbf{S}^1 \times \mathbf{S}^1, \mathcal{C})$ is a toroidal circle plane. If both $\mathcal{M}_1$ and $\mathcal{M}_2$ are flat Minkowski planes, then so is $\mathcal{M}$.*

The proof runs along the same lines as that of the respective results for spherical and cylindrical circle planes; see Propositions 3.3.9 and 5.3.8.

For a different method that does not quite fit the techniques described in this section also compare Subsection 4.3.6.

4.3.3 Integrals of $\mathbf{R}^2$*-Planes*

In Subsubsection 5.3.4.1 we describe a way to construct flat Laguerre planes by 'integrating' flat affine planes. It is also possible to integrate $\mathbf{R}^2$-planes in such a way that the resulting geometries can be turned into toroidal circle planes. Not much is known about these construction techniques. The known results are summarized at the end of Subsubsection 5.3.4.1.

4.3.4 The Generalized Hartmann Planes

The first examples of nonclassical Minkowski planes were given by Hartmann [1981].

For $r, s \in \mathbf{R}^+$ let $f_{r,s}$ be the orientation-preserving *semi-multiplicative homeomorphism* of $\mathbf{S}^1 \simeq \mathbf{R} \cup \{\infty\}$ to itself defined by

$$f_{r,s}(x) = \begin{cases} x^r & \text{for } x \geq 0, \\ -s|x|^r & \text{for } x < 0, \\ \infty & \text{for } x = \infty; \end{cases}$$

compare p. 442. With this notation, we construct the flat Minkowski plane $\mathcal{M}(r_1, s_1; r_2, s_2)$ for $r_1, s_1, r_2, s_2 \in \mathbf{R}^+$. The plane is in standard representation and the circle set of $\mathcal{M}(r_1, s_1; r_2, s_2)$ consists of all Euclidean lines extended by the point (∞, ∞) and the sets

$$\left\{ \left(x, \frac{a}{f_{r_1,s_1}(x-b)} + c \right) \middle| \, x \in \mathbf{S}^1 \right\},$$

where $a, b, c \in \mathbf{R}$, $a > 0$, and

$$\left\{ \left(x, \frac{a}{f_{r_2,s_2}(x-b)} + c \right) \,\middle|\, x \in \mathbf{S}^1 \right\},$$

where $a, b, c \in \mathbf{R}$, $a < 0$. As usual, we use the convention $\frac{1}{\infty} = 0$, $\frac{1}{0} = \infty$ and $a \cdot \infty + b = \infty$ for all $a, b \in \mathbf{R}$, $a \neq 0$.

The planes $\mathcal{M}(r_1, s_1; r_2, s_2)$ generalize Hartmann's [1981] construction of flat Minkowski planes which, in the above notation, are the planes $\mathcal{M}(r_1, 1; r_2, 1)$. In their generalized form these planes were introduced by Schenkel [1980].

Note that the map

$$\mathbf{S}^1 \times \mathbf{S}^1 \to \mathbf{S}^1 \times \mathbf{S}^1 : (x, y) \mapsto (rx + b, ry + c),$$

where $b, c, r \in \mathbf{R}$, $r > 0$, is an automorphism of $\mathcal{M}(r_1, s_1; r_2, s_2)$.

Clearly, $\mathcal{M}(1, 1; 1, 1)$ is the classical flat Minkowski plane. In fact, this is the only instance where the classical flat Minkowski plane occurs among the generalized Hartmann planes. In order to verify this claim we first look at the derived affine planes at the points $(\infty, 0)$ and $(0, \infty)$.

LEMMA 4.3.6 (Derived Planes of Hartmann Planes I) *The derived affine plane of a generalized Hartmann plane $\mathcal{M}(r_1, s_1; r_2, s_2)$ at $(0, \infty)$ is Desarguesian if and only if $r_1 = r_2$ and $s_1 = s_2$. If it is Desarguesian, then every derived affine plane at (x, ∞) for $x \in \mathbf{S}^1$ is Desarguesian.*

The derived plane at $(\infty, 0)$ is Desarguesian if and only if $r_1 = r_2$ and $s_1 s_2 = 1$. If it is Desarguesian, then every derived affine plane at (∞, y) for $y \in \mathbf{S}^1$ is Desarguesian.

Proof. Suppose that the derived affine plane at $(0, \infty)$ is Desarguesian. Under the coordinate transformation

$$(x, y) \mapsto \left(\frac{1}{f_{r_1,s_1}(x)}, y \right)$$

circles through $(0, \infty)$ give rise to lines of the form $\{(u, au + c) \mid u \in \mathbf{R}\}$ for $a, c \in \mathbf{R}$ and $a > 0$, and $\{(u, af_{r,s}(u) + c) \mid u \in \mathbf{R}\}$ for $a, c \in \mathbf{R}$ and $a < 0$, where $r = r_2/r_1$ and $s = s_1^r/s_2$. The projective extension of this plane is the dual of a semi-classical flat projective plane. Such a plane is Desarguesian if and only if $r = s = 1$; see Proposition 2.7.1. Thus $r_1 = r_2$ and $s_1 = s_2$.

Suppose that the derived affine plane at $(\infty, 0)$ is Desarguesian. We use the coordinate transformation $(x, y) \mapsto (f_{r_1,s_1}^{-1}(y), x)$. Then circles through $(\infty, 0)$ give rise to lines of the form $\{(u, au + c) \mid u \in \mathbf{R}\}$ for $a, c \in \mathbf{R}$, $a > 0$ and $\{(u, af_{r,s}(u) + c) \mid u \in \mathbf{R}\}$ for $a, c \in \mathbf{R}$, $a < 0$, where $r = r_1/r_2$ and $s = (s_1 s_2)^{(1/r_2)}$. Again, the projective extension of this plane is the dual of a semi-classical flat projective plane and such a plane is Desarguesian if and only if $r = s = 1$. We conclude that $r_1 = r_2$ and $s_1 s_2 = 1$. □

Combining the above two cases, we immediately obtain the following result.

COROLLARY 4.3.7 (Derived Planes of Hartmann Planes II) *Each derived affine plane of $\mathcal{M}(r, 1; r, 1)$ at the infinite points (x, ∞) or (∞, y) for $x, y \in \mathbf{S}^1$ is Desarguesian.*

Now we can show that if $\mathcal{M}(r_1, s_1; r_2, s_2)$ is the classical flat Minkowski plane, then $r_1 = s_1 = r_2 = s_2 = 1$. Since every derived affine plane of the classical flat Minkowski plane must be Desarguesian (see Subsection 4.2.1), the derived affine planes at the points $(\infty, 0)$, $(0, \infty)$ and $(0, 0)$, in particular, must be Desarguesian. From Lemma 4.3.6 we know that the first two conditions imply that $r_1 = r_2$, $s_1 = s_2$ and $r_1 = r_2$, $s_1 s_2 = 1$, respectively. We conclude that $r_1 = r_2 = r$ and $s_1 = s_2 = 1$ and we have a Minkowski plane $\mathcal{M}(r, 1; r, 1)$. Circles not passing through (∞, ∞) can be written in the form

$$\{(x, af_r(x - b) + c) \mid x \in \mathbf{R}\} \cup \{(\infty, \infty)\},$$

where $a, b, c \in \mathbf{R}$, $a \neq 0$ and $f_r(x) = x|x|^{-r-1}$. Note that f_r is multiplicative. Since $\mathcal{M}(r, 1; r, 1)$ is classical, the collection of circles representing homeomorphisms associated with the plane must be a group; see Theorem 4.4.17. In particular, $f_r \circ f_r$ represents a circle of $\mathcal{M}(r, 1; r, 1)$. But $f_r \circ f_r(x) = x|x|^{r^2-1}$ so that the corresponding circle of $\mathcal{M}(r, 1; r, 1)$ passes through the points (∞, ∞), $(0, 0)$, and $(1, 1)$. Of course, the unique circle through these three points is the Euclidean line $y = x$. Therefore $f_r \circ f_r(x) = x$ for all $x \in \mathbf{R}$ and $r = 1$. This concludes the proof of our claim and we have the following result.

THEOREM 4.3.8 (Classical Hartmann Planes) *A generalized Hartmann plane $\mathcal{M}(r_1, s_1; r_2, s_2)$ for $r_1, s_1, r_2, s_2 \in \mathbf{R}^+$ is classical if and only if $r_1 = s_1 = r_2 = s_2 = 1$.*

The planes $\mathcal{M}(r, 1; r, 1)$ for $r > 0$ can be characterized in terms of the configuration consisting of all points at which the Minkowski plane has Desarguesian derived planes (see Corollary 4.3.7 and Theorem 4.6.4) and in terms of their Klein–Kroll types; see Proposition 4.5.10. In fact, the generalized Hartmann planes can be obtained as a subfamily in a larger class of flat Minkowski planes, which will be introduced in the following subsection. See also Corollary 4.3.12 for the classification of generalized Hartmann planes with respect to their group dimensions.

4.3.5 The Artzy–Groh Planes

The planes we encounter in this section comprise the generalized Hartmann planes. The method of construction is to prescribe a large enough group that acts on the planes as a group of automorphisms. Artzy and Groh [1986] started off with the connected component of the group of affine similarities, that is, the group of transformations

$$\mathbf{R}^2 \to \mathbf{R}^2 : (x, y) \mapsto (rx + b, ry + c)$$

for $b, c, r \in \mathbf{R}$, $r > 0$, and all Euclidean lines. They then looked for curves C_1 and C_2 such that their images under the above group and the nonhorizontal and nonvertical Euclidean lines properly extended (as described in Subsection 4.2.2) yield a flat Minkowski plane.

Let $f, g : \mathbf{R} \setminus \{0\} \to \mathbf{R} \setminus \{0\}$ be two homeomorphisms. For $a, b, c \in \mathbf{R}$ let

$$C_{a,b,c} = \begin{cases} \{(x, bx + c) \mid x \in \mathbf{R}\} \cup \{(\infty, \infty)\} & \text{for } a = 0, \\ \{(x, af(\frac{x-b}{a}) + c) \mid x \in \mathbf{R}\} \cup \{(\infty, c), (b, \infty)\} & \text{for } a > 0, \\ \{(x, ag(\frac{x-b}{|a|}) + c) \mid x \in \mathbf{R}\} \cup \{(\infty, c), (b, \infty)\} & \text{for } a < 0, \end{cases}$$

and let

$$\mathcal{C}_{f,g} = \{C_{a,b,c} \mid a, b, c \in \mathbf{R}\}.$$

Artzy and Groh [1986] showed that $\mathcal{C}_{f,g}$ is the circle set of a flat Minkowski plane if and only if f and g satisfy the following four conditions.

(i) The functions f and g are differentiable.
(ii) The restrictions of f' and g' to both $\mathbf{R}^+$ and $\mathbf{R}^-$ are strictly monotonic.
(iii) The restrictions of f and g to both $\mathbf{R}^+$ and $\mathbf{R}^-$ have the x-axis and the y-axis as asymptotes.
(iv) We have $f'(x) < 0$ and $g'(x) < 0$ for all $x \in \mathbf{R} \setminus \{0\}$.

We denote this plane by $\mathcal{M}'(f, g)$. Note that we add the prime in order to distinguish these planes from the planes in Subsection 4.3.1.

Since the circles $C_{a,0,0}$ for $a \in \mathbf{R}$, $a \neq 0$, form the pencil of circles through the points $(\infty, 0)$ and $(0, \infty)$, there is exactly one circle in this pencil that passes through the point $(1, 1)$ and exactly one other circle that passes through the point $(1, -1)$. The former circle must be obtained for some $a > 0$ and the latter one for some $a < 0$. Of course, we can choose the corresponding functions as the ones that generate the plane. Therefore each Artzy–Groh plane is isomorphic to a plane $\mathcal{M}'(f, g)$, where f and g in addition satisfy $f(1) = g(1) = 1$. Using homeomorphisms $\mathbf{S}^1 \to \mathbf{S}^1$, the above conditions can be described more easily.

PROPOSITION 4.3.9 (Normalized Artzy–Groh Planes) *Up to isomorphisms the Artzy–Groh planes $\mathcal{M}'(f, g)$ can be obtained by only using functions f and g that are homeomorphisms $\mathbf{S}^1 \to \mathbf{S}^1$ whose restrictions to $\mathbf{R} \setminus \{0\}$ are differentiable and for 0, 1 and ∞ take on the values $f(\infty) = g(\infty) = 0$, $f(0) = g(0) = \infty$, $f(1) = g(1) = 1$.*

We obtain the generalized Hartmann planes $\mathcal{M}(r_1, s_1; r_2, s_2)$ (see Subsection 4.3.4) if $f = 1/f_{r_1,s_1}$ and $g = 1/f_{r_2,s_2}$, where $f_{r,s}$ is a semi-multiplicative function as defined at the beginning of Subsection 4.3.4. The classical flat Minkowski plane occurs for $f = g = 1/f_{1,1}$. In fact, assuming that f and g are both normalized, that is, $f(1) = g(1) = 1$, this is the only way to obtain the classical flat Minkowski plane. This characterization of the classical plane among the Artzy–Groh planes follows directly from Theorems 4.4.12 and 4.3.11.

THEOREM 4.3.10 (Classical Artzy–Groh Planes) *Under the assumptions of Proposition 4.3.9 one obtains the classical flat Minkowski plane if and only if f and g are both defined by $x \mapsto 1/x$.*

In their paper Artzy and Groh [1986] use Theorem 4.4.17 to prove Theorem 4.3.10. They show that $f \circ f$ is involutory and then derive a functional equation for f by determining the form of $f \circ f_a$ for $a > 0$, where $f_a(x) = af(x/a)$. A similar method is applied to the other homeomorphism g.

From the way the Artzy–Groh planes are constructed it becomes clear that this kind of flat Minkowski planes can be characterized by the type of their automorphism groups; compare Theorem 4.4.13.

THEOREM 4.3.11 (Group-Dimension Classification) *An Arzty–Groh plane $\mathcal{M}'(f,g)$, where f and g are homeomorphisms of $\mathbf{S}^1$ to itself whose restrictions to the set $\mathbf{R}\setminus\{0\}$ are differentiable and which satisfy the conditions $f(\infty)=g(\infty)=0$, $f(0)=g(0)=\infty$, $f(1)=g(1)=1$, has group dimension*

6 *if and only if f and g are both equal to the inversion $i : x \mapsto \frac{1}{x}$;*

4 *if and only if both f and g are inversely semi-multiplicative (see* *p. 442* *for a definition of inversely semi-multiplicative homeomorphisms) but at least one is not equal to i (then $\mathcal{M}'(f,g)$ is a nonclassical generalized Hartmann plane);*

3 *if and only if at least one of f or g is not inversely semi-multiplicative.*

The connected component of the automorphism group containing the identity of such a plane is $\mathrm{PSL}_2(\mathbf{R}) \times \mathrm{PSL}_2(\mathbf{R})$, $\mathrm{L}_2 \times \mathrm{L}_2$, *and*

$$\{(x,y) \mapsto (rx+b, ry+c) \mid b,c,r \in \mathbf{R}, r>0\},$$

respectively.

Proof. Every Artzy–Groh plane clearly has group dimension at least 3, because each such plane admits the group

$$\Phi = \{(x,y) \mapsto (rx+b, ry+c) \mid b,c,r \in \mathbf{R}, r>0\}$$

as group of automorphisms. This group fixes the point $p=(\infty,\infty)$ and acts transitively on $|p|_\pm \setminus \{p\}$ and $\mathbf{S}^1 \times \mathbf{S}^1 \setminus (|p|_+ \cup |p|_-)$.

We assume that the Artzy–Groh plane $\mathcal{M} = \mathcal{M}'(f,g)$ has group dimension at least 4. Let Γ^1 be the connected component of the automorphism group Γ of $\mathcal{M}$ and let $T^\pm$ be the kernel of the action of Γ^1 on the set $\Pi^\pm$ of $(\pm)$-parallel classes; see Subsection 4.4.3 for the definition of the kernels and their properties. From Theorem 4.4.12 we know that the group Γ is 4- or 6-dimensional, and that in the latter case $\mathcal{M}$ must be classical. We claim that in the former case Γ^1 fixes the distinguished point p. Assume otherwise. Then, using the orbits of the group Φ, we see that Γ^1 must be transitive on at least one of Π^+ and Π^-, the collections of $(+)$- and $(-)$-parallel classes, respectively. Without loss of generality, we may assume that Γ^1 is transitive on Π^+. Since Φ is transitive on $\Pi^+ \setminus \{|p|_+\}$, we see that Γ^1 is 2-transitive on Π^+. Hence Γ^1/T^+ is isomorphic to a finite covering group of $\mathrm{PSL}_2(\mathbf{R})$ by Brouwer's Theorem A2.3.8. The connected component Δ of the other kernel T^- is canonically isomorphic to a closed normal connected subgroup of Γ^1/T^+.

But Γ^1/T^+ is almost simple so that $\Delta = \{id\}$ or $\Delta = \Gamma^1/T^+$. In the former case, T^- is 0-dimensional and Γ^1/T^- is 4-dimensional in contradiction to Proposition 4.4.7. In the latter case, T^- is 3-dimensional and $\mathcal{M}$ is isomorphic to a flat Minkowski plane $\mathcal{M}(\tilde{f}, id)$ by Theorem 4.4.10. Furthermore, by Theorem 4.3.3, the homeomorphism $\tilde{f}$ must be projectively equivalent to a semi-multiplicative homeomorphism. But then $\mathcal{M}(\tilde{f}, id)$ is classical or the connected component of the automorphism group containing the identity of such a plane fixes two parallel classes of the same type. Both conclusions contradict our assumptions. This shows that Γ^1 fixes the point p.

Since Γ^1 fixes p, the group Γ^1 induces a 4-dimensional closed connected group of collineations of the derived projective plane $\mathcal{P}$ at p which, in addition, also fixes the two infinite points ω_+ and ω_- that come from parallel classes of the Minkowski plane. But $\mathcal{P}$ is Desarguesian so that the induced group, and thus Γ^1, is the group

$$\mathrm{L}_2 \times \mathrm{L}_2 = \{(x, y) \mapsto (rx + b, sy + c) \mid b, c, r, s \in \mathbf{R}, r, s > 0\}.$$

The stabilizer $\Psi = \Gamma^1_q$ of the point $q = (0, 0)$ consists of the transformations $(x, y) \mapsto (rx, sy)$ for $r, s \in \mathbf{R}^+$. This group operates on the bundle $\mathcal{B}$ of circles through the points $(\infty, 0)$ and $(0, \infty)$. Furthermore, this action is transitive on both connected components of $\mathcal{B}$, because no automorphism $(x, y) \mapsto (x, ry)$ for $r \in \mathbf{R}^+$, $r \neq 1$, can fix a circle in $\mathcal{B}$. Therefore, the stabilizer Ψ_C of a circle C in $\mathcal{B}$ is 1-dimensional. But a 1-dimensional closed connected subgroup of Ψ is of the form

(i) $\{(x, y) \mapsto (x, ry) \mid r \in \mathbf{R}^+\}$, or
(ii) $\{(x, y) \mapsto (rx, r^d y) \mid r \in \mathbf{R}^+\}$ for some $d \in \mathbf{R}$.

The group in (i) and likewise the group in (ii) for $d = 0$ cannot occur as the stabilizer of C (both are subgroups of the kernels of $\mathcal{M}$). Furthermore, Ψ_C must be transitive on each connected component of the set $C \setminus \{(\infty, 0), (0, \infty)\}$ by Lemma 4.4.2. If C is the unique circle in $\mathcal{B}$ through the point $(1, 1)$ and Ψ_C is as in (ii) for $d \neq 0$, then the connected component of $C \setminus \{(\infty, 0), (0, \infty)\}$ that contains $(1, 1)$ is the orbit of $(1, 1)$ under Ψ_C, that is, $\{(x, x^d) \mid x \in \mathbf{R}^+\}$ is contained in C. Since C passes through $(\infty, 0)$ and $(0, \infty)$, we see that $d < 0$. But C also passes through a point $(-1, c)$ for some $c < 0$ and the other connected component of $C \setminus \{(\infty, 0), (0, \infty)\}$ is the orbit of $(-1, c)$ under Ψ_C. Hence

$$C = \{(x, x^d) \mid x \in \mathbf{R}^+\} \cup \{(x, c|x|^d) \mid x \in \mathbf{R}, x < 0\} \cup \{(\infty, 0), (0, \infty)\},$$

that is, C is the graph of an inversely semi-multiplicative homeomorphism of $\mathbf{S}^1$ and thus f is inversely semi-multiplicative. Starting off with a circle in the other connected component of $\mathcal{B}$, say the one passing through $(1, -1)$, we similarly obtain that g is another inversely semi-multiplicative homeomorphism of $\mathbf{S}^1$. Since in this situation $\mathcal{M}'(f, g)$ is a generalized Hartmann plane, Theorems 4.3.8 and 4.4.12 guarantee that $f = g = id$ if the plane has group dimension at least 5. □

Since generalized Hartmann planes are special Artzy–Groh planes, we readily obtain the following.

COROLLARY 4.3.12 (Classification of Hartmann Planes) *A generalized Hartmann $\mathcal{M}(r_1, s_1; r_2, s_2)$ where $r_1, s_1, r_2, s_2 \in \mathbf{R}^+$ has group dimension*

6 *if and only if* $r_1 = s_1 = r_2 = s_2 = 1$;
4 *if and only if* $(r_1, s_1, r_2, s_2) \neq (1, 1, 1, 1)$.

The connected component of the automorphism group of such a plane containing the identity is $\mathrm{PSL}_2(\mathbf{R}) \times \mathrm{PSL}_2(\mathbf{R})$ *and* $\mathrm{L}_2 \times \mathrm{L}_2$, *respectively.*

Note that the above corollary also follows from Theorems 4.4.12 and 4.3.8 and the fact that each generalized Hartmann plane admits the 4-dimensional group $\mathrm{L}_2 \times \mathrm{L}_2$ as group of automorphisms.

4.3.6 Modified Classical Planes

The construction in Steinke [1985b] does not fit either of the two cut-and-paste methods described in Subsection 4.3.2. Let f be an orientation-preserving homeomorphism of the real line $\mathbf{R}$ that fixes 0. We construct a flat Minkowski plane $\mathcal{M}(f)$ whose circles are as follows.

- The nonhorizontal and nonvertical Euclidean lines extended by the point (∞, ∞)
$$\{(x, mx + t) \mid x \in \mathbf{R}\} \cup \{(\infty, \infty)\},$$
where $m, t \in \mathbf{R}$, $m \neq 0$.
- Unions of branches of Euclidean hyperbolas extended by two infinite points
$$\left\{(x, y) \in \mathbf{R}^2 \,\middle|\, (x-b)(y-c) = \begin{cases} a & \text{for } x > b, \\ f(a) & \text{for } x < b \end{cases}\right\} \cup \{(\infty, c), (b, \infty)\},$$
where $a, b, c \in \mathbf{R}$, $a \neq 0$.

Clearly, the derived affine plane of $\mathcal{M}(f)$ at (∞, ∞) is the Euclidean plane. Furthermore, the transformations

$$(x, y) \mapsto \left(rx + b, \frac{y}{r} + c\right),$$

where $b, c, r \in \mathbf{R}$, $r > 0$, are automorphisms of $\mathcal{M}(f)$.

Here circles are pasted together along the two parallel classes of the point $p = (\infty, \infty)$. Note, however, that $|p|$ is not a separating set of the point set $\mathbf{S}^1 \times \mathbf{S}^1$. Although the Euclidean lines form a separating set in the circle set, we do not exchange all the circles in one component with circles from another flat Minkowski plane. Instead, we first observe that every circle C not passing through p intersects $|p|$ in two points $p_+ = C \cap |p|_+$ and $p_- = C \cap |p|_-$ and that $C \setminus \{p_+, p_-\}$ has two connected components. Then we form a new circle by pasting together two complementary branches from different circles in the pencil of circles through the points p_+ and p_-. Since we basically swap branches of Euclidean hyperbolas, we call these flat Minkowski planes *modified classical Minkowski planes.*

It readily follows that the derived affine planes of $\mathcal{M}(f)$ at the two points (∞, c) or (b, ∞) for $b, c \in \mathbf{R}$ are generalized Moulton planes; see p. 65 for a definition of these planes. Using the above automorphisms one only has to consider the derived affine planes at $(\infty, 0)$ and $(0, \infty)$.

LEMMA 4.3.13 (Desarguesian Derived Affine Planes) *The derived affine plane of a modified classical Minkowski plane $\mathcal{M}(f)$ at $(0, \infty)$ is Desarguesian if and only if f is a multiplication μ_r, that is, $f(x) = rx$ for some $r > 0$. If it is Desarguesian, then every derived affine plane at (x, ∞) for $x \in \mathbf{S}^1$ is Desarguesian.*

The derived plane at $(\infty, 0)$ is Desarguesian if and only if f is semilinear of the form

$$f(x) = \begin{cases} rx & \text{for } x \geq 0, \\ x/r & \text{for } x \leq 0, \end{cases}$$

where $r > 0$. If it is Desarguesian, then every derived affine plane at (∞, y) for $y \in \mathbf{S}^1$ is Desarguesian.

Hence, the derived affine plane of $\mathcal{M}(f)$ at (∞, ∞) is the only Desarguesian derived affine plane if f is not of the form μ_r or

$$x \mapsto \begin{cases} rx & \text{for } x \geq 0, \\ x/r & \text{for } x \leq 0, \end{cases}$$

where $r > 0$.

From the above lemma we furthermore readily obtain that $f = \mu_r$ and $r = 1$, if a modified classical Minkowski plane is classical. We therefore can characterize the classical flat Minkowski plane as follows.

COROLLARY 4.3.14 (Classical Modified Planes) *A modified classical flat Minkowski plane $\mathcal{M}(f)$ is classical if and only if f is the identity on* $\mathbf{R}$.

All derived affine planes of $\mathcal{M}(f)$ at finite points (b, c) for $b, c \in \mathbf{R}$ are isomorphic to the one at $(0, 0)$. It turns out that the derived affine plane at $(0, 0)$ is Desarguesian if and only if $f = id$. More precisely, the derived projective plane at $(0, 0)$ is not (ω_-, W)-transitive, where W denotes the line at infinity and ω_- denotes the infinite point of lines that come from $(-)$-parallel classes. This then implies that the point (∞, ∞) must be fixed by every automorphism of $\mathcal{M}(f)$ if $f \neq id$.

PROPOSITION 4.3.15 (Fixed Point of Modified Planes) *Every automorphism of a nonclassical modified classical flat Minkowski plane fixes the point* (∞, ∞).

The above proposition allows us to determine the group dimensions of the modified classical planes. If $f \neq id$, then every automorphism of $\mathcal{M}(f)$ fixes the point (∞, ∞) and the derived affine plane at (∞, ∞) is the Euclidean plane. Thus every automorphism can be obtained as the extension of a collineation of the Euclidean plane of the form

$$(x, y) \mapsto (rx + b, sy + c), \text{ or}$$
$$(x, y) \mapsto (sy + c, rx + b)$$

for $b, c, r, s \in \mathbf{R}$, $r, s \neq 0$. Since the transformations

$$(x, y) \mapsto \left(rx + b, \frac{y}{r} + c\right)$$

for $a, b, r \in \mathbf{R}$, $r > 0$, are automorphisms in any case, we can restrict ourselves to analysing whether or not $(x, y) \mapsto (x, sy)$ for $s \in \mathbf{R}^+$ are automorphisms in order to determine the group dimensions of these planes. Indeed, if each of the above transformations is an automorphism of $\mathcal{M}(f)$, then f satisfies the functional equation $f(sa) = sf(a)$ for all $a, s \in \mathbf{R}$, $s > 0$, so that f is of the form

$$x \mapsto \begin{cases} rx & \text{if } x \geq 0, \\ sx & \text{if } x \geq 0, \end{cases}$$

where $r, s \in \mathbf{R}^+$. We call a homeomorphism $f : \mathbf{R} \to \mathbf{R}$ of this form semi-linear; see p. 441. With this notation one obtains the following.

THEOREM 4.3.16 (Group-Dimension Classification of $\mathcal{M}(f)$) *A modified classical Minkowski plane $\mathcal{M}(f)$, where f is a homeomorphism of $\mathbf{R}$ that fixes 0, has group dimension*

6 *if and only if $f = id$;*
4 *if and only if f is semi-linear but not the identity id;*
3 *if and only if f is not semi-linear.*

The connected component of the automorphism group of such a plane is $\mathrm{PSL}_2(\mathbf{R}) \times \mathrm{PSL}_2(\mathbf{R})$, $\mathrm{L}_2 \times \mathrm{L}_2$, *and*

$$\left\{(x, y) \mapsto \left(rx + b, \frac{y}{r} + c\right) \;\middle|\; b, c, r \in \mathbf{R}, r > 0\right\},$$

respectively.

4.3.7 Proper Toroidal Circle Planes

So far we have not encountered any proper toroidal circle planes, that is, toroidal circle planes that are not flat Minkowski planes. Remember that it is easy to construct proper spherical and cylindrical circle planes by modifying the ovoidal constructions of the respective classical examples. For example, the geometry of plane sections of a strictly convex sphere 'with a corner' is a proper spherical circle plane. A similar 'ovoidal' construction is not possible for proper toroidal circle planes, because the classical flat Minkowski plane is the only 'ovoidal' toroidal plane; see Buekenhout [1969].

A first construction of proper toroidal circle planes is given in Polster [1998b]. We start with the affine part of one of the Artzy–Groh planes $\mathcal{M}'(f, g)$ in Subsection 4.3.5, where the function f has the additional property that $f = f^{-1}$. This implies that the graph of f has the line with slope 1 through the origin of $\mathbf{R}^2$ as a symmetry axis. Consider all circles that have asymptotes the x- and y-axes and are completely contained in the first and third quadrants. Of course, all these curves are just the images of the graph of f under all positive dilatations; see the leftmost diagram in Figure 4.5. Now we 'collapse' the cross in the second diagram. This cross is the set of all points at distance less than 1 from the lines of slope 1 and -1 through the origin. In this way we obtain a new system of curves, as illustrated in the third diagram. Consider the set of all images of elements in this system under all possible Euclidean

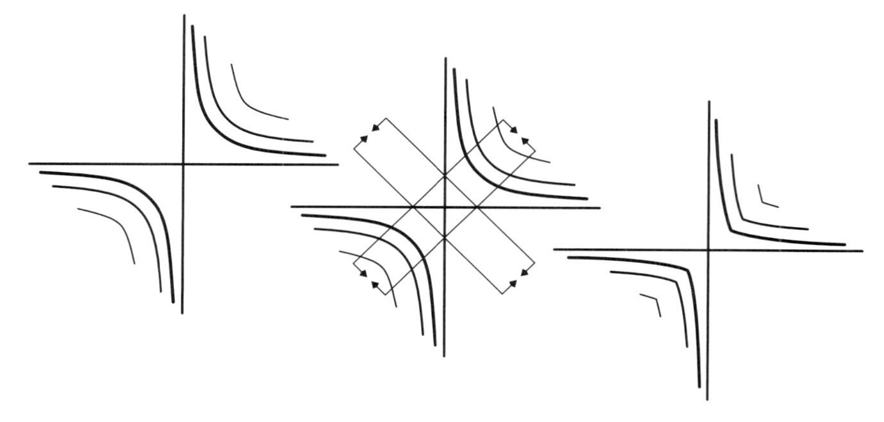

Fig. 4.5. Construction of a proper toroidal circle plane from an Artzy–Groh plane

plane translations. Then this set, together with the Euclidean lines of negative slope, can be turned into one half of a toroidal circle plane in the usual way. Together with the half of the flat Minkowski plane $\mathcal{M}'(f,g)$ generated from g, we obtain a new toroidal circle plane $\mathcal{M}''(f,g)$.

It is easy to see that $\mathcal{M}''(f,g)$ does not satisfy Axiom B2. Just consider one of the new circles and call it C. Such a circle has two 'corners' c_1 and c_2 in the affine part and there are infinitely many Euclidean lines that touch C at these points. In particular, this means that in the toroidal circle plane there are infinitely many circles through the point (∞,∞) that touch the circle C at c_1.

THEOREM 4.3.17 (Proper Toroidal Circle Planes) *Proper toroidal circle planes do exist. In particular, let f and g be functions as in Subsection 4.3.5, and let $f = f^{-1}$. Then $\mathcal{M}''(f,g)$ is a proper toroidal circle plane.*

4.4 Automorphism Groups and Groups of Projectivities

In this section we investigate automorphisms of toroidal circle planes and flat Minkowski planes and show that the collection of all automorphisms of a flat Minkowski plane is a Lie group when suitably topologized. The well-known structure of Lie groups of small dimensions can then be used to classify the most homogeneous flat Minkowski planes.

There clearly are two kinds of isomorphisms between two toroidal circle planes $\mathcal{M}_1$ and $\mathcal{M}_2$. If (+)-parallel classes are always taken to (+)-parallel classes, an isomorphism must have the form

$$(x, y) \mapsto (\alpha(x), \beta(y)),$$

where α and β are permutations of $\mathbf{S}^1$. In the other kind of isomorphism (+)-parallel classes of $\mathcal{M}_1$ are taken to (–)-parallel classes of $\mathcal{M}_2$ and such an isomorphism must have the form

$$(x, y) \mapsto (\alpha(y), \beta(x)),$$

where α and β are permutations of $\mathbf{S}^1$.

4.4.1 Automorphisms

If an automorphism γ of a toroidal circle plane fixes a point p, then clearly γ induces an automorphism of the derived $\mathbf{R}^2$-plane at this point. In particular, each such automorphism is continuous; see Theorem 2.4.2. More generally, an automorphism γ of a toroidal circle plane $\mathcal{M}$ induces an isomorphism from the derived $\mathbf{R}^2$-plane at any point p onto the derived $\mathbf{R}^2$-plane at $\gamma(p)$. Since isomorphisms between $\mathbf{R}^2$-planes are continuous, we see that γ must be continuous. In fact, the above argument applies to isomorphisms between toroidal circle planes; see Schenkel [1980] 5.1 for flat Minkowski planes.

THEOREM 4.4.1 (Isomorphisms Are Continuous) *Every isomorphism between toroidal circle planes is continuous. In particular every automorphism of a toroidal circle plane is continuous.*

As in the case of flat linear spaces and spherical circle planes an automorphism of a flat Minkowski plane that fixes sufficiently many points is the identity.

LEMMA 4.4.2 (Automorphism with Fixed Points) *If an automorphism γ of a flat Minkowski plane fixes three pairwise nonparallel points, then $\gamma^2 = id$.*

In particular, if such an automorphism γ takes (+)-parallel classes to (+)-parallel classes, then $\gamma = id$.

Proof. Suppose that γ fixes the three pairwise nonparallel points p_1, p_2 and p_3. In the derived projective plane $\mathcal{P}$ at p_3 the automorphism γ induces a collineation $\tilde{\gamma}$ that fixes the finite points p_1 and p_2 and leaves the set $\{\omega_+, \omega_-\}$ of infinite points corresponding to (+)- and (−)-parallel classes fixed. But then $\tilde{\gamma}^2$ fixes p_1, p_2, ω_+, and ω_-. These four points form a nondegenerate quadrangle in $\mathcal{P}$. Hence $\tilde{\gamma}^2 = id$ by Corollary 2.4.4. Therefore $\gamma^2 = id$ as well.

If γ takes (+)-parallel classes to (+)-parallel classes, then $\tilde{\gamma}$ already fixes ω_+ and ω_- so that $\tilde{\gamma} = id$ and thus $\gamma = id$. □

4.4.2 Groups of Automorphisms

In this subsection we establish that the automorphism group of a flat Minkowski plane is a Lie group and determine its maximum dimension. The crucial step is to show that if automorphisms converge on triodes, they converge to an automorphism globally; compare Schenkel [1980] 5.5.

LEMMA 4.4.3 (Convergence on Triodes) *If a sequence of automorphisms γ_n of a flat Minkowski plane $\mathcal{M}$ converges on the vertices of a triode towards points that also form the vertices of a triode, then the automorphisms γ_n converge pointwise to an automorphism γ of $\mathcal{M}$.*

The automorphism group Γ of a toroidal circle plane is an effective topological transformation group on the torus $\mathbf{S}^1 \times \mathbf{S}^1$ when Γ carries the compact-open topology (or, equivalently, the topology of uniform convergence on the torus). For Γ to be a Lie group we need that Γ is locally compact; see Theorem A2.3.5. This condition can be verified for flat Minkowski planes; see Schenkel [1980] 5.6 or Steinke [1984b]; compare also Subsection 3.4.1 for the corresponding result for spherical circle planes. Using the above lemma the automorphism group can be identified with a subset ('vertices of triodes') of $(\mathbf{S}^1 \times \mathbf{S}^1)^3$.

THEOREM 4.4.4 (Automorphism Group) *The collection Γ of all automorphisms of a flat Minkowski plane is a Lie group with respect to the compact-open topology (or, equivalently, the topology of uniform convergence) of dimension at most* 6.

The statement on the dimension of the automorphism group is an immediate consequence of the dimension formula A2.3.6 and Lemma 4.4.2 on the stabilizer of three pairwise nonparallel points because each point has an orbit of dimension at most 2.

As for flat linear spaces and other flat circle planes we say that a flat Minkowski plane $\mathcal{M}$ has *group dimension* n if the (full) automorphism group of $\mathcal{M}$ has dimension n.

In fact, if the group dimension of a flat Minkowski plane is at least 5, then the plane is classical; see Theorem 4.4.12. Examples in all other possible dimensions can be found among the planes $\mathcal{M}(f,g)$ described in Subsection 4.3.1; see Steinke [1994]. Let $f(x) = x^3$ and $g(x) = \sinh(x)$ for $x \in \mathbf{R}$ extended to $\mathbf{S}^1$ by $f(\infty) = g(\infty) = \infty$. Then we have the following flat Minkowski planes; compare Theorem 4.3.4.

- $\mathcal{M}(id, id)$ is classical. It has group dimension 6.
- $\mathcal{M}(f, id)$ has group dimension 4. This plane admits the transformations
$$(x,y) \mapsto \left(\frac{ax+b}{cx+d}, ry\right),$$
where $a, b, c, d, r \in \mathbf{R}$, $ad - bc = 1$, $r > 0$, as automorphisms.
- $\mathcal{M}(g, id)$ has group dimension 3. This plane admits the transformations
$$(x,y) \mapsto \left(\frac{ax+b}{cx+d}, y\right),$$
where $a, b, c, d \in \mathbf{R}$, $ad - bc = 1$, as automorphisms.
- $\mathcal{M}(f, f)$ has group dimension 2. This plane admits the transformations $(x, y) \mapsto (sx, ry)$, where $r, s \in \mathbf{R}^+$, as automorphisms.
- $\mathcal{M}(f, g)$ has group dimension 1. This plane admits the transformations $(x, y) \mapsto (x, ry)$, where $r \in \mathbf{R}^+$, as automorphisms.
- $\mathcal{M}(g, g)$ has group dimension 0.

4.4.3 The Kernels

The automorphism group Γ of a toroidal circle plane $\mathcal{M}$ has two distinguished normal subgroups, the *kernels* T^+ and T^- of the action of Γ on the sets of (+)- and (−)-parallel classes, that is, T^+ consists of all automorphisms of $\mathcal{M}$ that fix each (+)-parallel class globally, and similarly for T^-. If $\gamma \in T^\pm$ fixes a point p, then γ fixes each point in the ($\mp$)-parallel class of p. Furthermore, in the case of a flat Minkowski plane, γ induces a central collineation $\tilde{\gamma}$ in the derived flat projective plane at p with centre $\omega^\pm$. Note that $\omega^\mp$ is a fixed point of $\tilde{\gamma}$ so that the axis of $\tilde{\gamma}$ passes through $\omega^\mp$. In the Minkowski plane this means that γ fixes all points on a ($\mp$)-parallel class or no points.

In the case of an automorphism in one of the kernels Lemma 4.4.2 becomes the following.

LEMMA 4.4.5 (Kernel Automorphisms with Fixed Points) *If an automorphism $\gamma \in T^{\pm}$ of a toroidal circle plane fixes three pairwise non-$(\mp)$-parallel points, then $\gamma = id$. In particular, the stabilizer in $T^{\pm}$ of a circle is trivial and $T^{\pm}$ acts freely and effectively on the circle set.*

Proof. If $\gamma \in T^{\pm}$ fixes three pairwise non-$(\mp)$-parallel points, then γ fixes the circle C through these three points. However, the automorphism γ fixes each $(\pm)$-parallel class so that γ fixes C pointwise. This implies that γ fixes each $(\pm)$-parallel class and $\gamma = id$. □

The dimension of a kernel can again be calculated using the dimension formula A2.3.6 and Lemma 4.4.5. Note that now each point has an orbit of dimension at most 1.

PROPOSITION 4.4.6 (Dimensions of the Kernels) *Both kernels T^+ and T^- of a flat Minkowski plane are closed normal subgroups of Γ of dimension at most* 3.

Let Γ_0 be the set of all automorphisms that take $(+)$-parallel classes to $(+)$-parallel classes and $(-)$-parallel classes to $(-)$-parallel classes. Clearly, the group Γ_0 is a closed normal subgroup of Γ of index at most 2. Furthermore, the connected component Γ^1 of Γ that contains the identity and each kernel $T^{\pm}$ are contained in Γ_0.

The factor group $\Gamma_0/T^{\pm}$ operates effectively on the set $\Pi^{\pm}$ of $(\pm)$-parallel classes. Since $T^+ \cap T^- = \{id\}$, the kernel $T^{\mp}$ can be canonically identified with a closed normal subgroup of $\Gamma_0/T^{\pm}$. Furthermore, Γ_0 can be identified with a closed subgroup of $(\Gamma_0/T^+) \times (\Gamma_0/T^-)$. In fact, $\Gamma_0/T^{\pm}$ is at most 3-dimensional; see the proposition below. This yields another method to verify that a flat Minkowski plane has group dimension at most 6.

PROPOSITION 4.4.7 (Closed Subgroups) *Let Σ be a closed connected subgroup of the automorphism group of a flat Minkowski plane $\mathcal{M}$ and let $\Delta^{\pm}$ be the kernel of the action of Σ on the set $\Pi^{\pm}$ of $(\pm)$-parallel classes. Then $\Sigma/\Delta^{\pm}$ has dimension at most* 3.

Proof. We assume, to the contrary, that Σ/Δ^+ has dimension at least 4. Since Σ/Δ^+ is effective on Π^+, Corollary A2.3.9 shows that Σ/Δ^+ fixes at least two elements of Π^+, that is, Σ fixes at least two $(+)$-parallel classes π_1 and π_2. Choosing points on π_1 and π_2 the dimension formula A2.3.6 and Lemma 4.4.2 then show that Σ must be 4-dimensional so that Δ^+ is 0-dimensional. Furthermore, Σ must act 2-transitively on π_1. Brouwer's Theorem A2.3.8 tells us that Σ has a 1-dimensional kernel on π_1, which of course is Δ^-, and Σ/Δ^- is a finite covering group of $\mathrm{PSL}_2(\mathbf{R})$ and thus almost simple.

As a subgroup of Σ the kernel Δ^- fixes both π_1 and π_2, In fact, Δ^- fixes those parallel classes pointwise. But Δ^- is 1-dimensional and Lemma 4.4.5 shows that Δ^- must be transitive on each connected component C_i, $i = 1, 2$, of $\Pi^+ \setminus \{\pi_1, \pi_2\}$. Therefore Σ acts transitively on C_1 and C_2, too. Let K_i be the kernel of the action of Σ on C_i for $i = 1, 2$. By Corollary A2.3.9 we know that K_i is at least 1-dimensional. Moreover, $K_i \cap \Delta^- = \{id\}$.

The canonical homomorphism $\varphi : \Sigma \to \Sigma/\Delta^-$ takes K_i onto a closed normal subgroup of Σ/Δ^-. Hence $\varphi(K_i)$ is either discrete or 3-dimensional. But $K_i \cap \Delta^- = \{id\}$ so that K_i is isomorphic to $\varphi(K_i)$. As seen before, K_i is at least 1-dimensional and this implies that K_i is 3-dimensional.

The two closed subgroups K_1 and K_2 must then intersect in a subgroup of Σ of dimension at least 2. However, $K_1 \cap K_2 = \Delta^+$ in contradiction to Δ^+ being 0-dimensional. □

LEMMA 4.4.8 (Compact Subgroups of Kernels) *The kernel $T^\pm$ of a flat Minkowski plane contains no compact, connected subgroup of dimension at least* 2.

Proof. Let Δ be a compact, connected subgroup of $T^\pm$. Since every subgroup of $T^\pm$ acts effectively on $\Pi^\mp \simeq \mathbf{S}^1$, Theorem A2.3.13 shows that $\dim \Delta \leq 1$. □

From Lemmas 4.4.5, 4.4.8, the classification of Lie groups of dimension at most 3 and their possible actions on $\mathbf{S}^1$ or $\mathbf{R}$ (see Appendix 2) one obtains the following.

PROPOSITION 4.4.9 (Connected Components of Kernels) *The connected component of a kernel $T^\pm$ of a flat Minkowski plane is isomorphic to* $\mathrm{PSL}_2(\mathbf{R})$, L_2, $\mathrm{SO}_2(\mathbf{R})$, $\mathbf{R}$, *or the trivial group* $\{id\}$.

Proof. Let Δ be the connected component of $T^\pm$. If Δ is at most 2-dimensional, then the statement readily follows from Lemma 4.4.8 and Theorem A2.2.4. Furthermore, by Theorem A2.3.13, a maximally compact subgroup of Δ can be at most 1-dimensional. Hence, if Δ is 3-

dimensional, Δ cannot be compact and thus not isomorphic to $\mathrm{SO}_3(\mathbf{R})$ or $\mathrm{SU}_2(\mathbf{C})$.

We now assume that Δ is 3-dimensional. The dimension formula then shows that each $(\mp)$-parallel class has a 1-dimensional orbit so that Δ acts transitively on $\Pi^{\mp}$. In fact, repeating the argument, Δ must be 2-transitive on $\Pi^{\mp}$. Since Δ acts effectively on $\Pi^{\mp}$, Brouwer's Theorem A2.3.8 then shows that Δ is isomorphic to a finite covering group of $\mathrm{PSL}_2(\mathbf{R})$. The centre Z of Δ therefore is a finite cyclic group. Let $\delta \in Z$ and $\pi \in \Pi^{\mp}$. Since δ commutes with the stabilizer Δ_π of π, we see that Δ_π fixes $\delta(\pi)$ and thus that $\delta(\pi) = \pi$. But Δ is transitive on $\Pi^{\mp}$ and is centralized by δ so that δ fixes the whole orbit $\Pi^{\mp}$ elementwise. This shows that $\delta \in T^{\pm} \cap T^{\mp} = \{id\}$. Hence Z must be trivial and Δ is isomorphic to $\mathrm{PSL}_2(\mathbf{R})$. □

4.4.4 Planes Admitting 3-Dimensional Kernels

From Proposition 4.4.6 we know that a kernel of a flat Minkowski plane can be at most 3-dimensional. In this subsection we determine those flat Minkowski planes for which the maximum dimension occurs, that is, flat Minkowski planes that admit a 3-dimensional kernel. We show that the resulting planes are the planes $\mathcal{M}(f, id)$ introduced in Subsection 4.3.1; see Schenkel [1980] Satz 5.9.

Note that, by Proposition 4.4.9, a 3-dimensional kernel contains a subgroup isomorphic to the group $\mathrm{PSL}_2(\mathbf{R})$.

THEOREM 4.4.10 (Planes with 3-Dimensional Kernel) *Let $\mathcal{M}$ be a flat Minkowski plane such that the kernel T^+ is 3-dimensional. Then $\mathcal{M}$ is isomorphic to a plane $\mathcal{M}(f, id)$, where f is an orientation-preserving homeomorphism of $\mathbf{S}^1$. Without loss of generality, we may assume that f fixes the points ∞, 0, and 1.*

Proof. Because of Proposition 4.4.9 we know that the connected component Δ^+ of T^+ containing the identity is isomorphic to $\mathrm{PSL}_2(\mathbf{R})$. The group Δ^+ acts transitively and freely on each connected component of the circle space; see Lemma 4.4.5. We may assume that Δ^+ acts on the torus as $(x, y) \mapsto (x, \delta(y))$ for $\delta \in \mathrm{PSL}_2(\mathbf{R})$. Let f_1 and f_2 be two homeomorphisms of $\mathbf{S}^1$ that describe circles through (∞, ∞) and $(0, 0)$ in different connected components of the circle space of $\mathcal{M}$. Without loss of generality, say f_1 and f_2 are orientation-preserving and orientation-reversing, respectively. Then the graphs of the homeomorphisms in $\Delta^+ f_1 \cup \Delta^+ f_2$ form the circle set of $\mathcal{M}$. Changing the coordinates in $\mathcal{M}$ with the transformation $(x, y) \mapsto (x, f_1(y))$, we obtain the circle set $\Delta^+ \cup \Delta^+ f_2 f_1^{-1}$.

Let $\gamma(x) = sx$ for $x \in \mathbf{R} \cup \{\infty\}$ where $s = f_2 f_1^{-1}(1) < 0$. Then

$$\Delta^+ f_2 f_1^{-1} = \Delta^+ \gamma\gamma^{-1} f_2 f_1^{-1} = (\mathrm{PGL}_2(\mathbf{R}) \setminus \mathrm{PSL}_2(\mathbf{R}))f,$$

where $f = \gamma^{-1} f_2 f_1^{-1}$ is an orientation-preserving homeomorphism of $\mathbf{S}^1$ that fixes the points ∞, 0 and 1. Clearly, the flat Minkowski plane $\mathcal{M}$ is isomorphic to $\mathcal{M}(f, id)$. □

We now obtain a first characterization of the classical flat Minkowski plane as follows; see Schenkel [1980] p. 59.

COROLLARY 4.4.11 (3-Dimensional Kernel and Desarguesian Derived Plane Equals Classical) *A flat Minkowski plane is classical if and only if one of the kernels $T^\pm$ is 3-dimensional and at least one derived affine plane is Desarguesian.*

Proof. By Theorem 4.4.10 the Minkowski plane $\mathcal{M}$ must be isomorphic to a plane $\mathcal{M}(f, id)$ for some orientation-preserving homeomorphism f of $\mathbf{S}^1$. The derived projective plane at a point of $\mathcal{M}(f, id)$ is isomorphic to the dual of a semi-classical flat projective plane; see Subsection 2.7.2. Such a plane is Desarguesian if and only if $f \in \mathrm{PSL}_2(\mathbf{R})$; compare Proposition 2.7.1. Hence the plane is classical. □

4.4.5 Classification with Respect to the Group Dimension

Schenkel investigated flat Minkowski planes admitting an automorphism group of dimension at least 4 in her dissertation Schenkel [1980]. We give her results and include new proofs. From Theorem 4.4.4 we know that the automorphism group of a flat Minkowski plane $\mathcal{M}$ is a Lie group of dimension at most 6, that is, the group dimension of $\mathcal{M}$ is at most 6. We begin with the classification of flat Minkowski planes of group dimension at least 5 and show that such a Minkowski plane must be classical; see Schenkel [1980] Satz 5.10.

THEOREM 4.4.12 (At Least 5-Dimensional Equals Classical) *A flat Minkowski plane is isomorphic to the classical flat Minkowski plane if and only if its automorphism group has dimension at least 5.*

Proof. Let Σ be a closed connected subgroup of the automorphism group of a flat Minkowski plane $\mathcal{M}$ and assume that Σ has dimension at least 5. Let $\Delta^{\pm}$ be the kernel of the action of Σ on the set $\Pi^{\pm}$ of $(\pm)$-parallel classes. If Σ does not act transitively on either Π^+ or Π^-, then Σ fixes a $(+)$-parallel class π_+ and a certain $(-)$-parallel class π_- and thus the point $p = \pi_+ \cap \pi_-$. The dimension formula A2.3.6 implies that $\dim \Sigma \leq 4$, in contradiction to our assumption $\dim \Sigma \geq 5$.

We may therefore assume that Σ is transitive on Π^+. By Brouwer's Theorem A2.3.8 we then obtain that the factor group Σ/Δ^+ is 1- or 3-dimensional. The former case implies that $\dim \Delta^+ = \dim \Sigma - 1 \geq 4$ in contradiction to Proposition 4.4.6. Hence we must have that Σ/Δ^+ is 3-dimensional and, moreover, that Σ/Δ^+ is almost simple (a finite covering group of $\mathrm{PSL}_2(\mathbf{R})$). Since Δ^- can be identified with a closed normal subgroup of Σ/Δ^+, we obtain that Δ^- is either discrete or 3-dimensional.

The former cannot occur by Proposition 4.4.7 because then Σ/Δ^- is at least 5-dimensional. In the latter case $\mathcal{M}$ is isomorphic to a plane $\mathcal{M}(f, id)$ by Theorem 4.4.10. But then $\mathcal{M}$ must be classical by Theorem 4.3.3. □

Before we proceed with the classification of flat Minkowski planes of group dimension 4 we give a characterization of the Artzy–Groh planes (see Subsection 4.3.5) in terms of their automorphism groups.

THEOREM 4.4.13 (Characterization of Artzy–Groh Planes) *A flat Minkowski plane $\mathcal{M}$ in standard representation is isomorphic to an Artzy–Groh plane $\mathcal{M}'(f, g)$ if and only if $\mathcal{M}$ admits the group*

$$\{(x, y) \mapsto (rx + b, ry + c) \mid b, c, r \in \mathbf{R}, r > 0\}$$

as group of automorphisms.

The above theorem follows immediately from Artzy–Groh [1986]. For a flat Minkowski plane in standard representation to have the above group of automorphisms implies that the derived affine plane at (∞, ∞) is a translation plane and therefore Desarguesian. Hence we have all nonhorizontal and nonvertical Euclidean lines as circles. The form of the other circles was then derived by Artzy and Groh.

For the generalized Hartmann planes one then obtains the following characterization in terms of their automorphism groups.

THEOREM 4.4.14 (Characterization of Gen. Hartmann Planes) *A flat Minkowski plane $\mathcal{M}$ in standard representation is isomorphic to a generalized Hartmann plane $\mathcal{M}(r_1, s_1; r_2, s_2)$ for $r_1, s_1, r_2, s_2 \in \mathbf{R}^+$ if and only if $\mathcal{M}$ admits the group*

$$\{(x, y) \mapsto (rx + b, sy + c) \mid b, c, r, s \in \mathbf{R}, r, s > 0\}$$

as group of automorphisms.

The above theorem is an immediate consequence of Theorems 4.4.13 and 4.3.11, since the automorphism group Σ contains the transformations $(x, y) \mapsto (rx + b, ry + c)$ for all $b, c, r \in \mathbf{R}$, $r > 0$. Hence $\mathcal{M}$ is isomorphic to an Artzy–Groh plane $\mathcal{M}'(f, g)$ and, up to isomorphism, we may assume that f and g satisfy the assumptions made in Theorem 4.3.11. Therefore f and g must both be semi-multiplicative by Theorem 4.3.11 so that $\mathcal{M}'(f, g)$ is one of the generalized Hartmann planes $\mathcal{M}(r_1, s_1; r_2, s_2)$, where $r_1, s_1, r_2, s_2 \in \mathbf{R}^+$.

With these results the classification of flat Minkowski planes of group dimension 4 can be obtained as follows; see Schenkel [1980] Satz 5.11.

THEOREM 4.4.15 (Group-Dimension Classification) *Let $\mathcal{M}$ be a flat Minkowski plane of group dimension 4. Then $\mathcal{M}$ is isomorphic to one of the following planes.*

(i) *A nonclassical plane $\mathcal{M}(f, id)$, where f is a semi-multiplicative homeomorphism of the form $f_{d,s}$, $(d, s) \neq (1, 1)$; see p. 442 for a definition of semi-multiplicative homeomorphisms. This plane admits the 4-dimensional group of automorphisms*

$$\{(x, y) \mapsto (rx, \delta(y)) \mid r \in \mathbf{R}^+, \delta \in \mathrm{PSL}_2(\mathbf{R})\}.$$

(ii) *A nonclassical generalized Hartmann plane $\mathcal{M}(r_1, s_1; r_2, s_2)$, $r_1, s_1, r_2, s_2 \in \mathbf{R}^+$, $(r_1, s_1, r_2, s_2) \neq (1, 1, 1, 1)$. This plane admits the 4-dimensional group of automorphisms*

$$\{(x, y) \mapsto (rx + a, sy + b) \mid a, b, r, s \in \mathbf{R}, r, s > 0\}.$$

Proof. Let Σ be a closed connected group of automorphisms of the flat Minkowski plane $\mathcal{M}$ and let $\Delta^{\pm}$ be the kernel of the action of Σ on the set $\Pi^{\pm}$ of $(\pm)$-parallel classes.

We first assume that Σ is transitive on Π^+. Then Σ/Δ^+ is transitive and effective on Π^+. Hence Σ/Δ^+ is 1- or 3-dimensional by Brouwer's Theorem A2.3.8. In the former case, Δ^+ is 3-dimensional so that $\mathcal{M}$ is isomorphic to a plane $\mathcal{M}(f, id)$ for some orientation-preserving homeomorphism of $\mathbf{S}^1$; see Theorem 4.4.10. But $\mathcal{M}(f, id)$ admits a 4-dimensional group of automorphisms so that f must be projectively equivalent to a semi-multiplicative homeomorphism by Theorem 4.3.3. Hence, $\mathcal{M}$ is isomorphic to a plane $\mathcal{M}(f, id)$, where f is a semi-multiplicative homeomorphism $f_{d,s}$. The full automorphism group of such a plane has dimension 4 if and only if it is not classical, that is, if and only if $(d, s) \neq (1, 1)$; compare Theorem 4.3.3.

We now assume that Σ/Δ^+ is 3-dimensional. Then Σ/Δ^+ is almost simple (a finite covering group of $\mathrm{PSL}_2(\mathbf{R})$). Since Δ^- can be canonically identified with a closed normal subgroup of Σ/Δ^+, we see that Δ^- is either discrete or 3-dimensional. The former case cannot occur by Proposition 4.4.7. In the latter case, we see as above that $\mathcal{M}$ is isomorphic to a plane $\mathcal{M}(f_{d,s}, id)$.

We finally assume that Σ is not transitive on Π^+. By symmetry we further assume that Σ is not transitive on Π^- either. Then Σ fixes some $(+)$-parallel class and some $(-)$-parallel class and thus a point p. The dimension formula A2.3.6 and Lemma 4.4.2 on stabilizers show that Σ is transitive on $\mathbf{S}^1 \times \mathbf{S}^1 \setminus |p|$ and even that for any point q not parallel to p the stabilizer Σ_q must be transitive on each connected component of $\mathbf{S}^1 \times \mathbf{S}^1 \setminus (|p| \cup |q|)$. In particular, $\Sigma/\Delta^{\pm}$ is effective and transitive on $\Pi^{\pm} \setminus \{|p|_{\pm}\}$ and thus must be of dimension 2 or 3 by Brouwer's Theorem A2.3.8. Correspondingly, $\Delta^{\pm}$ is 2- or 1-dimensional.

If one kernel, say Δ^+, is 1-dimensional, then Σ/Δ^+ is 3-dimensional and, as seen before, Δ^- can only be 0- or 3-dimensional. This result contradicts our previous observation that Δ^- is 1- or 2-dimensional.

Hence we have shown that Δ^+ must be 2-dimensional. We likewise obtain that Δ^- is 2-dimensional, too. Therefore $\Sigma/\Delta^{\pm}$ is 2-dimensional and thus isomorphic to the group L_2 by Brouwer's Theorem A2.3.8. Furthermore, $\Delta^{\mp}$ is isomorphic to $\Sigma/\Delta^{\pm}$ so that $\Sigma \simeq \Delta^+ \times \Delta^-$. Since the action of $\Delta^{\pm}$ on $\Pi^{\mp}$ is equivalent to the usual affine action of L_2, we see that we can assume that Σ consists of all transformations of the form $(x, y) \mapsto (rx + b, sy + c)$ for all $b, c, r, s \in \mathbf{R}$, $r, s > 0$. Hence $\mathcal{M}$ is a generalized Hartmann plane by Theorem 4.4.14. Since $\mathcal{M}$ has group dimension 4, the plane $\mathcal{M}$ cannot be classical and we conclude that $(r_1, s_1, r_2, s_2) \neq (1, 1, 1, 1)$. □

No complete classification of flat Minkowski planes of group dimension 3 has been carried out yet. However, there are many different types of flat Minkowski planes of group dimension 3. For example, the Artzy–

Groh planes $\mathcal{M}'(f,g)$ for f or g not inversely semi-multiplicative (see Theorem 4.3.11), the modified classical Minkowski planes $\mathcal{M}(f)$ for f not semi-linear, that is, f is not of the form

$$x \mapsto \begin{cases} ax & \text{for } x \geq 0, \\ bx & \text{for } x \leq 0, \end{cases}$$

where $a, b \in \mathbf{R}$, $a, b > 0$, and the planes $\mathcal{M}(f, id)$ admitting a 3-dimensional kernel in case f is an orientation-preserving homeomorphism that is not projectively equivalent to a semi-multiplicative map (see Theorem 4.3.3) all have group dimension 3. These planes admit the three groups

$$\{(x,y) \mapsto (rx+b, ry+c) \mid r,b,c \in \mathbf{R}, r>0\},$$

$$\{(x,y) \mapsto (rx+b, \tfrac{1}{r}y+c) \mid r,b,c \in \mathbf{R}, r>0\},$$

$$\left\{(x,y) \mapsto \left(x, \frac{ay+b}{cy+d}\right) \;\middle|\; a,b,c,d \in \mathbf{R}, ad-bc=1\right\},$$

respectively, as groups of automorphisms.

4.4.6 Von Staudt's Point of View—Groups of Projectivities

Following von Staudt's point of view, the group of all projectivities of a circle plane was investigated by Kroll [1977b], [1977c] for general Minkowski planes and Löwen [1977], [1981d] for topological Minkowski planes. Here, as usual, a projectivity of a fixed circle C is a composition of finitely many perspectivities between circles, the first and last circle being C. Perspectivities between circles are defined in the familiar manner by the correspondence between the points of two circles via a pencil of circles with the obvious modifications for points of tangency or via parallel projections. These maps are continuous by coherence so that projectivities become homeomorphisms of the circle C to itself. For example, each circle C' gives rise to a projectivity of the distinguished circle C by projecting from C onto C' via $(+)$-parallel classes and then from C' back onto C via $(-)$-parallel classes. Explicitly, we have the function

$$\varphi_{C,C'} : C \to C : p \mapsto |(|p|_+ \cap C')|_- \cap C.$$

Given two triples of three distinct points of C we can find a projectivity of the form $\varphi_{C,C'}$ that takes the points in the first triple to the corresponding points in the second triple. More precisely, given the triples (p_1, p_2, p_3) and (q_1, q_2, q_3), let $r_i = |p_i|_+ \cap |q_i|_-$ for $i = 1, 2, 3$. Then r_1, r_2, r_3 are three pairwise nonparallel points and thus there is precisely one circle passing through these three points. This circle becomes the circle C' and, by construction, $\varphi_{C,C'}(p_i) = q_i$ for $i = 1, 2, 3$. This proves the following result.

LEMMA 4.4.16 (High Degree of Transitivity) *The group of projectivities of a circle C in a toroidal circle plane is* 3*-transitive on the points of C.*

Note that the group of projectivities Π_C of a circle C in a toroidal circle plane is independent of C, that is, the groups Π_C and $\Pi_{C'}$ are isomorphic for any circles C and C'. We can therefore simply speak of the group of projectivities without referring to any particular circle. The above lemma shows that the group of projectivities in a toroidal circle plane tends to be rather large. In fact, the classical flat Minkowski plane has the smallest group of projectivities, namely the group $\mathrm{PGL}_2(\mathbf{R})$. This is in marked contrast to the situation with respect to automorphism groups: The smaller the automorphism group of a flat Minkowski plane, the larger the group of projectivities.

Note that the projectivity $\varphi_{C,C'}$ completely describes the circle C' relative to C. Suppose we coordinatize the toroidal circle plane $\mathcal{M}$ in such a way that C is represented, as in Subsection 4.2.3, by the identity. Then $\varphi_{C,C'}$ is just the representing homeomorphism of the circle C'. Hence the set Σ of circle-representing homeomorphisms is contained in the group Π_C of projectivities of C. If we now assume that Π_C is sharply 3-transitive on C, then $\Sigma = \Pi_C$ and Σ is a group. Hence $\mathcal{M}$ is classical by Theorem 4.2.4.

THEOREM 4.4.17 (Sharply 3-Transitive Equals Classical) *A toroidal circle plane is isomorphic to the classical flat Minkowski plane if and only if the group of projectivities of a circle C is sharply* 3*-transitive on the points of C.*

Pursuing this line of investigation further, the following characterization of the classical flat Minkowski plane was obtained by Löwen [1981d].

THEOREM 4.4.18 (Locally Compact Equals Classical) *Let $\mathcal{M}$ be a flat Minkowski plane and let Π_C be the group of projectivities of a circle C. Then $\mathcal{M}$ is classical, if*

(i) *Π_C or its closure in the group of all homeomorphisms of C with respect to the compact-open topology τ is locally compact with respect to τ, or*

(ii) *Π_C acts ω-regularly on C, that is, there exists a finite set $F \subseteq C$ such that the subgroup in Π_C fixing F elementwise is discrete with respect to τ.*

See Polster [1996b] for examples of flat Minkowski planes whose groups of projectivities are homeomorphic to the group of piecewise projective homeomorphisms of the circle; see Subsection 2.9.2 for the most important properties of this group.

4.5 The Klein–Kroll Types

Similar to the Lenz–Barlotti classification of projective planes with respect to central collineations and the Hering classification of Möbius planes with respect to central automorphisms, Minkowski planes can be classified with respect to *central automorphisms*, that is, automorphisms that fix at least one point and induce central collineations in the derived projective plane at that fixed point. More precisely, one considers subgroups of central automorphisms which are linearly transitive, that is, the induced groups of central collineations are transitive on each central line except for the obvious fixed points, the centre and the point of intersection with the axis.

Unlike in the case of Möbius planes, where we have only two types of central automorphisms, central automorphisms of Minkowski planes come in a variety of types according to whether the axis of a central collineation in the derived projective plane at a fixed point is the line at infinity, a line that stems from a circle of the Minkowski plane or a line that comes from a parallel class of the Minkowski plane, and whether or not the centre is on the axis of the central collineation. Klein and Kroll therefore obtained far more types of Minkowski planes. They considered three types of central automorphisms, G-translations, q-translations and (p, q)-homotheties; see the following subsections. In fact, in their classification they considered groups of automorphisms and determined their types according to transitive subgroups of central automorphisms contained in them. In the following we only deal with the full automorphism group. We furthermore say that a flat Minkowski plane $\mathcal{M}$ is of Klein–Kroll type X if the full automorphism group of $\mathcal{M}$ is of type X; see the following subsections for the definitions of the various Klein–Kroll types.

4.5.1 *G-Translations*

Let G be a parallel class of a Minkowski plane $\mathcal{M}$. A *G-translation* of $\mathcal{M}$ is an automorphism of $\mathcal{M}$ that either fixes precisely the points of G or is the identity. A group of G-translations of $\mathcal{M}$ is called *G-transitive*, if it acts transitively on each parallel class H of type opposite to the type of G without the point of intersection with G. We say that the automorphism group Γ of $\mathcal{M}$ is G-transitive if Γ contains a G-transitive subgroup of G-translations. With respect to G-translations Klein and Kroll obtained six types of Minkowski planes, in fact, the more general hyperbola structures; see Klein–Kroll [1989] Theorem 3.4. A *hyperbola structure* is an incidence geometry that satisfies the Axioms B1, B3, and B4. The corresponding flat geometries are toroidal circle planes.

If $\mathcal{Z}$ denotes the set of all parallel classes G for which the hyperbola structure is G-transitive, then exactly one of the following holds.

A. $\mathcal{Z} = \emptyset$.
B. $|\mathcal{Z}| = 1$.
C. There is a point p such that $\mathcal{Z} = \{|p|_+, |p|_-\}$.
D. $\mathcal{Z}$ consists of all $(+)$-parallel classes or of all $(-)$-parallel classes.
E. $\mathcal{Z}$ consists of all $(+)$-parallel classes plus one $(-)$-parallel class or of all $(-)$-parallel classes plus one $(+)$-parallel class.
F. $\mathcal{Z}$ consists of all $(+)$- and all $(-)$-parallel classes.

Types D, E, and F all contain all $(+)$-parallel classes or all $(-)$-parallel classes. In this situation we can explicitly describe the flat Minkowski planes. These are the planes $\mathcal{M}(f, id)$ from Subsection 4.3.1.

LEMMA 4.5.1 (Types D, E, F Equal $\mathcal{M}(f, id)$) *If the set $\mathcal{Z}$ of all parallel classes G for which the automorphism group of the flat Minkowski plane $\mathcal{M}$ is G-transitive contains all $(-)$-parallel classes, then the kernel T^+ is 3-dimensional and $\mathcal{M}$ is isomorphic to a plane $\mathcal{M}(f, id)$ (see Subsection 4.3.1) for some orientation-preserving homeomorphism f of the unit circle $\mathbf{S}^1$.*

The lemma follows from Theorem 4.4.10 and the fact that T^+ is 3-dimensional by Brouwer's Theorem A2.3.8 and because T^+ must be 2-transitive on each (+)-parallel class; see Steinke [20XXd].

Since a 3-dimensional kernel contains a subgroup isomorphic to the group $\mathrm{PSL}_2(\mathbf{R})$ by Proposition 4.4.9, we readily obtain the following.

COROLLARY 4.5.2 (Type D) *A flat Minkowski plane is of Klein–Kroll type at least D if and only if the connected component of the identity in one kernel is isomorphic to* $\mathrm{PSL}_2(\mathbf{R})$.

An immediate consequence of the above Lemma 4.5.1, Corollary 4.4.11 and Proposition 2.4.9 (every flat translation plane is Desarguesian) is that type E cannot occur in flat Minkowski planes.

COROLLARY 4.5.3 (Type E) *If the set $\mathcal{Z}$ of all parallel classes G for which the automorphism group of the flat Minkowski plane $\mathcal{M}$ is G-transitive contains all (−)-parallel classes plus one (+)-parallel class, then $\mathcal{M}$ is the classical flat Minkowski plane and thus of type F.*

We can now characterize the flat Minkowski planes of Klein–Kroll type D as follows.

PROPOSITION 4.5.4 (Type D) *A flat Minkowski plane is of Klein–Kroll type D if and only if it is isomorphic to a plane $\mathcal{M}(f, id)$ for some orientation-preserving homeomorphism f of the unit circle* $\mathbf{S}^1$, *where* $f \notin \mathrm{PSL}_2(\mathbf{R})$.

In summary we obtain the following possible Klein–Kroll types with respect to G-translations.

PROPOSITION 4.5.5 (Possible Types) *A flat Minkowski plane is of Klein–Kroll type A, B, C, D, or F.*

For examples for each of the Klein–Kroll types A, B, C, D, and F see Subsection 4.5.4.

Clearly, a type F plane is the classical flat Minkowski plane, and, by Proposition 4.5.4, Klein–Kroll type D comprises, up to isomorphism, precisely the nonclassical flat Minkowski planes of the form $\mathcal{M}(f, id)$.

In a type C plane the derived affine plane at the distinguished point p is Desarguesian and all translations are induced from automorphisms of the Minkowski plane. This type is realized in the Hartmann planes. Type A occurs in planes of group dimension 0.

4.5.2 q-*Translations*

Let q be a point of a Minkowski plane $\mathcal{M} = (P, \mathcal{C})$. A *$q$-translation* of $\mathcal{M}$ is an automorphism of $\mathcal{M}$ that either is the identity, or fixes precisely the point q and induces a translation of the derived affine plane $\mathcal{A}_q$ at q. More precisely, let C be a circle passing through q. As for flat Möbius planes let $B(q, C)$ denote the *touching pencil with support* q, that is, $B(q, C)$ consists of all circles that touch the circle C at the point q. In the derived affine plane at q the touching pencil represents a parallel class of lines and we can look at translations in this direction. Then a *$(q, B(q, C))$-translation* of $\mathcal{M}$ is a q-translation that fixes C (and thus each circle in $B(q, C)$) globally. A group of $(q, B(q, C))$-translations of $\mathcal{M}$ is called *$(q, B(q, C))$-transitive*, if it acts transitively on $C \setminus \{q\}$. A group of q-translations is called *q-transitive*, if it acts transitively on $P \setminus |q|$. We say that the automorphism group Γ of $\mathcal{M}$ is $(q, B(q, C))$-transitive or q-transitive if Γ contains a $(q, B(q, C))$-transitive subgroup of $(q, B(q, C))$-translations or a q-transitive subgroup of q-translations, respectively.

With respect to q-translations Klein and Kroll obtained seven types of Minkowski planes; see Klein–Kroll [1989] Theorem 4.9.

If $\mathcal{T}$ denotes the set of all points q for which the Minkowski plane $\mathcal{M}$ with automorphism group Γ is $(q, B(q, C))$-transitive for some touching pencil $B(q, C)$ with support q, then exactly one of the following statements is valid.

I. $\mathcal{T} = \emptyset$.
II. There is one point q such that $\mathcal{T} = \{q\}$ and there is exactly one touching pencil with support q such that Γ is $(q, B(q, C))$-transitive.
III. There is a point q such that $\mathcal{T} = \{q\}$ and Γ is q-transitive.
IV. $\mathcal{T}$ consists of the points on a circle.
V. $\mathcal{T}$ consists of the points on a parallel class.
VI. $\mathcal{T} = P$ and for each point q there is exactly one touching pencil $B(q, C)$ with support q such that $\mathcal{M}$ is $(q, B(q, C))$-transitive.
VII. $\mathcal{T} = P$ and Γ is q-transitive for every point q.

Types V and VI cannot occur in flat Minkowski planes.

LEMMA 4.5.6 (Type VII) *If the set $\mathcal{T}$ of a flat Minkowski plane $\mathcal{M}$ contains all points of a parallel class, then $\mathcal{M}$ is the classical flat Minkowski plane and thus of type VII.*

To prove the above lemma one shows that the automorphism group Γ of such a flat Minkowski plane $\mathcal{M}$ is at least 4-dimensional. One then uses the classification of flat Minkowski planes of group dimension at least 4 (see Theorems 4.4.12 and 4.4.15) to show that nonclassical planes cannot occur; see Steinke [20XXd]. Of course, the classical flat Minkowski plane is of Klein–Kroll type VII.

Since in Minkowski planes of types V and VI the set $\mathcal{T}$ contains all points on a parallel class, we readily obtain the following possible types for flat Minkowski planes.

PROPOSITION 4.5.7 (Possible Types) *A flat Minkowski plane is of Klein–Kroll type I, II, III, IV, or VII.*

For examples for types I, III, and VII see Subsection 4.5.4.

Again, type VII is the classical flat Minkowski plane. In a type III plane the derived affine plane at the distinguished point q is Desarguesian and all translations extend to automorphisms of the Minkowski plane. This type is realized in the Hartmann planes. Type I occurs in planes of group dimension 0.

By combining both the classifications with respect to G-translations and q-translations Klein–Kroll [1989] Theorem 4.12 obtained ten types of Minkowski planes, of which only seven types can possibly occur in flat Minkowski planes.

PROPOSITION 4.5.8 (Combined Types w.R.t. Translations) *A flat Minkowski plane is of Klein–Kroll type I.A, I.B, I.D, II.A, III.C, IV.A, or VII.F.*

For examples for types I.A, I.B, I.D, III.C, and VII.F see Subsection 4.5.4.

4.5.3 (p, q)*-Homotheties*

Finally, a third kind of central automorphisms has been used in the classification in Klein [1992]. Let p and q be two nonparallel points of a Minkowski plane $\mathcal{M}$. An automorphism γ of $\mathcal{M}$ is a (p, q)-*homothety* if γ fixes p and q and induces a homothety with centre q in the derived affine plane at p. Note that each (p, q)-homothety is also a (q, p)-homothety. A group of (p, q)-homotheties is called (p, q)-*transitive* if it acts transitively on each circle through p and q minus the two points p and q. We say that the automorphism group Γ of $\mathcal{M}$ is (p,q)-transitive if Γ contains a (p,q)-transitive subgroup of (p,q)-homotheties.

With respect to (p,q)-homotheties Klein [1992] obtained 23 types of Minkowski planes. Some of the types are known to occur only in finite Minkowski planes or only as the type of a proper subgroup of the full automorphism group; see Klein [1992], Klein–Kroll [1994], Kroll [1995], Kroll–Matrás [1997], and Jakóbowski–Kroll–Matrás [20XX]. In the following we list only those types for which we have examples in flat Minkowski planes. In these types the set $\mathcal{H}$ of all unordered pairs of points $\{p,q\}$ for which the automorphism group of the flat Minkowski plane is (p,q)-transitive is of the following form.

1. $\mathcal{H} = \emptyset$.
3. There are two nonparallel points p and q such that
$$\mathcal{H} = \{\{p,q\}, \{|p|_+ \cap |q|_-, |p|_- \cap |q|_+\}\}.$$
19. There is a point p such that
$$\begin{aligned}\mathcal{H} &= \{\{p,q\} \mid q \text{ not parallel to } p\} \\ &\cup \{\{r,s\} \mid r \in |p|_+ \setminus \{p\}, s \in |p|_- \setminus \{p\}\}.\end{aligned}$$
23. $\mathcal{H}$ consists of all unordered pairs of nonparallel points.

PROPOSITION 4.5.9 (Possible Types) *A flat Minkowski plane is of type 1, 2, 3, 10, 11, 12, 13, 14, 15, 17, 18, 19, 20, or 23.*

For examples for types 1, 3, 19, and 23 see Subsection 4.5.4.

Type 23 describes the classical flat Minkowski plane. Type 19 yields the proper Hartmann planes $\mathcal{M}(r,1;r,1)$ in the notation of Subsection 4.3.4; see Klein–Kroll [1994] Theorem 4.1.

PROPOSITION 4.5.10 (Type 19) *A flat Minkowski plane $\mathcal{M}$ is of type 19 if and only if $\mathcal{M}$ is a proper Hartmann planes $\mathcal{M}(r,1;r,1)$ for some $r \in \mathbf{R}^+$, $r \neq 1$.*

Combining all three classifications Klein [1992] Theorem 2.16 obtained 32 types of Minkowski planes. Of those only 20 types can possibly occur in flat Minkowski planes.

THEOREM 4.5.11 (Combined Types) *A flat Minkowski plane is of Klein–Kroll type I.A.1, I.A.2, I.A.3, I.B.1, I.B.10, I.B.11, I.B.12, I.D.1, I.D.13, I.D.14, II.A.1, II.A.13, II.A.14, III.C.1, III.C.17, III.C.18, III.C.19, IV.A.1, IV.A.20, or VII.A.23*

For examples of the types I.A.1, I.A.3, I.B.1, I.D.1, III.C.1, III.C.19 and VII.F.23 see the following subsection.

4.5.4 Some Examples

So far no complete classification of those types that can occur in flat Minkowski planes has been carried out. Examples for some of the possible Klein–Kroll types of flat Minkowski planes are as follows.

I.A.1

A plane of group dimension 0. In particular, a rigid plane; see Subsection 4.7.1. Clearly, such a plane must be of Klein–Kroll type I.A.1. For example, the plane $\mathcal{M}(f, f)$, where $f(\infty) = \infty$ and $f(x) = \sinh x$ for $x \in \mathbf{R}$ is of Klein–Kroll type I.A.1; compare Theorem 4.3.4.

I.A.3

The plane $\mathcal{M}(f, f)$, where $f(x) = x^3$ for $x \in \mathbf{R}$ and $f(\infty) = \infty$; compare Theorem 4.3.4. The distinguished points p and q are the points (∞, ∞) and $(0, 0)$. For each $r \in \mathbf{R}$, $r \neq 0$, the transformation $(x, y) \mapsto (rx, ry)$, where $r \cdot \infty = \infty$, is an $\{(\infty, \infty), (0, 0)\}$-homothety. Likewise, the transformation $(x, y) \mapsto (rx, \frac{y}{r})$ is an $\{(\infty, 0), (0, \infty)\}$-homothety.

I.B.1

The plane obtained from a modified classical Minkowski plane $\mathcal{M}(g)$ by swapping its negative half with the negative half of the plane $\mathcal{M}(f, id)$, where $f \notin \mathrm{PGL}_2(\mathbf{R})$ and g is not the identity. For example $f(x) = x^3$ and $g(x) = 2x$ for $x \in \mathbf{R}$ and $f(\infty) = g(\infty) = \infty$ yields a flat Minkowski plane of Klein–Kroll type I.B.1. Each translation in the y-direction extends to an automorphism of the Minkowski plane.

I.D.1

The planes $\mathcal{M}(f, id)$, where $f \notin \mathrm{PGL}_2(\mathbf{R})$. For example $f(x) = x^3$ for $x \in \mathbf{R}$ and $f(\infty) = \infty$ yields a flat Minkowski plane of Klein–Kroll type I.D.1. The transformations $(x, y) \mapsto (x, \delta(y))$, where $\delta \in \mathrm{PSL}_2(\mathbf{R})$ has precisely one fixed point, are G-translations.

III.C.1

The generalized Hartmann planes $\mathcal{M}(r_1, s_1; r_2, s_2)$, $r_1, r_2, s_1, s_2 \in \mathbf{R}^+$ and $r_1 \neq r_2$. The distinguished point is the point $p = (\infty, \infty)$. Each translation of the derived affine plane at p extends to an automorphism of the flat Minkowski plane.

III.C.19

The Hartmann planes $\mathcal{M}(d, 1; d, 1)$, $d \in \mathbf{R}^+$, $d \neq 1$. The distinguished point is the point $p = (\infty, \infty)$. Each translation and each homothety of the (Desarguesian) derived affine plane at p extend to an automorphism of the Minkowski plane. The transformation $(x, y) \mapsto (rx, r|r|^{-d-1}y)$ for $r \in \mathbf{R}$, $r \neq 0$, is an $\{(\infty, 0), (0, \infty)\}$-homothety. Using translations one obtains $\{(\infty, c), (b, \infty)\}$-homotheties for any $b, c \in \mathbf{R}$.

VII.F.23

The classical flat Minkowski plane. Here all admissible subgroups of central automorphisms are linearly transitive.

4.6 Characterizations of the Classical Plane

In this section we characterize the classical flat Minkowski plane with respect to several geometric properties. We only state the characterizations. For the proofs the reader is referred to the same sources as for the proofs of the respective results in Subsection 3.6.2.

4.6.1 The Locally Classical Plane

Being classical is a local property of a flat Minkowski plane, that is, if a flat Minkowski plane looks like the classical flat Minkowski plane around each point, then the plane is classical.

THEOREM 4.6.1 (Locally Classical Equals Classical) *A locally classical flat Minkowski plane is classical, that is, isomorphic to the classical flat Minkowski plane.*

A global isomorphism is constructed via a monodromy argument by extending local isomorphisms; see Steinke [1983b].

4.6.2 The Miquelian Plane

Just like the classical flat Möbius plane the classical flat Minkowski plane is geometrically characterized by the fact that Miquel's configuration (locally) closes in it; see Kaerlein [1970] for the global version of this result and Steinke [1984a] for the local version.

THEOREM 4.6.2 (Locally Miquelian Equals Classical) *The classical flat Minkowski plane is the only flat Minkowski plane in which Miquel's theorem is (locally) valid.*

4.6.3 The Plane with Many Desarguesian Derivations

For each point p of a flat Minkowski plane $\mathcal{M}$ we obtain a derived affine plane at p. This flat affine plane may be Desarguesian or not; there are many flat affine planes that may possibly occur. Let $\mathcal{D}$ be the collection of all points of $\mathcal{M}$ for which the derived affine plane is Desarguesian. With this notation Steinke [1990b] characterized the classical flat Minkowski plane as follows.

THEOREM 4.6.3 (Classical Derived Planes Equals Classical) *A flat Minkowski plane is classical if and only if the set $\mathcal{D}$ contains all points on one $(+)$-parallel class or one $(-)$-parallel class and at least three pairwise nonparallel points.*

In fact, if there are enough derived affine planes that are Desarguesian, the flat Minkowski planes are completely determined; see Steinke [1990b].

THEOREM 4.6.4 (Many Classical Derived Planes) *A flat Minkowski plane is isomorphic to one of the Hartmann planes $\mathcal{M}(r,1;r,1)$ for some $r>0$ if and only if the set $\mathcal{D}$ contains all points on one $(+)$-parallel class or one $(-)$-parallel class and at least one further point.*

4.6.4 The Plane in Which the Rectangle Configuration Closes

We say that four points p_1, p_2, p_3, and p_4 in a flat Minkowski plane $\mathcal{M}$ form a *rectangle* $[p_1,p_2;p_3,p_4]$ if p_1 and p_3 are not parallel and, furthermore, $p_1||_-p_2||_+p_3||_-p_4||_+p_1$ (the sides of the rectangle are formed by parallel classes).

We then say that the *rectangle configuration closes in $\mathcal{M}$* if for any four rectangles $[a_1,a_2;a_3,a_4]$, $[b_1,b_2;b_3,b_4]$, $[c_1,c_2;c_3,c_4]$, and $[d_1,d_2;d_3,d_4]$ such that a_i, b_i, c_i and d_i are on a circle C_i for $i=1,2,3$, the remaining four points a_4, b_4, c_4, and d_4 are on a circle C_4, too. Figure 4.6 illustrates the rectangle configuration in the geometry of Euclidean lines and hyperbolas. The four circles in the axiom correspond to the four skew lines. In particular, the grey line corresponds to the circle C_4.

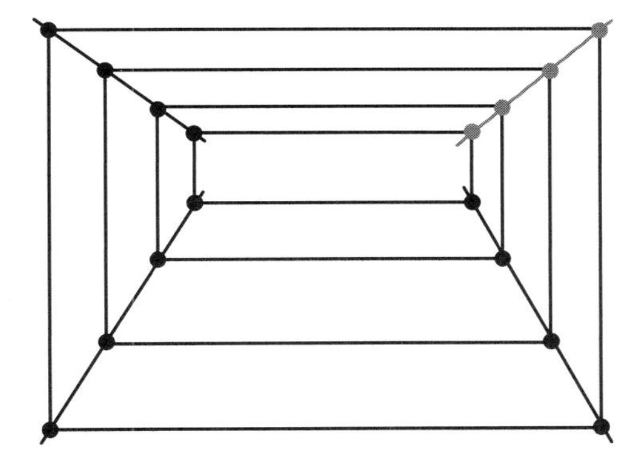

Fig. 4.6. The rectangle configuration

If the rectangle configuration closes the associated sharply 3-transitive set of circle-defining homeomorphisms is a group; see Wefelscheid [1977].

PROPOSITION 4.6.5 (Rectangle Equals Classical) *The rectangle configuration closes in a flat Minkowski plane $\mathcal{M}$ if and only the associated sharply* 3*-transitive set of homeomorphisms of* $\mathbf{S}^1$ *is a group.*

The above proposition together with Theorem 4.4.17 yields the following characterization of the classical flat Minkowski plane.

THEOREM 4.6.6 (Rectangle Equals Classical) *A flat Minkowski plane is classical if and only if the rectangle configuration closes in it.*

4.6.5 The Symmetric Plane

An *inversion at a circle C* of a flat Minkowski plane $\mathcal{M}$ is an automorphism σ of $\mathcal{M}$ that fixes precisely the points of C. Such an automorphism exchanges (+)- and (−)-parallel classes and has the property that a, $\sigma(a)$, b, and $\sigma(b)$ are on a circle for any points a and b with b parallel to neither a nor $\sigma(a)$. We say that a 2-set $\{p, q\}$ of nonparallel points p and q in $\mathcal{M}$ is *Miquelian* if the collection of all inversions of $\mathcal{M}$ that interchange p and q operates transitively on each circle through the points p and q minus the two points p and q. For a subgroup of the automorphism group of $\mathcal{M}$ the collection of all Miquelian pairs of points such that the corresponding inversions are in the subgroup are determined in Kroll–Matraś [1997]. There are nine types; see Kroll–Matraś [1997] Theorem 3.6. Only five of these types can occur in infinite Minkowski planes as the type of the full automorphism group. In the following we list those types, where M denotes the set of all Miquelian unordered pairs of points in the Minkowski plane.

M_0. $M = \emptyset$.
M_1. $|M| = 1$.
M_2. There are two nonparallel points p and q such that
$M = \{\{p,q\}, \{|p|_+ \cap |q|_-, |p|_- \cap |q|_+\}\}$.
M_3. There is a circle C such that $M = \{\{p,q\} \mid p,q \in C, p \neq q\}$.
M_4. M consists of all pairs of nonparallel points.

The classical flat Minkowski plane is the only plane of type M_4; see Percsy [1983] Theorem 1.3. Furthermore, if $\mathcal{M}$ is (p,q)-Miquelian, then $\mathcal{M}$ is also (p,q)-transitive; see Percsy [1983] 4.1(ii). In particular, Klein–Kroll type 1 implies type M_0 and a type M_i plane must be of Klein–Kroll type at least 2, 3, and 20 for $i = 1,2,3$, respectively. Indeed we have the following examples.

- The plane $\mathcal{M}(f,f)$, where $f(x) = \sinh x$ for $x \in \mathbf{R}$ and $f(\infty) = \infty$ is of type M_0; see Subsection 4.5.4.
- The plane $\mathcal{M}(f,f)$, where $f(x) = x^3$ for $x \in \mathbf{R}$ and $f(\infty) = \infty$ is of type M_2; see Subsection 4.5.4. The special points p and q are the points (∞,∞) and $(0,0)$. For each $m \in \mathbf{R}$, $m \neq 0$, the map $(x,y) \mapsto (y/m, mx)$ is an inversion at the circle $y = mx$ that exchanges $(\infty,0)$ and $(0,\infty)$. Furthermore, the collection of these inversions for $m \neq 0$ is transitive on each circle $y = -a/x$ through the two points $(\infty,0)$ and $(0,\infty)$ minus these two points. Likewise, the map $(x,y) \mapsto (-a/y, -a/x)$ for $a \neq 0$ is an inversion at the circle $y = -a/x$ that exchanges (∞,∞) and $(0,0)$.
- The classical flat Minkowski plane is of type M_4.

In fact, it suffices for a Minkowski plane to have two suitable Miquelian pairs in order to be Miquelian; see Kroll–Matrás [1997] Theorem 4.3.

PROPOSITION 4.6.7 (Two Miquelian Pairs Equals Classical) *Let $\mathcal{M}$ be a flat Minkowski plane that has two Miquelian pairs $\{p,q\}$ and $\{p',q'\}$ such that p' is $(+)$-parallel to p and $q' \neq q$ is $(-)$-parallel to q. Then $\mathcal{M}$ is the classical flat Minkowski plane.*

We say that a Minkowski plane is *symmetric* if it admits an inversion at every circle; see Percsy [1983], Dienst [1977], and Heise–Karzel [1973] for different definitions of a symmetric Minkowski plane. It readily follows that a Minkowski plane is symmetric as defined above if and only if each unordered pair of nonparallel points is Miquelian.

THEOREM 4.6.8 (Symmetric Equals Classical) *A flat Minkowski plane is isomorphic to the classical flat Minkowski plane if and only if each circle is the axis of an inversion.*

4.6.6 The Plane with Flag-Transitive Group

In Subsections 4.1.5 and 4.1.6 we have already seen that the automorphism group of the classical flat Minkowski plane is transitive on the set of all flags, that is, incident point–circle pairs. Conversely, the classical flat Minkowski plane is characterized by this property.

THEOREM 4.6.9 (Flag-Transitive Equals Classical) *A flat Minkowski plane is isomorphic to the classical flat Minkowski plane if and only if the plane is flag-transitive.*

Proof. Since, by Theorem 4.2.16, the flag space $\mathcal{F}$ of a flat Minkowski plane $\mathcal{M}$ is a 4-dimensional space, a group of automorphisms that is transitive on $\mathcal{F}$ must be of dimension at least 4. From the classification of flat Minkowski planes of group dimension at least 4 (see Theorems 4.4.12 and 4.4.15) we see that $\mathcal{M}$ must be isomorphic to a plane $\mathcal{M}(f, id)$, where f is semi-multiplicative or a generalized Hartmann plane $\mathcal{M}(r_1, s_1; r_2, s_2)$. However, the automorphism groups of the nonclassical planes of the former kind fix the parallel classes $\{\infty\} \times \mathbf{R}$ and $\{0\} \times \mathbf{R}$ so that they cannot be flag-transitive.

The automorphism groups of the nonclassical Hartmann planes fix the point (∞, ∞), so that they cannot be flag-transitive either. This shows that of all the flat Minkowski planes of group dimension at least 4 only the classical flat Minkowski plane has a flag-transitive automorphism group. □

4.6.7 The Plane of Klein–Kroll Type at Least V, E, or 21

We say that a flat Minkowski plane $\mathcal{M}$ is of type at least X if the automorphism group of $\mathcal{M}$ is of type X or one of the following types as given in the three lists of possible types in Section 4.5. For example, type at least V means that with respect to q-translations the plane is of type V, VI, or VII. Using the results in Section 4.5, we obtain the following characterizations of the classical flat Minkowski plane.

PROPOSITION 4.6.10 (At Least Type V Equals Classical) *A flat Minkowski plane $\mathcal{M}$ is of Klein–Kroll type at least V if and only if $\mathcal{M}$ is the classical flat Minkowski plane.*

PROPOSITION 4.6.11 (At Least Type E Equals Classical) *A flat Minkowski plane $\mathcal{M}$ is of Klein–Kroll type at least E if and only if $\mathcal{M}$ is the classical flat Minkowski plane.*

PROPOSITION 4.6.12 (At Least Type 21 Equals Classical) *A flat Minkowski plane $\mathcal{M}$ is of Klein–Kroll type at least 21 if and only if $\mathcal{M}$ is the classical flat Minkowski plane.*

4.6.8 Summary

We summarize the various characterizations of the classical flat Minkowski plane obtained in the previous subsections.

THEOREM 4.6.13 (Characterizations of the Classical Plane) *A flat Minkowski plane $\mathcal{M}$ in standard representation is isomorphic to the classical flat Minkowski plane if and only if any one of the following holds.*

(i) *The associated sharply 3-transitive set of homeomorphisms of $\mathbf{S}^1$ is a group.*
(ii) *The automorphism group Γ of $\mathcal{M}$ is flag-transitive.*
(iii) *Each circle is the axis of an inversion.*
(iv) *Each pair of distinct points occur as centres of a nontrivial central automorphism.*
(v) *Γ is at least 5-dimensional.*
(vi) *A kernel is 3-dimensional and at least one derived affine plane is Desarguesian.*
(vii) *$\mathcal{M}$ is Miquelian.*
(viii) *$\mathcal{M}$ is locally Miquelian.*
(ix) *$\mathcal{M}$ is locally classical.*
(x) *The rectangle configuration closes in $\mathcal{M}$.*
(xi) *$\mathcal{M}$ is of Klein–Kroll type at least V.*
(xii) *$\mathcal{M}$ is of Klein–Kroll type at least E.*
(xiii) *$\mathcal{M}$ is of Klein–Kroll type at least 21.*

(xiv) *$\mathcal{M}$ contains two Miquelian pairs $\{p, q\}$ and $\{p', q'\}$ such that p' is (+)-parallel to p and $q' \neq q$ is (−)-parallel to q.*

(xv) *$\mathcal{M}$ is symmetric.*

(xvi) *The group of projectivities Π_C of a circle C in $\mathcal{M}$ is sharply 3-transitive on the points of C.*

(xvii) *The group Π_C or its closure in the group of all homeomorphisms of the circle C with respect to the compact-open topology τ is locally compact with respect to τ.*

(xviii) *Π_C acts ω-regularly on C.*

4.7 Planes with Special Properties

In the following two subsections we consider flat Minkowski planes that do not admit any nontrivial automorphisms and differentiable flat Minkowski planes.

4.7.1 Rigid Planes

From the classification of flat Minkowski planes of group dimension at least 4 we see that the most homogeneous planes are of rather special form. As for flat linear spaces and spherical circle planes, we expect that 'most' flat Minkowski planes have small automorphism groups and that the vast majority of planes has 0-dimensional or even finite automorphism groups.

Choosing the homeomorphisms f and g suitably in the Minkowski planes $\mathcal{M}(f, g)$ constructed in Subsection 4.3.1 one even obtains flat Minkowski planes whose only automorphism is the identity. Again, we call such a plane *rigid.*

For example, let $a, c \in \mathbf{R}^+$ and let $h_{a,c}$ be the homeomorphism of $\mathbf{S}^1$ defined by

$$h_{a,c}(x) = \begin{cases} 1 - a(1 - x) & \text{if } x \in \mathbf{R},\ x \geq 1, \\ x & \text{if } 0 \leq x \leq 1, \\ cx & \text{if } x \in \mathbf{R},\ x \leq 0, \\ \infty & \text{if } x = \infty. \end{cases}$$

One can use this kind of homeomorphism to obtain rigid Minkowski planes. For each $a \in \mathbf{R}^+$, $a \neq 1$ the plane $\mathcal{M}(h_{a,a^2}, h_{a,a^3})$ is rigid. Furthermore, two planes $\mathcal{M}(h_{a,a^2}, h_{a,a^3})$ and $\mathcal{M}(h_{b,b^2}, h_{b,b^3})$ are nonisomorphic if $b \neq a$; see Steinke [1994].

4.7.2 Differentiable Planes

The point set $\mathbf{S}^1 \times \mathbf{S}^1$ and the circle set $\mathrm{PGL}_2(\mathbf{R})$ of a toroidal circle plane naturally carry smooth differentiable structures which make them into smooth manifolds. In the classical flat Minkowski plane each parallel class and each circle then are submanifolds of $\mathbf{S}^1 \times \mathbf{S}^1$ and the geometric operations are not only continuous (see Theorem 4.2.10) but even smooth maps with respect to the appropriate smooth manifold structures. We now can replace the topological conditions in the definition of a toroidal circle plane or flat Minkowski plane by differentiability assumptions and ask what planes can occur. More precisely, a *smooth (differentiable) toroidal circle plane or flat Minkowski plane* is a toroidal circle plane or flat Minkowski plane such that parallel classes and circles are submanifolds of the point space and such that the geometric operations are smooth on their respective domains of definition.

Since $\mathbf{S}^1$ has only one differentiable manifold structure and because parallel classes are submanifolds, the point space $\mathbf{S}^1 \times \mathbf{S}^1$ carries the usual differentiable product manifold structure. Each circle of a smooth toroidal circle plane is now represented (in the sense of Subsection 4.2.3) by a smooth diffeomorphism of $\mathbf{S}^1$.

Note that the flat Minkowski planes of group dimension 4 discussed in Theorem 4.4.15 are not differentiable because a semi-multiplicative map $f_{r,s}$ is differentiable in ∞ if and only if $r = s = 1$, and similarly for inversely semi-multiplicative maps.

If we assume that the homeomorphisms f and g in the flat Minkowski planes of the form $\mathcal{M}(f, g)$ (see Subsection 4.3.1) are even smooth diffeomorphisms, then, of course, each circle is represented by a diffeomorphism. This then implies that parallel projection is smooth. Since the two connected components of the circle space are totally independent of each other and because each component essentially consists of circles of the classical flat Minkowski plane, the negative component being modified by f and g, the operation of joining is smooth. Furthermore, because circles that touch each other must be in the same component, the operation of forming tangent circles basically can be performed in the classical flat Minkowski plane and thus is smooth, too. Finally, the operation of intersecting nondisjoint circles is smooth in pairs of circles in the same connected component of the circle space; we are again essentially in the classical flat Minkowski plane. If the circles involved are in different components, they intersect in two points. This operation is also smooth.

THEOREM 4.7.1 (Nonclassical Smooth Planes) *A flat Minkowski plane $\mathcal{M}(f,g)$ is smooth if and only if f and g are both smooth.*

For 4-dimensional compact Minkowski planes Scholz [1980] showed that such a plane is complex-analytic, that is, the point space is a complex-analytic manifold and the geometric operations are complex-analytic, if and only if it is isomorphic to the classical Minkowski plane over the complex numbers.

4.8 Subgeometries and Lie Geometries

Apart from the constructions in this section, a large number of constructions of other types of geometries from flat Minkowski planes are described in detail in Chapter 6. Also included in Chapter 6 is the solution of the Apollonius problem for flat Minkowski planes.

4.8.1 Flocks and Resolutions

A *flock* of a toroidal circle plane $\mathcal{M}$ is a partition of the point set $\mathbf{S}^1 \times \mathbf{S}^1$ into pairwise disjoint circles of $\mathcal{M}$.

Examples of flocks for the classical flat Minkowski plane can be obtained as follows. Let $\mathcal{Q}$ be the nondegenerate ruled quadric in 3-dimensional projective space $\mathrm{PG}(3,\mathbf{R})$, as in Subsection 4.1.1, and let L be a line of $\mathrm{PG}(3,\mathbf{R})$ that has no points in common with $\mathcal{Q}$. Since a tangent plane to $\mathcal{Q}$ intersects $\mathcal{Q}$ in two lines, we see that a plane through L cannot be tangent to $\mathcal{Q}$. Each plane through L of $\mathrm{PG}(3,\mathbf{R})$ therefore intersects $\mathcal{Q}$ in a circle of the Minkowski plane. The collection of circles obtained in this way yields a flock of the classical flat Minkowski plane $\mathcal{M}(\mathcal{Q})$. This kind of flock is called *linear*.

Note, however, that there are nonlinear flocks of $\mathcal{M}(\mathcal{Q})$ as well. For example, using the same setup as before, one can choose three exterior lines to $\mathcal{Q}$ through a common point p which are not in a plane, say L_1, L_2 and L_3. Any two of these lines determine a plane and dually, any two of these planes intersect in one of the three lines. One then obtains a 'trilinear' flock by taking the intersections of $\mathcal{Q}$ with planes through the lines L_i between the two planes determined by L_i and one of the other two lines. If the lines L_1, L_2, and L_3 are chosen suitably one obtains in this way a flock of $\mathcal{M}(\mathcal{Q})$ that is made of three sections of linear flocks.

In the case of a linear flock of $\mathcal{M}(\mathcal{Q})$ the circles in the flock correspond to the planes through a line. Furthermore, the collection of all planes through a line topologically forms a 1-sphere. More generally, let $\mathcal{F}$ be a flock of a toroidal circle plane $\mathcal{M}$. Let G be one (+)- or (−)-parallel class. Every circle C in $\mathcal{F}$ intersects G in exactly one point, and vice versa, every point of G determines a unique circle in $\mathcal{F}$. This shows that, as a subset of the circle space, $\mathcal{F}$ is homeomorphic to $\mathbf{S}^1$.

PROPOSITION 4.8.1 (Flocks) *Each flock of a toroidal circle plane is homeomorphic to* $\mathbf{S}^1$.

Recall from Subsection 4.2.3 that a toroidal circle plane $\mathcal{M}$ can be described by a sharply 3-transitive set Σ of homeomorphisms of $\mathbf{S}^1$. The circles then are just the graphs of the homeomorphisms in Σ. If we use this standard representation of a toroidal circle plane, a flock $\mathcal{F}$ of $\mathcal{M}$ corresponds to a sharply transitive subset of Σ; we just take the homeomorphisms in Σ whose graphs are circles in $\mathcal{F}$. Conversely, every sharply transitive subset of Σ gives rise to a flock of $\mathcal{M}$; the circles in the flock are the graphs of the homeomorphisms in the subset.

PROPOSITION 4.8.2 (Flocks and Sharply Transitive Sets) *In a toroidal circle plane* $\mathcal{M}$ *represented by a sharply* 3*-transitive set* Σ *of homeomorphisms of* $\mathbf{S}^1$ *the sharply transitive subsets of* Σ *are precisely the flocks of* $\mathcal{M}$.

Given two disjoint circles C_1 and C_2 in a flat Minkowski plane $\mathcal{M}$ and a point $p \notin C_1 \cup C_2$ one can always find a circle C_3 through p that is disjoint to both C_1 and C_2. In fact, starting from two disjoint circles one can inductively construct a *partial flock*, that is, a collection of circles of the plane that are pairwise disjoint, that covers a dense subset of the torus. This partial flock can further be completed to a flock of $\mathcal{M}$; see Rosehr [1998] Theorem 1.9.

THEOREM 4.8.3 (Existence of Flock) *Every flat Minkowski plane admits a flock.*

Let $\mathcal{F}$ be a flock of a toroidal circle plane $\mathcal{M}$. An *automorphism of* $\mathcal{F}$ is an automorphism of the plane $\mathcal{M}$ that leaves $\mathcal{F}$ invariant, that is, circles in $\mathcal{F}$ are taken to circles in $\mathcal{F}$ under the automorphism. Clearly, the collection of all automorphisms of $\mathcal{F}$ forms a group with respect to composition, the *automorphism group of* $\mathcal{F}$.

In the case that $\mathcal{M}$ is a flat Minkowski plane, the automorphism group of a flock $\mathcal{F}$ of $\mathcal{M}$ is a closed subgroup of the automorphism group of $\mathcal{M}$

and thus is itself a Lie group. The automorphism group of a flock of the classical flat Minkowski plane is at most 3-dimensional, the linear flocks attaining that maximum dimension. Furthermore, all flocks of the classical flat Minkowski plane that admit a nondiscrete automorphism group are all classified; see Rosehr [1998] Theorem 3.11. In fact, the linear flocks are distinguished by their automorphism groups; see Rosehr [1998] Lemma 3.7 and Corollary 3.10.

THEOREM 4.8.4 (Linear Flock) *Let $\mathcal{F}$ be a flock of the classical flat Minkowski plane and let Σ be the automorphism group of $\mathcal{F}$. Then each of the following conditions implies that $\mathcal{F}$ is linear.*

(i) Σ *is at least 2-dimensional;*

(ii) Σ *contains an automorphism $\sigma \neq id$ that fixes three circles of $\mathcal{F}$.*

A *resolution* of a toroidal circle plane $\mathcal{M}$ is a partition of the circle space of $\mathcal{M}$ into flocks of $\mathcal{M}$. We call the Minkowski plane $\mathcal{M}$ *resolvable* if it admits a resolution. For example, the classical flat Minkowski plane admits a resolution into linear flocks. A nice geometric construction is as follows; see Thas [1975] for finite Minkowski planes. Let $\mathcal{Q}$ be the nondegenerate ruled quadric as before and let $\mathcal{R}$ denote one of the reguli of lines on $\mathcal{Q}$, that is, all parallel classes of $\mathcal{Q}$ of one type. It is possible to embed $\mathcal{R}$ into a spread $\mathcal{S}$ of $\mathrm{PG}(3, \mathbf{R})$. Recall that a *spread* $\mathcal{S}$ of $\mathrm{PG}(3, \mathbf{R})$ is a partition of $\mathrm{PG}(3, \mathbf{R})$ into lines such that any two lines in the spread are skew, that is, they span the entire projective space. The linear flocks of the classical flat Minkowski plane $\mathcal{M}(\mathcal{Q})$ that are defined by the lines of $\mathcal{S} \setminus \mathcal{R}$ determine a resolution of $\mathcal{M}(\mathcal{Q})$. Note that a line in $\mathcal{S} \setminus \mathcal{R}$ cannot intersect $\mathcal{Q}$ since each point of $\mathcal{Q}$ is on a line in $\mathcal{R}$. Furthermore, the flocks are pairwise disjoint, because the lines in $\mathcal{S}$ are pairwise skew.

Using the standard representation of a toroidal circle plane in terms of a sharply 3-transitive set Σ of homeomorphisms of $\mathbf{S}^1$ to itself from Subsection 4.2.3, we can obtain resolutions of certain toroidal circle planes as follows; see Bonisoli [1989] Proposition 1 for finite Minkowski planes. Let $\Sigma_\infty = \{\sigma \in \Sigma \mid \sigma(\infty) = \infty\}$ be the set of all homeomorphisms in Σ that fix the point ∞. Then it is clear that Σ_∞ is sharply 2-transitive on $\mathbf{R} = \mathbf{S}^1 \setminus \{\infty\}$. Let further Φ be a sharply transitive subset of Σ and assume that $\Phi\Sigma_\infty \subseteq \Sigma$. Since it readily follows that $\Phi\Sigma_\infty$ is 3-transitive, the condition $\Phi\Sigma_\infty \subseteq \Sigma$ implies that $\Phi\Sigma_\infty = \Sigma$. Obviously each 'right coset' $\Phi\sigma = \{\varphi\sigma \mid \varphi \in \Phi\}$, where $\sigma \in \Sigma_\infty$, is a sharply transitive subset of Σ and thus determines a flock. Furthermore, every $\sigma \in \Sigma$

has a unique decomposition as $\sigma = \varphi\sigma_\infty$ for $\varphi \in \Phi$ and $\sigma_\infty \in \Sigma_\infty$. Hence $\{\Phi\sigma \mid \sigma \in \Sigma_\infty\}$ corresponds to a resolution.

4.8.2 Double Covers of Disk Möbius Strip Planes

In Polster [1998a] the fixed-circle sets of special involutory automorphisms of toroidal circle planes were shown to be closely related to parts of flat projective planes.

Let C be a circle in a toroidal circle plane $\mathcal{M}$. If p is a point, let $C(p)$ be the set of all points on C that are parallel to p. This set contains one or two points depending on whether p is contained in C or not. In the first case, let $\gamma(p)$ be p and, in the second case, let $\gamma(p)$ be the unique second point q such that $C(q) = C(p)$. This defines an involutory homeomorphism γ of the point set to itself that exchanges (+)- and (−)-parallel classes and whose fixed point set is C. Note that, in general, we do not obtain an automorphism of $\mathcal{M}$. (In the case of a flat Minkowski plane and an automorphism, we have an inversion at C; see Subsection 4.6.5.) We call γ an *exterior involution* if every circle that intersects C and contains two distinct points that are exchanged by γ is globally fixed by γ. We call an exterior involution *strong* if it fixes all circles that contain any two distinct points that get exchanged by γ. If γ is an automorphism it can be shown that it is an exterior involution (which may or may not be strong); see Polster [1998a].

The definition of a *bundle involution* of a hyperbolic quadric in real projective 3-space mirrors that of an oval in a projective plane or that of an ovoid in projective 3-space; compare Subsection 2.10.2. A bundle involution is an automorphism of the classical flat Minkowski plane whose fixed-point set is a circle in the classical plane. Furthermore, such a bundle involution is a strong exterior involution of this plane.

Assume that γ is an exterior involution. For every $p \in C$ consider the union of the two parallel classes containing p and let $T(\gamma)$ be the set of all such unions. Let $S(\gamma)$ be the set of all circles that intersect the circle C in exactly two points and are globally fixed by γ. Finally, let $E(\gamma)$ be the set of all circles that do not intersect the circle C and are globally fixed by γ.

Using these three sets, we can define some point–line geometries. Let $D'(\gamma)$ be the geometry whose point set is the torus and whose line set is $T(\gamma) \cup S(\gamma) \cup E(\gamma)$ or $T(\gamma) \cup S(\gamma)$ depending on whether or not γ is strong. Let $D(\gamma)$ be the geometry we arrive at by identifying points of D that get exchanged by γ. This means that $D(\gamma)$ is a 'double cover' of $D'(\gamma)$.

THEOREM 4.8.5 (Subgeometries of Flat Projective Planes) *Let γ be an exterior involution of a toroidal circle plane that fixes the circle C pointwise. Then there exists a topological oval $\mathcal{O}$ in some flat projective plane $\mathcal{P}$ such that the geometry $D(\gamma)$ is topologically isomorphic to the geometry whose point set is the complement of the interior of $\mathcal{O}$. If γ is strong its lines are the restrictions of the lines of $\mathcal{P}$ to the complement. If γ is not strong the lines of the geometry are the restrictions of the tangents and secants of $\mathcal{O}$ to this complement.*

Furthermore, the sets $T(\gamma)$, $S(\gamma)$, and $E(\gamma)$ correspond to the sets of tangent, secant, and exterior lines of $\mathcal{O}$, respectively, and the circle C to the oval $\mathcal{O}$.

This means that in the case of a strong exterior involution the geometry of globally fixed circles is a 'double cover' of a disk Möbius strip plane 'with boundary'.

In the case that γ is an automorphism, this theorem is a corollary of Theorem 2.10.6; see the remarks following this result. To make the necessary connection with Theorem 2.10.6, we remark that, given an inversion γ at the circle C of a toroidal circle plane, there is a representation of this circle plane as an invertible sharply 3-transitive set of homeomorphisms $\mathbf{S}^1 \to \mathbf{S}^1$ such that C corresponds to the identity in this set. Translated into the language of sharply 3-transitive sets the proofs for the above special case in Polster [1998a] also work for the more general case that γ is a (strong) exterior involution.

4.9 Open Problems

PROBLEM 4.9.1 *What $\mathbf{R}^2$-planes and flat affine planes can occur as derived planes of toroidal circle planes and flat Minkowski planes? What disk Möbius strip planes and abstract ovals arise from toroidal circle planes as described in Subsection 4.8.2?*

Some more constructions of flat linear spaces from flat Minkowski planes via generalized quadrangles (see Chapter 6) correspond to some more similar questions.

PROBLEM 4.9.2 *Develop the theory of flat Minkowski planes that are integrals of* $\mathbf{R}^2$*-planes; see Subsection 5.3.4.1.*

PROBLEM 4.9.3 *Are there any further essentially new ways to construct flat linear spaces from toroidal circle planes apart from the ones that come about as combinations of the links between the different classes of geometries described in this book?*

PROBLEM 4.9.4 *Are there toroidal circle planes all of whose derived planes are* $\mathbf{R}^2$*-planes that are not flat affine planes?*

PROBLEM 4.9.5 *Show that the automorphism group of a toroidal circle plane is a Lie group and develop a classification of toroidal circle planes with respect to their group dimension.*

PROBLEM 4.9.6 *Classify all flat Minkowski planes of group dimension* 3.

A complete classification of the flat Minkowski planes/toroidal circle planes of group dimension 3 should still be possible, whereas a complete classification of the planes of group dimension 2 seems to be beyond reach.

PROBLEM 4.9.7 *Develop a classification of flat Minkowski planes with respect to Klein–Kroll types.*

PROBLEM 4.9.8 *Does every toroidal circle plane contain a flock or, even stronger, admit a resolution into flocks?*

PROBLEM 4.9.9 *In general, extend some more results about flat Minkowski planes listed in this chapter to results about toroidal circle planes.*

More problems include the counterparts of some of the problems listed at the end of the chapter on spherical circle planes.

5
Cylindrical Circle Planes

Cylindrical circle planes and, in particular, flat Laguerre planes were first investigated by Groh [1968], [1969]. For information about general Laguerre planes we refer to Delandtsheer [1995], Hartmann [1982b], Kleinewillinghöfer [1980], and references given there.

A cylindrical circle plane is a point–circle geometry whose point set is (homeomorphic to) the cylinder $\mathbf{S}^1 \times \mathbf{R}$. Its point set is equipped with one nontrivial parallelism. The parallel classes of this parallelism are the verticals (generators) on the cylinder. The circles of the cylindrical circle plane are graphs of continuous functions $\mathbf{S}^1 \to \mathbf{R}$ that form a system of topological circles on the cylinder such that the Axiom of Joining B1 (see p. 7) is satisfied, that is, any three pairwise nonparallel points determine exactly one curve in the system. A cylindrical circle plane is a flat Laguerre plane if and only if it also satisfies Axiom B2.

Cylindrical circle planes are just one of the infinitely many types of tubular circle planes that we will investigate in the last chapter of this book. On the other hand, among these different types of circle planes only the cylindrical circle planes allow an exposition that parallels those of the spherical and toroidal circle planes. Also, we will see in the next chapter that flat Laguerre planes are really the most important, the best understood, and the most general among the nested flat circle planes of rank 3. In fact, it turns out that for every flat Möbius and Minkowski plane there is a flat Laguerre plane from which it can be reconstructed in a natural manner. It is for these reasons that we dedicate a separate chapter to cylindrical circle planes whose organization mirrors that of the previous two chapters. The emphasis is on special features of flat Laguerre planes. Proofs for results that are part of a general result for all flat circle planes living on the cylinder or the Möbius strip will be deferred to the last chapter of this book.

5.1 Models of the Classical Flat Laguerre Plane

In this first section we describe a number of models of the classical flat Laguerre plane. For detailed information about most of these models see Benz [1973].

5.1.1 The Geometry of Plane Sections

The point set of the classical flat Laguerre plane is the vertical cylinder over the unit circle in the xy-plane

$$\mathbf{S}^1 \times \mathbf{R} = \{(x, y, z) \in \mathbf{R}^3 \mid x^2 + y^2 = 1\}$$

in 3-dimensional Euclidean space. Its circles are the intersections of the cylinder $\mathbf{S}^1 \times \mathbf{R}$ with the nonvertical planes

$$\{(x, y, z) \in \mathbf{R}^3 \mid z = ax + by + c\},$$

where $a, b, c \in \mathbf{R}$. The parallel classes of the Laguerre plane are the vertical Euclidean lines $\{(x_0, y_0, z) \mid z \in \mathbf{R}\}$, $x_0^2 + y_0^2 = 1$, contained in the cylinder.

More generally, let C be a cone with vertex v and base an ellipse in real projective 3-space $\mathrm{PG}(3, \mathbf{R})$. Then a plane intersects C in either v, one line through v, two lines through v, or a nondegenerate conic, that is, an ellipse. We call the intersection of a plane with the cone *nontrivial* if it intersects the cone in an ellipse. Then a model of the classical flat Laguerre plane has as point set the set $C \setminus \{v\}$, as circles the nontrivial plane intersections of the cone, and as parallel classes the lines on C through v that have been punctured at v.

5.1.2 The Geometry of Euclidean Lines and Parabolas

We fix a point p on the cylinder and let V denote the parallel class through p. We now consider the derived plane at the point p. Remember that its points are all points not on V. Its lines are all circles passing through p that have been punctured at p and all parallel classes different from V. The derived plane can be identified with the Euclidean plane via a stereographic projection (see below) such that the lines corresponding to the parallel classes are the verticals. Under this identification circles not passing through the point p induce conics in the affine plane. More precisely, these conics are Euclidean parabolas with vertical axes. In fact, any Euclidean parabola with a vertical axis occurs. Hence we are dealing with a system of curves consisting of the Euclidean lines and parabolas

$$\{(x, y) \in \mathbf{R}^2 \mid y = ax^2 + bx + c\},$$

where $a, b, c \in \mathbf{R}$. Two Euclidean points (x_1, y_1) and (x_2, y_2) are parallel if and only if $x_1 = x_2$. Thus the vertical Euclidean lines form the parallel classes of points.

Conversely, we can start off with the Euclidean plane together with all nonvertical Euclidean lines and all Euclidean parabolas with vertical axes. This essentially describes the classical flat Laguerre plane. To reconstruct this geometry, we augment the point space $\mathbf{R}^2$ of the Euclidean plane by one *parallel class at infinity* $\{\infty\} \times \mathbf{R}$ and augment every one of the curves that correspond to quadratic polynomials in which a is the coefficient of x^2 by the point (∞, a). In particular, the nonvertical Euclidean lines are augmented by the point $(\infty, 0)$.

Now, for the stereographic projection mentioned above we may choose, without loss of generality, the point p we project from to be $(0, 1, 0)$ and the plane we project onto the xz-plane; see Figure 5.1.

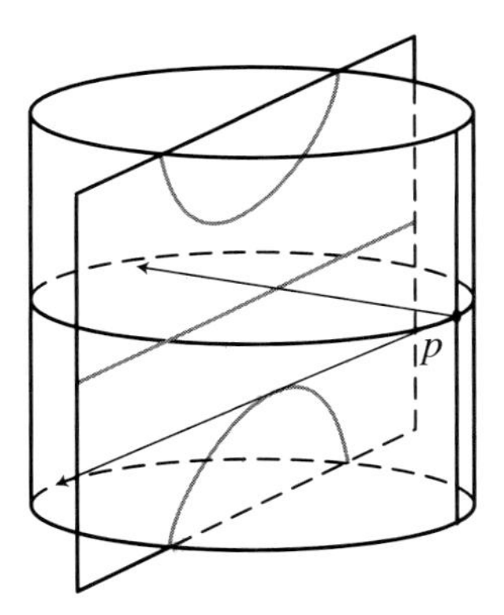

Fig. 5.1. Stereographic projection

Then the stereographic projection is given by the map

$$(x, y, z) \mapsto \left(\frac{x}{1-y}, \frac{z}{1-y}\right).$$

We are interested in the image under this map of the intersection of the cylinder with the nonvertical plane

$$E = \{(x, y, z) \in \mathbf{R}^3 \mid z = vx + uy + w\},$$

where $v, u, w \in \mathbf{R}$. This image is

$$\left\{ (x,z) \in \mathbf{R}^2 \;\middle|\; z = \frac{u+w}{2}x^2 + vx + \frac{w-u}{2} \right\}.$$

Clearly, the plane E passes through the point $(0, 1, a)$ of the parallel class through p if and only if $u + w = a$. This means that the circles of the flat Laguerre plane through the point $(0, 1, a)$ on the parallel class at infinity correspond to the parabolas/quadratic polynomials with the same leading coefficient $a/2$.

5.1.3 The Geometry of Trigonometric Polynomials

We again start with the geometry of nontrivial plane sections of the vertical cylinder over the unit circle $\mathbf{S}^1$ in the xy-plane and parametrize the circle $\mathbf{S}^1$ by $[0, 2\pi)$ in the usual way

$$[0, 2\pi) \to \mathbf{R}^2 : t \mapsto (\sin t, \cos t).$$

Let E be the nonvertical plane

$$\{(x, y, z) \in \mathbf{R}^3 \mid z = ax + by + c\},$$

where $a, b, c \in \mathbf{R}$. If we view the cylinder as the strip $[0, 2\pi] \times \mathbf{R}$ whose left and right boundaries have been identified in the obvious way, then the circle of the classical flat Laguerre plane that corresponds to the plane E is the graph of the periodic function

$$[0, 2\pi] \to \mathbf{R} : t \mapsto a \sin t + b \cos t + c.$$

This means that the set T of trigonometric polynomials of degree at most 1 is another model of the classical flat Laguerre plane. The fact that our geometry satisfies Axiom B1 is equivalent to this set of continuous periodic functions solving the Lagrange interpolation problem of order 3.

In Chapter 7 we will investigate the different kinds of geometries that correspond to sets of continuous functions that solve the Lagrange interpolation problem of some order.

5.1.4 The Geometry of Oriented Lines and Circles

The following model of the classical flat Laguerre plane is constructed completely in terms of the objects in the affine part of the classical flat Möbius plane; see also Subsection 6.5.2.

Consider the abstract geometry $\mathcal{L}$ whose points are the oriented Euclidean lines and whose circles are the points and the oriented Euclidean circles in $\mathbf{R}^2$. This means that every Euclidean line or circle corresponds to two objects in $\mathcal{L}$. A point and a circle of $\mathcal{L}$ are incident if and only if their corresponding objects touch, that is, have exactly one point in common, and share the same orientation if they are both oriented; see Figure 5.2.

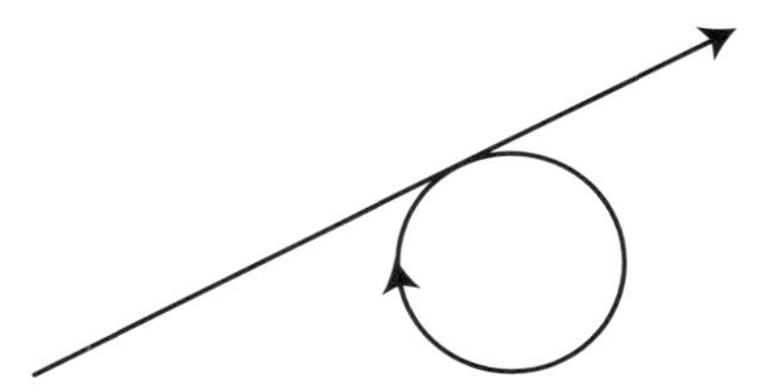

Fig. 5.2. An incident point–circle pair in the geometry of oriented Euclidean lines and circles

It is easy to see that two points of $\mathcal{L}$ are parallel if and only if they correspond to parallel Euclidean lines that are oriented in the same way. Figure 5.3 shows the unique connecting circle of three pairwise nonparallel points in the geometry $\mathcal{L}$. The diagram on the cover of this book shows an example of Miquel's configuration in $\mathcal{L}$.

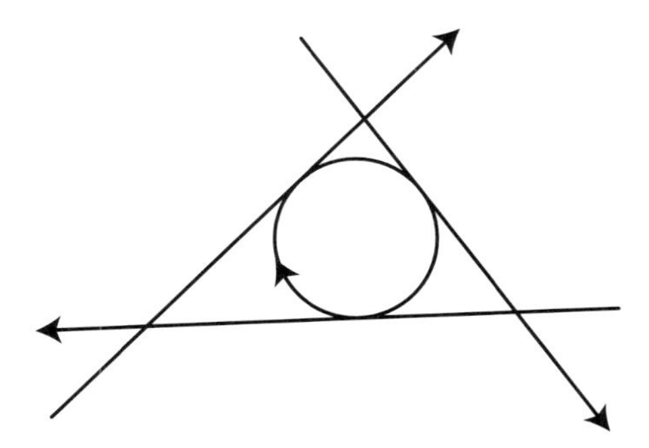

Fig. 5.3. Three pairwise nonparallel points and their connecting circle in the geometry of oriented Euclidean lines and circles

We want to show that $\mathcal{L}$ is really isomorphic to the classical flat Laguerre plane. To this end we first assign homogeneous coordinates to both points and circles of $\mathcal{L}$. Let L be a point in $\mathcal{L}$ that is carried by the Euclidean line

$$L' = \{(x, y) \mid a + bx + cy = 0\}.$$

Then $V = (b, c)$ is a vector that is perpendicular to L'. Let $W = (b', c')$ be a vector in $\mathbf{R}^2$ whose orientation coincides with that of L. Finally, let

$$d = \pm\sqrt{b^2 + c^2}$$

depending on whether the determinant

$$\begin{vmatrix} b & b' \\ c & c' \end{vmatrix}$$

is positive or negative, that is, depending on whether V and W define a right- or left-handed coordinate system. Now the point in real projective 3-space corresponding to L is $(a : b : c : d)$. Convince yourself of the fact that this assignment of homogeneous coordinates is well defined. Also a point $(a : b : c : d)$ in $\mathrm{PG}(3, \mathbf{R})$ corresponds to a point of $\mathcal{L}$ if and only if

$$b^2 + c^2 = d^2 \neq 0.$$

Furthermore, two points $(a : b : c : d)$ and $(a'' : b'' : c'' : d'')$ correspond to parallel points of $\mathcal{L}$ if and only if

$$\frac{b}{d} = \frac{b''}{d''} \text{ and } \frac{c}{d} = \frac{c''}{d''}.$$

Let C be a circle of $\mathcal{L}$. If C is a point of $\mathbf{R}^2$, we consider it as a circle of radius 0. Let C' be the Euclidean circle or point that carries C and let r be its radius and (a, b) the affine coordinates of its centre. Finally, let $\mathrm{rad}(C) = r$ or $-r$ depending on whether C is oriented in the anticlockwise or clockwise direction, respectively. The point

$$(w : x : y : z)$$

corresponds to C if

$$w \neq 0, \frac{x}{w} = a, \frac{y}{w} = b, \frac{z}{w} = \mathrm{rad}(C).$$

Again, convince yourself that this assignment of homogeneous coordinates to circles of $\mathcal{L}$ is well defined.

Furthermore it can be shown that the point in $\mathcal{L}$ with the coordinates $(a : b : c : d)$ is incident with the circle with the coordinates $(w : x : y : z)$ if and only if

$$aw + bx + cy + dz = 0.$$

If $\mathcal{C}$ is the set of all points of $\mathcal{L}$ given in the form $(a, b, c, 1)$, then

$$\mathcal{C} \to \mathbf{R}^3 : (a, b, c, 1) \mapsto (-c, b, -a)$$

defines an isomorphism from $\mathcal{L}$ to the classical flat Laguerre plane; see Benz [1973] Section I.2, for more details about this construction.

5.1.5 Pentacyclic Coordinates

In this subsection we introduce a certain higher-dimensional space such that points and circles of the classical flat Laguerre plane are both represented by certain points of this space.

Let $\mathcal{Q}$ be the nondegenerate quadric in $\mathrm{PG}(4, \mathbf{R})$ defined by

$$\mathcal{Q} = \{(x_0 : x_1 : x_2 : x_3 : x_4) \in \mathrm{PG}(4, \mathbf{R}) \mid x_0^2 + x_1^2 - x_3^2 + 2x_2x_4 = 0\}.$$

We intersect $\mathcal{Q}$ with hyperplanes

$$\begin{aligned} E_{\mathbf{a}} &= \{(x_0 : x_1 : x_2 : x_3 : x_4) \in \mathrm{PG}(4, \mathbf{R}) \mid \\ &\qquad a_0x_0 + a_1x_1 + a_2x_2 + a_3x_3 + a_4x_4 = 0\}, \end{aligned}$$

where $\mathbf{a}$ is a point $\mathbf{a} = (a_0 : a_1 : a_2 : a_3 : a_4)$ of the quadric $\mathcal{Q}$. For the point $\mathbf{e} = (0 : 0 : 0 : 0 : 1)$ the intersection of $E_{\mathbf{e}}$ and $\mathcal{Q}$ is

$$\begin{aligned} E_{\mathbf{e}} \cap \mathcal{Q} &= \{(x_0 : x_1 : x_2 : x_3 : 0) \in \mathrm{PG}(4, \mathbf{R}) \mid x_0^2 + x_1^2 - x_3^2 = 0\} \\ &\simeq \{(x_0 : x_1 : x_2 : x_3) \in \mathrm{PG}(3, \mathbf{R}) \mid x_0^2 + x_1^2 - x_3^2 = 0\}, \end{aligned}$$

that is, an elliptic cone in 3-dimensional projective space $E_{\mathbf{e}} \simeq \mathrm{PG}(3, \mathbf{R})$ with vertex $\mathbf{v} = (0 : 0 : 1 : 0 : 0)$. We thus have, except for the vertex $\mathbf{v}$, the point space of the classical flat Laguerre plane. Introducing affine coordinates in such a way that $x_3 = 0$ describes the plane at infinity of the projective space $E_{\mathbf{e}}$ (this plane intersects the cone only in the vertex $\mathbf{v}$) we just obtain the standard cylinder $\mathbf{S}^1 \times \mathbf{R}$ in $\mathbf{R}^3$.

$$\begin{aligned} (E_{\mathbf{e}} \cap \mathcal{Q})_{aff} &= (E_{\mathbf{e}} \cap \mathcal{Q}) \setminus \{\mathbf{v}\} \\ &= \{(x_0 : x_1 : x_2 : 1 : 0) \in \mathrm{PG}(4, \mathbf{R}) \mid x_0^2 + x_1^2 - 1 = 0\} \\ &\simeq \{(x_0, x_1, x_2) \in \mathbf{R}^3 \mid x_0^2 + x_1^2 = 1\}. \end{aligned}$$

We further intersect $E_{\mathbf{e}} \cap \mathcal{Q}$ with hyperplanes $E_{\mathbf{a}}$ for $\mathbf{a} \in \mathcal{Q} \setminus E_{\mathbf{v}}$ to obtain the circles. We have to exclude hyperplanes that contain $\mathbf{v}$ because such hyperplanes intersect the cone in $\mathbf{v}$, one generator, or a pair of generators.

Clearly, $E_{\mathbf{e}} \cap \mathcal{Q} \cap E_{\mathbf{a}}$ is the intersection of the cone $E_{\mathbf{e}} \cap \mathcal{Q} \subset E_{\mathbf{e}}$ in

3-dimensional projective space $E_{\mathbf{e}} \simeq \mathrm{PG}(3,\mathbf{R})$ with the plane $E_{\mathbf{e}} \cap E_{\mathbf{a}}$ of $E_{\mathbf{e}}$. Furthermore, we can put $a_2 = 1$ because $\mathbf{a} \notin E_{\mathbf{v}}$. Then

$$\begin{aligned} E_{\mathbf{e}} \cap \mathcal{Q} \cap E_{\mathbf{a}} &= \{(x_0 : x_1 : x_2 : x_3) \in \mathrm{PG}(3,\mathbf{R}) \mid \\ &\qquad x_0^2 + x_1^2 - x_3^2 = 0, a_0x_0 + a_1x_1 + x_2 + a_3x_3 = 0\} \\ &\simeq \{(x_0, x_1, x_2) \in \mathbf{R}^3 \mid \\ &\qquad x_0^2 + x_1^2 = 1, a_0x_0 + a_1x_1 + x_2 + a_3 = 0\} \\ &= \{(x_0, x_1, -a_0x_0 - a_1x_1 - a_3) \mid \\ &\qquad x_0, x_1 \in \mathbf{R}, x_0^2 + x_1^2 = 1\}. \end{aligned}$$

Moreover, all planes of $E_{\mathbf{e}}$ that intersect $\mathbf{S}^1 \times \mathbf{R}$ in a circle are covered; compare Subsection 5.1.1.

Note that

$$\begin{aligned} \mathcal{Q} \setminus E_{\mathbf{v}} &= \{(x_0 : x_1 : 1 : x_3 : x_4) \in \mathrm{PG}(4,\mathbf{R}) \mid x_0^2 + x_1^2 - x_3^2 + x_4 = 0\} \\ &\simeq \{(x_0, x_1, x_3, x_4) \in \mathbf{R}^4 \mid x_0^2 + x_1^2 - x_3^2 + x_4 = 0\} \\ &\simeq \{(x_0, x_1, x_3, x_3^2 - x_0^2 - x_1^2 x_4) \mid x_0, x_1, x_3 \in \mathbf{R}\} \\ &\simeq \mathbf{R}^3. \end{aligned}$$

More generally, one can use any point $\mathbf{p}$ on $\mathcal{Q}$ instead of $\mathbf{e}$. Thus the point space is $(E_{\mathbf{p}} \cap \mathcal{Q}) \setminus \{\varphi(p)\}$ where φ is the involutory collineation of $\mathrm{PG}(4,\mathbf{R})$ given by $\varphi(x_0 : x_1 : x_2 : x_3 : x_4) = (x_0 : x_1 : x_4 : -x_3 : x_2)$ that leaves the quadric $\mathcal{Q}$ invariant. Circles are of the form $E_{\mathbf{p}} \cap \mathcal{Q} \cap E_{\mathbf{q}}$ for $\mathbf{q} \in \mathcal{Q} \setminus E_{\varphi(\mathbf{p})}$. Hence the quadric $\mathcal{Q}$ contains many copies of the classical flat Laguerre plane.

Points and circles of the classical flat Laguerre plane $\mathcal{L}$ are both represented by points of $\mathcal{Q}$. Points of $\mathcal{L}$ are the points of $\mathcal{Q}$ that are orthogonal to a fixed point $\mathbf{p} \in \mathcal{Q}$ but different from $\varphi(\mathbf{p})$. Circles of $\mathcal{L}$ correspond precisely to the points of $\mathcal{Q}$ not collinear with $\varphi(\mathbf{p})$. A point $\mathbf{u} \in \mathcal{Q}$ is incident with a circle $\mathbf{v} \in \mathcal{Q}$ if and only if $\mathbf{u}$ and $\mathbf{v}$ are orthogonal; see also Section 6.1.

5.1.6 The Geometry of Chains

In this subsection we construct the classical flat Laguerre plane as a geometry of chains. As in the case of the respective models for the classical flat Möbius and Minkowski planes, we start with a 2-dimensional ring extension of $\mathbf{R}$. This ring extension is isomorphic to the ring of *dual numbers*

$$\mathbf{D} = \{a + b\varepsilon \mid a, b \in \mathbf{R}\},$$

where $\varepsilon^2 = 0$. The algebraic operations in $\mathbf{D}$ are given by

$$(a + b\varepsilon) + (c + d\varepsilon) = (a + c) + (b + d)\varepsilon$$

and

$$(a + b\varepsilon)(c + d\varepsilon) = (ac) + (ad + bc)\varepsilon.$$

Clearly, $\mathbf{D}$ is commutative and contains $\mathbf{R}$ as a subring. As in the case of the classical flat Möbius and Minkowski planes the chain geometry associated with $\mathbf{D}$ has as points the points of the projective line $\mathrm{PG}(1, \mathbf{D})$ and as blocks the images of $\mathrm{PG}(1, \mathbf{R}) \subset \mathrm{PG}(1, \mathbf{D})$ under the projective group of $\mathrm{PG}(1, \mathbf{D})$. Here are some of the details of this construction.

Just as the complex numbers, $\mathbf{D}$ is a commutative ring with identity 1. A map $\varphi : \mathbf{D} \to \mathbf{D}$ is an automorphism of $\mathbf{D}$ that fixes $\mathbf{R}$ globally if and only if it is one of the maps

$$\varphi_l : \mathbf{D} \to \mathbf{D} : a + b\varepsilon \mapsto a + bl\varepsilon,$$

where $l \in \mathbf{R} \setminus \{0\}$. Furthermore, $\mathbf{D}$ admits precisely two involutory ring automorphisms that fix $\mathbf{R}$ globally. These are the identity and the involution

$$\varphi_{-1} : \mathbf{D} \to \mathbf{D} : a + b\varepsilon \mapsto \overline{a + b\varepsilon} = a - b\varepsilon.$$

Apart from $\{0\}$ and $\mathbf{D}$ itself, the ring $\mathbf{D}$ has just one more ideal, namely

$$\mathcal{J} = \mathbf{R}\varepsilon = \{r\varepsilon \mid r \in \mathbf{R}\}.$$

Let $\mathcal{R} = \mathbf{D} \setminus \mathcal{J}$. Then $\mathcal{R}$ is the set of invertible elements of $\mathbf{D}$. The inverse of $a + b\varepsilon$, $a \neq 0$, is given by $\frac{1}{a^2}(\overline{a + b\varepsilon}) = \frac{1}{a^2}(a - b\varepsilon)$.

Let P be the set of all pairs (x, y) such that not both of x and y are contained in the ideal $\mathcal{J}$ and let two pairs (u, v) and (x, y) in P be equivalent if there is an invertible element $s \in \mathcal{R}$ such that $(u, v) = (sx, sy)$. This really defines an equivalence relation on P. The projective line $\mathrm{PG}(1, \mathbf{D})$ consists of the equivalence classes of this equivalence relation. As usual, if $(x, y) \in P$, then $(x : y)$ denotes its corresponding equivalence class.

The invertible 2×2 matrices with entries in $\mathbf{D}$ form a group $\mathrm{GL}_2(\mathbf{D})$ with respect to the usual matrix multiplication. A 2×2 matrix

$$\begin{pmatrix} a & b \\ c & d \end{pmatrix}$$

is invertible if and only if its determinant $ad - bc$ is an element of $\mathcal{R}$. If S is the subgroup consisting of all elements of the form

$$\begin{pmatrix} a & 0 \\ 0 & a \end{pmatrix}$$

for $a \in \mathcal{R}$, then $\mathrm{PGL}_2(\mathbf{D})$ is the quotient group $\mathrm{GL}_2(\mathbf{D})/S$ which acts in a natural way on $\mathrm{PG}(1,\mathbf{D})$. We define a geometry $\Sigma(\mathbf{R},\mathbf{D})$ whose points are the elements of $\mathrm{PG}(1,\mathbf{D})$ and whose circles are the real projective line $\mathrm{PG}(1,\mathbf{R})$, which is naturally embedded in $\mathrm{PG}(1,\mathbf{D})$, and all its images under $\mathrm{PGL}_2(\mathbf{D})$.

As in Subsection 5.1.2, we consider $Z = (\mathbf{R} \cup \{\infty\}) \times \mathbf{R}$ to be the point set of the classical flat Laguerre plane. Then

$$Z \to \mathrm{PG}(1,\mathbf{D}) : (x,y) \mapsto \begin{cases} (x + y\varepsilon, 1) & \text{for } x \in \mathbf{R}, \\ (1, -y\varepsilon) & \text{for } x = \infty \end{cases}$$

defines an isomorphism between the classical flat Laguerre plane and the (chain) geometry $\Sigma(\mathbf{R},\mathbf{D})$. Here the circle

$$\{(x,y) \in \mathbf{R}^2 \mid y = ax^2 + bx + c\} \cup \{(\infty, a)\}$$

is mapped to the set of all $(x : y)$ such that

$$(x,y)\begin{pmatrix} -a\varepsilon & \frac{1-b\varepsilon}{2} \\ -\frac{1+b\varepsilon}{2} & -c\varepsilon \end{pmatrix}\begin{pmatrix} \overline{x} \\ \overline{y} \end{pmatrix} = 0.$$

We only remark that two points $(u : v)$ and $(x : y)$ in $\Sigma(\mathbf{R},\mathbf{D})$ are parallel if and only if $uy - vx \in \mathcal{J}$.

The automorphisms φ_l of $\mathbf{D}$ also extend in the natural way to automorphisms of $\mathrm{PG}(1,\mathbf{D})$. (With the identifications made above, φ_l gives rise to the transformation $(x,y) \mapsto (x, ly)$ in the usual model of the classical flat Laguerre plane on $(\mathbf{R} \cup \{\infty\}) \times \mathbf{R}$.) Together with the elements of $\mathrm{PGL}_2(\mathbf{D})$ these automorphisms generate the full group of automorphisms $\mathrm{P\Gamma L}_2(\mathbf{D})$ of $\Sigma(\mathbf{R},\mathbf{D})$.

By construction, the group $\mathrm{PGL}_2(\mathbf{D})$ is transitive on the set of chains. The stabilizer of the chain $\mathrm{PG}(1,\mathbf{R})$ is obtained from the matrices in $\mathrm{GL}_2(\mathbf{R})$ (canonically embedded into $\mathrm{GL}_2(\mathbf{D})$). This stabilizer is transitive on $\mathrm{PG}(1,\mathbf{R})$.

THEOREM 5.1.1 (Flag-Transitive Automorphism Group) *The group* $\mathrm{P\Gamma L}_2(\mathbf{D})$ *is the automorphism group of* $\Sigma(\mathbf{R},\mathbf{D})$. *Furthermore, this group is a 7-dimensional Lie group that acts flag-transitively on the projective line* $\mathrm{PG}(1,\mathbf{D})$.

In fact, the stabilizer of $\mathrm{PG}(1,\mathbf{R})$ is 2-transitive on $\mathrm{PG}(1,\mathbf{R})$. This readily implies the following.

PROPOSITION 5.1.2 (2-Transitive Automorphism Group) *The group* $\mathrm{P\Gamma L}_2(\mathbf{D})$ *is* 2-*transitive on the set of parallel classes of* $\Sigma(\mathbf{R},\mathbf{D})$.

5.2 Derived Planes and Topological Properties

In this section we define the two standard representations of cylindrical circle planes. The first generalizes the representation of the classical plane as a geometry of Euclidean lines and parabolas, the second the representation as a set of continuous functions $\mathbf{S}^1 \to \mathbf{R}$ that solves the Lagrange interpolation problem of order 3. Following this we consider the continuity of the geometric operations, some questions of coherence, and the spaces of lines and flags of a cylindrical circle plane.

5.2.1 Derived R^2-Planes

The idea of representing the classical flat Laguerre plane in the Euclidean plane can be extended as follows.

Let p be a point of a cylindrical circle plane and let V be the parallel class containing p. Remember that the point set of the derived plane at the point p consists of all points not parallel to p. Its lines are the circles through p that have been punctured at p, and all parallel classes, except V. This is an $\mathbf{R}^2$-plane. In the case of a flat Laguerre plane we even obtain a flat affine plane, the *derived affine plane at* p. This flat affine plane extends uniquely to a flat projective plane, which we call the *derived projective plane at* p.

THEOREM 5.2.1 (Characterization via Derived Planes) *Let $\mathcal{L}$ be a geometry whose point set the cylinder $\mathbf{S}^1 \times \mathbf{R}$ is equipped with one nontrivial parallelism, the parallel classes of which are the verticals on the cylinder. Furthermore, the circles of $\mathcal{L}$ are graphs of continuous functions $\mathbf{S}^1 \to \mathbf{R}$. Then $\mathcal{L}$ is a cylindrical circle plane or a flat Laguerre plane if and only if all its derived planes are $\mathbf{R}^2$-planes or flat affine planes, respectively.*

Each derived affine plane of the classical flat Laguerre plane is isomorphic to the Euclidean plane.

Just as in the case of spherical and toroidal circle planes, it is possible to use results about the derived planes of cylindrical circle planes to prove similar results for cylindrical circle planes. Here is an example.

THEOREM 5.2.2 (Intersection of Circles) *Two distinct circles in a cylindrical circle plane intersect in at most two points. If they intersect in exactly one point, then they touch topologically. If they intersect in two points, then they intersect transversally at these points.*

5.2.2 Affine Parts

A circle C not passing through the distinguished point p intersects the parallel class through p in one point. We call the restriction of the circle C to the point set of the derived plane at p a *parabolic curve.* Clearly, this parabolic curve is a topological arc in the derived plane. In the case of a flat Laguerre plane, a closer look at the derived projective plane at p reveals that the parabolic curve can be extended to a topological oval by adding to the parabolic arc the point at infinity of the lines that correspond to the parallel classes in the Laguerre plane.

A cylindrical circle plane can thus be described in one derived plane $\mathcal{A}$ at a point p by the lines of $\mathcal{A}$ and a collection of parabolic curves. We call the induced geometry the *affine part at p.* Note that as geometries the affine parts at all points of one parallel class coincide.

As demonstrated at the end of Subsection 5.1.2, the affine part (at any point) of the classical flat Laguerre plane can be identified with the Euclidean plane together with its lines and all Euclidean parabolas that have a vertical symmetry axis.

Let $\mathcal{L} = (P, \mathcal{C})$ be a geometry whose point set is equipped with one nontrivial parallelism and which satisfies Axiom B1. We say that $\mathcal{L}$ has a *flat affine part with respect to one of its parallel classes* V if it has the following properties.

- $P \setminus V = \mathbf{R}^2$.
- The parallel classes different from V are the verticals in $\mathbf{R}^2$.
- The restrictions of circles in $\mathcal{C}$ to $P \setminus V = \mathbf{R}^2$ are graphs of continuous functions $\mathbf{R} \to \mathbf{R}$.

Of course, this implies that the derived planes at all points of V are $\mathbf{R}^2$-planes. In the case that $\mathcal{L}$ is a Laguerre plane all derived planes are flat affine plane.

THEOREM 5.2.3 (Affine Parts) *Let $\mathcal{L} = (P, \mathcal{C})$ be a geometry whose point set is equipped with one nontrivial parallelism. If $\mathcal{L}$ satisfies Axiom B1 and $\mathcal{L}$ has a flat affine part with respect to one of its parallel classes V, then there are a cylindrical circle plane $\mathcal{L}' = (P', \mathcal{C}')$ that is isomorphic to $\mathcal{L}$ and a parallel class V' in $\mathcal{L}'$ such that the restriction of $\mathcal{L}'$ to $P' \setminus V'$ and the restriction of $\mathcal{L}$ to $P \setminus V$ coincide. The cylindrical circle plane $\mathcal{L}'$ is uniquely determined up to isomorphism.*

If $\mathcal{L}$ is a Laguerre plane, then $\mathcal{L}'$ is a flat Laguerre plane.

This result is a straightforward generalization of Polster–Steinke [1994] Proposition 2; see also Theorem 7.2.4.

In the following, a flat Laguerre plane will often be given as an abstract Laguerre plane on the set $(\mathbf{R}\cup\{\infty\})\times\mathbf{R}$ with parallel classes the verticals in this set and circles that are graphs of functions $\mathbf{R}\cup\{\infty\}\to\mathbf{R}$ that are continuous when restricted to $\mathbf{R}$. Every flat Laguerre plane given in this form has a flat affine part with respect to the parallel class $\{\infty\}\times\mathbf{R}$. We will refer to this distinguished affine part as *the affine part* of the Laguerre plane.

5.2.3 Continuity of the Geometric Operations

As in the case of the geometries that we considered in previous chapters, the topology on the circle set of a cylindrical circle plane is uniquely determined by the topology on its point set; for proofs of the following results in the case of flat Laguerre planes that generalize to the respective results for cylindrical circle planes see Groh [1968] 2.10 and [1970] 3.10.

THEOREM 5.2.4 (Topology Cylinder → Topology Circle Space) *The natural topology of the circle space of a cylindrical circle plane is the finest topology with respect to which the map α of joining three pairwise nonparallel points by a circle is continuous. In particular, a basis for the topology of the circle space is formed by the sets $\alpha(U_1,U_2,U_3)$, where the sets U_1,U_2,U_3 are nonempty open subsets of the cylinder such that any three points $p_i\in U_i$, $i=1,2,3$, are pairwise nonparallel.*

In the following we always assume that both point and circle spaces of a cylindrical circle plane carry their natural topologies.

The topology of the point space is, of course, the usual Euclidean topology of the cylinder $\mathbf{S}^1\times\mathbf{R}$ but one can also describe it geometrically. Let C_1, C_2, C_3 be three circles that pairwise intersect in two points and such that $C_1\cap C_2\cap C_3$ consists of a single point q. Let the *triode* $\mathcal{D}(C_1,C_2,C_3;q)$ be the collection of all points in the interior of the triangle formed by the lines induced by C_1, C_2, C_3 in the derived plane at q.

PROPOSITION 5.2.5 (Triodes Form Neighbourhood Basis) *The triodes for a fixed q that contain a point p that is not parallel to q form a basis for the neighbourhoods of p.*

In fact, one can generalize the notion of a triode and replace one of the circles by a parallel class such that in the derived plane one of the sides of the triangle is vertical (that is, it comes from a parallel class of the cylindrical circle plane).

It now follows that the other geometric operations are also continuous with respect to the topologies defined above.

THEOREM 5.2.6 (Continuity) *In a cylindrical circle plane the operations of*

- *joining three pairwise nonparallel points by a circle,*
- *intersecting a parallel class and a circle in a point,*
- *intersecting two distinct circles*

are continuous on their respective domains of definition.

In a flat Laguerre plane, forming the tangent circle to a circle through two nonparallel points as in Axiom B2 is also a continuous operation.

5.2.4 Topological Laguerre Planes

A topological Laguerre plane is a Laguerre plane whose sets of points and circles carry nonindiscrete topologies such that the geometric operations of joining, touching, intersecting distinct circles, and intersecting circles and parallel classes are continuous on their domains of definition. The results of the previous subsections imply that the flat Laguerre planes are topological Laguerre planes.

As usual, we call a topological Laguerre plane an X Laguerre plane if its point space is a topological space of type X. The following result characterizes the flat Laguerre planes among the topological Laguerre planes; see Groh [1968], [1970].

THEOREM 5.2.7 (Topological Characterization) *The flat Laguerre planes are precisely the 2-dimensional locally compact connected Laguerre planes.*

A topological locally compact finite-dimensional Laguerre plane has dimension 0, 2, or 4; see Buchanan–Hähl–Löwen [1980], Löwen [1981d]. For example, one obtains a 4-dimensional Laguerre plane as in Subsections 5.1.1 and 5.1.2 by replacing the real numbers $\mathbf{R}$ by the complex numbers $\mathbf{C}$.

As in the case of topological Möbius and Minkowski planes, a number of coherence axioms for topological Laguerre planes have to be investigated to decide whether or not the derived affine planes of a given topological Laguerre plane are topological, too. In the case of a flat Laguerre plane the answer clearly is 'Yes'; compare Theorem 2.3.4.

To a large extent the different coherence axioms that need to be considered coincide with those investigated for topological Möbius and Minkowski planes. One of these coherence axioms reflects the fact that ovals induced in derived projective planes must be topological. In flat Laguerre planes this becomes that touching is the limit of proper intersection. Specializing points and circles yields the continuity of forming parallel lines in derived affine planes. Other coherence axioms deal with the intersection of circles and parallel classes as the circles pass through points that converge to parallel points.

A topological Laguerre plane is called *coherent* if it satisfies all coherence axioms. Groh [1970] 2.7 proved the following result.

THEOREM 5.2.8 (Coherence) *Every flat Laguerre plane is coherent.*

See the references given above and at the end of Subsection 3.2.4 and Iversen [1970] for further information about topological Laguerre planes.

THEOREM 5.2.9 (Derived Planes, Arcs, and Ovals) *Each derived plane of a cylindrical circle at a point p is an $\mathbf{R}^2$-plane. The restriction of the natural topology on the line set of the $\mathbf{R}^2$-plane to the set of lines that arise from circles through p is the restriction of the natural topology on the circle set to the set of circles through p.*

A circle not passing through the point p turns into a topological arc in the derived plane at p. In a flat Laguerre plane such an arc extends to a topological oval in the derived flat projective plane at p that touches the line at infinity of the derived affine plane at p at the point at infinity of the vertical lines.

5.2.5 Circle Space and Flag Space

From the description of parallel classes as verticals of the cylinder $\mathbf{S}^1 \times \mathbf{R}$ it is clear that each parallel class is homeomorphic to $\mathbf{R}$ and that the set Π of all parallel classes is homeomorphic to $\mathbf{S}^1$.

THEOREM 5.2.10 (The Space of Parallel Classes) *The set of parallel classes of a cylindrical circle plane is homeomorphic to* $\mathbf{S}^1$.

Using the fact that each circle in a cylindrical circle plane is uniquely determined by its intersection with three distinct parallel classes, the circle space of a cylindrical circle plane can be explicitly described topologically.

THEOREM 5.2.11 (The Space of Circles) *The set of circles of a cylindrical circle plane is homeomorphic to* $\mathbf{R}^3$.

Remember that a *flag* in a cylindrical circle plane $\mathcal{L}$ is an incident point–circle pair, that is, a pair (p, C), where p is a point and C a circle of $\mathcal{L}$ such that $p \in C$. Since parallel classes and circles always intersect in unique points, it readily follows that the map $(p, C) \mapsto (|p|, C)$ provides a homeomorphism from the flag set $\mathcal{F}$ onto the Cartesian product $\Pi \times \mathcal{C}$ of the set Π of all parallel classes and the circle set $\mathcal{C}$.

THEOREM 5.2.12 (The Space of Flags) *The flag space* $\mathcal{F}$ *of a cylindrical circle plane is homeomorphic to the topological space* $\mathbf{S}^1 \times \mathbf{R}^3$.

5.3 Constructions

In this section we present some of the most important construction methods for nonclassical flat Laguerre planes and cylindrical circle planes.

5.3.1 Ovoidal Planes

Let C be a strictly convex topological circle in the xy-plane. Then the geometry of nonvertical plane sections of the vertical cylinder over this curve is a cylindrical circle plane. Cylindrical circle planes constructed like this are called *ovoidal planes.* The parallel classes in such a plane are the vertical Euclidean lines of $\mathbf{R}^3$ contained in the cylinder. The ovoidal plane associated with C is a flat Laguerre plane if and only if C is differentiable. It is classical if and only if C is an ellipse.

We can also describe this construction within real projective 3-space. Here we start with C and a point that is not contained in the plane C is contained in. Then the cylindrical circle plane associated with C and p is the geometry of nontrivial plane sections of the cone K with base C and vertex p. Here the point set of the plane is $K \setminus \{p\}$, the parallel classes are the lines in K that have been punctured at p, and a plane section is nontrivial if and only if it is a topological circle. The resulting circle planes for fixed C and variable p are isomorphic via automorphisms of the projective space that fix C pointwise. In fact, the group of collineations of $\mathrm{PG}(3, \mathbf{R})$ that fix a plane E pointwise is transitive on the complement of E.

Let $f : \mathbf{R} \to \mathbf{R}$ be a continuous function. Then f is called *parabolic* if it satisfies the following conditions:

- f is strictly convex (or concave);
- $\lim_{x \to \pm\infty} f(x) - cx = +\infty$ (or $-\infty$) for all $c \in \mathbf{R}$.

Recall that parabolic functions generate shift planes, that is, special flat affine planes; see Subsection 2.7.4 for a summary of results about these special functions. A parabolic function is called *strictly parabolic* if it is differentiable. This is the case if and only if f is a differentiable function whose derivative is a homeomorphism $\mathbf{R} \to \mathbf{R}$. A strictly parabolic function is *normalized* if and only if it is convex and $f(0) = f'(0) = 0$ and $f(1) = 1$.

Given a parabolic function f, we obtain a closed topological arc C_f in $\mathrm{PG}(2, \mathbf{R})$ by augmenting the graph of f by the infinite point of the y-axis, that is

$$C_f = \{(x : f(x) : 1) \mid x \in \mathbf{R}\} \cup \{(0 : 1 : 0)\},$$

where $(x : y : z)$ denotes the point in 2-dimensional projective space spanned by the vector $(x, y, z) \in \mathbf{R}^3$. The ovoidal cylindrical circle plane obtained from this closed topological arc will be denoted by $\mathcal{L}(f)$. Moreover, every ovoidal cylindrical circle plane is isomorphic to one such plane and every ovoidal flat Laguerre plane is isomorphic to a plane $\mathcal{L}(f)$ for a suitable normalized strictly parabolic function f; see Löwen–Pfüller [1987a] Lemma 2.1.

We represent $\mathcal{L}(f)$ on the cylinder $Z = (\mathbf{R} \cup \{\infty\}) \times \mathbf{R}$ using a stereographic projection. On Z the circles of $\mathcal{L}(f)$ are the sets

$$\{(x, y) \in \mathbf{R}^2 \mid y = af(x) + bx + c)\} \cup \{(\infty, a)\},$$

where $a, b, c \in \mathbf{R}$.

A special class of normalized strictly parabolic functions are the *skew parabola functions*; see p. 443. These functions are of the form

$$f_{d,r}(x) = \begin{cases} x^d & \text{for } x \geq 0, \\ r|x|^d & \text{for } x \leq 0, \end{cases}$$

where $d > 1$ and $r > 0$. A *skew parabola* is the graph of a skew parabolic function. Note that $d = 2$, $r = 1$ corresponds to a Euclidean parabola.

An ovoidal plane $\mathcal{L}(f)$ represented on Z as above admits the following automorphisms:

$$Z \to Z : (x, y) \mapsto \begin{cases} (x, ry + af(x) + bx + c) & \text{for } x \in \mathbf{R}, \\ (\infty, ry + a) & \text{for } x = \infty, \end{cases}$$

where $a, b, c, r \in \mathbf{R}$, $r \neq 0$. These automorphisms form a 4-dimensional group of automorphisms. If f is skew parabolic, then the maps

$$\mathbf{R}^2 \to \mathbf{R}^2 : (x, y) \mapsto (sx, y),$$

where $s \in \mathbf{R}^+$ also extend to automorphisms of $\mathcal{L}(f)$. In fact, it turns out that the ovoidal flat Laguerre planes over proper skew parabolas are precisely the flat Laguerre planes of group dimension 5; see Theorem 5.4.12.

5.3.2 Semi-classical Flat Laguerre Planes

In general, the construction of semi-classical flat Laguerre planes can be imagined as two halves of a classical flat Laguerre plane being pasted together along one circle or two distinct parallel classes. We describe these two essentially different methods in the following two subsubsections in more detail.

5.3.2.1 Gluing along Two Parallel Classes

Let $\mathcal{H}$ be the set of all orientation-preserving homeomorphisms of $\mathbf{R}$ to itself. Let $h_0, h_1, h_2 \in \mathcal{H}$ and for every $x \in (-\infty, 0]$ let

$$F_x : \mathbf{R} \to \mathbf{R} : y \mapsto \begin{cases} (1 + x)h_0^{-1}(y) - xh_2^{-1}(y) & \text{for } x \in [-1, 0], \\ h_2^{-1}(y) & \text{for } x \in (-\infty, -1]. \end{cases}$$

Then the sets

$$\begin{aligned} & \{(x, y) \in \mathbf{R}^2 \mid y = ax^2 + bx + c, x \geq 0\} \\ \cup \ & \{(x, y) \in \mathbf{R}^2 \mid y = F_x(h_2(a)x^2 + h_1(b)x + h_0(c)), x \leq 0\} \\ \cup \ & \{(\infty, a)\}, \end{aligned}$$

where $a, b, c \in \mathbf{R}$, form the circle set of the *semi-classical flat Laguerre plane* $\mathcal{L}(h_0, h_1, h_2)$ *of type 1* on the cylinder $(\mathbf{R} \cup \{\infty\}) \times \mathbf{R}$.

The functions F_x are used to obtain topological circles in the usual topology of the cylinder $(\mathbf{R} \cup \{\infty\}) \times \mathbf{R}$. An isomorphic model with a slightly different topology but circles that are easier to visualize and deal with is one where circles have the form

$$\begin{array}{ll} & \{(x,y) \in \mathbf{R}^2 \mid y = ax^2 + bx + c, x \geq 0\} \\ \cup & \{(x,y) \in \mathbf{R}^2 \mid y = h_0^{-1}(h_2(a)x^2 + h_1(b)x + h_0(c)), x \leq 0\} \\ \cup & \{(\infty, a)\}, \end{array}$$

where $a, b, c \in \mathbf{R}$. One only has to fit in the parallel class at infinity topologically in a suitable way. The topology on $\mathbf{R}^2$ is the usual one, that is, the affine parts of the two models are (topologically) isomorphic. We use this description when specifying automorphisms of these planes.

If $h_0 = id$ and $h_1 = h_2$ are of the form $x \mapsto x$ and $x \mapsto kx$ for $x \geq 0$ and $x \leq 0$, respectively, the resulting planes are called *Moulton–Laguerre planes* in Hartmann [1976]; see also Mäurer [1972].

Let V and W be two distinct parallel classes in the classical flat Laguerre plane. One *type 1 half* of this plane is the restriction of the plane to one of the two connected components of the complement of $V \cup W$ in the point set. Since the automorphism group of the classical plane acts 2-transitively on parallel classes (see Proposition 5.1.2), all type 1 halves of the classical plane are isomorphic.

THEOREM 5.3.1 (Characterization of Semi-classical Planes) *Let $\mathcal{L} = (Z, \mathcal{C})$ be a flat Laguerre plane and let V and W be two parallel classes such that the restrictions of $\mathcal{L}$ to the two connected components of $Z \backslash (V \cup W)$ are both isomorphic to one type 1 half of the classical plane. Then $\mathcal{L}$ is isomorphic to one of the semi-classical planes $\mathcal{L}(h_0, h_1, h_2)$ of type 1 defined above.*

Semi-classical flat Laguerre planes of type 1 have two distinguished parallel classes, which in the above representation are the two parallel classes $\{\infty\} \times \mathbf{R}$ and $\{0\} \times \mathbf{R}$. We summarize some properties of $\mathcal{L}(h_0, h_1, h_2)$ in the following theorem. Recall that a homeomorphism in $\mathcal{H}$ is called a *multiplication* if it is of the form $x \mapsto rx$ for $r \in \mathbf{R}^+$.

THEOREM 5.3.2 (Special Properties) *Let $\mathcal{L} = \mathcal{L}(h_0, h_1, h_2)$ be a semi-classical flat Laguerre plane of type 1. Then the following hold.*

(i) *The derived planes at points of the two distinguished parallel classes are semi-classical flat affine planes; see Subsection 2.7.2.*

(ii) *The flat Laguerre plane $\mathcal{L}$ is ovoidal if and only if all the defining homeomorphisms are multiplications.*

(iii) *The flat Laguerre plane $\mathcal{L}$ is classical if and only if all the defining homeomorphisms are multiplications and $h_1^2 = h_2 h_0$.*

Since both classical subgeometries of a semi-classical flat Laguerre plane can be uniquely embedded, up to automorphisms of the classical flat Laguerre plane, into the classical flat Laguerre plane, automorphisms of semi-classical flat Laguerre planes can be explicitly determined.

We say that $\mathcal{L}(h_0, h_1, h_2)$ is of group/kernel dimensions (m, n) if it is of group dimension m and its kernel has dimension n; see Subsection 5.4.2 for the definition of the kernel of a flat Laguerre plane.

We define some special classes of orientation-preserving homeomorphisms $\mathbf{R} \to \mathbf{R}$ that have properties which make them particularly useful when it comes to constructing semi-classical flat Laguerre planes of large group dimensions.

Let

$$h_{d,r} : \mathbf{R} \to \mathbf{R} : x \mapsto \begin{cases} x^d & \text{for } x \geq 0, \\ -r|x|^d & \text{for } x < 0, \end{cases}$$

and let

$$h : \mathbf{R} \to \mathbf{R} : x \mapsto \delta(h_{d,r}(x+t) - h_{d,r}(t)),$$

where $\delta, d, r \in \mathbf{R}^+, t \in \mathbf{R}$. This means that $h_{d,r}$ is a *semi-multiplicative homeomorphism*; compare the end of Section A1.4.

It turns out that h is also a homeomorphism. Let $\widetilde{\mathcal{H}}_M$ be the set of all homeomorphisms of this form, let $\widetilde{\mathcal{H}}_{M,d}$ be the homeomorphisms in $\widetilde{\mathcal{H}}_M$ with fixed d, and let $\mathcal{H}_{sa} = \widetilde{\mathcal{H}}_{M,1}$ (*semi-affine homeomorphisms*). Finally, let $\mathcal{H}_m$ be the set of all multiplications as defined above.

THEOREM 5.3.3 (Group/Kernel-Dimension Classification) *Let $\mathcal{L} = \mathcal{L}(h_0, h_1, h_2)$ be a semi-classical flat Laguerre plane of type 1. Then $\mathcal{L}$ has group/kernel dimensions*

7,4 *if and only if $h_0, h_1, h_2 \in \mathcal{H}_m$ and $h_0 h_2 = h_1^2$; in this case $\mathcal{L}$ is the classical flat Laguerre plane;*

5,4 *if and only if $h_0, h_1, h_2 \in \mathcal{H}_m$ and $h_0 h_2 \neq h_1^2$; in this case $\mathcal{L}$ is a nonclassical ovoidal flat Laguerre plane;*

4,3 *if and only if two of the defining homeomorphisms are in $\mathcal{H}_m$ and the third is in $\mathcal{H}_{sa} \setminus \mathcal{H}_m$;*

3,2 *if and only if one of the defining homeomorphisms is in $\mathcal{H}_m$ and the other two are in $\mathcal{H}_{sa} \setminus \mathcal{H}_m$, or two of the defining homeomorphisms are in $\mathcal{H}_m$ and the third is not in $\mathcal{H}_{sa}$;*

2,1 *if and only if all three defining homeomorphisms are in $\widetilde{\mathcal{H}}_{M,d} \setminus \mathcal{H}_m$ for some $d > 0$, or $h_i \in \mathcal{H}_m, h_j \in \widetilde{\mathcal{H}}_{M,d} \setminus \mathcal{H}_m, h_k \in \widetilde{\mathcal{H}}_{M,d} \setminus \mathcal{H}_{sa}$, such that $\{i, j, k\} = \{0, 1, 2\}$;*

1,1 *if and only if exactly one of the defining homeomorphisms h_i is in $\mathcal{H}_m$ and at least one is not in $\widetilde{\mathcal{H}}_M$;*

1,0 *if and only if exactly two of the defining homeomorphisms are in $\mathcal{H}_{M,d}$ for some $d > 0$, or if they are all contained in $\widetilde{\mathcal{H}}_M \setminus \mathcal{H}_m$ but not all contained in $\widetilde{\mathcal{H}}_{M,d}$ for the same $d > 0$;*

0,0 *in all other cases, that is, none of the homeomorphisms is in $\mathcal{H}_m$, for every $d > 0$ at most one of them is contained in $\widetilde{\mathcal{H}}_{M,d}$ and at least one of them is not contained in $\widetilde{\mathcal{H}}_M$.*

By choosing the determining homeomorphisms carefully, we arrive at *rigid* flat Laguerre planes, that is, flat Laguerre planes that do not admit any nontrivial automorphisms. For example, let

$$f_{r,m} : \mathbf{R} \to \mathbf{R} : x \mapsto \left(1 - \frac{r}{3}\right) x^{2m+1} + r\left(\frac{1}{3}x^3 - x^2 + x\right),$$

where $m \in \mathbf{N}$, $m \geq 2$, and $r \in \mathbf{R}$, $0 < r < 3$. It is easily verified that each $f_{r,m}$ is a strictly increasing homeomorphism. Furthermore, if (r_i, m_i), $i = 1, 2, 3$, are three pairwise distinct pairs such as this, then $\mathcal{L}(f_{r_1,m_1}, f_{r_2,m_2}, f_{r_3,m_3})$ is a rigid flat Laguerre plane.

For these and other information about semi-classical flat Laguerre planes of this type see Steinke [1988]. In Polster–Rosehr–Steinke [1998] the results summarized in this subsubsection have been extended. In it all flat Laguerre planes have been determined that have been glued together from two halves of ovoidal flat Laguerre planes.

5.3.2.2 Gluing along a Circle

In the following let $\mathcal{H}_{0,1}$ be the set of all orientation-preserving homeomorphisms of $\mathbf{R}$ to itself that fix 0 and 1, and let $\mathcal{H}_s$ be the subset of $\mathcal{H}_{0,1}$ consisting of all those homeomorphisms h such that $h(x) = -h(-x)$ for all $x \in \mathbf{R}$. Let $\varphi \in \mathcal{H}_s$ and $\psi \in \mathcal{H}_{0,1}$.

The circles of the *semi-classical flat Laguerre plane of type 2*, denoted by $\mathcal{L}(\varphi, \psi)$, on the cylinder $(\mathbf{R} \cup \{\infty\}) \times \mathbf{R}$ consist of the following sets.

$$\{(x, c) \in \mathbf{R}^2 \mid x \in \mathbf{R}\} \cup \{(\infty, 0)\},$$

where $c \in \mathbf{R}$,

$$\begin{aligned} & \{(x, y) \in \mathbf{R}^2 \mid x = my + t, y \geq 0\} \\ \cup \quad & \{(x, y) \in \mathbf{R}^2 \mid x = \psi^{-1}(\varphi(m)y + \psi(t)), y \leq 0\} \cup \{(\infty, 0)\}, \end{aligned}$$

where $m, t \in \mathbf{R}$, $m \neq 0$, and

$$\begin{array}{ll} & \{(x,y) \in \mathbf{R}^2 \mid y = ax^2 + bx + c\} \\ \cup & \{(\infty, a)\} \qquad \text{for } b^2 \leq 4ac, a > 0, \end{array}$$

$$\begin{array}{ll} & \{(x,y) \in \mathbf{R}^2 \mid y = a(\psi(x))^2 + b\psi(x) + c\} \\ \cup & \{(\infty, a)\} \qquad \text{for } b^2 \leq 4ac, a < 0, \end{array}$$

$$\begin{array}{ll} & \{(x,y) \in \mathbf{R}^2 \mid y = ax^2 + bx + c, y \geq 0\} \\ \cup & \{(x,y) \in \mathbf{R}^2 \mid y = \overline{a}(\psi(x))^2 + \overline{b}\psi(x) + \overline{c}, y \leq 0\} \\ \cup & \{(\infty, a)\} \qquad \text{for } b^2 > 4ac, a > 0, \end{array}$$

$$\begin{array}{ll} & \{(x,y) \in \mathbf{R}^2 \mid y = ax^2 + bx + c, y \geq 0\} \\ \cup & \{(x,y) \in \mathbf{R}^2 \mid y = \overline{a}(\psi(x))^2 + \overline{b}\psi(x) + \overline{c}, y \leq 0\} \\ \cup & \{(\infty, \overline{a})\} \qquad \text{for } b^2 > 4ac, a < 0, \end{array}$$

where in the last two cases

$$\begin{aligned} \tfrac{1}{\overline{a}} &= \varphi\left(\tfrac{1}{\sqrt{b^2-4ac}}\right)\left(\psi\left(\tfrac{-b+\sqrt{b^2-4ac}}{2a}\right) - \psi\left(\tfrac{-b-\sqrt{b^2-4ac}}{2a}\right)\right), \\ \overline{b} &= -\overline{a}\left(\psi\left(\tfrac{-b+\sqrt{b^2-4ac}}{2a}\right) + \psi\left(\tfrac{-b-\sqrt{b^2-4ac}}{2a}\right)\right), \\ \overline{c} &= \overline{a}\psi\left(\tfrac{-b+\sqrt{b^2-4ac}}{2a}\right)\psi\left(\tfrac{-b-\sqrt{b^2-4ac}}{2a}\right), \end{aligned}$$

and $a, b, c \in \mathbf{R}$, $a \neq 0$.

Let C be a circle in the classical flat Laguerre plane. One *type 2 half* of this plane is the restriction of the plane to one of the two connected components of the complement of C in the point set. Since the automorphism group of the classical plane acts transitively on the set of circles (see Theorem 5.1.1) and because each circle admits a reflection at it, all type 2 halves of the classical plane are isomorphic.

THEOREM 5.3.4 (Characterization of Semi-classical Planes) *Let $\mathcal{L} = (Z, \mathcal{C})$ be a flat Laguerre plane and a let C be a circle such that the restrictions of $\mathcal{L}$ to the two connected components of $Z \setminus C$ are both isomorphic to one type 2 half of the classical plane. Then $\mathcal{L}$ is isomorphic to one of the semi-classical planes $\mathcal{L}(\varphi, \psi)$ of type 2 defined above.*

We summarize some properties of $\mathcal{L}(\varphi, \psi)$ in the following theorem.

THEOREM 5.3.5 (Special Properties) *Let $\mathcal{L} = \mathcal{L}(\varphi, \psi)$ be a semi-classical flat Laguerre plane of type 2. Then the derived plane at $(\infty, 0)$ is a semi-classical flat affine plane. Furthermore, $\mathcal{L}$ is ovoidal if and only if $\varphi = \psi = id$. In this case the Laguerre plane is classical.*

Since both classical halves of a semi-classical Laguerre plane of type 2 can be uniquely embedded, up to automorphisms of the classical flat Laguerre plane, into the classical flat Laguerre plane, the automorphisms of the semi-classical Laguerre planes can be explicitly determined; see Steinke [1987]. One can therefore completely classify the semi-classical planes of type 2 with respect to the group/kernel dimensions. In order to be able to state this classification, we define $\psi_\infty = \psi$ and for $c \in \mathbf{R}$

$$\psi_c : \mathbf{R} \to \mathbf{R} : x \mapsto \begin{cases} \dfrac{\psi(c+1) - \psi(c)}{\psi(x + \frac{1}{x}) - \psi(x)} & \text{for } x \neq 0, \\ 0 & \text{for } x = 0. \end{cases}$$

This is an orientation-preserving homeomorphism that fixes 0 and 1. We say that two homeomorphism $\psi, \psi' \in \mathcal{H}$ are *equivalent* if and only if there are affine bijections $\alpha, \beta : \mathbf{R} \to \mathbf{R}$ and a coefficient $c \in \mathbf{R} \cup \{\infty\}$ such that $\psi \circ \alpha = \beta \circ \psi'_c$. We remark that ψ is equivalent to id if and only if $\psi = id$. Let $\mathcal{H}_M$ be the set of all semi-multiplicative homeomorphisms and let $\mathcal{H}_{s,M} = \mathcal{H}_s \cap \mathcal{H}_M$, that is, the set of all $h_{d,1}$ in the notation of the previous subsubsection. (These are multiplicative homeomorphisms of $\mathbf{R}$ onto itself.)

THEOREM 5.3.6 (Group/Kernel-Dimension Classification) *Let $\mathcal{L} = \mathcal{L}(\varphi, \psi)$ be a semi-classical flat Laguerre plane of type 2. Then $\mathcal{L}$ has group/kernel dimensions*

- 7,4 *if and only if $\varphi = \psi = id$; in this case $\mathcal{L}$ is the classical flat Laguerre plane;*
- 4,1 *if and only if $\psi = id$ and $\varphi \in \mathcal{H}_{s,M} \setminus \{id\}$;*
- 3,0 *if and only if $\psi = id$ and $\varphi \notin \mathcal{H}_{s,M}$;*
- 2,1 *if and only if the function ψ is equivalent to some $h \in \mathcal{H}_M \setminus \{id\}$ and $\varphi \in \mathcal{H}_{s,M}$;*
- 1,1 *if and only if the function ψ is not equivalent to any $h \in \mathcal{H}_M$ and $\varphi \in \mathcal{H}_{s,M}$;*
- 1,0 *if and only if the function ψ is equivalent to some $h \in \mathcal{H}_M \setminus \{id\}$ and $\varphi \notin \mathcal{H}_{s,M}$;*
- 0,0 *if and only if the function ψ is not equivalent to any $h \in \mathcal{H}_M \setminus \{id\}$ and $\varphi \notin \mathcal{H}_{s,M}$.*

Choose for φ and ψ any of the functions $f_{r,m}$ defined at the end of the previous subsubsection. Then the semi-classical flat Laguerre plane $\mathcal{L}(\varphi, \psi)$ is a rigid flat Laguerre plane, that is, a flat Laguerre plane that does not admit any nontrivial automorphisms.

For these and other information about semi-classical flat Laguerre planes of this type see Steinke [1987].

5.3.3 Different Ways to Cut and Paste

In this subsection we look at different ways to construct new cylindrical circle planes and flat Laguerre planes from old ones using a variety of cut-and-paste techniques. The results in this subsection are very similar to the cut-and-paste results about spherical circle planes. There are two essentially different types of techniques.

One type uses a separating set of the point set $Z = \mathbf{S}^1 \times \mathbf{R}$. Examples for flat Laguerre planes that arise from this kind of cut and paste are the semi-classical flat Laguerre planes discussed in the previous subsection. More generally, one can start with two cylindrical circle planes or flat Laguerre planes $\mathcal{L}_1$ and $\mathcal{L}_2$ and cut and paste either along two parallel classes or along a common circle. In both cases one then needs a rule that matches circles of $\mathcal{L}_1$ that intersect the separating set in two points with circles of $\mathcal{L}_2$ that intersect the separating set in the same two points. However, a general rule for a matching such as this is difficult to specify, may even be impossible in general, and has only been carried out so far for classical and ovoidal flat Laguerre planes.

In the following we concentrate on a second type of cut-and-paste method, namely cutting and pasting along separating sets in the circle space. A formal definition of (strong) (X-embedded) separating sets of cylindrical circle planes can be modelled after the definition of such sets in the case of spherical circle planes; see Subsection 3.3.4. For example, essentially this type of cut-and-paste method is used to obtain the flat Laguerre planes of generalized shear type; see Subsection 5.3.5 for this kind of plane.

The following general result can be used to prove that certain 'nice' sets of circles are separating sets. It also suggests that there may be pre-separating sets that do not work for all but only a proper subset of the cylindrical circle planes and flat Laguerre planes they are contained in. Others may only work for cylindrical circle planes they are contained in but not for flat Laguerre planes.

THEOREM 5.3.7 (General Cut and Paste) *Let $\mathcal{L}_i = (Z, \mathcal{C}_i)$, $i = 1, 2$, be two cylindrical circle planes. Suppose that there is a set of circles $S \subseteq \mathcal{C}_1 \cap \mathcal{C}_2$ such that $\mathcal{C}_i \setminus S$ has precisely two (nonempty) connected components $\mathcal{C}_i^+$ and $\mathcal{C}_i^-$ and that each of these components is path-connected. Furthermore, the labelling of the components $\mathcal{C}_1^+$ and $\mathcal{C}_2^+$ agrees, that is, there are three pairwise nonparallel points p_1, p_2, and p_3 such that the circles joining these three points in the respective cylindrical circle planes belong to $\mathcal{C}_1^+$ and $\mathcal{C}_2^+$, respectively. Then $(Z, S \cup \mathcal{C}_1^+ \cup \mathcal{C}_2^-)$ is a cylindrical circle plane.*

If $\mathcal{L}_1$ and $\mathcal{L}_2$ are flat Laguerre planes, then so is $(Z, S \cup \mathcal{C}_1^+ \cup \mathcal{C}_2^-)$.

The first part of this result is part of the more general Theorem 7.3.7. For the second part, see Polster–Steinke [20XXc].

To make this result really applicable we construct some nice separating sets. The following considerations suggest a way to find such sets.

We have a look at the ovoidal cylindrical circle plane over a closed topological arc $\mathcal{O}$. Circles of the cylindrical circle plane are the nontrivial plane intersections of a cone with basis $\mathcal{O}$ and vertex v in projective 3-dimensional space $\mathrm{PG}(3, \mathbf{R})$. We now choose a point $p \neq v$ of $\mathrm{PG}(3, \mathbf{R})$ and let L be the line through p and v. Since L is homeomorphic to $\mathbf{S}^1$, the complement $L \setminus \{p, v\}$ has precisely two connected components C_1 and C_2. Every plane of $\mathrm{PG}(3, \mathbf{R})$ that does not contain L intersects L in precisely one point. Clearly, the collection of all planes of $\mathrm{PG}(3, \mathbf{R})$ that pass through the point p separate the space of all planes of $\mathrm{PG}(3, \mathbf{R})$ that intersect the cone nontrivially into two connected components. These components are the planes that intersect L in precisely one point of the components C_1 and C_2, respectively. Similarly, the set $\mathcal{C}^p$ of all circles that arise from planes through p separate the circle set $\mathcal{C}$ of the ovoidal cylindrical circle plane into two connected components. Moreover, each of these components is path-connected. The point p is a point of the cone, an inner point of the cone, or an outer point of the cone. This suggests looking for three different kinds of separating sets in general cylindrical circle planes that correspond to these three different kinds of points. In the first case this is no problem.

Let $\mathcal{L} = (Z, \mathcal{C})$ be a cylindrical circle plane. We fix one point p and consider the collection $\mathcal{C}^p$ of all circles that contain p. The point p separates the parallel class V it is contained in into two different components V_+ and V_-, and the set $\mathcal{C} \setminus \mathcal{C}^p$ has two open components $\mathcal{C}^{p+}$ and $\mathcal{C}^{p-}$ consisting of all those circles that intersect V in V_+ and V_-, respectively. The following result is a corollary of Theorem 5.3.7; see also Polster–Steinke [1995] Theorem 5.1. It shows that $\mathcal{C}^p$ is a strong separating set.

PROPOSITION 5.3.8 (Cut and Paste I) *Let $\mathcal{L}_i = (Z, \mathcal{C}_i)$, $i = 1, 2$, be cylindrical circle planes. Suppose that $\mathcal{C}_1 \cap \mathcal{C}_2 \supseteq \mathcal{C}_1^p$ for some point p. Let $\mathcal{C} = \mathcal{C}_1^p \cup \mathcal{C}_1^{p+} \cup \mathcal{C}_2^{p-}$. Then $\mathcal{L} = (Z, \mathcal{C})$ is a cylindrical circle plane. If both $\mathcal{L}_1$ and $\mathcal{L}_2$ are flat Laguerre planes, then $\mathcal{L}$ is a flat Laguerre plane, too.*

It does not make sense to speak of inner and outer points of general cylindrical circle planes but, just like in the case of spherical circle planes, certain involutory homeomorphisms of the cylinder to itself can play the role of such points.

Consider the cylinder associated with an ovoidal cylindrical circle plane $\mathcal{L}$. The *bundle involution* γ of $\mathcal{L}$ associated with a point p not contained in the cylinder is defined by using the lines through p just like the bundle involution of a topological oval in a flat projective plane associated with a point off the oval; see p. 120.

Let $Fix(\gamma)$ be the set of all circles in $\mathcal{L}$ that are fixed by γ. Then γ has the following properties.

(i) If p is an inner point of $\mathcal{O}$, then γ is fixed-point-free.

(ii) If p is an exterior point of $\mathcal{O}$, then the fixed-point set F of γ is the union of two parallel classes and γ exchanges the two connected components of $Z \setminus F$.

(iii) A circle is contained in $Fix(\gamma)$ if and only if it contains a point q and its image $\gamma(q)$, where $q \neq \gamma(q)$.

(iv) The involution γ maps parallel classes to parallel classes.

Note that γ is not necessarily an automorphism of $\mathcal{L}$.

Now let $\mathcal{L} = (Z, \mathcal{C})$ be a cylindrical circle plane and let γ be an involutory homeomorphism of Z to itself. We call γ an *inner involution* of $\mathcal{L}$ if it satisfies the properties listed in (i), (iii), and (iv) above. We call it an *outer involution* of $\mathcal{L}$ if it satisfies the properties listed in (ii), (iii), and (iv). Inner and outer involutions are virtual inner and outer points that can play the same role in the construction of separating sets as the points of the projective space $\mathrm{PG}(3, \mathbf{R})$ in the ovoidal case. See also Subsection 5.8.2 for fix-geometries of inner and outer involutions.

We start by summarizing some useful properties of inner involutions in the following lemma.

LEMMA 5.3.9 (Properties of Inner Involutions) *Let γ be an inner involution of a cylindrical circle plane $\mathcal{L} = (Z, \mathcal{C})$. Then the following hold.*

(i) *Two distinct circles in $Fix(\gamma)$ have exactly two points in common. These two points get exchanged by γ.*

(ii) *Let C be a circle not in $Fix(\gamma)$. Then C is disjoint from its image under the involution.*

(iii) *Let r, s, $\gamma(r)$, and $\gamma(s)$ be four distinct points on a circle C in $Fix(\gamma)$. Then r and $\gamma(r)$ are contained in different connected components of $C \setminus \{s, \gamma(s)\}$.*

For a proof of this lemma see Polster–Steinke [1997] Lemma 4.

From property (iii) we see that the set $\mathcal{C} \setminus Fix(\gamma)$ has two connected components $\mathcal{C}^{\gamma+}$ and $\mathcal{C}^{\gamma-}$. Here $\mathcal{C}^{\gamma+}$ consists of all circles C such that $\gamma(C)$ is situated above C on the cylinder. The other connected component $\mathcal{C}^{\gamma-}$ consists of all circles C such that $\gamma(C)$ is situated below C on the cylinder. Both components are path-connected.

It is easy to show that γ is an inner involution of any cylindrical circle plane that contains $Fix(\gamma)$. This observation plus the following corollary of Theorem 5.3.7 shows that $Fix(\gamma)$ is a strong separating set; see also Polster–Steinke [1997] Propositions 2 and 2*.

PROPOSITION 5.3.10 (Cut and Paste II) *Let $\mathcal{L}_i = (Z, \mathcal{C}_i)$, $i = 1, 2$, be two cylindrical circle planes that share the set $Fix(\gamma)$, where γ is some inner involution of $\mathcal{L}_1$. Let $\mathcal{C} = Fix(\gamma) \cup \mathcal{C}_1^{\gamma+} \cup \mathcal{C}_2^{\gamma-}$. Then $\mathcal{L} = (Z, \mathcal{C})$ is a cylindrical circle plane. If both $\mathcal{L}_1$ and $\mathcal{L}_2$ are flat Laguerre planes, then $\mathcal{L}$ is a flat Laguerre plane, too.*

We now consider an outer involution γ with fixed-point set F. Some useful properties of such an involution are summarized in the following lemma.

LEMMA 5.3.11 (Properties of Outer Involutions) *Let γ be an outer involution of a cylindrical circle plane $\mathcal{L} = (Z, \mathcal{C})$ with fixed-point set F (this set is a union of two parallel classes). Then the following hold.*

(i) *Let $C \in \mathcal{C} \setminus Fix(\gamma)$. Then $C \cap \gamma(C) = C \cap F$.*

(ii) *Let r, s, $\gamma(r)$, and $\gamma(s)$ be four distinct points on a circle C in $Fix(\gamma)$. Then r and $\gamma(r)$ are contained in the same connected component of $C \setminus \{s, \gamma(s)\}$.*

For a proof of this lemma see Polster–Steinke [1997] Lemma 4.

Let H be one of the connected components of $Z \setminus F$. From property (i) listed in the above lemma we conclude that $\mathcal{C} \setminus Fix(\gamma)$ has two connected components $\mathcal{C}^{\gamma+}$ and $\mathcal{C}^{\gamma-}$. Here $\mathcal{C}^{\gamma+}$ consists of all circles C such that $\gamma(C)$ is situated above C on the half H. The other connected component $\mathcal{C}^{\gamma-}$ consists of all circles C such that $\gamma(C)$ is situated below C on H. Both components are path-connected.

Again it is easy to see that $Fix(\gamma)$ is a strong separating set; see also Polster–Steinke [1997] Propositions 3 and 3*.

PROPOSITION 5.3.12 (Cut and Paste III) *Let $\mathcal{L}_i = (Z, \mathcal{C}_i)$, $i = 1, 2$, be two cylindrical circle planes that share the set $Fix(\gamma)$, where γ is some inner involution of $\mathcal{L}_1$. Let $\mathcal{C} = Fix(\gamma) \cup \mathcal{C}_1^{\gamma+} \cup \mathcal{C}_2^{\gamma-}$. Then $\mathcal{L} = (Z, \mathcal{C})$ is a cylindrical circle plane. If both $\mathcal{L}_1$ and $\mathcal{L}_2$ are flat Laguerre planes, then $\mathcal{L}$ is a flat Laguerre plane, too.*

For further cut-and-paste constructions involving separating sets associated with points and virtual points of cylindrical circle planes, see Polster–Steinke [1995], [1997], and [20XXa]. We only note that separating sets have been constructed whose complement in the circle set falls into more than two connected components. These separating sets allow us to simultaneously combine more than two cylindrical circle planes.

5.3.4 Integrals of Flat Linear Spaces

Let F be an n-unisolvent set on the interval I, that is, a set of continuous functions $I \to \mathbf{R}$ that solve the Lagrange interpolation problem of order n; see Chapter 7. For example, the set $\mathrm{Poly}(n, \mathbf{R})$ of all polynomials of degree at most $n-1$ is an n-unisolvent set on $\mathbf{R}$. The integral $S(F)$ of F is the set consisting of the functions

$$I \to \mathbf{R} : x \mapsto \int_a^x f(t)dt + b,$$

where $a \in I, b \in \mathbf{R}$. Clearly, $S(\mathrm{Poly}(n, \mathbf{R})) = \mathrm{Poly}(n+1, \mathbf{R})$. Since the 2-unisolvent set $\mathrm{Poly}(2, \mathbf{R})$ essentially corresponds to the Euclidean plane and the 3-unisolvent set $\mathrm{Poly}(3, \mathbf{R})$ to the affine part of the classical flat Laguerre plane, this means that this affine part 'is' the integral of the Euclidean plane. This observation can be generalized in a number of ways.

5.3.4.1 Integrals of Flat Affine Planes

Recall from Subsection 2.7.9 the following necessary and sufficient condition that the integral of a 1-unisolvent set F on $\mathbf{R}$ gives rise to a flat affine plane on $\mathbf{R}^2$ by adding the vertical Euclidean lines.

$$\lim_{x\to\pm\infty}\left|\int_0^x (f(t)-g(t))dt\right| = \infty$$

for all $f, g \in F$, $f \neq g$. Let F be a 2-unisolvent set on $\mathbf{R}$ that corresponds to a flat affine plane. Then the 1-unisolvent subsets of F correspond to the nonvertical parallel classes of the affine plane.

THEOREM 5.3.13 (Integrals of Flat Affine Planes) *Let F be a 2-unisolvent set on $\mathbf{R}$. Then $S(F)$ corresponds to the affine part of a flat Laguerre plane if and only if F corresponds to a flat affine plane and the integral of every parallel class of F corresponds to a flat affine plane.*

Let F be as in the above theorem. Then there are so-called *coordinatizing functions* $\mathbf{R}^3 \to \mathbf{R}$ for F. Such a coordinatizing function, let us call it F as well, is continuous and has the property that every nonvertical parallel class corresponds to an $a \in \mathbf{R}$ and consists of the functions $\mathbf{R} \to \mathbf{R} : x \mapsto F(a, b, x)$ for $b \in \mathbf{R}$. Starting with a 2-unisolvent set F as in the theorem let

$$F(a, b, x) = f(x),$$

where $f \in F$ corresponds to the parallel through the point $(0, b)$ of the line/function in F that contains the points $(0, 0)$ and $(1, a)$. For example, if F is $\mathrm{Poly}(2, \mathbf{R})$, then $(a, b, x) \mapsto ax + b$ defines such an associated coordinatizing function. Represented on the cylinder $(\mathbf{R} \cup \{\infty\}) \times \mathbf{R}$, the circles of the flat Laguerre plane whose affine part corresponds to $S(F)$ are the sets

$$\left\{(x, y) \in \mathbf{R}^2 \;\middle|\; y = \int_0^x F(a, b, t)dt + c\right\} \cup \{(\infty, a)\},$$

where $a, b, c \in \mathbf{R}$. This means that the derived planes at points of the parallel class at infinity are the integrals of the (nonvertical) parallel classes in F. If $G \subset F$ corresponds to the pencil of lines through the point (x_0, y_0), then for a fixed $b \in \mathbf{R}$ the set of functions

$$\mathbf{R} \to \mathbf{R} : x \mapsto \int_{x_0}^x g(t)dt + b,$$

where $g \in G$, corresponds to a pencil of touching circles of the integral of F through the point (x_0, b). Every pencil of touching circles at a point of $\mathbf{R}^2$ arises like this. Finally, note that all Euclidean translations in the vertical direction are automorphisms of the integral of F.

Let us have a look at some examples of flat Laguerre planes that are integrals of flat affine planes. We first deform the Euclidean plane by the homeomorphism

$$\mathbf{R}^2 \to \mathbf{R}^2 : (x, y) \to (x, h(y)),$$

where $h : \mathbf{R} \to \mathbf{R}$ is a nonlinear homeomorphism. Then

$$F : \mathbf{R}^3 \to \mathbf{R} : (a, b, x) \mapsto h(ax + b)$$

is a coordinatizing function of the deformed Euclidean plane. Now $S(F)$ consists of the functions

$$\mathbf{R} \to \mathbf{R} : x \mapsto \frac{1}{a}(p(ax + b) - p(b)) + c$$

and

$$\mathbf{R} \to \mathbf{R} : x \mapsto h(b)x + c,$$

where $a, b, c \in \mathbf{R}$, $a \neq 0$, and $p : \mathbf{R} \to \mathbf{R}$ is a strictly parabolic function such that $p' = h$. It can be shown that $S(F)$ corresponds to the affine part of a nonclassical flat Laguerre plane; see Artzy–Groh [1986] or Subsection 5.3.7 for more information on these kind of planes.

As a second example let

$$F : \mathbf{R}^3 \to \mathbf{R} : (a, b, x) \mapsto \begin{cases} ax + b & \text{for } x \leq 1, \\ \frac{a}{x} + \frac{b}{x^2} & \text{for } x \geq 1. \end{cases}$$

This is the coordinatizing function of yet another model of the Euclidean plane, namely, the classical model whose point set has been deformed by the homeomorphism

$$\mathbf{R}^2 \to \mathbf{R}^2 : (x, y) \mapsto \begin{cases} (x, y) & \text{for } x \leq 1, \\ (x, \frac{y}{x^2}) & \text{for } x > 1. \end{cases}$$

However, the integral of this model is no longer a flat Laguerre plane since

$$\lim_{x \to \infty} \int_0^x [F(0, 1, t) - F(0, 0, t)]dt = 2 < \infty.$$

Thirdly, remember that the affine part of an ovoidal flat Laguerre plane corresponds to the set of functions

$$\mathbf{R} \to \mathbf{R} : x \mapsto ap(x) + bx + c,$$

where $a, b, c \in \mathbf{R}$ and $p : \mathbf{R} \to \mathbf{R}$ is a fixed strictly parabolic function. The derivative of the differentiable function p defines a homeomorphism h and

$$(a, b, x) \mapsto \frac{\partial}{\partial x}(ap(x) + bx + c)) = ah(x) + b$$

is a coordinatizing function of a flat affine plane. Hence, we obtain the following.

PROPOSITION 5.3.14 (Ovoidal Planes Are Integrals) *Every ovoidal flat Laguerre plane is an integral of a flat affine plane.*

We now describe how the classical flat Möbius and Minkowski planes can be represented as 'integrals'. Let I be the open interval $I = (-1, 1)$. The map

$$s : \mathbf{R} \to I : x \mapsto \frac{x}{\sqrt{1 + x^2}}$$

is a homeomorphism with inverse

$$s^{-1} : I \to \mathbf{R} : x \mapsto \frac{x}{\sqrt{1 - x^2}}.$$

Let

$$S : \mathbf{R} \times \mathbf{R} \to \mathbf{R} \times I : (x, y) \mapsto (x, s(y)).$$

This is a homeomorphism that compresses $\mathbf{R}^2$ into the strip $\mathbf{R} \times I$.

The image of the Euclidean line given by the linear function

$$\mathbf{R} \to \mathbf{R} : x \mapsto ax + b$$

under S is the graph of

$$\mathbf{R} \to \mathbf{R} : x \mapsto s(ax + b)).$$

Clearly, $\int s(x)dx = \sqrt{1 + x^2} + c$. The graph of

$$H : \mathbf{R} \to \mathbf{R} : x \mapsto \sqrt{1 + x^2}$$

is the branch of a hyperbola with asymptotes having slope ± 1 and

$$\int s(ax + b)dx = \begin{cases} s(b)x + c & \text{for } a = 0, \\ \frac{1}{a}H(ax + b) + c & \text{for } a \neq 0. \end{cases}$$

Hence the integral of the compressed Euclidean plane consists of all lines with slope less than 1 and greater than -1, and all branches of hyperbolas that have lines with slope 1 and -1 as asymptotes and that are graphs of convex or concave functions. This means that the integral of

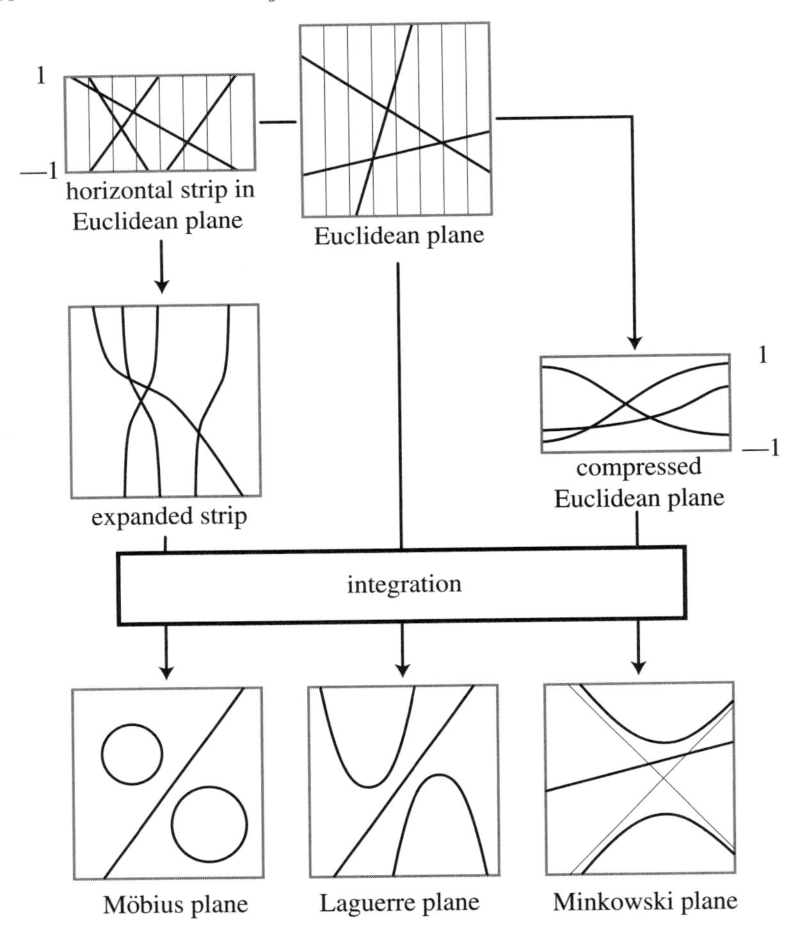

Fig. 5.4. The classical flat Möbius, Laguerre, and Minkowski planes as integrals of (parts of) the Euclidean plane

the compressed Euclidean plane is one half of the classical flat Minkowski plane. We get the other half by just rotating this half around 90 degrees. All this corresponds to the right column of Figure 5.4.

Let us concentrate on the strip $\mathbf{R} \times I$ in the Euclidean plane, where I denotes the open interval $I = (-1, 1)$, and expand it by the inverse S^{-1} of S. Under this map the line segment which is the graph of the function $I \to I : x \mapsto x$ turns into the graph of the function $I \to \mathbf{R} : x \mapsto s^{-1}(x)$. The integral of this function is defined by $K(x) = -\sqrt{1-x^2}, x \in I$, and the graph of K is the bottom half of the unit circle. By integrating all other expanded line segments in the strip, we get all other top and bottom parts of the Euclidean circles. By 'integrating' the horizontal lines we get all nonvertical Euclidean lines. We conclude that the integral of the expanded (Euclidean) strip is the classical flat Möbius plane. All this corresponds to the left column of Figure 5.4.

We only remark that there is a second way in which the affine part of the classical flat Laguerre plane can be seen to be an integral. Just rotate the middle picture in the bottom row of Figure 5.4 around 90 degrees, think of every parabola as being split into a top and a bottom part and start arguing as in the case of flat Möbius planes. The resulting way of integrating seems to be a mix of the ways in which the classical flat Möbius and Minkowski planes were represented as integrals. This makes sense considering that the classical flat Laguerre plane is the limiting case of both the classical flat Möbius plane and the classical flat Minkowski plane (an elliptic quadric in real 3-dimensional projective space can be continuously deformed into a hyperbolic quadric via elliptic quadrics, one cone and hyperbolic quadrics).

Many of the known examples of flat Möbius, Laguerre and Minkowski planes can be interpreted as integrals of flat affine planes and $\mathbf{R}^2$-planes as described above. For details see Polster [1997], [1998c].

5.3.4.2 Integrals of Point Möbius Strip Planes

In this subsection we work with the following normalized representations of point Möbius strip planes and flat Laguerre planes.

The point set of a point Möbius strip plane is the strip $\mathcal{M} = [0, \pi] \times \mathbf{R}$ whose left and right boundaries have been identified via the map

$$\{0\} \times \mathbf{R} \to \{\pi\} \times \mathbf{R} : (0, y) \mapsto (\pi, -y).$$

This is a Möbius strip. The set of nonvertical lines in the point Möbius strip plane corresponds to a 2-unisolvent set of continuous half-periodic functions on $[0, \pi]$. The classical example of such a set consists of the functions

$$[0, \pi] \to \mathbf{R} : x \mapsto a \sin x + b \cos x,$$

where $a, b \in \mathbf{R}$.

The point set of a flat Laguerre plane is the strip $\mathcal{Z} = [0, 2\pi] \times \mathbf{R}$ whose left and right boundaries have been identified via the map

$$\{0\} \times \mathbf{R} \to \{2\pi\} \times \mathbf{R} : (0, y) \mapsto (2\pi, y).$$

This is a cylinder. Its circle set corresponds to a 3-unisolvent set of continuous periodic functions on $[0, 2\pi]$. The classical example of such a set consists of the functions

$$[0, 2\pi] \to \mathbf{R} : x \mapsto a \sin x + b \cos x + c,$$

where $a, b, c \in \mathbf{R}$.

Note that the second set of functions is the integral of the periodic extension of the first set to the interval $[0, 2\pi]$. In this way, the classical flat Laguerre plane is the integral of the classical point Möbius strip plane. This observation generalizes as follows.

Let F be a 2-unisolvent set of continuous half-periodic functions on the interval $[0, \pi]$. Such a set automatically corresponds to a point Möbius strip plane. Let $\overline{F}$ be the set of functions

$$\overline{f} : [0, 2\pi] \to \mathbf{R} : x \mapsto \begin{cases} f(x) & \text{for } x \in [0, \pi], \\ -f(x - \pi) & \text{for } x \in [\pi, 2\pi], \end{cases}$$

where $f \in F$. Furthermore, let $S(F)$ be the set of functions

$$[0, 2\pi] \to \mathbf{R} : x \mapsto \int_0^x g(t)dt + c,$$

where $g \in \overline{F}$ and $c \in \mathbf{R}$. It turns out that $S(F)$ is a set of continuous periodic functions the set of whose graphs is the circle set of a flat Laguerre plane on the cylinder $\mathcal{Z}$. We summarize these considerations in the following theorem.

THEOREM 5.3.15 (Integrals of Point Möbius Strip Planes) *Let F be a 2-unisolvent set of continuous half-periodic functions on $[0, \pi]$. Then $S(F)$ corresponds to a flat Laguerre plane.*

Note that the translations of the cylinder in the vertical direction are automorphisms of the resulting flat Laguerre planes. This means that a nonvertical line in a point Möbius strip plane corresponds to a circle and all its vertical translates in the Laguerre plane. This set of circles is a so-called *flock*, that is, a set of circles that partitions the cylinder, and the set of all these flock is a partition, or *resolution*, of the circle set into flocks; see Subsection 5.8.3 for more information about flocks and resolutions of flat Laguerre planes.

The integral of a pencil of lines in the Möbius strip plane corresponds to tangent pencils of circles in the Laguerre plane, and every tangent pencil of circles arises in this manner.

By construction, the map

$$\mathcal{Z} \to \mathcal{Z} : (x, y) \mapsto (x + \pi \bmod 2\pi, -y)$$

defines a fixed-point-free involutory automorphism of the Laguerre plane. We will see in the next chapter that this automorphism is a so-called Möbius involution and that from every flat Laguerre plane that admits a Möbius involution a flat Möbius plane can be constructed.

If the point Möbius strip plane $\mathcal{M}$ we started with admits the involutory automorphism

$$\mathcal{M} \to \mathcal{M} : (x, y) \mapsto (-x + \pi, y),$$

then its integral admits also admits the involutory automorphism

$$\mathcal{Z} \to \mathcal{Z} : (x, y) \mapsto (-x + 2\pi, y)$$

whose fixed points are the points of the verticals $\{0\} \times \mathbf{R}$ and $\{\pi\} \times \mathbf{R}$. We will see in the next chapter that this automorphism is a so-called Minkowski involution and that from every flat Laguerre plane that admits a Minkowski involution a flat Minkowski plane can be constructed.

Let f be one of the special functions in Subsection 2.7.3 that we used to construct the radial Moulton planes. By removing the point that is fixed by all automorphisms of such a flat projective plane, we arrive at a point Möbius strip plane whose nonvertical lines are the graphs of the functions

$$[0, \pi] \to \mathbf{R} : x \mapsto a\overline{f}(x + b),$$

where $a \in [0, +\infty)$ and $b \in [0, \pi]$. This implies that the integral of such a point Möbius strip plane has an at least 3-dimensional point-transitive group of automorphisms that consists of the homeomorphisms

$$\mathcal{Z} \to \mathcal{Z} : (x, y) \mapsto (x + r \bmod 2\pi, sy + t),$$

where $r \in [0, 2\pi]$, $s \in \mathbf{R}^+$, $t \in \mathbf{R}$. In fact, such a plane has group dimension 3 unless it is classical; see Theorem 5.4.13. Furthermore, f can be chosen in such a way that the integral admits the special Möbius and Minkowski involutions defined above.

For more information about this special way of integrating point Möbius strip planes see Polster [2000].

5.3.5 Planes of Generalized Shear Type

Given two functions f and g on $\mathbf{R}$ we define the binary operation $*_{f,g}$ on $\mathbf{R}$ by

$$a *_{f,g} x = \begin{cases} af(x) & \text{for } a \geq 0, \\ ag(x) & \text{for } a < 0. \end{cases}$$

Furthermore, let id and 1 denote the identity function $x \mapsto x$ and the constant function $x \mapsto 1$, respectively. With this notation the following family of flat *Laguerre planes of generalized shear type* was constructed in Polster–Steinke [2000].

Let $f, g, h, k : \mathbf{R} \to \mathbf{R}$ be four continuous functions on $\mathbf{R}$ such that the following conditions are satisfied.

- The function k is positive, that is, $k(x) > 0$ for all $x \in \mathbf{R}$.
- The function h is an orientation-preserving homeomorphism of $\mathbf{R}$ to itself.
- The two functions φ and ψ defined by $\varphi(x) = \frac{x}{k(x)}$ and $\psi(x) = \frac{h(x)}{k(x)}$ for $x \in \mathbf{R}$ are homeomorphisms of $\mathbf{R}$.
- The functions f, g, $f \circ h^{-1}$, $g \circ h^{-1}$, $\frac{f}{k} \circ \varphi^{-1}$, $\frac{g}{k} \circ \varphi^{-1}$, $\frac{f}{k} \circ \psi^{-1}$, $\frac{g}{k} \circ \psi^{-1}$ are strictly parabolic.

Let $\mathcal{C}(f, g, h, k)$ be the collection of all circles of the form

$$\{(x, a *_{f,g} x + b *_{id,h} x + c *_{1,k} x) \mid x \in \mathbf{R}\} \cup \{(\infty, a)\},$$

where $a, b, c \in \mathbf{R}$. Then $\mathcal{C}(f, g, h, k)$ is the circle set of a flat Laguerre plane $\mathcal{L}(f, g, h, k)$ represented on the cylinder $Z = (\mathbf{R} \cup \{\infty\}) \times \mathbf{R}$.

One can think of $\mathcal{L}(f, g, h, k)$ as a Laguerre plane whose circle space is made up of pieces of the circle spaces from up to eight different ovoidal Laguerre planes that have certain circles in common. The ovoidal planes involved are represented by the parabolic functions f, g, $f \circ h^{-1}$, $g \circ h^{-1}$, $\frac{f}{k} \circ \varphi^{-1}$, $\frac{g}{k} \circ \varphi^{-1}$, $\frac{f}{k} \circ \psi^{-1}$, $\frac{g}{k} \circ \psi^{-1}$. The 'basic' circles are graphs of the functions f, g, id, h, 1, and k augmented by the infinite points $(\infty, 1)$ or $(\infty, 0)$. Then linear combinations are formed. To verify that one indeed obtains a Laguerre plane several cut-and-paste methods as described in Subsection 5.3.3 (in particular, Propositions 5.3.8 and 5.3.12) are employed.

In fact, one can generalize $\mathcal{L}(f, g, h, k)$ a bit further by replacing the identity id by another orientation-preserving homeomorphism and the constant function 1 by another positive function. However, no new Laguerre planes arise.

The planes $\mathcal{L}(f, g, id, 1)$ are the flat Laguerre planes of shear type in Löwen–Pfüller [1987a]. These planes comprise all ovoidal Laguerre planes; see Löwen–Pfüller [1987a] Theorem 3. Indeed, $\mathcal{L}(f, f, id, 1)$ is the ovoidal plane $\mathcal{L}(f)$. In fact, ovoidal Laguerre planes can be characterized among the flat Laguerre planes of generalized shear type as follows.

THEOREM 5.3.16 (Ovoidal Planes of Gen. Shear Type) *A Laguerre plane $\mathcal{L}(f, g, h, k)$ is ovoidal if and only if g is related to f by $g(x) = af(x) + bx + c$ for some $a, b, c \in \mathbf{R}$, $a > 0$, h is an affine function, and k is constant.*

A Laguerre plane $\mathcal{L}(f, g, h, k)$ is classical if and only if f and g are quadratic polynomials, h is an affine function, and k is constant.

Note that for all $r > 0$ the maps $(x, y) \mapsto (x, ry)$, $x \in \mathbf{R} \cup \{\infty\}$, are automorphisms of $\mathcal{L}(f, g, h, k)$. Steinke [2000] showed that the family of Laguerre planes $\mathcal{L}(f, g, h, k)$ comprises, up to isomorphisms, all flat Laguerre planes whose kernels are at least 3-dimensional; they are obtained for $h = id$, $k = 1$ or $f = g$, $k = 1$ or $f = g$, $h = id$. For more information see Theorem 5.4.11. The flat Laguerre planes $\mathcal{L}(f, g, id, 1)$ are called *Laguerre planes of skew parabola type* by Löwen–Pfüller [1987a] and the flat Laguerre planes $\mathcal{L}(f, f, h, 1)$ where f is a skew parabola function and h is semi-multiplicative are called *Laguerre planes of modified skew parabola type* by Schroth [2000]. Furthermore, the Laguerre planes of generalized shear type also comprise all flat Laguerre planes that admit 4-dimensional groups of automorphisms fixing at least two parallel classes; see Theorem 5.4.15.

5.3.6 Planes of Translation Type

A function f is called *strongly parabolic* if it is normalized strictly parabolic (see p. 305), twice differentiable on $\mathbf{R} \setminus \{0\}$, $\log |f'|$ is strictly concave on the open intervals $(-\infty, 0)$ and $(0, +\infty)$, and $\lim_{x \to \pm\infty} \frac{f(x+b)}{f(x)} = 1$ for each $b \in \mathbf{R}$.

Let f and g be two strongly parabolic functions and let $\mathcal{C}(f, g)$ be the collection of the sets

$$\{(x, mx + t) \mid x \in \mathbf{R}\} \cup \{(\infty, 0)\},$$

where $m, t \in \mathbf{R}$ and

$$\{(x, a *_{f,g} (x - b) + c) \mid x \in \mathbf{R}\} \cup \{(\infty, a)\},$$

where $a, b, c \in \mathbf{R}, a \neq 0$. Then $\mathcal{C}(f, g)$ is the circle set of a flat Laguerre plane $\mathcal{L}(f, g)$ represented on the cylinder $Z = (\mathbf{R} \cup \{\infty\}) \times \mathbf{R}$.

Note that the derived affine plane at the point $(\infty, 0)$ is the Euclidean plane. We call a flat Laguerre plane of *translation type* if it is isomorphic to one of the planes $\mathcal{L}(f, g)$; see Löwen–Pfüller [1987a] for this kind of flat Laguerre planes. These planes also generalize a construction of Hartmann [1979]. The ovoidal planes among the flat Laguerre planes of translation type can be characterized as follows; see Löwen–Pfüller [1987a] Theorem 5.

THEOREM 5.3.17 (Ovoidal Flat Laguerre Planes) *A Laguerre plane $\mathcal{L}(f, g)$ is ovoidal if and only if $f(x) = g(x) = x^2$. In this case, the flat Laguerre plane $\mathcal{L}(f, g)$ is classical.*

The following homeomorphisms of $\mathbf{R}^2$ to itself extend to automorphisms of the plane $\mathcal{L}(f, g)$ that fix the point $(\infty, 0)$ and the parallel class at infinity (globally).

$$\mathbf{R}^2 \to \mathbf{R}^2 : (x, y) \mapsto (x + b, ay + c),$$

where $a, b, c \in \mathbf{R}$, $a > 0$.

If $f = g$ is the skew parabola function $f_{d,s}$ (see p. 443), then the homeomorphisms

$$\mathbf{R}^2 \to \mathbf{R}^2 : (x, y) \mapsto (rx, r^d y)$$

also extend to automorphisms of $\mathcal{L}(f, g)$. Hence in this case the Laguerre plane $\mathcal{L}(f, g)$ has an at least 4-dimensional group of automorphisms. Furthermore, the point $(\infty, 0)$ is fixed under all these automorphisms. In fact, a flat Laguerre plane whose automorphism group is 4-dimensional and fixes a point is isomorphic to a Laguerre plane of shear type over a pair of different skew parabolas or to a Laguerre plane of translation type over a pair of skew parabolas. See Theorem 5.4.14 and Löwen–Pfüller [1987b] Theorem 2, for this and other results about the planes of translation type.

5.3.7 The Artzy–Groh Planes

The planes that we will encounter in this subsection are constructed in the same manner as the flat Minkowski planes from Subsection 4.3.5 and are similar to the planes of generalized translation type from the previous subsection. As for flat Minkowski planes, Artzy and Groh [1986] started their construction with the group of affine similarities, that is, the group of transformations

$$\mathbf{R}^2 \to \mathbf{R}^2 : (x, y) \mapsto (rx + b, ry + c),$$

where $b, c, r \in \mathbf{R}$, $r \neq 0$, and all Euclidean lines. They then looked for a curve C such that its images under the above group and the nonvertical Euclidean lines properly extended (in the sense of Subsection 5.2.2) yield a flat Laguerre plane.

Let $f : \mathbf{R} \to \mathbf{R}$ be a continuous function. For $a, b, c \in \mathbf{R}$ let

$$C_{a,b,c} = \begin{cases} \{(x, bx + c) \mid x \in \mathbf{R}\} \cup \{(\infty, 0)\} & \text{for } a = 0, \\ \{(x, af(\frac{x-b}{a}) + c) \mid x \in \mathbf{R}\} \cup \{(\infty, a)\} & \text{for } a \neq 0, \end{cases}$$

and let

$$\mathcal{C}_f = \{C_{a,b,c} \mid a, b, c \in \mathbf{R}\}.$$

Artzy and Groh [1986] showed that $\mathcal{C}_f$ is the circle set of a flat Laguerre plane if and only if f is strictly parabolic, that is, f is differentiable and f' is a homeomorphism $\mathbf{R} \to \mathbf{R}$. We denote this plane by $\mathcal{L}'(f)$.

Since the generating curve C can be moved around, one can assume that C passes through the point $(0,0)$ and has the Euclidean line $C_{0,0,0}$ as a tangent at this point. This means that $f(0) = f'(0) = 0$. Moreover, because the circles $C_{a,0,0}$ for $a \in \mathbf{R}$, $a \neq 0$, then form the tangent pencil of circles through the point $(0,0)$, there is exactly one circle in this pencil that passes through the point $(1,1)$. This circle must be obtained for some $a > 0$. Of course, we can choose the corresponding function as the one that generates the plane. Therefore each Artzy–Groh plane is isomorphic to a plane $\mathcal{L}'(f)$, where f is a normalized strictly parabolic function, that is, the function f is strictly parabolic and $f(0) = f'(0) = 0$ and $f(1) = 1$; see also p. 305.

PROPOSITION 5.3.18 (Normalized Artzy–Groh Planes) *Every Artzy–Groh plane is isomorphic to a plane $\mathcal{L}'(f)$ where f is a normalized strictly parabolic function.*

THEOREM 5.3.19 (Classical Artzy–Groh Planes) *Under the assumptions of Proposition 5.3.18 one obtains the classical flat Laguerre plane if and only if f is equal to $x \mapsto x^2$.*

From the way the Artzy–Groh planes are constructed it becomes clear that these flat Laguerre planes can be characterized by the type of their automorphism groups. Furthermore, using the cut-and-paste method described in Proposition 5.3.8, these planes can be generalized as follows.

Let $f, g : \mathbf{R} \to \mathbf{R}$ be two normalized strictly parabolic functions of $\mathbf{R}$. For $a, b, c \in \mathbf{R}$ let

$$C_{a,b,c} = \begin{cases} \{(x, bx + c) \mid x \in \mathbf{R}\} \cup \{(\infty, 0)\} & \text{for } a = 0, \\ \{(x, af(\frac{x-b}{a}) + c) \mid x \in \mathbf{R}\} \cup \{(\infty, a)\} & \text{for } a > 0, \\ \{(x, ag(\frac{x-b}{|a|}) + c) \mid x \in \mathbf{R}\} \cup \{(\infty, a)\} & \text{for } a < 0. \end{cases}$$

Then

$$\mathcal{C}_{f,g} = \{C_{a,b,c} \mid a, b, c \in \mathbf{R}\}$$

is the circle set of a flat Laguerre plane $\mathcal{L}(f, g)$.

Note that if f and g are both skew parabola functions, then the generalized Artzy–Groh plane $\mathcal{L}(f, g)$ is isomorphic to a Laguerre plane of translation type; see Subsection 5.3.6.

5.3.8 Planes of Shift Type

Let h be a differentiable homeomorphism of $\mathbf{R}$ such that the derivative h' of h is planar, that is, for each $a \in \mathbf{R}$, $a \neq 0$, the function

$$x \mapsto h'(x + a) - h'(x)$$

is bijective; see also Subsection 2.7.4 for a different description of planar functions and a characterization of those functions on $\mathbf{R}$. Then the collection of all sets

$$\{(x, h(x - a) + bx + c) \mid x \in \mathbf{R}\} \cup \{(\infty, a)\},$$

where $a, b, c \in \mathbf{R}$, is the circle set of a flat Laguerre plane $\mathcal{L}(h)$ represented on the cylinder $Z = (\mathbf{R} \cup \{\infty\}) \times \mathbf{R}$. Note that derived affine planes at points (∞, a) are classical.

This kind of Laguerre plane can be obtained by integrating the shift plane generated by h'; see Subsubsection 5.3.4.1 for the general construction principle and Steinke [20XXa] for automorphisms of the flat Laguerre planes of shift type.

Note that $h(x) = x^3$ yields the classical flat Laguerre plane and the map $(x, y) \mapsto (x, y - x^3)$ is an isomorphism onto the parabola model. Conversely, this essentially is the only way to obtain the classical flat Laguerre plane in this family of flat Laguerre planes.

THEOREM 5.3.20 (Ovoidal Planes of Shift Type) *A Laguerre plane $\mathcal{L}(h)$ is ovoidal if and only if h is a polynomial of degree 3. In this case $\mathcal{L}(h)$ is classical.*

Clearly, each transformation

$$\mathbf{R}^2 \to \mathbf{R}^2 : (x, y) \mapsto (x + a, y + bx + c),$$

where $a, b, c \in \mathbf{R}$, extends to an automorphism of $\mathcal{L}(h)$. Furthermore, if h is the semi-multiplicative homeomorphism $h_{d,r}$ (see p. 442), then

$$\mathbf{R}^2 \to \mathbf{R}^2 : (x, y) \mapsto (x, t^d y),$$

where $t \in \mathbf{R}^+$, also extends to an automorphism of $\mathcal{L}(h)$. Hence in this case the Laguerre plane $\mathcal{L}(h)$ has an at least 4-dimensional group of automorphisms. Furthermore, the parallel class $\{\infty\} \times \mathbf{R}$ is fixed under all these automorphisms. In fact, a flat Laguerre plane whose automorphism group is 4-dimensional and fixes precisely one parallel class is isomorphic to a Laguerre plane of shift type over a semi-multiplicative function or to a Laguerre plane of translation type over a pair of skew parabolas. See Theorem 5.4.16 and Steinke [20XXa] for this and other results about the planes of shift type.

5.4 Automorphism Groups and Groups of Projectivities

In this section we investigate automorphisms of cylindrical circle planes and flat Laguerre planes and show that the collection of all automorphisms of a flat Laguerre plane is a Lie group when suitably topologized. Then the well-known structure of Lie groups of small dimensions can be used to classify the most homogeneous flat Laguerre planes. The results and their proofs to a large extent mirror those of their counterparts for flat Möbius and Minkowski planes. We only summarize the main results and refer to the literature for proofs and to the preceding chapters for the ideas behind these proofs.

5.4.1 Automorphisms

Using the fact that isomorphisms of $\mathbf{R}^2$-planes are continuous, the respective result can be proved for cylindrical circle planes.

THEOREM 5.4.1 (Isomorphisms Are Continuous) *Every isomorphism between cylindrical circle planes is continuous. In particular, every automorphism of a cylindrical circle plane is continuous.*

Recall from Corollary 2.4.4 that a collineation of a flat projective plane that fixes the points of a nondegenerate quadrangle is the identity. The corresponding result for flat Laguerre planes is as follows.

LEMMA 5.4.2 (Automorphism with Fixed Points) *If an automorphism γ of a flat Laguerre plane fixes four nonconcircular points of which at least three are pairwise nonparallel, then $\gamma = id$.*

Proof. Suppose that γ fixes the four nonconcircular points p_1, p_2, p_3, and p_4 of which p_1, p_2, and p_3 are pairwise nonparallel and p_4 is not parallel to p_2 or p_3. In the derived projective plane $\mathcal{P}$ at p_3 the automorphism γ induces a collineation $\tilde{\gamma}$ that fixes the finite points p_1, p_2 and p_4 and the infinite point ω corresponding to parallel classes. Furthermore, the point q of intersection of the line at infinity with the line through p_1 and p_2 is also fixed by $\tilde{\gamma}$. The four points p_4, p_2, ω, and q form a nondegenerate quadrangle in $\mathcal{P}$. Hence $\tilde{\gamma} = id$ by Corollary 2.4.4. Therefore $\gamma = id$ as well. □

The automorphism group Γ of a cylindrical circle plane is an effective topological transformation group on the cylinder $\mathbf{S}^1 \times \mathbf{R}$ when Γ carries the compact-open topology (or, equivalently, the topology of uniform convergence on compact sets). For Γ to be a Lie group we essentially need that Γ is locally compact; compare Theorem A2.3.5. This condition can be verified for flat Laguerre planes; see Steinke [1986b].

THEOREM 5.4.3 (Automorphism Group) *The collection Γ of all automorphisms of a flat Laguerre plane is a Lie group with respect to the compact-open topology of dimension at most 7.*

The statement on the dimension of the automorphism group is an immediate consequence of the dimension formula A2.3.6 and Lemma 5.4.2 on the stabilizer of four nonconcircular points because each of the three pairwise nonparallel points has an orbit of dimension at most 2 and the fourth point, which can be chosen to be parallel to one of the first three points, has an orbit of dimension at most 1 under the stabilizer of the other three points.

As for flat linear spaces and the other flat circle planes, we say that a flat Laguerre plane $\mathcal{L}$ has *group dimension* n if the (full) automorphism group of $\mathcal{L}$ has dimension n.

In fact, if the group dimension of a flat Laguerre plane is at least 6, then the plane is classical; see Theorem 5.4.12. Examples in all other possible dimensions can be found among the semi-classical flat Laguerre planes $\mathcal{L}(f, g, h)$ of type 1; see Subsection 5.3.2 and Steinke [1988].

Let $f, g, h, k : \mathbf{R} \to \mathbf{R}$ be homeomorphisms defined by

$$\begin{aligned} f(x) &= 2x, \\ h(x) &= x^3, \\ k(x) &= \sinh(x), \\ g(x) &= \begin{cases} x & \text{for } x \geq 0, \\ 2x & \text{for } x \leq 0, \end{cases} \end{aligned}$$

where $x \in \mathbf{R}$. Then we have the following flat Laguerre planes; compare Theorem 5.3.3.

- $\mathcal{L}(id, id, id)$ is the classical flat Laguerre plane. It has group dimension 7.
- $\mathcal{L}(f, id, id)$ has group dimension 5. The transformations

$$\mathbf{R}^2 \to \mathbf{R}^2 : (x, y) \mapsto \begin{cases} \left(sx, ry + ax^2 + bx + c\right) & \text{for } x \geq 0, \\ \left(sx, ry + f(a)x^2 + bx + c\right) & \text{for } x \leq 0, \end{cases}$$

  where $a, b, c, r, s \in \mathbf{R}$, $r, s > 0$, extend to automorphisms of this plane.
- $\mathcal{L}(g, id, id)$ has group dimension 4. The transformations

$$\mathbf{R}^2 \to \mathbf{R}^2 : (x, y) \mapsto (sx, ry + bx + c),$$

  where $b, c, r, s \in \mathbf{R}$, $r, s > 0$, extend to automorphisms of this plane.
- $\mathcal{L}(h, id, id)$ has group dimension 3. The transformations

$$\mathbf{R}^2 \to \mathbf{R}^2 : (x, y) \mapsto \begin{cases} (x, ry + bx + c) & \text{for } x \geq 0, \\ (x, h(r)y + bx + c) & \text{for } x \leq 0, \end{cases}$$

  where $b, c, r \in \mathbf{R}$, $r > 0$, extend to automorphisms of this plane.
- $\mathcal{L}(h, g, id)$ has group dimension 2. The transformations

$$\mathbf{R}^2 \to \mathbf{R}^2 : (x, y) \mapsto \left(sx, s^2 y + c\right),$$

  where $c, s \in \mathbf{R}$, $s > 0$, extend to automorphisms of this plane.
- $\mathcal{L}(k, h, id)$ has group dimension 1. The transformations

$$\mathbf{R}^2 \to \mathbf{R}^2 : (x, y) \mapsto (x, y + c),$$

  where $c \in \mathbf{R}^+$, extend to automorphisms of this plane.
- $\mathcal{L}(k, k, k)$ has group dimension 0.

5.4.2 The Kernel

The automorphism group Γ of a flat Laguerre plane $\mathcal{L}$ has a distinguished subgroup, the *kernel* T of the action of Γ on the set of parallel classes, that is, T consists of all automorphisms that fix each parallel class globally. This collection of automorphisms is a closed normal subgroup of the automorphism group of the Laguerre plane. If $\gamma \in T$ fixes a point p, then γ induces a central collineation $\tilde{\gamma}$ in the derived projective plane at p with centre ω, the infinite point of the lines that correspond to parallel classes.

For the dimension of the kernel one again uses the dimension formula A2.3.6 and Lemma 5.4.2. Note that now each point has an orbit of dimension at most 1.

THEOREM 5.4.4 (The Dimension of the Kernel) *The dimension of the kernel of a flat Laguerre plane is at most* 4.

The maximum dimension 4 is realized in ovoidal flat Laguerre planes. If v is the vertex of the associated cone in 3-dimensional real projective space, then every collineation of $\mathrm{PG}(3, \mathbf{R})$ that leaves each line through v invariant induces an automorphism of the Laguerre plane that belongs to the kernel. Moreover, the group of central collineations with fixed centre in $\mathrm{PG}(3, \mathbf{R})$ is 4-dimensional. From this description we see that there are no compact subgroups of positive dimension in the kernel. This is true for arbitrary flat Laguerre planes; see Steinke [1990a] Lemma 2.4.

LEMMA 5.4.5 (Compact Subgroups) *The kernel T of a flat Laguerre plane contains no compact, connected nontrivial subgroup. More precisely, a compact subgroup of T is finite and has order at most* 2.

From the structure theory of Lie groups and the classification of almost simple Lie groups of dimension at most 4 one obtains the following.

PROPOSITION 5.4.6 (Kernel Is Solvable) *The kernel of a flat Laguerre plane is solvable.*

In an ovoidal flat Laguerre plane with automorphism group Γ and kernel T one sees that the factor Γ/T can be at most 3-dimensional. This follows from the dimension formula for quotients A2.2.2 and the fact that the group Γ is at most 7-dimensional and the kernel T is 4-dimensional.

For arbitrary flat Laguerre planes we still find that Γ/T is at most 3-dimensional, but there are nonovoidal semi-classical flat Laguerre planes where this maximum dimension is attained; see Theorem 5.3.6.

THEOREM 5.4.7 (Closed Connected Subgroups) *Let Σ be a closed connected subgroup of the automorphism group of a flat Laguerre plane $\mathcal{L}$ and let Δ be the kernel of the action of Σ on the set Π of parallel classes. Then Σ/Δ has dimension at most 3. Furthermore, if $\dim(\Sigma/\Delta) = 3$, then $\Sigma/\Delta \simeq \mathrm{PSL}_2(\mathbf{R})$ and Σ acts transitively on Π.*

Either Σ is solvable, or Δ is the radical of Σ and Σ/Δ is isomorphic to $\mathrm{PSL}_2(\mathbf{R})$. In the latter case Σ contains a locally simply connected Lie subgroup (Levi complement) isomorphic to either $\mathrm{PSL}_2(\mathbf{R})$ or its simply connected covering group.

For the proof see Steinke [1990a]. In particular, if Γ is the automorphism group of a flat Laguerre plane $\mathcal{L}$ and T denotes the kernel of $\mathcal{L}$, then we deduce the inequalities $m \leq n \leq m + 3$ and $m \leq 4$ for the dimensions n and m of Γ and T, respectively.

5.4.3 Planes Admitting an at Least 3-Dimensional Kernel

Recall that the kernel of an ovoidal flat Laguerre plane is 4-dimensional. Every central collineation of $\mathrm{PG}(3, \mathbf{R})$ with centre the vertex of the associated cone induces an automorphism of the Laguerre plane that belongs to the kernel. Conversely, Groh [1969] showed that ovoidal flat Laguerre planes are characterized by this property.

THEOREM 5.4.8 (Ovoidal Equals 4-Dimensional Kernel) *A flat Laguerre plane $\mathcal{L}$ is ovoidal if and only if the kernel T of $\mathcal{L}$ is 4-dimensional. In this case $\mathcal{L}$ has a representation as one of the flat Laguerre planes $\mathcal{L}(f)$ on $Z = (\mathbf{R} \cup \{\infty\}) \times \mathbf{R}$, where f is a parabolic function on $\mathbf{R}$. The kernel of $\mathcal{L}(f)$ consists of all maps*

$$Z \to Z : (x, y) \mapsto \begin{cases} (x, ry + af(x) + bx + c) & \text{for } x \in \mathbf{R}, \\ (\infty, ry + a) & \text{for } x = \infty, \end{cases}$$

where $a, b, c, r \in \mathbf{R}$, $r \neq 0$.

The central collineations of $\mathrm{PG}(3, \mathbf{R})$ with fixed centre v that fix each point on a plane through v form a closed normal subgroup of the group of all central collineations with centre v. In fact, this normal subgroup is isomorphic to $\mathbf{R}^3$. Moreover, such a collineation fixes no plane but the distinguished plane and planes passing through v. In the ovoidal Laguerre plane we obtain automorphisms in the kernel that fix no circle except the identity. We call an automorphism in the kernel of a flat Laguerre plane $\mathcal{L}$ an *elation* if it is the identity or if it fixes no circle.

It is not clear whether or not the collection of all elations of $\mathcal{L}$ forms a group. If it does we call it the *elation group* of $\mathcal{L}$. The corresponding construction for finite Laguerre planes and 4-dimensional Laguerre planes always yields a group; see Steinke [1991a] and [1991b], respectively. The elation group of a flat Laguerre plane can be at most 3-dimensional, and in this case it must be isomorphic to $\mathbf{R}^3$. Such a group again characterizes ovoidal flat Laguerre planes.

THEOREM 5.4.9 (Ovoidal Equals 3-Dim. Elation Group) *A flat Laguerre plane $\mathcal{L}$ is ovoidal if and only if the kernel of $\mathcal{L}$ contains a subgroup isomorphic to $\mathbf{R}^3$.*

The proof of this result is implicitly contained in Groh [1969]; see also Steinke [1993] Lemma 2.8 and Löwen [1994] Corollary 6.8.

Under the assumptions of the above theorem, the elation group of $\mathcal{L}$ acts sharply transitively on the circle set $\mathcal{C}$. Hence the kernel of $\mathcal{L}$ is transitive on $\mathcal{C}$. Conversely, the existence of such a kernel implies ovoidal; see Steinke [2000a] Proposition 1. The kernel must be at least 3-dimensional. If it is 4-dimensional, $\mathcal{L}$ is ovoidal by Theorem 5.4.8, or the connected component containing the identity is sharply transitive on the circle set and thus isomorphic to $\mathbf{R}^3$.

THEOREM 5.4.10 (Ovoidal Equals Circle-Transitive Kernel) *The kernel of a flat Laguerre plane $\mathcal{L}$ is transitive on the circle set if and only if $\mathcal{L}$ is ovoidal.*

The flat Laguerre planes admitting a 3-dimensional kernel are also completely classified; see Steinke [2000a]. All these planes are of generalized shear type; see Subsection 5.3.5.

Recall that a semi-multiplicative homeomorphism of $\mathbf{R}$ to itself is of the form

$$x \mapsto \begin{cases} x^d & \text{for } x \geq 0, \\ s|x|^d & \text{for } x \leq 0, \end{cases}$$

where $d > 0 > s$, written $h_{d,s}$, and that one obtains a skew parabola function, written $f_{d,s}$, if $d > 1$ and $s > 0$; see p. 442 for the definition of semi-multiplicative homeomorphisms and skew parabola functions.

THEOREM 5.4.11 (Planes with 3-Dimensional Kernel) *The family of Laguerre planes $\mathcal{L}(f, g, h, k)$ of generalized shear type comprises, up to isomorphisms, all 2-dimensional Laguerre planes whose kernels are at least 3-dimensional. These special planes are obtained for $h = id$, $k = 1$ or $f = g$, $k = 1$ or $f = g$, $h = id$. The kernel of such a plane is 3-dimensional unless the Laguerre plane is ovoidal.*

More precisely, a 2-dimensional Laguerre plane whose kernel is at least 3-dimensional is isomorphic to one of the following planes.

(i) *A Laguerre plane $\mathcal{L}(f, g, id, 1)$ where f and g are normalized strictly parabolic functions. This plane admits the 3-dimensional group*

$$\left\{ (x,y) \mapsto \begin{cases} (x, ry + bx + c) & \text{for } x \in \mathbf{R} \\ (\infty, ry) & \text{for } x = \infty \end{cases} \;\middle|\; b, c, r \in \mathbf{R}, r > 0 \right\}$$

as group of automorphisms in the kernel. The flat Laguerre plane $\mathcal{L}(f, g, id, 1)$ is ovoidal if and only if $f = g$. The automorphism group of $\mathcal{L}(f, g, id, 1)$ is

- *7-dimensional if and only if $f(x) = g(x) = x^2$;*
- *5-dimensional if and only if $f = g$ describes a skew parabola but not a conic;*
- *4-dimensional if and only if either $f = g$ does not describe a skew parabola or $f \neq g$ and both describe skew parabolas;*
- *3-dimensional if and only if $f \neq g$ and at least one of f, g does not describe a skew parabola.*

(ii) *A Laguerre plane $\mathcal{L}(f, f, h, 1)$ where h is an orientation-preserving homeomorphism of $\mathbf{R}$ fixing 0 and 1 and f and $f \circ h^{-1}$ are strictly parabolic functions. This plane admits the 3-dimensional group*

$$\left\{ (x,y) \mapsto \begin{cases} (x, ry + af(x) + c) & \text{for } x \in \mathbf{R} \\ (\infty, ry + a) & \text{for } x = \infty \end{cases} \;\middle|\; a, c, r \in \mathbf{R}, r > 0 \right\}$$

as group of automorphisms in the kernel. The flat Laguerre plane $\mathcal{L}(f, f, h, 1)$ is ovoidal if and only if $h = id$. The automorphism group of $\mathcal{L}(f, f, h, 1)$ for $h \neq id$ is

- *4-dimensional if and only if f describes a skew parabola and h is semi-multiplicative; in this case $\mathcal{L}(f, f, h, 1)$ also admits the group*

$$\left\{ (x,y) \mapsto \begin{cases} (sx, y) & \text{for } x \in \mathbf{R} \\ (\infty, \frac{y}{f(s)}) & \text{for } x = \infty \end{cases} \;\middle|\; s \in \mathbf{R}^+ \right\}$$

as group of automorphisms not contained in the kernel;

- *3-dimensional if and only if f does not describe a skew parabola or h is not semi-multiplicative.*

(iii) *A Laguerre plane $\mathcal{L}(f, f, id, k)$ where k is a positive continuous function such that $k(0) = 1$ and $\varphi : x \mapsto \frac{x}{k(x)}$ is a homeomorphism of $\mathbf{R}$ and f and $\frac{f}{k} \circ \varphi^{-1}$ are parabolic functions such that $f(0) = -1$ and $f'(0) = 0$. This plane admits the 3-dimensional group*

$$\left\{ (x,y) \mapsto \begin{cases} (x, ry + af(x) + bx) & \text{for } x \in \mathbf{R} \\ (\infty, ry + a) & \text{for } x = \infty \end{cases} \middle| a, b, r \in \mathbf{R}, r > 0 \right\}$$

as group of automorphisms in the kernel. $\mathcal{L}(f, f, id, k)$ is ovoidal if and only if $k = 1$. The automorphism group of $\mathcal{L}(f, f, id, k)$ is 3-dimensional for $k \neq 1$.

Note that in the above theorem the planes $\mathcal{L}(f, g, id, 1)$, $\mathcal{L}(f, f, h, 1)$, and $\mathcal{L}(f, f, id, k)$ are ovoidal for $f = g$, $h = id$ and $k = 1$, respectively, so that they have 4-dimensional kernels.

For the proof one considers a 3-dimensional closed connected subgroup in the kernel. Such a group must be isomorphic to $\mathbf{R}^3$, and then the plane is ovoidal by Theorem 5.4.9, or is the semi-direct product of $\mathbf{R}$ by $\mathbf{R}^2$. In the latter case, $\mathbf{R}^2$ is in the elation group of the Laguerre plane and $\mathbf{R}$ fixes a circle. An analysis of the possible actions of these groups on parallel classes yields the planes of generalized shear type.

5.4.4 Classification with Respect to the Group Dimension

In this subsection we give the classification of the flat Laguerre planes with respect to the group dimension so far as it is known at present. As for the other circles planes, the classical plane has the largest group dimension among all flat Laguerre planes. In the next possible lower dimension, that is, group dimension 5, one still finds close relatives of the classical flat Laguerre plane. In particular, all these planes are ovoidal over 'nice' ovals.

THEOREM 5.4.12 (Group-Dimension Classification I) *Let $\mathcal{L}$ be a flat Laguerre plane.*

(i) *The maximal possible group dimension of $\mathcal{L}$ is that of the classical flat Laguerre plane. The classical plane has group dimension 7 and kernel dimension 4.*

(ii) *The plane $\mathcal{L}$ has group dimension at least* 6 *if and only if it is classical.*

(iii) *The plane $\mathcal{L}$ has group dimension* 5 *if and only if it is a plane of shear type over a proper skew parabola. These flat Laguerre planes are ovoidal and have kernel dimension* 4*; see p. 304.*

See Löwen–Pfüller [1987a] for the proof of the above theorem.

We already encountered a variety of flat Laguerre planes of group dimension 4 so that the classification cannot be as nice as for group dimension 5. Flat Laguerre planes of group dimension 4 were investigated by Steinke in a series of papers. There are various possible ways a closed connected 4-dimensional group of automorphisms can act on the point set Z, the set of parallel classes Π, and the circle set $\mathcal{C}$. It can fix

- neither a parallel class nor a circle;
- a point;
- at least two parallel classes, being transitive on each of them;
- precisely one parallel class, being transitive on it;
- no parallel class (and thus being transitive on Π) but not being transitive on Z.

The first case leads to the classical flat Laguerre plane and the group must be transitive on the point set; see Steinke [1993].

THEOREM 5.4.13 (4-Dimensional + Point-Transitive) *Let $\mathcal{L}$ be a flat Laguerre plane with automorphism group Γ and let $\Sigma \leq \Gamma$ be a closed connected 4-dimensional group of automorphisms that fixes neither a parallel class nor a circle. Then Σ is point-transitive, and $\mathcal{L}$ is isomorphic to the classical flat Laguerre plane.*

The kernel Δ of the action of Σ on Π is 3-dimensional, Σ/Δ is isomorphic to $\mathrm{SO}_2(\mathbf{R})$, and Σ is conjugate to one of the following groups.

$$\{(x,y,z) \mapsto (x\cos t - y\sin t, x\sin t + y\cos t, rz + ay + bx) | a,b,t \in \mathbf{R}, r>0\},$$

$$\{(x,y,z) \mapsto (x\cos t - y\sin t, x\sin t + y\cos t, z + ay + bx + c) \mid a,b,c,t \in \mathbf{R}\},$$

where we use the spatial description of the classical flat Laguerre plane on the cyclinder $Z = \{(x,y,z) \in \mathbf{R}^3 \mid x^2 + y^2 = 1\}$.

The second case was dealt with in Löwen–Pfüller [1987a] Theorem 2. The derived plane at the distinguished fixed point is Desarguesian and the group can be identified with a subgroup of the 5-dimensional group of collineations

$$\{(x,y) \mapsto (sx+a, ry+bx+c) \mid a,b,c,r,s \in \mathbf{R}, r,s>0\}.$$

Depending on the form of this subgroup one obtains a Laguerre plane of translation type or one of shear type.

THEOREM 5.4.14 (4-Dimensional + 1 Fixed Point) *A flat Laguerre plane whose automorphism group is 4-dimensional and fixes a point is isomorphic to a Laguerre plane of shear type over a pair of different skew parabolas or to a nonclassical Laguerre plane of translation type over a pair of skew parabolas. In the former planes precisely two parallel classes are fixed and in the latter planes precisely one parallel class is fixed.*

If the 4-dimensional group fixes at least two parallel classes, it turns out that the kernel must be relatively large and the resulting flat Laguerre planes are covered by Theorem 5.4.11; see Steinke [20XXb]. In particular, all such flat Laguerre planes are of generalized shear type.

THEOREM 5.4.15 (4-Dimensional + 2 Fixed Parallel Classes) *Let $\mathcal{L}$ be a flat Laguerre plane with automorphism group Γ and let $\Sigma \leq \Gamma$ be a closed connected 4-dimensional group of automorphisms that fixes at least two parallel classes. Then the kernel of $\mathcal{L}$ is at least 3-dimensional.*

If Γ is 4-dimensional, then $\mathcal{L}$ is isomorphic to one of the following planes represented on $Z = (\mathbf{R} \cup \{\infty\}) \times \mathbf{R}$.

(i) *An ovoidal Laguerre plane $\mathcal{L}(f)$ where f is a normalized strictly parabolic function such that the associated oval $\mathcal{O}_f$ is not projectively equivalent to a skew parabola. In this case Γ^1 consists of all automorphisms*

$$Z \to Z : (x,y) \mapsto \begin{cases} (x, ry+af(x)+bx+c) & \text{for } x \in \mathbf{R}, \\ (\infty, ry+a) & \text{for } x = \infty, \end{cases}$$

where $a,b,c,r \in \mathbf{R}$, $r>0$ and fixes every parallel class. Two such planes $\mathcal{L}(f)$ and $\mathcal{L}(g)$ are isomorphic if and only if the ovals $\mathcal{O}_f$ and $\mathcal{O}_g$ are projectively equivalent.

(ii) *A Laguerre plane $\mathcal{L}(f_{d_1,r_1}, f_{d_2,r_2}, id, 1)$ of shear type over two different skew parabolas, that is, $(d_1,r_1) \neq (d_2,r_2)$; see p. 443. In this case Γ^1 fixes precisely the parallel classes $\pi_\infty = \{\infty\} \times \mathbf{R}$*

and $\pi_0 = \{0\} \times \mathbf{R}$ *and each transformation*

$$\mathbf{R}^2 \to \mathbf{R}^2 : (x, y) \mapsto (sx, ry + bx + c)$$

where $r, s, b, c \in \mathbf{R}$, $r, s > 0$ *extends to an automorphism. Such a flat Laguerre plane is isomorphic to exactly one of the planes with parameters* $d_1, d_2 > 1$, $r_1 > r_2 > 0$ *or* $d_1 > d_2 > 1$, $r_1 = r_2 \geq 1$.

(iii) *A Laguerre plane* $\mathcal{L}(f, f, h, 1)$ *where* $f = f_{d_1,r_1}$*, describes a skew parabola and* $h = h_{d_2,r_2} \neq id$ *is semi-multiplicative. Here* Γ^1 *fixes precisely the two parallel classes* π_∞ *and* π_0 *and each transformation*

$$\mathbf{R}^2 \to \mathbf{R}^2 : (x, y) \mapsto (sx, ry + af(x) + c)$$

where $r, s, a, c \in \mathbf{R}$, $r, s > 0$ *extends to an automorphism. Such a Laguerre plane is isomorphic to exactly one of the flat Laguerre planes* $\mathcal{L}(f_{d_1,r_1}, f_{d_1,r_1}, h_{d_2,r_2}, 1)$*, where*

- $1 < d_2 < d_1 - 1$, $r_1 > 1$, $r_2 > 0$, *or*
- $1 < d_2 < d_1 - 1$, $r_1 = 1$, $r_2 \geq 1$, *or*
- $d_2 = 1$, $d_1 > 2$, $r_1 \geq 1$, $r_2 > 1$, *or*
- $d_1 - d_2 = 1$, $d_2 > 1$, $r_1, r_2 \geq 1$, *or*
- $d_2 = 1$, $d_1 = 2$, $r_1 \geq 1$, $r_2 > 1$.

In the second to last case, where precisely one parallel class is fixed, one looks at the kernel Δ of the action of the 4-dimensional group on Π. This kernel has dimension 2 or 3. If Δ is 3-dimensional, then we are in the situation of Theorem 5.4.11 and the Laguerre plane is, in fact, classical. If Δ is 2-dimensional, one further uses the classification of flat projective planes of group dimension at least 3 that admit a closed connected 3-dimensional group of collineations that fix a flag. One then obtains that the Laguerre plane must be a flat Laguerre plane of shift type over a semi-multiplicative homeomorphism; see Steinke [20XXc].

THEOREM 5.4.16 (4-Dimensional + 1 Fixed Parallel Class) *Let* $\mathcal{L}$ *be a flat Laguerre plane whose automorphism group* Γ *is 4-dimensional and such that the connected component* Γ^1 *fixes precisely one parallel class. Then* $\mathcal{L}$ *is isomorphic to one of the following planes represented on* $Z = (\mathbf{R} \cup \{\infty\}) \times \mathbf{R}$.

(i) *A nonclassical Laguerre plane* $\mathcal{L}(f_{d_1,r_1}, f_{d_2,r_2})$ *of translation type over a pair of skew parabolas. Here* Γ^1 *fixes the point* $(\infty, 0)$ *and precisely the parallel class* $\pi_\infty = \{\infty\} \times \mathbf{R}$ *and consists of all automorphisms*

$$Z \to Z : (x, y) \mapsto \begin{cases} (sx + b, ry + c) & \text{for } x \in \mathbf{R}, \\ \left(\infty, \frac{r}{f_{d_1,r_1}(s)} y\right) & \text{for } x = \infty, y \geq 0, \\ \left(\infty, \frac{r}{f_{d_2,r_2}(s)} y\right) & \text{for } x = \infty, y \leq 0, \end{cases}$$

where $b, c, r, s \in \mathbf{R}$, $r, s > 0$. *Such a plane is isomorphic to exactly one of the planes corresponding to parameters* $r_1, r_2 > 1$, $d_1 > d_2 > 1$ *or* $r_1 \geq r_2$, $d_1 = d_2 > 1$.

(ii) *A Laguerre plane* $\mathcal{L}(h_{m,q})$ *of shift type over a semi-multiplicative function* $h_{m,q}$ *where* $m > 2$, $q > 0$, $(m, q) \neq (3, 1)$. *The group* Γ^1 *fixes precisely the parallel class* $\pi_\infty = \{\infty\} \times \mathbf{R}$, *acts transitively on* π_∞ *and consists of all automorphisms*

$$Z \to Z : (x, y) \mapsto \begin{cases} (sx + a, s^m y + bx + c) & \text{for } x \in \mathbf{R}, \\ (\infty, sy + a) & \text{for } x = \infty, \end{cases}$$

where $a, b, c, s \in \mathbf{R}$, $s > 0$. *Such a Laguerre plane is isomorphic to exactly one plane where* $m > 2$ *and* $q \geq 1$.

In the last case, where no parallel class is fixed but the group is not transitive on Z, one always has a circle that is fixed under the group, a 1-dimensional kernel and an almost simple subgroup isomorphic to $\mathrm{PSL}_2(\mathbf{R})$ or its simply connected covering group $\widetilde{\mathrm{PSL}_2}(\mathbf{R})$. In the former case one obtains a semi-classical flat Laguerre plane of type 2. The latter case is not completely worked through.

The various cases can now be summarized as follows.

THEOREM 5.4.17 (Group-Dimension Classification II) *Let* $\mathcal{L}$ *be a flat Laguerre plane. Then* $\mathcal{L}$ *is of group dimension* 4 *if and only if it has*

(i) *kernel dimension* 4 *and is an ovoidal flat Laguerre plane of shear type over a parabolic function that is not skew-parabolic; see p. 324;*

(ii) *kernel dimension* 3 *and is one of the following planes–*

  (a) *a plane of shear type over a pair of different skew parabolas; see p. 325;*

  (b) *a plane of modified shear type; see p. 325;*

(iii) *kernel dimension* 2 *and is one of the following planes–*

  (a) *a plane of translation type over two different skew parabolas; see p. 325;*

(b) *a plane of shift type over a skew parabola; see p. 328;*

(iv) *kernel dimension* 1 *and is one of the following planes–*

(a) *a semi-classical flat Laguerre plane* $\mathcal{L}(\varphi, id)$ *of type 2 obtained by gluing along a circle where* $\varphi \neq id$ *is a multiplicative homeomorphism of* $\mathbf{R}$ *onto itself;*

(b) *a plane that admits the universal covering group* $\widetilde{\mathrm{PSL}_2(\mathbf{R})}$ *of* $\mathrm{PSL}_2(\mathbf{R})$ *as a group of automorphisms and the connected component of the automorphism group containing the identity fixes a circle.*

5.4.5 Von Staudt's Point of View—Groups of Projectivities

Following von Staudt's point of view, the group of all projectivities of a circle plane was investigated by Kroll [1977c] and Löwen [1977], [1981d]. Here, as in the case of the other circle planes, a projectivity of a fixed circle C is a composition of finitely many perspectivities between circles, the first and last circle being C. Perspectivities between circles are defined in the familiar manner by the correspondence between the points of two circles via a pencil of circles with the obvious modifications for parallel points and points of tangency or via parallel projections; compare Subsections 2.9.2 and 3.4.4 for the corresponding constructions in flat projective planes and flat Möbius planes, respectively. These maps are continuous by coherence so that projectivities become homeomorphisms of the circle C to itself; see also Groh [1968] 3.2.

Given two triples of three distinct points of C we can find a projectivity that takes the points in the first triple to the corresponding points in the second triple. For example, if (p_1, p_2, p_3) and (q_1, q_2, q_3) are two such triples on a circle C, we can choose a point $p \notin C$ and a circle D through p. If D_i denotes the tangent circle to D at p through the point p_i for $i = 1, 2, 3$, let $r_i = D_i \cap |q_i|$ and let C' be the circle through r_1, r_2 and r_3. If we project C onto C' via the tangent pencil to D at p and then project C' back onto C using parallel classes, we obtain a projectivity that takes (p_1, p_2, p_3) to (q_1, q_2, q_3); see also Groh [1968] 3.3.

LEMMA 5.4.18 (High Degree of Transitivity) *The group of projectivities of a circle C in a flat Laguerre plane is 3-transitive on the points of C.*

Note that the group of projectivities Π_C of a circle C in a flat Laguerre plane is independent of C, that is, the groups Π_C and $\Pi_{C'}$ are isomorphic for any circles C and C'. We can therefore simply speak of the group of projectivities without referring to any particular circle. The above lemma shows that the group of projectivities in a flat Laguerre plane tends to be rather large. In fact, the classical flat Laguerre plane has the smallest group of projectivities, namely the group $\mathrm{PGL}_2(\mathbf{R})$. This is in marked contrast to the situation with respect to automorphism groups: the smaller the automorphism group of a flat Laguerre plane, the larger the group of projectivities.

THEOREM 5.4.19 (Sharply 3-Transitive Equals Classical) *A flat Laguerre plane is isomorphic to the classical flat Laguerre plane if and only if the group of projectivities of a circle C is sharply 3-transitive on the points of C.*

Pursuing this line of investigation further, the following characterization of the classical flat Laguerre plane was obtained by Löwen [1981d].

THEOREM 5.4.20 (Locally Compact Equals Classical) *Let $\mathcal{L}$ be a flat Laguerre plane and let Π_C be the group of projectivities of a circle C. Then $\mathcal{L}$ is classical, if*

(i) *Π_C or its closure in the group of all homeomorphisms of C with respect to the compact-open topology τ is locally compact with respect to τ, or*

(ii) *Π_C acts ω-regularly on C, that is, there exists a finite set $F \subseteq C$ such that the subgroup in Π_C fixing F elementwise is discrete with respect to τ.*

5.5 The Kleinewillinghöfer Types

Similarly to the Lenz–Barlotti classification of projective planes, the Hering classification of Möbius planes, and the Klein–Kroll classification of Minkowski planes (see Subsection 2.9.1 and Sections 3.5 and 4.5, respectively), Laguerre planes can be classified with respect to *central automorphisms*, that is, automorphisms that fix at least one point and induce a central collineation in the derived projective plane at that fixed point. More precisely, one considers subgroups of central automorphisms that are linearly transitive, that is, the induced groups of central collineations are transitive on each central line except for the obvious fixed points, the centre and the point of intersection with the axis.

As in the case of Minkowski planes central automorphisms of Laguerre planes come in a variety of types according to whether the axis of a central collineation in the derived projective plane at a fixed point is the line at infinity, a line that stems from a circle of the Laguerre plane, or a line that comes from a parallel class of the Laguerre plane, and whether or not the centre is on the axis of the central collineation. Kleinewillinghöfer therefore obtained a multitude of types of Laguerre planes. She considered four kinds of central automorphisms, G-translations, $(G, B(q, C))$-translations, C-homologies, and (p, q)-homotheties; see the following subsections. In fact, in her classification she considered groups of automorphisms and determined their types according to transitive subgroups of central automorphisms contained in them. In the following we only deal with the full automorphism group.

5.5.1 C-Homologies

Let C be a circle of a Laguerre plane $\mathcal{L}$. A *C-homology* of $\mathcal{L}$ is an automorphism of $\mathcal{L}$ that either is the identity or fixes precisely the points of C. A C-homology induces a homology of the derived projective plane $\mathcal{P}_q$ at each $q \in C$ with infinite centre ω (the infinite point associated with lines coming from parallel classes of the Laguerre plane). Note that every C-homology is in the kernel of $\mathcal{L}$. A group of C-translations of $\mathcal{L}$ is called *C-transitive*, if it acts transitively on each parallel class minus its point of intersection with C. We say that the automorphism group Γ of $\mathcal{L}$ is C-transitive if Γ contains a C-transitive subgroup of C-homologies.

With respect to C-homologies Kleinewillinghöfer obtained seven types of Laguerre planes; see Kleinewillinghöfer [1979] Satz 3.1.

If $\mathcal{Z}$ denotes the set of all circles C for which the automorphism group of the Laguerre plane is C-transitive, then exactly one of the following statements is valid.

I. $\mathcal{Z} = \emptyset$.

II. $|\mathcal{Z}| = 1$.

III. There are a point p and a circle C with $p \in C$ such that $\mathcal{Z}$ is the tangent bundle $B(p, C)$.

IV. There are nonparallel points p and q such that the set $\mathcal{Z}$ is the bundle $\{C \in \mathcal{C} \mid p, q \in C\}$ of circles through p and q.

V. $\mathcal{Z}$ consists of a flock of $\mathcal{L}$, that is, the circles in $\mathcal{Z}$ partition the point set of $\mathcal{L}$; see Subsection 5.8.3 for more information about flocks.

VI. $\mathcal{Z}$ has the property that for each point p of $\mathcal{L}$ the collection of all circles in $\mathcal{Z}$ that pass through p equals $B(p,C)$ for some circle C or the bundle of circles through p and q for some point $q \nparallel p$.

VII. $\mathcal{Z} = \mathcal{C}$.

In fact, Type VI cannot occur in flat Laguerre planes; see Polster–Steinke [20XXb].

LEMMA 5.5.1 (Type VII) *If the set $\mathcal{Z}$ of a flat Laguerre plane $\mathcal{L}$ contains at least three circles that neither are pairwise disjoint nor pass through the same point, then $\mathcal{L}$ is ovoidal and thus of type VII.*

Since in Laguerre planes of type VI the set $\mathcal{Z}$ contains at least one circle through each point of $\mathcal{L}$, we readily obtain the following possible types for flat Laguerre planes.

PROPOSITION 5.5.2 (Possible Types) *A flat Laguerre plane is of type I, II, III, IV, V, or VII.*

For flat Laguerre planes of types I, II, III, or VII see the examples in Subsection 5.5.4.

An immediate consequence of Lemma 5.5.1 is the following characterization of ovoidal flat Laguerre planes; see also Mäurer [1977] Satz 6.

COROLLARY 5.5.3 (Ovoidal Equals Type VII) *A flat Laguerre plane $\mathcal{L}$ is of type VII if and only if $\mathcal{L}$ is ovoidal.*

5.5.2 Laguerre Translations

A *Laguerre translation* of a Laguerre plane $\mathcal{L}$ is an automorphism of $\mathcal{L}$ that either is the identity or fixes precisely the points of one parallel class. More precisely, there are two kinds of Laguerre translations. Let G be a parallel class of $\mathcal{L}$. A *G-translation* of $\mathcal{L}$ is an automorphism in the kernel of $\mathcal{L}$ that either is the identity or fixes precisely the points of G. A group of G-translations of $\mathcal{L}$ is called *G-transitive*, if it acts transitively on each parallel class $H \neq G$. We say that the automorphism group Γ of $\mathcal{L}$ is G-transitive if Γ contains a G-transitive subgroup of G-translations.

The second kind of Laguerre translations uses a circle C passing through $p \in G$. As for flat Möbius planes let $B(p,C)$ denote the *touching pencil with support p*, that is, $B(p,C)$ consists of all circles that touch the circle C at the point p. In the derived affine plane at p the touching pencil represents a bundle of parallel lines and we can look at translations in this direction. Then a $(G, B(p, C))$-*translation* of $\mathcal{L}$ is a Laguerre translation that fixes C (and thus each circle in $B(p, C)$) globally. Note that G is fixed pointwise by each $(G, B(p, C))$-translation. A group of $(G, B(p, C))$-translations of $\mathcal{L}$ is called $(G, B(p, C))$-*transitive*, if it acts transitively on $C \setminus \{p\}$. We say that the automorphism group Γ of $\mathcal{L}$ is $(G, B(p, C))$-transitive if Γ contains a $(G, B(p, C))$-transitive subgroup of $(G, B(p, C))$-translations.

With respect to Laguerre translations Kleinewillinghöfer obtained 11 types of Laguerre planes; see Kleinewillinghöfer [1979] Satz 3.3, or [1980] Satz 2.

Let Π be the set of all parallel classes of $\mathcal{L}$. If $\mathcal{E} \subseteq \Pi$ denotes the set of all parallel classes G for which the automorphism group Γ of the Laguerre plane is G-transitive, and $\mathcal{B}$ denotes the set of all touching pencils $B(p, C)$ with support p for which Γ is $(|p|, B(p, C))$-transitive, then exactly one of the following statements is valid.

A. $\mathcal{E} = \emptyset$; $\mathcal{B} = \emptyset$.
B. $|\mathcal{E}| = 1$; $\mathcal{B} = \emptyset$.
C. $|\mathcal{E}| = 2$; $\mathcal{B} = \emptyset$.
D. $|\mathcal{E}| \geq 3$; $\mathcal{B} = \emptyset$.
E. $\mathcal{E} = \emptyset$; $|\mathcal{B}| = 1$.
F. $\mathcal{E} = \emptyset$; there are a parallel class G, a subset $U \subseteq G$, $|U| \geq 2$, and an injective map $\varphi : U \to \mathcal{C}$ such that $q \in \varphi(q)$ and

$$\mathcal{B} = \{B(q, \varphi(q)) \mid q \in U\}.$$

G. $\mathcal{E} = \emptyset$; there is a circle C such that

$$\mathcal{B} = \{B(q, C) \mid q \in C\}.$$

H. There is a point p such that $\mathcal{E} = \{|p|\}$,

$$\mathcal{B} = \{B(p, C) \mid p \in C \in \mathcal{C}\}.$$

I. There is a parallel class G such that

$$\mathcal{E} = \{G\}, \quad \mathcal{B} = \{B(C \cap G, C) \mid C \in \mathcal{C}\}.$$

J. $\mathcal{E} = \Pi$; there is a parallel class G such that

$$\mathcal{B} = \{B(C \cap G, C) \mid C \in \mathcal{C}\}.$$

K. $\mathcal{E} = \Pi$; $\mathcal{B} = \{B(q, C) \mid q \in C \in \mathcal{C}\}$.

In types J and K the set $\mathcal{E}$ contains all parallel classes. In fact, $\mathcal{E}$ also contains all parallel classes in type D; see Polster–Steinke [20XXb]. This situation is realized in the ovoidal flat Laguerre planes.

LEMMA 5.5.4 (Types D, J, K Imply Ovoidal) *If the set $\mathcal{E}$ of all parallel classes G for which the automorphism group of the flat Laguerre plane $\mathcal{L}$ is G-transitive contains at least three parallel classes, then the kernel T contains a 3-dimensional subgroup isomorphic to $\mathbf{R}^3$ and $\mathcal{L}$ is ovoidal.*

The groups of G-translations for three distinct parallel classes G_1, G_2, and G_3 in $\mathcal{E}$ generate an abelian subgroup isomorphic to $\mathbf{R}^3$ and the Laguerre plane is ovoidal by Theorem 5.4.9.

Since a type K plane is the classical flat Laguerre plane and because $\mathcal{E} = \Pi$ in ovoidal Laguerre planes, we can characterize the type D flat Laguerre planes as follows.

COROLLARY 5.5.5 (Type D) *A flat Laguerre plane is of type D if and only if it is a nonclassical ovoidal flat Laguerre plane.*

A further consequence of the above Lemma 5.5.4 and the classification of ovoidal flat Laguerre planes of group dimension at least 5 (see Theorem 5.4.12) is that type J cannot occur in flat Laguerre planes. In fact, Type I cannot occur either.

LEMMA 5.5.6 (Type K) *If there is a parallel class G in a flat Laguerre plane $\mathcal{L}$ such that the automorphism group of $\mathcal{L}$ is $(G, B(p, C))$-transitive for each $p \in G$ and $C \in \mathcal{C}$, $p \in C$, then $\mathcal{L}$ is the classical flat Laguerre plane and thus of type K.*

Each derived affine plane at points of the parallel class G is a translation plane and thus must be the Euclidean plane by Proposition 2.4.9. The translation group of the derived plane at one point $p \in G$ acts as a group of shifts in the derived plane at $q \neq p$, $q \in G$. However, a shift plane is classical if and only if the associated planar function is a quadratic polynomial. This then leads to the classical flat Laguerre plane. We can further exclude Type F in flat Laguerre planes.

LEMMA 5.5.7 (Type F) *If there are a parallel class G and at least two points $p_1, p_2 \in G$ and circles $C_1, C_2 \in \mathcal{C}$, $p_i \in C_i$, for which the automorphism group of a flat Laguerre plane $\mathcal{L}$ is $(G, B(p_i, C_i))$-transitive, then $\mathcal{L}$ is $(G, B(p_1, C))$-transitive for all circles C through p_1.*

From the classification of types with respect to Laguerre translations it then follows that under the assumptions of the above lemma the set $\mathcal{B}$ must contain all bundles $B(C \cap G, C)$ for $C \in \mathcal{C}$.

In summary, we obtain the following possible types with respect to Laguerre translations.

PROPOSITION 5.5.8 (Possible Types) *A flat Laguerre plane is of type A, B, C, D, E, G, H, or K.*

For flat Laguerre planes of types A, B, C, D, G, H, and K see the examples in Subsection 5.5.4.

5.5.3 (p, q)*-Homotheties*

Let p and q be two nonparallel points of a Laguerre plane $\mathcal{L}$. An automorphism γ of $\mathcal{L}$ is a (p, q)*-homothety* if γ fixes p and q and induces a homothety with centre q in the derived affine plane $\mathcal{A}_p$ at p. Note that each (p, q)-homothety is also a (q, p)-homothety. A group of (p, q)-homotheties is called (p, q)*-transitive* if it acts transitively on each circle through p and q minus the two points p and q. We say that the automorphism group Γ of $\mathcal{L}$ is (p, q)-transitive if Γ contains a (p, q)-transitive subgroup of (p, q)-homotheties.

With respect to (p, q)-homotheties Kleinewillinghöfer [1979] Satz 3.2, or [1980] Satz 1, obtained 13 types of Laguerre planes.

If $\mathcal{H}$ denotes the set of all unordered pairs of points $\{p, q\}$ for which the automorphism group of the flat Laguerre plane is (p, q)-transitive, then exactly one of the following statements is valid.

1. $\mathcal{H} = \emptyset$.
2. $|\mathcal{H}| = 1$.
3. There are a point p and a parallel class G with $p \notin G$ such that
$$\mathcal{H} = \{\{p, q\} \mid q \in G\}.$$
4. There are a point p and a circle C through p such that
$$\mathcal{H} = \{\{p, q\} \mid q \in C \setminus \{p\}\}.$$
5. There are a circle C and a fixed-point-free involution $\varphi : C \to C$ such that
$$\mathcal{H} = \{\{p, \varphi(p)\} \mid p \in C\}.$$

6. There is a circle C such that
$$\mathcal{H} = \{\{p, q\} \mid p, q \in C, p \neq q\}.$$
7. There are two distinct parallel classes F and G and a bijection $\varphi : F \to G$ such that
$$\mathcal{H} = \{\{p, \varphi(p)\} \mid p \in F\}.$$
8. There are two distinct parallel classes F and G such that
$$\mathcal{H} = \{\{p, q\} \mid p \in F, q \in G\}.$$
9. Each point of $\mathcal{L}$ is in exactly one pair in $\mathcal{H}$.
10. There are a parallel class G and a bijection $\varphi : G \to \Pi \setminus \{G\}$ such that
$$\mathcal{H} = \{\{p, q\} \mid p \in G, q \in \varphi(p)\}.$$
11. There is a point p such that
$$\mathcal{H} = \{\{p, q\} \mid q \in Z \setminus |p|\}.$$
12. There is a parallel class G such that
$$\mathcal{H} = \{\{p, q\} \mid p \in G, q \in Z \setminus G\}.$$
13. $\mathcal{H}$ consists of all unordered pairs of nonparallel points.

Type 13 describes the classical flat Laguerre plane. In fact, it suffices for $\mathcal{H}$ to contain sufficiently many unordered pairs in order to obtain the classical flat Laguerre plane.

LEMMA 5.5.9 (Type 13) *If the set $\mathcal{H}$ of a flat Laguerre plane $\mathcal{L}$ contains all unordered pairs of nonparallel points p, q for all points p on a parallel class G and all points q off G, then $\mathcal{L}$ is the classical flat Laguerre plane and thus of Type 13.*

A set $\mathcal{H}$ as in the lemma implies that each derived affine plane at a point of the distinguished parallel class is a translation plane and thus the Euclidean plane by Proposition 2.4.9, and each translation of the derived plane is induced by a Laguerre translation. It then follows that $\mathcal{E} = \Pi$, and $\mathcal{L}$ is classical by Lemma 5.5.6.

The above lemma readily implies that type 12 cannot occur in flat Laguerre planes; compare Hartmann [1982b] Satz 8. Apart from that, little is known about the possible Kleinewillinghöfer types with respect to (p, q)-homologies.

PROPOSITION 5.5.10 (Possible Types) *A flat Laguerre plane is of Type 1, 2, 3, 4, 5, 6, 7, 8, 9, 10, 11, or 13.*

For flat Laguerre planes of types 1, 8, 11, or 13 see the examples in Subsection 5.5.4.

Combining all three classifications Kleinewillinghöfer [1979] Satz 3.4 obtained 46 types of Laguerre planes. Of those only the types I.A.1, I.A.2, I.A.5, I.A.7, I.A.9, I.B.1, I.B.3, I.B.10, I.C.1, I.C.8, I.E.1, I.E.4, I.G.1, I.G.6, I.H.1, I.H.11, I.K.1, I.K.13, II.A.1, II.A.2, II.A.5, II.E.1, II.E.4, II.G.1, II.G.6, III.B.1, III.B.3, III.H.1, III.H.11, IV.A.1, IV.A.2, V.A.1, VII.D.1, VII.D.8, VII.K.13 can possibly occur in flat Laguerre planes. Note that Kleinewillinghöfer's labelling of the combined types differs from the above one where the combined type just refers to the respective single types. For example, type III.B.3 refers to type III with respect to C-homotheties, type B with respect to Laguerre translations, and type 3 with respect to (p, q)-homotheties. (In Kleinewillinghöfer [1979] Satz 3.4, this combined type is labelled as III.A.1.)

Kleinewillinghöfer furthermore coordinatizes derived affine planes at suitable points. She investigates what algebraic properties such a coordinatizing ternary ring has and derives functional equations for circle-describing functions of circles not passing through the distinguished point of derivation. This gives more information about the possible forms of circles of certain types. See also Hartmann [1982b] for characterizations of ovoidal or Miquelian Laguerre planes in terms of transitivity properties of central automorphisms.

5.5.4 Some Examples

So far no complete classification of those types that can occur in flat Laguerre planes has been carried out. Examples for some of the possible Kleinewillinghöfer types of flat Laguerre planes are as follows.

I.A.1

A plane of group dimension 0. In particular, a rigid plane; see Subsection 5.3.2. Clearly, such a plane must be of type I.A.1.

I.B.1

A semi-classical plane $\mathcal{L}(f, g, id)$ of type 1 where f and g are the homeomorphisms given by $f(x) = \sinh x$ and $g(x) = x^3$ for $x \in \mathbf{R}$. The distinguished parallel class is $\{\infty\} \times \mathbf{R}$. The transformations $(x, y) \mapsto (x, y+c)$ for $c \in \mathbf{R}$ extend to G-translations of the Laguerre plane.

I.C.1

A semi-classical plane $\mathcal{L}(id, g, id)$ of type 1 where $g(x) = x^3$ for $x \in \mathbf{R}$. The distinguished parallel classes are $\{\infty\} \times \mathbf{R}$ and $\{0\} \times \mathbf{R}$ and the transitive groups of G-translations consist of the maps $(x, y) \mapsto (x, y+c)$, $(\infty, y) \mapsto (\infty, y)$ for $c \in \mathbf{R}$ and $(x, y) \mapsto (x, y+ax^2)$, $(\infty, y) \mapsto (\infty, y+a)$ for $a \in \mathbf{R}$, respectively.

I.G.1

A semi-classical plane $\mathcal{L}(f, id)$ of type 2 with $f(x) = x^3$ for $x \in \mathbf{R}$. The distinguished circle C is the circle $y = 0$. The transformations of the form $(x, y) \mapsto (x + t, y)$ for $t \in \mathbf{R}$ form a transitive group of $(\{\infty\} \times \mathbf{R}, B((\infty, 0), C))$-translations. Likewise, the transformations

$$(x, y) \mapsto \left(\frac{x}{sx+1}, \frac{y}{(sx+1)^2}\right),$$

for $s \in \mathbf{R}$ form a transitive group of $(\{0\} \times \mathbf{R}, B((0, 0), C))$-translations. The other transitive groups of translations can be found by forming compositions. (Each transformation

$$(x, y) \mapsto \left(\frac{ax+b}{cx+d}, \frac{y}{(cx+d)^2}\right),$$

where $a, b, c, d \in \mathbf{R}$, $ad - bc = 1$, extends to an automorphism of the Laguerre plane.)

II.A.1

A semi-classical plane $\mathcal{L}(h, h, h)$ of type 1 where $h \neq id$ is a multiplicative homomorphism; for example, $h(x) = x^3$ for $x \in \mathbf{R}$. The distinguished circle is the circle $y = 0$ and $\{(x, y) \mapsto (x, sy) \mid s \in \mathbf{R}, s \neq 0\}$ is a transitive group of C-homologies.

III.H.1

A flat Laguerre plane $\mathcal{L}(f, f)$ of translation type over f, where f is not a skew parabola function. The distinguished point is $p = (\infty, 0)$ and the distinguished circle C is the circle $y = 0$. In the derived affine plane at p (this is the Euclidean plane) each of the translations $(x, y) \mapsto (x+b, y+c)$ for $b, c \in \mathbf{R}$ and each homology $(x, y) \mapsto (x, a(y-c)+c)$, $a \in \mathbf{R}$, $a \neq 0$, at the parallel $y = c$ of C extend to automorphisms of the Laguerre plane.

III.H.11

A flat Laguerre plane $\mathcal{L}(f_d, f_d)$ of translation type over a skew parabola function f_d of the form $f_d(x) = x|x|^{d-1}$, $d > 2$. The distinguished point is $p = (\infty, 0)$ and the distinguished circle C is the circle $y = 0$. Each translation and each homology at a parallel of C of the derived affine plane at p (this is the Euclidean plane) extend to an automorphism of the Laguerre plane. Furthermore, for each $b, c \in \mathbf{R}$ the homotheties $(x, y) \mapsto (a(x - b) + b, a(y - c) + c)$ for $a \in \mathbf{R}$, $a \neq 0$, form a transitive group of $((\infty, \infty), (b, c))$-homotheties.

VII.D.1

An ovoidal flat Laguerre plane $\mathcal{L}(f)$, where f is not projectively equivalent to a skew parabola. For example, $f(x) = \cosh x$ yields such a plane. Note that the transformations $(x, y) \mapsto (x, y + t(af(x) + bx + c))$, where $t \in \mathbf{R}$, form a transitive group of G-translations whenever the circle $\{(x, af(x) + bx + c) \mid x \in \mathbf{R}\} \cup \{(\infty, a)\}$ touches the circle $y = 0$ at a point.

VII.D.8

An ovoidal flat Laguerre plane $\mathcal{L}(f)$, where f is a skew parabola function different from the squaring function $x \mapsto x^2$. The two distinguished parallel classes are $\{\infty\} \times \mathbf{R}$ and $\{0\} \times \mathbf{R}$. For each $a, c \in \mathbf{R}$ the transformations $(x, y) \mapsto (sx, sy + (1 - s)af(x) + (1 - s)c)$ for $s \in \mathbf{R}$, $s \neq 0$, form a transitive group of $((\infty, a), (0, c))$-homotheties.

VII.K.13

The classical flat Laguerre plane. Here all admissible subgroups of central automorphisms are linearly transitive.

5.6 Characterizations of the Classical Plane

In this section we summarize various characterizations of the classical flat Laguerre plane. Most of these characterizations parallel those of the classical flat Möbius und Minkowski planes; see Theorems 3.6.9 and 4.6.8. We only state the characterizations. For most of the proofs the reader is referred to the same sources as for the proofs of the respective results in Subsection 3.6.2. For characterization (vi) see Mäurer [1978]. The characterizations in terms of the group dimension, the Kleinewillinghöfer types, etc. have been proved earlier on in this chapter.

THEOREM 5.6.1 (Characterizations of the Classical Plane) *A flat Laguerre plane $\mathcal{L}$ with automorphism group Γ is isomorphic to the classical flat Laguerre plane if and only if any one of the following holds.*

(i) *Γ is flag-transitive.*
(ii) *Γ is at least 5-dimensional.*
(iii) *$\mathcal{L}$ is Miquelian.*
(iv) *$\mathcal{L}$ is locally Miquelian.*
(v) *$\mathcal{L}$ is locally classical.*
(vi) *For any two nonparallel points p and q the plane $\mathcal{L}$ admits a unique reflection at $\{p, q\}$.*
(vii) *$\mathcal{L}$ is of Kleinewillinghöfer type at least I.*
(viii) *$\mathcal{L}$ is of Kleinewillinghöfer type at least 12.*
(ix) *The group of projectivities of $\mathcal{L}$ is sharply 3-transitive on the points of a circle C.*
(x) *The group of projectivities Π_C of a circle C or its closure in the group of all homeomorphisms of C with respect to the compact-open topology τ is locally compact with respect to τ.*
(xi) *Π_C acts ω-regularly on C.*

5.7 Planes with Special Properties

In the following two subsections we consider flat Laguerre planes that do not admit any nontrivial automorphisms and differentiable flat Laguerre planes.

5.7.1 Rigid Planes

Examples of *rigid* flat Laguerre planes, that is, flat Laguerre planes that do not admit any nontrivial automorphisms, can be found among the semi-classical flat Laguerre planes; see Section 5.3.2.

5.7.2 Differentiable Planes

The point set $\mathbf{S}^1 \times \mathbf{R}$ and the circle set $\mathbf{R}^3$ of a cylindrical circle plane naturally carry smooth differentiable structures that make them into (differentiable) smooth manifolds. In the classical flat Laguerre plane each parallel class and each circle are then submanifolds of $\mathbf{S}^1 \times \mathbf{R}$ and the geometric operations are not only continuous (see Theorem 5.2.6) but even smooth (differentiable) maps with respect to the appropriate smooth manifold structures. We can now replace the topological conditions in the definition of a cylindrical circle plane or flat Laguerre plane by differentiability assumptions and ask what planes can occur. More precisely, a *smooth cylindrical circle plane or flat Laguerre plane* is a cylindrical circle plane or flat Laguerre plane such that parallel classes and circles are submanifolds of the point space and such that the geometric operations are smooth on their respective domains of definition.

The classical flat Laguerre plane is smooth according to this definition. More generally, any ovoidal flat Laguerre plane over a smooth oval is smooth.

THEOREM 5.7.1 (Smooth Ovoidal Planes) *An ovoidal flat Laguerre plane over an oval $\mathcal{O}$ in $\mathbf{R}^3$ is smooth if and only if $\mathcal{O}$ is a smooth submanifold of $\mathbf{R}^2$.*

5.8 Subgeometries and Lie Geometries

Apart from the constructions in this section, a large number of constructions of other types of geometries from flat Laguerre planes are described in detail in Chapter 6. Also included in Chapter 6 is the solution of the Apollonius problem for flat Laguerre planes.

5.8.1 Recycled Flat Projective Planes

In Subsection 3.8.1 we introduced a way to 'recycle' flat projective planes from spherical circle planes. We did this by constructing disk models of flat projective planes on disks bounded by circles. In cylindrical circle planes it is both circles and pairs of parallel classes that can be used to construct flat projective planes.

5.8.1.1 Recycling along Pairs of Parallel Classes

Let V and W be two distinct parallel classes of a cylindrical circle plane $\mathcal{L} = (Z, \mathcal{C})$, let H be one of the connected components of the set $Z \setminus (V \cup W)$, and let $\overline{H} = H \cup V \cup W$. Given an orientation-reversing homeomorphism $\gamma : V \to W$ we define a geometry $\mathcal{R}$ whose point set is $\overline{H}$ and whose lines are the restrictions to $\overline{H}$ of the circles in $\mathcal{C}$ that pass through pairs of points $(p, \gamma(p))$, $p \in V$. We construct a geometry $\mathcal{R}'$ from $\mathcal{R}$ by identifying V and W via γ. Furthermore, we add all parallel classes of lines contained in $\overline{H}$ to the line set. Consequently, the point set of $\mathcal{R}'$ is a Möbius strip and all lines that do not arise from parallel classes are topological circles.

THEOREM 5.8.1 (Strip Model) *Let $\mathcal{R}'$ be the geometry constructed as above from a strip H bounded by two distinct parallel classes in a cylindrical circle plane $\mathcal{L}$ and let A be the restriction to H of a circle in $\mathcal{L}$ that does not give rise to a line in $\mathcal{R}'$. Then $\mathcal{R}'$ is a point Möbius strip plane and A is a topological arc in $\mathcal{R}'$.*

This result is a corollary of a more general result that we will prove in Theorem 7.1.8. See also Polster [1996a] Section 2.3.

5.8.1.2 Recycling along a Circle

Let C be a circle in a cylindrical circle plane $\mathcal{L} = (Z, \mathcal{C})$. We compactify the cylinder Z by two points u, d (one for each of the two open ends). The resulting topological space $\overline{Z}$ is homeomorphic to the 2-sphere. Let H be the connected component of $\overline{Z} \setminus C$ that contains the compactifying point u and let $\overline{H} = H \cup C$. Clearly, $\overline{H}$ is a closed disk. Starting with a fixed-point-free involutory homeomorphism γ of C to itself we construct a disk model $(D, \gamma, \mathcal{G})$ of a flat projective plane (as defined in Subsection 3.8.1). If p is a point on the circle C, let $V(p)$ be the parallel class through p. Now $D = \overline{H}$ and $\mathcal{G}$ consists of the restrictions to D of the sets

$$V(p) \cup \{u\} \cup V(\gamma(p)),$$

where $p \in C$, and circles in $\mathcal{L}$ through pairs of points $(p, \gamma(p))$ on C.

THEOREM 5.8.2 (Disk Model) *Let $\mathcal{L}$ be a cylindrical circle plane, let C be one of its circles, and let $\gamma : C \to C$ be a fixed-point-free continuous involution. Then $(D, \gamma, \mathcal{G})$, as defined above, is a disk model of a flat projective plane $\mathcal{P}$.*

Let I be the set of all circles completely contained in the interior of D, let T be the set of all circles completely contained in D and touching C at exactly one point, and let S be the restrictions to D of all circles that intersect C in two points that do not get exchanged by γ. Then the elements of I, T, and S are topological arcs in the flat projective plane $\mathcal{P}$. If $\mathcal{L}$ is a flat Laguerre plane, then the elements of I and T are topological ovals in $\mathcal{P}$.

For more information about this construction see Polster [1996a] Section 2.2.

5.8.2 Double Covers of $\mathbb{R}^2$-Planes and Flat Projective Planes

In Subsection 5.3.3 on cut-and-paste constructions we introduced inner and outer involutions of cylindrical circle planes and used them to construct separating sets in such planes. In the following we consider the fixed-circle sets of these involutions.

5.8.2.1 Inner Involutions

Let γ be an inner involution of a cylindrical circle plane $\mathcal{L} = (Z, \mathcal{C})$ with fixed-circle set $Fix(\gamma)$. This means that γ is a fixed-point-free involutory homeomorphism $Z \to Z$ mapping parallel classes to parallel classes such that a circle is contained in $Fix(\gamma)$ if and only if it contains some point p and its image $\gamma(p)$. For further useful properties of inner involutions see Lemma 5.3.9.

It readily follows that the complete bundle of circles through a point and its image under γ is contained in $Fix(\gamma)$. Hence any two points of Z whose parallel classes are not exchanged by γ are contained in exactly one circle in $Fix(\gamma)$. Let P be the quotient space Z/γ and let $\mathcal{G}$ be the set consisting of all C/γ, $C \in Fix(\gamma)$, and all images of parallel classes under the identification. We conclude that $\mathcal{R} = (P, \mathcal{G})$ is a geometry on the Möbius strip all of whose lines are either topological circles or, if they arise from parallel classes, topological lines. We conclude that this geometry is a point Möbius strip plane; see Subsubsection 1.2.1.3. We express this by saying that $(Z, Fix(\gamma))$ is a *double cover* of the point Möbius strip plane $\mathcal{R}$. Furthermore, it is easy to see that $\mathcal{R}$ is isomorphic to one of the point Möbius strip planes constructed in Subsubsection 5.8.1.1 and that the flat projective plane corresponding to $\mathcal{R}$ is isomorphic to one of the flat projective planes constructed in Subsubsection 5.8.1.2.

THEOREM 5.8.3 (Double Covers of Möbius Strip Planes) *Let γ be an inner involution of a cylindrical circle plane $\mathcal{L} = (Z, \mathcal{C})$. Then the geometry $(Z, Fix(\gamma))$ is the double cover of a point Möbius strip plane $\mathcal{R}$.*

Let H be a strip on Z whose two bounding parallel classes V and W are exchanged by γ. Then γ defines an orientation-reversing homeomorphism $V \to W$ and $\mathcal{R}$ is isomorphic to the point Möbius strip plane constructed in Subsubsection 5.8.1.1.

Let C be a circle in $Fix(\gamma)$. Then the restriction of γ to C is a fixed-point-free continuous involution and the flat projective plane constructed in Subsubsection 5.8.1.2 is isomorphic to the flat projective plane that extends $\mathcal{R}$.

5.8.2.2 Outer Involutions

Let γ be an outer involution of a cylindrical circle plane $\mathcal{L} = (Z, \mathcal{C})$ and let $Fix(\gamma)$ consist of all circles that are globally fixed by γ. This means that the fixed-point set F of γ is the union of two parallel classes V and W such that the two connected components of $Z \setminus (V \cup W)$ get exchanged by γ. Furthermore, a circle is contained in $Fix(\gamma)$ if and only if it contains some point p and its image $\gamma(p)$, where $p \neq \gamma(p)$. For further useful properties of outer involutions see Lemma 5.3.11.

Now it follows that the complete bundle of circles through a point and its image under γ is contained in $Fix(\gamma)$ provided that $p \neq \gamma(p)$. Hence any two points of Z whose parallel classes do not get exchanged by γ are contained in exactly one circle in $Fix(\gamma)$. Let H_i, $i = 1, 2$, be the two connected components of $Z \setminus (V \cup W)$ and let $\mathcal{G}_i$ consist of all the parallel classes contained in H_i and all restrictions of circles in $Fix(\gamma)$ to D_i. Then $\mathcal{R}_i = (H_i, \mathcal{G}_i)$ is an $\mathbf{R}^2$-plane. This $\mathbf{R}^2$-plane cannot be a flat affine plane. Furthermore, γ defines an isomorphism between $\mathcal{R}_1$ and $\mathcal{R}_2$. We express all this by saying that $(Z, Fix(\gamma))$ is a *double cover* of an $\mathbf{R}^2$-plane.

If $\mathcal{L}$ is the classical flat Laguerre plane on the cylinder Z and γ is the bundle involution of Z associated with an exterior point, then this $\mathbf{R}^2$-plane is isomorphic to the restriction of the Euclidean plane to the strip bounded by two parallel Euclidean lines.

We summarize our considerations in the following theorem.

THEOREM 5.8.4 (Double Covers of $\mathbf{R}^2$-Planes) *Let γ be an outer involution of a cylindrical circle plane $\mathcal{L} = (Z, \mathcal{C})$. Then $(Z, Fix(\gamma))$ is the double cover of an $\mathbf{R}^2$-plane.*

5.8.3 Flocks and Resolutions

A *flock* of a cylindrical circle plane $\mathcal{L}$ is a partition of its point set into pairwise disjoint circles of $\mathcal{L}$.

Examples of flocks in ovoidal cylindrical circle planes can be obtained as follows. Let $\mathcal{L}$ be an ovoidal cylindrical circle plane that is the geometry of nontrivial plane sections of a cone C in $\mathrm{PG}(3, \mathbf{R})$ with vertex v. Let L be a line of $\mathrm{PG}(3, \mathbf{R})$ that has no points in common with C. Each plane through L that does not pass through v intersects the point set $Z = C \setminus \{v\}$ in a circle of the ovoidal plane. The collection of circles obtained in this way yields a flock in this plane. This kind of flock is called *linear*. Nonlinear flocks can be constructed as follows. Using the same setup, let K be a second line different from L that intersects L but does not intersect the cone. Let P be the plane determined by both lines. Then P dissects the cylinder Z into an upper and a lower part. Now a nonlinear flock consists of the circles that correspond to planes through L that intersect the cylinder in the upper part, the planes through K that intersect the cylinder in the lower part, and the plane P. This kind of flock is called *bilinear*; see Biliotti–Johnson [1999] for bilinear flocks of Miquelian Laguerre planes over ordered fields.

Intersection of circles in a flock with a fixed parallel class in a cylindrical circle plane defines a bijection between the parallel class and the flock.

PROPOSITION 5.8.5 (Flocks) *Each flock of a cylindrical circle plane is homeomorphic to* $\mathbf{R}$.

As for flat Minkowski planes, one can define partial flocks and the automorphism group of a flock; see Subsection 4.8.1. Starting with two disjoint circles of a flat Laguerre plane, one can again inductively construct a partial flock such that the intersections of its circles with each parallel class form a dense subset of that parallel class. This partial flock can then be completed to a flock of the Laguerre plane; see Rosehr [1998] Theorem 1.9.

THEOREM 5.8.6 (Existence of Flock) *Every flat Laguerre plane admits a flock.*

The automorphism group of a flock of the classical flat Laguerre plane is at most 3-dimensional, the linear flocks attaining that maximum dimension. In fact, the linear flocks are characterized by this property.

THEOREM 5.8.7 (Linear Flock) *A flock of the classical flat Laguerre plane has an automorphism group of dimension* 3 *if and only if the flock is linear.*

Rosehr [1998] Section 3.3 derives further characterizations of linear and bilinear flocks.

A *resolution* of a cylindrical circle plane $\mathcal{L}$ is a partition of the circle space of $\mathcal{L}$ into flocks of $\mathcal{L}$. We call a cylindrical circle plane *resolvable* if it admits a resolution. For example, every ovoidal cylindrical circle plane $\mathcal{L}$ given as above is resolvable as follows. Start with a plane in $\mathbf{R}^3$

that does not intersect the cylinder $\mathcal{L}$ is living on. Then the set of all linear flocks that correspond to nonvertical lines in this plane are a resolution of $\mathcal{L}$.

Further examples of flat Laguerre planes that are resolvable are given in Polster–Steinke [1997] Section 4.

5.9 Open Problems

PROBLEM 5.9.1 *What $\mathbf{R}^2$-planes and flat affine planes can occur as derived planes of cylindrical circle planes and flat Laguerre planes? What flat projective planes and point Möbius strip planes do we arrive at by recycling along circles or pairs of parallel classes of cylindrical circle planes, respectively? In particular, classify the flat projective planes and point Möbius strip planes that we arrive at by recycling along a circle or two parallel classes of the classical flat Laguerre plane, respectively. Classify the flat Laguerre planes that are integrals of the Euclidean plane.*

Some more constructions of flat linear spaces from flat Laguerre planes via generalized quadrangles (see Chapter 6) correspond to some further similar questions.

PROBLEM 5.9.2 *Assume that all derived affine planes of a flat Laguerre plane are Desarguesian. Does this imply that the Laguerre plane is ovoidal?*

In the case of a spherical circle plane the answer to the respective question is unknown. On the other hand, we know that a flat Minkowski plane all of whose derived planes are classical is classical itself; see Theorem 4.6.3. For finite Laguerre planes of odd order the answer is 'Yes'. In fact, in this case the Laguerre plane must be Miquelian.

PROBLEM 5.9.3 *Are there any further essentially new ways to construct flat linear spaces from cylindrical circle planes apart from the ones that come about as combinations of the links between the different classes of geometries described in this book?*

PROBLEM 5.9.4 *Show that the automorphism group of a cylindrical circle plane is a Lie group and develop a classification of cylindrical circle planes with respect to their group dimensions.*

PROBLEM 5.9.5 *Classify all flat Laguerre planes of group dimension* 3.

A complete classification of the flat Laguerre planes/cylindrical circle planes of group dimension 3 should still be possible, whereas a complete classification of the planes of group dimension 2 seems to be beyond reach.

PROBLEM 5.9.6 *Are there flat Laguerre planes whose collection of elations does not form a group?*

For finite Laguerre planes and 4-dimensional Laguerre planes the answer to the respective question is negative.

PROBLEM 5.9.7 *Develop a classification of flat Laguerre planes with respect to Kleinewillinghöfer types.*

PROBLEM 5.9.8 *If α is an automorphism of a Laguerre plane that fixes a point and induces a translation in the derived affine plane at p, does α necessarily fix the parallel class of p pointwise?*

In the definition of Laguerre translations one requires that such translations fix all the points of one parallel class. Is this restriction necessary? It is often easy to see that an automorphism that fixes a point p induces a translation but it is usually harder to deduce that all the points on the parallel class of p are indeed fixed.

PROBLEM 5.9.9 *Does every cylindrical circle plane contain a flock or, even stronger, is every such plane resolvable?*

PROBLEM 5.9.10 *In general, extend some more results about flat Laguerre planes listed in this chapter to results about cylindrical circle planes.*

More problems include the counterparts of some of the problems listed at the ends of the chapters on spherical and toroidal circle planes.

6
Generalized Quadrangles

Although the main objects of this chapter are not geometries on surfaces the theory of antiregular 3-dimensional generalized quadrangles as developed in Schroth [1995a] ties the flat Möbius, Laguerre, and Minkowski planes into a tight knot full of beautiful connections and unexpected relationships. In this chapter we give a summary of Schroth's theory and state the most important results about flat Laguerre, Möbius, and Minkowski planes that are best expressed in the language of generalized quadrangles.

A *generalized quadrangle* is a geometry $\mathcal{G} = (P, \mathcal{L})$ with point set P, line set $\mathcal{L}$, and set of flags $\mathcal{F}$ satisfying the following axioms.

Axioms for Generalized Quadrangles

(Q1) Any two distinct points have at most one joining line.

(Q2) For every antiflag $(p, K) \in (P \times \mathcal{L}) \setminus \mathcal{F}$ there exists exactly one flag $(q, L) \in \mathcal{F}$ such that $(p, L) \in \mathcal{F}$ and $(q, K) \in \mathcal{F}$.

Two points p, q are said to be *collinear*, denoted by $p \perp q$, if they can be joined by a line. The set of points collinear with a given point p is called the *perp* of p and is denoted by $p^\perp$. Given two points p and q, the set $p^\perp \cap q^\perp$ is called the *trace* of p and q. Note that if p and q are distinct and collinear, then $p^\perp \cap q^\perp$ is just the line connecting the two points.

A generalized quadrangle is called *antiregular* if for any three pairwise noncollinear points a, b, and c the *centre* $a^\perp \cap b^\perp \cap c^\perp$ either is empty or contains precisely two points. A set of three pairwise noncollinear points is called a *triad.*

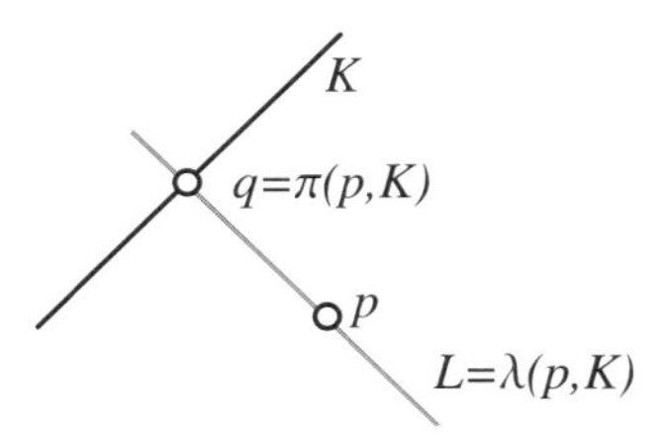

Fig. 6.1.

Axiom Q2 yields two maps $\pi : (P \times \mathcal{L}) \backslash \mathcal{F} \to P$ and $\lambda : (P \times \mathcal{L}) \backslash \mathcal{F} \to \mathcal{L}$ with $\pi(p, K) = q$ and $\lambda(p, K) = L$; see Figure 6.1. A *topological (generalized) quadrangle* is a generalized quadrangle in which P and $\mathcal{L}$ carry Hausdorff topologies such that the maps π and λ are continuous. A topological quadrangle is called a *compact* quadrangle if P and $\mathcal{L}$ carry compact topologies. If $\mathcal{G}$ is a compact quadrangle whose lines and line pencils are topological manifolds of positive dimensions, then there are parameters $m, n \in \mathbf{N}$ such that all lines are homeomorphic to the n-dimensional sphere $\mathbf{S}^n$ and all line pencils to the m-dimensional sphere $\mathbf{S}^m$. In this case we speak of a compact quadrangle $\mathcal{G}$ with *topological parameters* (m, n). Compact quadrangles with parameters (1,1) are also known as 3-dimensional quadrangles. It turns out that, given a 3-dimensional quadrangle, either it or its dual is antiregular; see p. 21 for the definition of the dual of a geometry. Note that the dual of a generalized quadrangle is also a generalized quadrangle.

In the following we will only be dealing with antiregular 3-dimensional generalized quadrangles, for short A3GQs.

In every A3GQ there exist triads of points whose centres are empty and there exist triads of points whose centres contain exactly two points.

For proofs of the results mentioned above and more background information about topological generalized quadrangles see Schroth [1995a], Forst [1981], and Kramer [1994]. For more information about generalized quadrangles in general see Payne–Thas [1984] and Thas [1995].

6.1 The Classical Antiregular 3-Dimensional Quadrangle

The orthogonal quadrangle $Q(4, \mathbf{R})$ over $\mathbf{R}$ is the classical A3GQ. It can be described as follows. Consider the symmetric bilinear form on $\mathbf{R}^5$

defined by the matrix

$$\begin{pmatrix} 1 & & & & \\ & 1 & & & \\ & & -1 & & \\ & & & & 1 \\ & & & 1 & \end{pmatrix}.$$

Then the point set P and line set $\mathcal{L}$ of the quadrangle consist of the totally isotropic 1- and 2-dimensional subspaces of $\mathbf{R}^5$, respectively, that is, the 1- and 2-dimensional subspaces that are completely contained in the quadric associated with the symmetric bilinear form. Note that this construction is very similar to the construction of the classical flat projective plane as the geometry of 1- and 2-dimensional subspaces of $\mathbf{R}^3$.

Two points of the quadrangle are collinear if and only if the correponding 1-dimensional isotropic subspaces are perpendicular with respect to the symmetric form. In this case the 2-dimensional subspace spanned by the two 1-dimensional isotropic subspaces is totally isotropic. This construction works for every field. However, the generalized quadrangle is antiregular if and only if the field is not of characteristic 2.

The Euclidean plane $\mathbf{R}^2$ together with its lines contains most of the information about the classical flat projective plane. Similarly, we now show that the 3-dimensional Euclidean space $\mathbf{R}^3$ together with an easily described subset of its lines contains most of the information about the classical quadrangle $Q(4, \mathbf{R})$.

We start by describing the orthogonal quadrangle using homogeneous coordinates in $\mathrm{PG}(4, \mathbf{R})$.

Two points $p_i = (a_i : b_i : c_i : d_i : e_i) \in \mathrm{PG}(4, \mathbf{R})$, $i = 1, 2$, are orthogonal with respect to the above symmetric form if and only if

$$a_1a_2 + b_1b_2 - c_1c_2 + d_1e_2 + e_1d_2 = 0.$$

This implies that

$$P = \{(a : b : c : d : e) \in \mathrm{PG}(4, \mathbf{R}) \mid a^2 + b^2 - c^2 + 2de = 0\}.$$

Let $p_\infty = (0 : 0 : 0 : 0 : 1) \in P$. Then $(a : b : c : d : e) \in p_\infty^\perp$ if and only if $d = 0$. Consequently, $(a : b : c : d : e) \in P \setminus p_\infty^\perp$ if and only if $d \neq 0$ and $a^2 + b^2 - c^2 + 2de = 0$. Because $d \neq 0$ we may assume that $d = 1$. Then $e = -(a^2 + b^2 - c^2)/2$. Therefore, the space $P \setminus p_\infty^\perp$ can be identified with $\mathbf{R}^3$ via the correspondence

$$(x, y, z) \longleftrightarrow (x : y : z : 1 : (z^2 - x^2 - y^2)/2).$$

Next we determine within $\mathbf{R}^3$ the set of points collinear with a given point $(a, b, c) \in \mathbf{R}^3$.

$$
\begin{aligned}
& (x, y, z) \perp (a, b, c) \\
\Leftrightarrow \quad & (x : y : z : 1 : (z^2 - x^2 - y^2)/2) \perp (a : b : c : 1 : (c^2 - a^2 - b^2)/2) \\
\Leftrightarrow \quad & (x - a)^2 + (y - b)^2 = (z - c)^2.
\end{aligned}
$$

So $(a, b, c)^\perp \cap \mathbf{R}^3 = \{ (x, y, z) \in \mathbf{R}^3 \mid (x - a)^2 + (y - b)^2 = (z - c)^2 \}$. This is a double cone with vertex (a, b, c); see Figure 6.2.

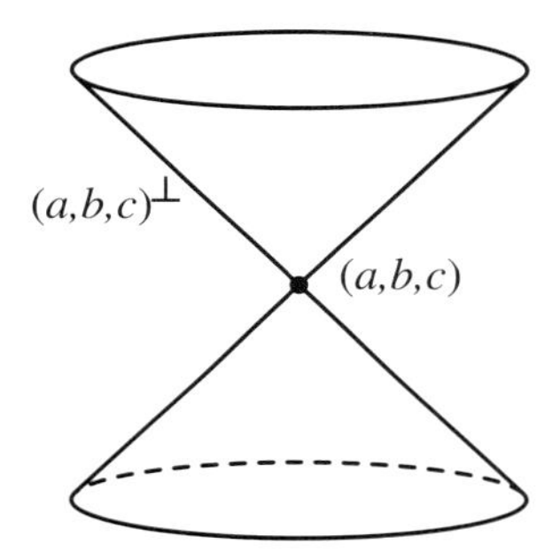

Fig. 6.2. The points collinear to a given point form a double cone

The group of those automorphisms of $\mathrm{PG}(4, \mathbf{R})$ that fix P globally acts transitively on P. This means that none of the points of P is distinguished among the rest. This in turn means that in order to verify that our geometry really satisfies Axioms Q1 and Q2 we may assume that the points and lines in the axioms are contained in $\mathbf{R}^3$. It is clear that two points of $\mathbf{R}^3$ are contained in at most one element of the line set $\mathcal{L}$. Hence Axiom Q1 is satisfied. Let $\mathcal{L}'$ be the set of all lines in $\mathbf{R}^3$ that are part of a line in $\mathcal{L}$. Given a point p of $\mathbf{R}^3$ and a line $K \in \mathcal{L}'$ that does not contain p, consider the unique Euclidean plane in $\mathbf{R}^3$ containing p and K. Then this plane intersects the double cone at p in two or one generator so that this plane contains either two or one line in $\mathcal{L}'$ through p. In the first case one of the two lines is parallel to K and the other one and its point of intersection with K are what we are looking for in Axiom Q2. In the second case K meets the unique line in a point outside $\mathbf{R}^3$. Also try to visualize what antiregularity means in this affine part of the quadrangle.

6.2 Basic Properties

We summarize some important geometrical and topological properties of A3GQs in the following two theorems.

THEOREM 6.2.1 (Topology) *Let $\mathcal{G}$ be an A3GQ with point set P and line set $\mathcal{L}$. Then the following hold.*

(i) *The trace $p^\perp \cap q^\perp$ of two distinct points p and q is homeomorphic to $\mathbf{S}^1$; see Figure 6.3.*

(ii) *The perp $p^\perp$ of a point p is homeomorphic to a quadratic cone in 3-dimensional projective space, where p corresponds to the vertex of the cone. The set $p^\perp \setminus \{p\}$ is homeomorphic to a cylinder in $\mathbf{R}^3$.*

(iii) *In the dual of $\mathcal{G}$ the perp of a point q is homeomorphic to the real projective plane and the set $q^\perp \setminus \{q\}$ is homeomorphic to the Möbius strip.*

(iv) *The sets P and $\mathcal{L}$ are homeomorphic to the point and line sets of the classical A3GQ, respectively. In particular, the set of points is homeomorphic to $(\mathbf{S}^2 \times \mathbf{S}^1)/\sim$, where $(a, b) \sim (c, d)$ if and only if $(c, d) \in \{(a, b), (-a, -b)\}$. The set $P \setminus p^\perp$ for $p \in P$ is homeomorphic to $\mathbf{R}^3$. The set of lines is homeomorphic to the real projective space $\mathrm{PG}(3, \mathbf{R})$.*

For proofs of (i), (ii), and (iii) see Schroth–Van Maldeghem [1994] 2.2 and Schroth [1995a] Lemma 2.3. For a proof of (iv) see Schroth [20XX] Theorem 4.6.

THEOREM 6.2.2 (Automorphisms) *Let $\mathcal{G}$ be an A3GQ with point set P. Then the following hold.*

(i) *Every automorphism of an A3GQ is continuous, that is, is induced by a homeomorphism of the point set P to itself.*

(ii) *If $\mathcal{G}$ has an automorphism group that acts transitively on P, then $\mathcal{G}$ is classical.*

(iii) *The connected component of the automorphism group of $\mathcal{G}$ is a finite-dimensional real Lie group of dimension ≤ 10 (the classical example has group dimension 10).*

(iv) *If $\mathcal{G}$ is nonclassical, then its group dimension is at most 5. If its group dimension is 5, then it arises from a flat Laguerre plane of skew parabola type (as described in the following section).*

For a proofs of (i) and (iii) see Forst [1981] 4.1 and 4.6. For a proof of (ii) see Kramer [1994] Corollary 5.2.7. Finally (iv) is proved in Schroth [2000] Corollary 4.9 and Theorem 4.10. Schroth's paper is the first step in the direction of a group-dimension classification of A3GQs. It is based on the group-dimension classification of flat Laguerre planes by Steinke; see Chapter 5.

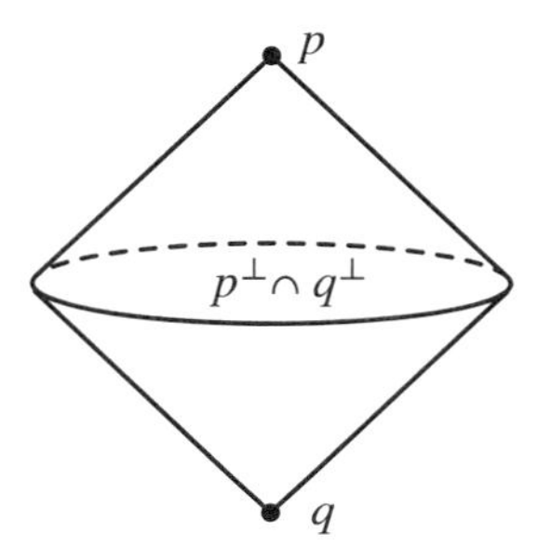

Fig. 6.3. The trace of two noncollinear points in $Q(4, \mathbf{R})$ is a conic

6.3 From Circle Planes to Generalized Quadrangles and Back

The main relationships between circle planes and A3GQs are summarized in Figure 6.4. We will give detailed descriptions of all these relationships following the exposition in Schroth [1995a] Chapters 3, 4, and 5.

6.3.1 Flat Laguerre Planes

The connection between flat Laguerre planes and A3GQs is very strong. In fact, in a way, an A3GQ is just a special representation of a flat Laguerre plane, and vice versa.

Let $\mathcal{G}$ be an A3GQ with point set P. Then for every $p \in P$ the *derived structure*

$$Lp(\mathcal{G}, p) = (p^{\perp} \setminus \{p\}, \{p^{\perp} \cap q^{\perp} \mid q \in P \setminus p^{\perp}\})$$

turns out to be a flat Laguerre plane. Note that our use of the word 'derived' is different from the way we usually use this word.

The proof that the geometry $Lp(\mathcal{G}, p)$ is really a flat Laguerre plane is not difficult; see Schroth [1995a] Theorem 3.1. Note that Theorem 6.2.1

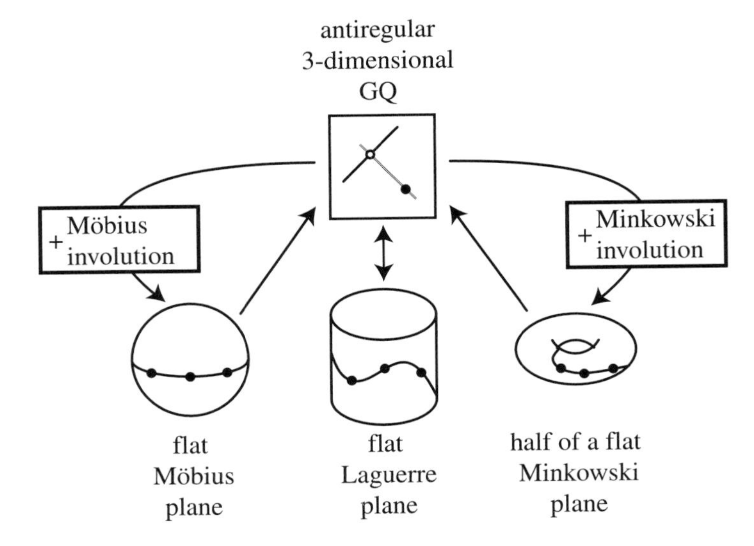

Fig. 6.4. The network of relationships between A3GQs and flat Möbius, Laguerre, and Minkowski planes

guarantees that the point set $p^\perp \setminus \{p\}$ is a cylinder. The parallel classes of points in the Laguerre plane $Lp(\mathcal{G}, p)$ are the lines in the quadrangle through p that have all been punctured at p.

Let $\mathcal{L}$ be a flat Laguerre plane and let S, $\mathcal{C}$, and E be its point set, circle set, and set of parallel classes, respectively. Let ∞ denote an additional point. An *extended parallel class* is a parallel class to which the point ∞ has been added. An *extended tangent pencil* is the set consisting of a point p, a circle through this point, and all circles touching the given circle at p. Let B and G denote the sets of all extended tangent pencils and parallel classes, respectively.

Then

$$Lie(\mathcal{L}) = (S \cup \mathcal{C} \cup \{\infty\}, B \cup G)$$

is the so-called *Lie geometry* of $\mathcal{L}$. This geometry turns out to be an antiregular generalized quadrangle. Furthermore, it is possible to make this generalized quadrangle into an A3GQ by equipping its point and line sets with topologies that induce the topologies that the four object sets S, $\mathcal{C}$, B, and G inherit from the flat Laguerre plane $\mathcal{L}$. It turns out that such topologies exist and that they are unique. We call them *natural topologies*.

The following theorem summarizes the relationship between flat Laguerre planes and A3GQs.

THEOREM 6.3.1 (Flat Laguerre Plane and A3GQ) *Let $\mathcal{L}$ be a flat Laguerre plane. Then the Lie geometry $Lie(\mathcal{L})$ equipped with its natural topology is an A3GQ and*

$$Lp(Lie(\mathcal{L}), \infty) \simeq \mathcal{L}.$$

Conversely, let $\mathcal{G}$ be an A3GQ and let ∞ be one of its points. Then the geometry $Lp(\mathcal{G}, \infty)$ is a flat Laguerre plane and

$$Lie(Lp(\mathcal{G}, \infty)) \simeq \mathcal{G}.$$

If p and q are two points of $\mathcal{G}$, then $Lp(\mathcal{G}, p)$ is isomorphic to $Lp(\mathcal{G}, q)$ if and only if there is an automorphism of $\mathcal{G}$ mapping the point p to the point q.

The group of automorphisms of $\mathcal{G}$ fixing the point p is the full automorphism group of the flat Laguerre plane $Lp(\mathcal{G}, p)$.

One surprising consequence of this result is that it allows us to construct new flat Laguerre planes from a given flat Laguerre plane. Just observe that the derived structure $Lp(Lie(\mathcal{L}), q)$ at any point $q \neq \infty$ is a new flat Laguerre plane that may not be isomorphic to the one we started with; see Subsection 6.5.1 for more details. Of course, the derived structure of the classical A3GQ at any point is isomorphic to the classical flat Laguerre plane.

In general it can be shown that the derived structures of any antiregular generalized quadrangle are Laguerre planes; see again Schroth [1995a] Theorem 3.1. On the other hand, the Lie geometry of an abstract Laguerre plane is not necessarily a generalized quadrangle. The topology of a flat Laguerre plane is used in an essential way to obtain a generalized quadrangle.

For more details and proofs we refer to Schroth [1995a] Chapter 3.

6.3.2 Flat Möbius Planes

Let $\mathcal{M}$ be a flat Möbius plane with point set S and circle set $\mathcal{C}$. A *cycle* of $\mathcal{M}$ is a pair (C, X) where C is a circle of $\mathcal{M}$ and X is one of the connected components of $S \setminus C$. This means that to every circle correspond two cycles, or equivalently, that every circle can be *oriented* in two different ways. Two cycles (C, X) and (D, Y) *touch* if C touches D

and $X \subseteq Y$ or $Y \subseteq X$. In the following we will depict a cycle (C, X) by drawing the circle C together with an arrow. Here X is the connected component of $S \setminus C$ that is on your left when you travel along C in the direction of the arrow. An *extended cycle pencil* in the point $p \in S$ consists of the point p, a cycle containing p, and all other cycles touching the given one at p; see Figure 6.5.

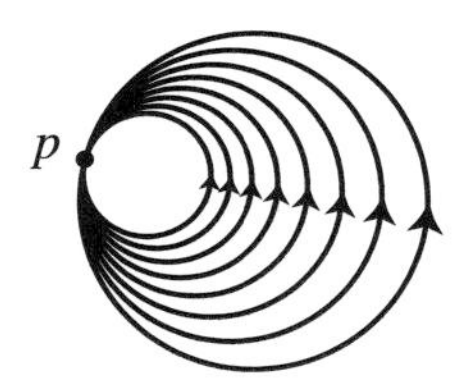

Fig. 6.5. An extended cycle pencil in the classical flat Möbius plane

Let $\overline{\mathcal{C}}$ be the set of all cycles and B the set of all extended cycle pencils. Then the so-called *lifted Lie geometry*

$$Lie(\mathcal{M}) = (S \cup \overline{\mathcal{C}}, B)$$

turns out to be an antiregular quadrangle.

Again, as in the case of the Lie geometry of a flat Laguerre plane, there are unique natural topologies on the point and line sets of this antiregular quadrangle that turn it into an A3GQ and extend the natural topologies on S, $\overline{\mathcal{C}}$, and B.

Note that we could also define a Lie geometry for flat Möbius planes as described for flat Laguerre planes. This Lie geometry is not a generalized quadrangle, though.

Unlike in the case of flat Laguerre planes, not all A3GQs arise as lifted Lie geometries of flat Möbius planes. In fact, only those do that admit a *Möbius involution*, that is, an involutory automorphism that does not fix any lines and whose fixed-point set is homeomorphic to $\mathbf{S}^2$. It turns out that every line in an A3GQ intersects the fixed-point set of a Möbius involution in exactly one point and that every three distinct points of the fixed-point set form a triad; see Schroth [1995a] Proposition 4.16 for these results and a number of other characterizations of Möbius involutions. In the A3GQ $Lie(\mathcal{M})$ the involution $\tau_{\mathcal{M}}$ that fixes all elements of S and exchanges the two cycles belonging to the same circle is a Möbius involution. Let $\mathcal{G}$ be an A3GQ with point set P that admits a Möbius involution τ with fixed point set S. We define

$$Mo(\mathcal{G},\tau) = (S, \{\, S \cap q^\perp \mid q \in P \setminus S \,\}).$$

The following result summarizes the relationship between A3GQs and flat Möbius planes.

THEOREM 6.3.2 (Flat Möbius Plane and A3GQ) *Let $\mathcal{M}$ be a flat Möbius plane. Then the lifted Lie geometry $Lie(\mathcal{M})$ is an A3GQ with Möbius involution $\tau_{\mathcal{M}}$ and*

$$Mo(Lie(\mathcal{M}), \tau_{\mathcal{M}}) \simeq \mathcal{M}.$$

Conversely, let $\mathcal{G}$ be an A3GQ with Möbius involution τ. Then the geometry $Mo(\mathcal{G},\tau)$ is a flat Möbius plane and

$$Lie(Mo(\mathcal{G},\tau)) \simeq \mathcal{G}.$$

If υ is another Möbius involution of $\mathcal{G}$, then the Möbius plane $Mo(\mathcal{G},\upsilon)$ is isomorphic to $Mo(\mathcal{G},\tau)$ if and only if there is an automorphism γ of $\mathcal{G}$ such that $\tau = \gamma^{-1}\upsilon\gamma$.

The automorphism group of the Möbius plane $Mo(\mathcal{G},\tau)$ is the quotient group of the group consisting of all automorphisms of $\mathcal{G}$ commuting with τ by the group consisting of τ and the identity.

We give examples of Möbius involutions of the real orthogonal quadrangle $Q(4, \mathbf{R})$, as described in Section 6.1, at the end of the following section. The flat Möbius plane constructed from this classical A3GQ and one of its Möbius involutions is classical.

6.4 Flat Minkowski Planes

The connection between flat Minkowski planes and A3GQs is similar to the connection between flat Möbius planes and A3GQs described in the previous section. There are two main differences, though.

First, remember that the circle space of every flat Minkowski plane has two geometrically completely independent connected components. In fact, it is possible to combine any two halves taken from (possibly) different flat Minkowski planes into a new flat Minkowski plane; see Chapter 4. We will associate one A3GQ each with these two components.

Second, there is no easy way to define orientation for circles that will yield a lifted Lie geometry that is an antiregular quadrangle. In particular, just imitating what we did in the case of flat Möbius planes does not give anything usable. Here is a brief outline of the solution of this problem and how it is used to define working lifted Lie geometries for flat Minkowski planes.

Let $\mathcal{M}$ be one half of a flat Minkowski plane with point set S, circle set $\mathcal{C}$, and sets G^+ and G^- of $(+)$- and $(-)$-parallel classes, respectively. This means that S, G^+, and G^- are the full point set and the full sets of parallel classes of a flat Minkowski plane. Only $\mathcal{C}$ is just one of the two connected components of the circle set of this Minkowski plane.

As a topological space the circle set $\mathcal{C}$ is homeomorphic to $\mathrm{PSL}_2(\mathbf{R})$ which, in turn, is homeomorphic to $\mathbf{S}^1 \times \mathbf{R}^2$. This space has a unique 2-fold covering space which is homeomorphic to $\mathrm{SL}_2(\mathbf{R})$. We denote this space by $\overline{\mathcal{C}}$ and the covering map $\overline{\mathcal{C}} \to \mathcal{C}$ by κ. If S denotes the point set of $\mathcal{M}$, we can extend κ to a map $\overline{\mathcal{C}} \cup S \to \overline{\mathcal{C}} \cup S$ which is the identity on S. This extended map induces a 2-fold covering $\overline{B}$ of the space B of all extended tangent pencils of $\mathcal{M}$.

Then the so-called *lifted Lie geometry*

$$Lie(\mathcal{M}) = (S \cup \overline{\mathcal{C}}, \overline{B} \cup G^+ \cup G^-)$$

turns out to be an antiregular quadrangle.

Again, there are unique natural topologies on the point and line sets of this antiregular quadrangle that turn it into an A3GQ.

As in the case of flat Möbius planes, the Lie geometries of halves of flat Minkowski planes, defined as for flat Laguerre planes, are not generalized quadrangles.

Furthermore, not all A3GQs arise as lifted Lie geometries of halves of flat Minkowski planes. In fact, only those do that admit a *Minkowski involution*, that is, an involutory automorphism having the following properties: the fixed-point set is homeomorphic to the torus $\mathbf{S}^1 \times \mathbf{S}^1$, the fixed lines are all contained in this torus, and the set of fixed lines is, just like the set of parallel classes of a flat Minkowski plane, a disjoint union of two sets G^+ and G^- such that the elements of either set partition the torus and a line in G^+ intersects a line in G^- in exactly one point.

In the A3GQ $Lie(\mathcal{M})$ the involution $\tau_{\mathcal{M}}$ that fixes all elements of S and exchanges the two elements of $\overline{\mathcal{C}}$ belonging to the same circle is a Minkowski involution.

Let $\mathcal{G}$ be an A3GQ with point set P that admits a Minkowski involution τ with fixed-point set S. We define

$$Mi(\mathcal{G}, \tau) = (S, \{\, S \cap q^\perp \mid q \in P \setminus S \,\}).$$

The following result summarizes the relationship between A3GQs and flat Minkowski planes.

THEOREM 6.4.1 (Flat Minkowski Plane and A3GQ) *Let $\mathcal{M}$ be one half of a flat Minkowski plane. Then the lifted Lie geometry $Lie(\mathcal{M})$ is an A3GQ with Minkowski involution $\tau_{\mathcal{M}}$ and*

$$Mi(Lie(\mathcal{M}), \tau_{\mathcal{M}}) \simeq \mathcal{M}.$$

Conversely, let $\mathcal{G}$ be an A3GQ that has a Minkowski involution τ. Then $Mi(\mathcal{G}, \tau)$ is one half of a flat Minkowski plane where the sets G^+ and G^- of fixed lines of τ form the parallel classes of points of all flat Minkowski planes that $Mi(\mathcal{G}, \tau)$ is a half of. Furthermore,

$$Lie(Mi(\mathcal{G}, \tau)) \simeq \mathcal{G}.$$

If υ is another Minkowski involution of $\mathcal{G}$, then the half $Mi(\mathcal{G}, \upsilon)$ of a flat Minkowski plane is isomorphic to $Mi(\mathcal{G}, \tau)$ if and only if there is an automorphism γ of $\mathcal{G}$ such that $\tau = \gamma^{-1}\upsilon\gamma$.

The automorphism group of the half of a Minkowski plane $Mi(\mathcal{G}, \tau)$ is the quotient group of the group consisting of all automorphisms of $\mathcal{G}$ commuting with τ by the group consisting of τ and the identity.

In the real orthogonal quadrangle $Q(4, \mathbf{R})$, as described in Section 6.1, Möbius and Minkowski involutions are rather natural maps.

For $q = \{a : b : c : d : e\} \in \mathrm{PG}(4, \mathbf{R})$ let $r_q = a^2 + b^2 - c^2 + 2de$ and

$$H_q = \{\, \{s : t : u : v : w\} \mid as + bt - cu + dw + ev = 0 \,\}.$$

Reflection at H_q is a Möbius involution if $r_q < 0$ and a Minkowski involution if $r_q > 0$.

Let $p_\infty = \{0 : 0 : 0 : 0 : 1\}$ and identify $P \setminus p_\infty^\perp$ with $\mathbf{R}^3$ as in Section 6.1. Then every reflection at a plane perpendicular to the z-axis corresponds to a Möbius involution fixing p_∞. Every reflection at a plane parallel to the z-axis corresponds to a Minkowski involution fixing p_∞.

A flat Möbius plane or half of a flat Minkowski plane constructed from the classical A3GQ and one of its Möbius or Minkowski involutions is classical.

6.5 Sisters of Circle Planes

The results of the last three sections imply that, given any flat Laguerre, Möbius, or Minkowski plane, it is possible to construct many new flat Laguerre planes by first constructing the associated A3GQ and then deriving this generalized quadrangle at different points. Sometimes it is also possible to construct new flat Möbius or Minkowski planes depending on whether the associated A3GQ admits any (new) Möbius or Minkowski involutions.

A flat Laguerre plane derived from an A3GQ $\mathcal{G}$ is called a *daughter* of $\mathcal{G}$. A flat Möbius or Minkowski plane associated with $\mathcal{G}$ is called a *son*. It then makes sense to speak of sisters and brothers of circle planes. Certain types of sisters of a given circle plane have attractive characterizations entirely within the given circle plane. In the following we give a summary of these characterizations.

6.5.1 Sisters of Flat Laguerre Planes

Sisters of a flat Laguerre plane $\mathcal{L}$, with point set S and circle set $\mathcal{C}$, are themselves flat Laguerre planes. In fact, by Theorem 6.3.1 these sisters are the Laguerre planes

$$\mathcal{L}(q) = Lp(Lie(\mathcal{L}), q)),$$

where $q \in S \cup \mathcal{C}$.

By Theorem 6.3.1, the sisters corresponding to the points q and r are isomorphic if and only if there is an automorphism of the generalized quadrangle $Lie(\mathcal{L})$ that maps q to r.

It is possible to describe the sisters of $\mathcal{L}$ within $\mathcal{L}$. We distinguish between two essentially different 'embeddings' depending on whether q is a point or circle of $\mathcal{L}$.

Case where q is a point of $\mathcal{L}$: The points of the sister $\mathcal{L}(q)$ are all points parallel to but different from p, plus all circles incident with p, plus the additional point ∞. The circles of the sister are all points of $\mathcal{L}$ not parallel to p, and all circles not incident with p. A flag of the sister consists of either a flag of $\mathcal{L}$, a pair of tangent circles of $\mathcal{L}$, or a pair of the form (∞, r), where r is a point not parallel to p.

Case where q is a circle of $\mathcal{L}$: The points of the sister of $\mathcal{L}$ are all points on this circle, plus all circles that *touch q properly*, that is, at exactly one point. The circles are all points not on q, the additional point ∞, plus all circles not tangent to q. A flag of the sister consists of either a pair of parallel points, a flag of $\mathcal{L}$, a pair of tangent circles of $\mathcal{L}$, or a pair (r, ∞), where r is a point incident with q.

We have already noted before, using different language, that all sisters of the classical flat Laguerre plane are isomorphic. Recently, it was proved by Schroth [1995b] that the flat Laguerre planes of 'shear type' constructed in Löwen–Pfüller [1987a] (see also Subsection 5.3.5) are sisters of the flat Laguerre planes of 'translation type' described in Subsection 5.3.6. For further results that relate different families of flat Laguerre planes via sisterhood see Schroth [2000]. First results towards a classification of the A3GQs of large group dimension suggest that this classification can be roughly described as the classification of flat Laguerre planes modulo sisterhood; see again Schroth [1995b].

6.5.2 Sisters of Flat Möbius Planes

The sisters of a flat Möbius plane $\mathcal{M}$ with point set S and circle set $\mathcal{C}$ are the flat Laguerre planes

$$\mathcal{M}(q) = Lp(Lie(\mathcal{M}), q)),$$

where $q \in S \cup \overline{\mathcal{C}}$.

It is possible to describe these sisters within $\mathcal{M}$. We distinguish between two essentially different embeddings.

Case where q is a point of $\mathcal{M}$: Then the points of the sister $\mathcal{M}(q)$ are the cycles containing q and the circles are all the cycles not containing q plus all elements of S. Two points are parallel if and only if they correspond to cycles that touch (necessarily at q). A point is incident with a circle if, considered as objects of the Möbius plane, they form a flag or touch. In the affine part of the Möbius plane with respect to the point q, the points of $\mathcal{M}(q)$ are just the oriented lines in the derived affine plane of $\mathcal{M}$ at q. Two points of the Laguerre plane are parallel if and only if the corresponding oriented lines in the affine plane are parallel and share the same orientation. This means that the classical flat Laguerre plane is the geometry whose points are the oriented Euclidean lines (these are also called *spears*), and whose circles are the points of the Möbius plane plus the oriented Euclidean circles; see Subsection 5.1.4 for details about this model of the classical flat Laguerre plane. We note that Groh was the first to completely establish this particular way of constructing flat Laguerre planes from flat Möbius planes; see Groh [1974a]

Case where q is a cycle of $\mathcal{M}$: Points of the sister $\mathcal{M}(q)$ are the cycles that touch q properly plus all points incident with q. Circles are all points of the Möbius plane not incident with q plus all cycles not touching q. A point of the Laguerre plane is incident with a circle if and only if the corresponding objects in the Möbius plane form a flag or touch.

A *Möbius involution* of a flat Laguerre plane is an inner involution that is also an automorphism of the plane; see Subsection 5.3.3. This means that a Möbius involution of a flat Laguerre plane is fixed-point-free, that any two fixed circles intersect in two points that get exchanged by the involution, and that a circle that is not fixed by the involution does not intersect its image under the involution.

The reflection $(x, y, z) \mapsto (-x, -y, -z)$ through the origin of $\mathbf{R}^3$ induces a Möbius involution on the classical Laguerre plane in its representation as the geometry of nontrivial plane sections of the vertical cylinder over the unit circle in the xy-plane. The circles fixed by this involution are the intersections of the cylinder with the nonvertical planes that contain the origin of $\mathbf{R}^3$

Back to $\mathcal{M}$ and its sister $\mathcal{M}(q)$, $q \in S$. The Möbius involution $\tau_{\mathcal{M}}$ of the quadrangle $Lie(\mathcal{M})$ fixes the point q. Therefore $\tau_{\mathcal{M}}$ induces an involutory automorphism $\tau_{\mathcal{M},q}$ on the Laguerre plane $\mathcal{M}(q)$. Conversely, any involutory automorphism τ of a flat Laguerre plane $\mathcal{L}$ induces an involutory automorphism $\tau_{\mathcal{L}}$ of its associated quadrangle.

THEOREM 6.5.1 (Möbius and Laguerre) *Let $\mathcal{M}$ be a flat Möbius plane with point set S. Then $\tau_{\mathcal{M},q}$ is a Möbius involution of the flat Laguerre plane $\mathcal{M}(q)$, $q \in S$.*

Conversely, let $\mathcal{L}$ be a flat Laguerre plane with Möbius involution τ. Then $\tau_{\mathcal{L}}$ is a Möbius involution of its associated A3GQ $Lie(\mathcal{L})$.

For details see Schroth [1995a] Section 6.3, as well as Groh [1974a] and [1974b]. Note that the two constructions 'flat Möbius plane to flat Laguerre plane with Möbius involution' and 'flat Laguerre plane with Möbius involution to flat Möbius plane' are inverse to each other.

In a flat Laguerre plane $\mathcal{L}$ with Möbius involution τ, the points of the associated flat Möbius plane are just the circles fixed by τ plus the extra point ∞ in $Lie(\mathcal{L})$. Its circles are pairs of points and pairs of circles in the Laguerre plane that get exchanged by τ. A point and circle of the Möbius plane are incident if they arise from collinear points in $Lie(\mathcal{L})$.

6.5.3 Sisters of (Halves of) Flat Minkowski Planes

The sisters of one half of a flat Minkowski plane $\mathcal{M}$ with point set S and circle set $\mathcal{C}$ are the flat Laguerre planes

$$\mathcal{M}(q) = Lp(Lie(\mathcal{M}), q)),$$

where $q \in S \cup \overline{\mathcal{C}}$. A description of these sisters within $\mathcal{M}$ is very cumbersome and we do not give one here.

A *Minkowski involution* of a flat Laguerre plane is an outer involution that is also an automorphism of the plane; see Subsection 5.3.3. Remember that the reflection $(x, y, z) \mapsto (x, -y, z)$ through the xz-plane of $\mathbf{R}^3$ induces a Minkowski involution on the classical Laguerre plane in its representation as the geometry of nontrivial plane sections of the vertical cylinder over the unit circle in the xy-plane. The circles fixed by this involution are the intersections of the cylinder with planes that are perpendicular to the xz-plane. Just like this prototype, every Minkowski involution fixes two parallel classes pointwise. Furthermore, every pair of nonparallel fixed points is contained in exactly one fixed circle.

Consider $\mathcal{M}(q)$, $q \in S$. The Minkowski involution $\tau_{\mathcal{M}}$ of the quadrangle $Lie(\mathcal{M})$ fixes the point q. Therefore $\tau_{\mathcal{M}}$ induces an involutory automorphism $\tau_{\mathcal{M},q}$ on the Laguerre plane $\mathcal{M}(q)$. Conversely, any involutory automorphism τ of a flat Laguerre plane $\mathcal{L}$ induces an involutory automorphism $\tau_{\mathcal{L}}$ of its associated quadrangle.

THEOREM 6.5.2 (Minkowski and Laguerre) *Let $\mathcal{M}$ be one half of a flat Minkowski plane with point set S. Then $\tau_{\mathcal{M},q}$ is a Minkowski involution of the flat Laguerre plane $\mathcal{M}(q)$, $q \in S$.*

Conversely, let $\mathcal{L}$ be a flat Laguerre plane with Minkowski involution τ. Then $\tau_{\mathcal{L}}$ is a Minkowski involution of its associated A3GQ $Lie(\mathcal{L})$.

The constructions 'half of a flat Minkowski plane to flat Laguerre plane with Minkowski involution' and 'flat Laguerre plane with Minkowski involution to half of a flat Minkowski plane' are inverse to each other.

Given a flat Laguerre plane $\mathcal{L}$ with Minkowski involution τ, the points of the associated half of a flat Minkowski plane are just the circles fixed by τ plus the extra point ∞ in $Lie(\mathcal{L})$. Its circles are pairs of points and pairs of circles in the Laguerre plane that get exchanged by τ. A point and circle of the half are incident if they arise from collinear points in $Lie(\mathcal{L})$. The circles through a fixed point of τ form a parallel class of points in the half.

6.6 Flat Biaffine Planes and Flat Homology Semibiplanes

Let $\mathcal{G}$ be an A3GQ with point set P. Recall that the flat Laguerre plane associated with a point $p \in P$ is defined as

$$Lp(\mathcal{G}, p) = \{p^\perp \setminus \{p\}, \{\, p^\perp \cap q^\perp \mid q \in P \setminus p^\perp\}\}.$$

Let q be a point of this Laguerre plane, that is, a point in $\mathcal{G}$ distinct from but collinear with p. Then the derived geometry of $Lp(\mathcal{G}, p)$ at the point q is

$$\mathcal{G}(p, q) = \{p^\perp \setminus (\{p\} \cup q^\perp), q^\perp \setminus (\{q\} \cup p^\perp), \perp\}.$$

This is a *flat biaffine plane*, that is, a flat affine plane from which all lines in a certain parallel class have been removed. This means that the point and line sets of a flat biaffine plane carry one nontrivial parallelism each. Furthermore, the dual of a flat biaffine plane is also a flat biaffine plane. Note that incidence in $\mathcal{G}(p, q)$ is just 'being collinear' in the quadrangle. We also call $\mathcal{G}(p, q)$ the *derived geometry of* $\mathcal{G}$ *at the two points* p *and* q.

The definition of the geometry $\mathcal{G}(p, q)$ still makes sense if the two points p and q are no longer collinear. We first describe this geometry within the Laguerre plane $Lp(\mathcal{G}, p)$. As usual, let S and $\mathcal{C}$ denote the sets of points and circles of this circle plane. That p and q are not collinear in the quadrangle implies the following.

- The point q is a circle in the Laguerre plane.
- The point set of $\mathcal{G}(p, q)$ is $S \setminus q$. This means that the point set has two connected components S_1 and S_2 that are both homeomorphic to the cylinder $\mathbf{S}^1 \times \mathbf{R}$.
- The block set B of the geometry is the set of all circles in $\mathcal{C}$ that touch the circle q properly, that is, at one point each. Again, the block set has two connected components B_1 and B_2, where B_i consists of all those elements of B that are contained in $\mathcal{S}_i \cup q$.
- Incidence between points and blocks is containment.

Since no point of S_i, $i = 1, 2$, is contained in any block of B_{3-i}, we conclude that the point–block geometry $\mathcal{G}(p, q)$ is the disjoint union of the two geometries $\mathcal{G}_i(p, q) = (S_i, B_i)$, $i = 1, 2$. Note that $\mathcal{G}(p, q)$ is the dual of $\mathcal{G}(q, p)$ and that, similarly, $\mathcal{G}_i(p, q)$ is the dual of $\mathcal{G}_i(q, p)$.

THEOREM 6.6.1 (A3GQ to Semibiplane I) *Let* $\mathcal{G}$ *be an A3GQ and* p *and* q *two distinct points of* $\mathcal{G}$.

If p *and* q *are collinear, then the derived geometry* $\mathcal{G}(p, q)$ *is a flat biaffine plane.*

If p *and* q *are not collinear, then* $\mathcal{G}(p, q)$ *is the disjoint union of the two geometries* $\mathcal{G}_i(p, q)$, $i = 1, 2$. *In this case* $\mathcal{G}_i(p, q)$ *is a flat homology semibiplane (see Subsection 2.10.3 for the definition of flat homology semibiplanes).*

A flat biaffine plane has a unique extension to a flat projective plane and every flat homology semibiplane arises from a flat projective plane with homology involution in a unique manner. This means that associated with any two points of an A3GQ are one or two flat projective planes depending on whether the points are collinear or not.

Proof of theorem. Keeping in mind our discussion above and Theorem 2.10.7, which states that flat homology semibiplanes are basically the same thing as cylinder semibiplanes, it suffices to show that $\mathcal{G}_i(p,q)$ is a cylinder semibiplane. From what we said before, we know that $\mathcal{G}_i(p,q)$ certainly looks like a cylinder semibiplane in that its point space and the embeddings of the blocks in the point space are of the right form. It remains to check that the following three properties are satisfied.

(i) Two points are parallel if and only if they are parallel in the Laguerre plane.
(ii) Two blocks are parallel if and only if they touch the circle q at the same point.
(iii) Axioms S1 and S2 for semibiplanes.

We check that Axiom S1 and (i) are satisfied. Let r and s be two nonparallel (in the Laguerre plane) points in S_i. Then the solution of the Apollonius problem for flat Laguerre planes (see Theorem 6.8.1) guarantees that there are exactly two circles in the Laguerre plane that touch the circle q and contain the two points r and s. Both these circles are contained in B_i. Clearly, if r and s are parallel, then there is no block containing both of them. Since $\mathcal{G}_i(p,q)$ is the dual of $\mathcal{G}_i(q,p)$, we can simply exchange the roles of p and q to conclude that Axiom S2 and (ii) are also satisfied. □

Following our approach in the last three sections, we can now try to identify the biaffine planes and semibiplanes that we encountered above as attractive substructures of the various flat Laguerre and Möbius planes associated with $\mathcal{G}$. For more details about semibiplanes constructed from A3GQs and circle planes see Polster–Schroth [1997a], [1997b], [1998a], [1998b], and [20XX].

6.6.1 Flat Biaffine Planes in Flat Laguerre Planes

Let $\mathcal{G} = Lie(\mathcal{L})$, where $\mathcal{L}$ is a flat Laguerre plane. Let p and q be two collinear points in $\mathcal{G}$, let L be the line connecting p and q, and let r denote the extra point (at infinity) of the generalized quadrangle. We interpret the flat biaffine plane $\mathcal{G}(p,q)$ in terms of the objects that make up $Lie(\mathcal{L})$. Then $\mathcal{G}(p,q)$ can be viewed as a 'subgeometry' of the Laguerre plane $\mathcal{L}$. We discuss the essentially different possibilities for r relative to p and q. Remember that by exchanging the roles of p and q, we switch from $\mathcal{G}(p,q)$ to its dual (which is also a flat biaffine plane); we do not really get anything new.

Case where $r = p$: Here q has to be a point of the Laguerre plane and $\mathcal{G}(p,q)$ is the derived geometry of the Laguerre plane $\mathcal{L}$ at the point q.

Case where $r \in L \setminus \{p,q\}$: Here p and q correspond to parallel points of the Laguerre plane. The points and lines of $\mathcal{G}(p,q)$ are the following objects.

points:	circles containing p
lines:	circles containing q

Incidence between points and lines is touching.

As an example, choose p and q to be the points $(\infty, 1)$ and $(\infty, -1)$ on the parallel class at infinity of the classical flat Laguerre plane. Then $\mathcal{G}(p,q)$ lives on $\mathbf{R}^2$. Its points and lines correspond to the translates of the parabolas $\{(x,x^2) \mid x \in \mathbf{R}\}$ and $\{(x,-x^2) \mid x \in \mathbf{R}\}$, respectively. Two points or lines are parallel if one is the vertical translate of the other. Figure 6.6 shows two points and their connecting (grey) line in this model of $\mathcal{G}(p,q)$.

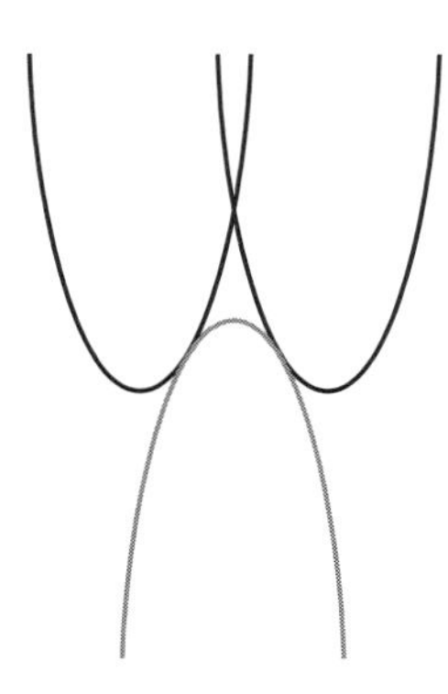

Fig. 6.6. Two points and their connecting (grey) line in a flat biaffine plane embedded in the affine part of the classical flat Laguerre plane

Case where $r \notin L$ and $p \in r^\perp$: In this case p is a point of the Laguerre plane and q is a circle containing this point. The points and lines of $\mathcal{G}(p,q)$ are the following objects.

points:	points parallel to but different from p
	the extra point r (not a point of the Laguerre plane)
	circles through p but not touching q
lines:	points on q different from p
	circles that touch q properly at points different from p

A point is incident with a line if, as objects of the Laguerre plane, they form a flag or are two touching circles. The extra point r is incident with all lines that correspond to points of the Laguerre plane.

Case where $p, q \notin r^\perp$: In this case p and q are circles of the Laguerre plane that touch at a point s. The points and lines of $\mathcal{G}(p,q)$ are the following objects.

points:	points contained in the circle p different from s
	circles that touch p properly at points different from s
lines:	points contained in the circle q different from s
	circles that touch q properly at points different from s

A point is incident with a line if, as objects of the Laguerre plane, they form a flag, are parallel points, or are two touching circles.

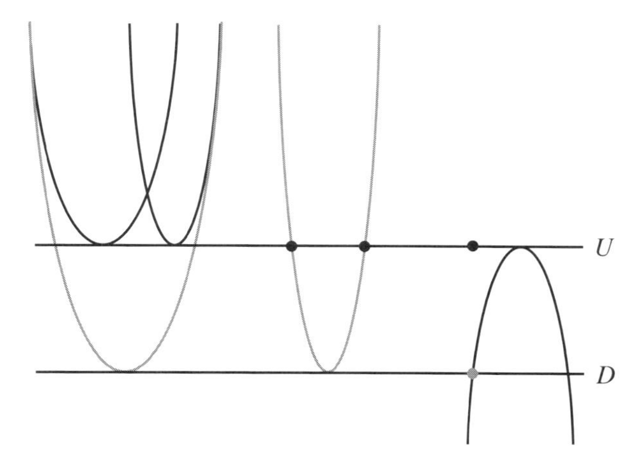

Fig. 6.7. Three pairs of points and their connecting (grey) lines in a flat biaffine plane embedded in the affine part of the classical flat Laguerre plane

As an example, choose s to be the point $(\infty, 0)$ on the parallel class at infinity of the affine part of the classical flat Laguerre plane and let p and q correspond to the horizontal lines U and D in $\mathbf{R}^2$. Then $\mathcal{G}(p,q)$ lives on $\mathbf{R}^2$. Its points are all points of U and all Euclidean parabolas that touch U. Its lines are all points of D and all Euclidean parabolas that touch D; see Figure 6.7.

6.6.2 Flat Biaffine Planes in Flat Möbius Planes

Assume that $\mathcal{G} = Lie(\mathcal{M})$, where $\mathcal{M} = (\mathcal{S}, \mathcal{C})$ is a flat Möbius plane and, as above, let p and q be two collinear points in $\mathcal{G}$. We interpret the flat biaffine plane $\mathcal{G}(p,q)$ in terms of the objects that make up $Lie(\mathcal{M})$, that is, in terms of the points and cycles of the flat Möbius plane $\mathcal{M}$.

Case where p and q are two cycles of $\mathcal{M}$ that touch in a point s of $\mathcal{M}$: The points and lines of $\mathcal{G}(p,q)$ are the following objects.

points:	points of the cycle p different from s
	cycles that touch p in points different from s
lines:	points of the cycle q different from s
	cycles that touch q in points different from s

A point is incident with a line if, as objects of the Möbius plane, they form a flag or are two touching cycles.

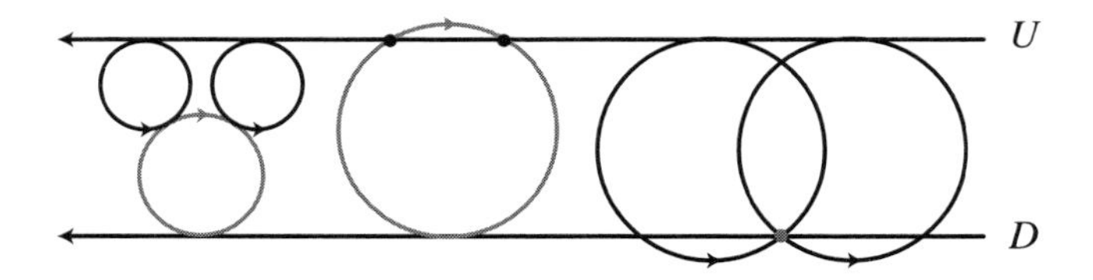

Fig. 6.8. Three pairs of points and their connecting (grey) lines in a flat biaffine plane embedded in the affine part of the classical flat Möbius plane

As an example, choose s to be the infinite point of the affine part of the classical flat Möbius plane. Then p and q correspond to two parallel oriented Euclidean lines U and D and $\mathcal{G}(p,q)$ lives on $\mathbf{R}^2$. Its points are all points of U and all oriented Euclidean circles that touch U. Its lines are all points of D and all oriented Euclidean circles that touch D; see Figure 6.8. Compare this model of our geometry to the one introduced at the end of the previous subsection.

Case where q is a cycle of $\mathcal{M}$ that contains the point p of $\mathcal{M}$: The points and lines of $\mathcal{G}(p,q)$ are the following objects.

points:	cycles containing p but not touching q
lines:	points of the cycle q different from p
	cycles that touch q at points different from p

A point is incident with a line if, as objects of the Möbius plane, they form a flag or are two touching cycles.

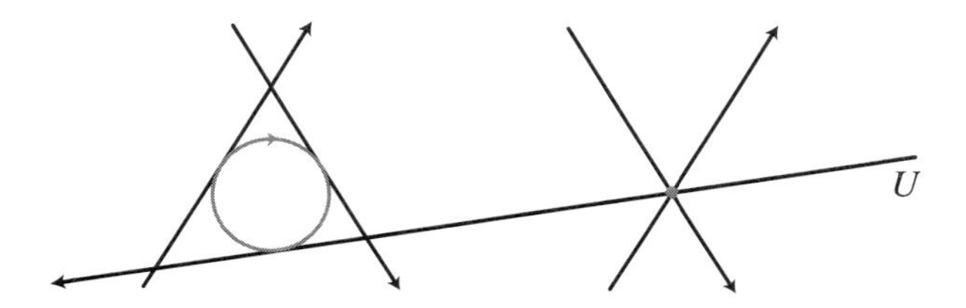

Fig. 6.9. Two pairs of points and their connecting (grey) lines in a flat biaffine plane embedded in the affine part of the classical flat Möbius plane

As an example, choose p to be the infinite point of the affine part of the classical flat Möbius plane. Then q corresponds to an oriented Euclidean line U and $\mathcal{G}(p,q)$ lives on $\mathbf{R}^2$. Its points are all oriented Euclidean lines that are not parallel to U. Its lines are all points of U and all oriented Euclidean circles that touch U; see Figure 6.9.

6.6.3 Flat Homology Semibiplanes in Flat Laguerre Planes

Let $\mathcal{G} = Lie(\mathcal{L})$, where $\mathcal{L}$ is a flat Laguerre plane. Let p and q be two noncollinear points in $\mathcal{G}$ and let r denote the extra point (at infinity) of the generalized quadrangle. We interpret the flat homology semibiplane $\mathcal{G}(p,q)$ in terms of the objects that make up $Lie(\mathcal{L})$. We discuss the essentially different possibilities for r relative to p and q. Remember that by exchanging the roles of p and q we switch from $\mathcal{G}(p,q)$ to its dual (which is a geometry of the same type).

In every single case, the division of $\mathcal{G}(p,q)$ into its two components corresponds to the natural division of its points and blocks by p and q (within the Laguerre plane).

Case where $r = p$: Here we get the model of $\mathcal{G}(p,q)$ that we discussed at the beginning of this section.

Case where $r \in p^\perp \cap q^\perp$: Here p and q correspond to nonparallel points of the Laguerre plane. The points and blocks of $\mathcal{G}(p, q)$ are the following objects.

points:	points parallel to but different from p
	circles through p but not through q
blocks:	points parallel to but different from q
	circles through q but not through p

A point is incident with a block if, as objects of the Laguerre plane, they form a flag or are two touching circles in the Laguerre plane.

Case where $r \notin p^\perp \cup q^\perp$: Here p and q correspond to circles that do not touch in the Laguerre plane. This means that the two circles are disjoint or intersect in two points. The points and blocks of $\mathcal{G}(p, q)$ are the following objects.

points:	points contained in p but not in q
	circles that touch p properly but do not touch q
blocks:	points contained in q but not in p
	circles that touch q properly but do not touch p

A point is incident with a block if, as objects of the Laguerre plane, they form a flag, are parallel, or are two touching circles.

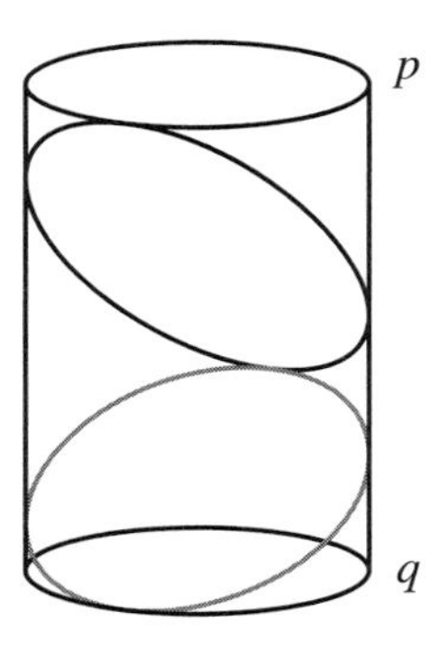

Fig. 6.10. An incident point–block pair of a flat homology semibiplane embedded in the classical flat Laguerre plane

Suppose, as circles of the Laguerre plane, p and q do not intersect. Then one of the two flat homology semibiplanes contained in $\mathcal{G}(p, q)$ is particularly attractive. Let Y be the closed cylinder bounded by p and q.

Then the points of the semibiplane are the circles of the Laguerre plane that touch p properly, do not touch q, and are contained in Y. Similarly, the blocks of the semibiplane are the circles of the Laguerre plane that touch q properly, do not touch p, and are contained in Y; see Figure 6.10.

Case where $r \in p^\perp \setminus (\{p\} \cup q^\perp)$: Here p is a point and q a circle of the Laguerre plane. The point is not incident with the circle. The points and blocks of $\mathcal{G}(q,p)$ are the following objects.

points:	points not in q that are parallel to but different from p
	circles incident with p that do not touch q
	the extra point r (not a point of the Laguerre plane)
blocks:	points incident with q but not parallel to p
	circles not incident with p that touch q at one point

A point is incident with a block if, as objects of the Laguerre plane, they form a flag or are two touching circles. The extra point is incident with those blocks that correspond to points of the circle q.

6.6.4 Flat Homology Semibiplanes in Flat Möbius Planes

Assume that $\mathcal{G} = Lie(\mathcal{M})$ where $\mathcal{M}$ is a flat Möbius plane. We interpret the flat homology semibiplane $\mathcal{G}(p,q)$ in terms of the objects that make up $Lie(\mathcal{M})$, that is, in terms of the points and cycles of the flat Möbius plane $\mathcal{M}$. In fact, it turns out that the semibiplanes that make up $\mathcal{G}(p,q)$ can sometimes be described in terms of points and circles of $\mathcal{M}$ without our having to refer to orientation of those circles at all. There are many cases to consider and we only concentrate on those cases that give the most attractive models for our semibiplanes.

Case where both p and q are points of the Möbius plane: Consider the point–block geometry with the following object sets.

points:	circles through p but not through q
blocks:	circles through q but not through q

Incidence is touching. This is a flat homology semibiplane. There are two ways to orient circles in this geometry consistently such that touching circles turn into touching cycles. By orienting the circles in these two different ways, we arrive at the two flat homology semibiplanes that make up $\mathcal{G}(p,q)$. This means that reversing orientation of the circles induces an isomorphism between these two semibiplanes.

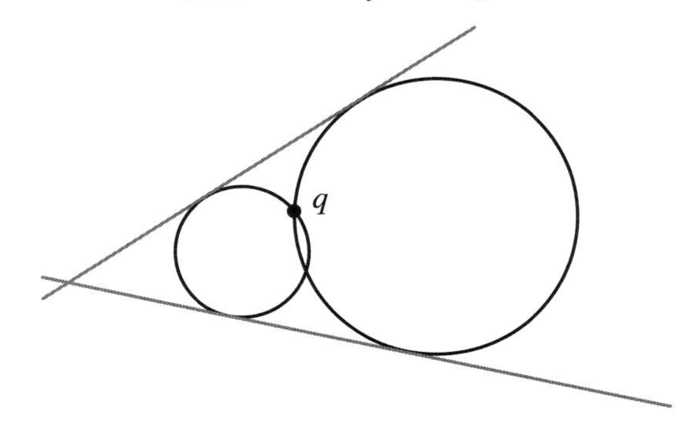

Fig. 6.11. Two points and their connecting (grey) blocks of a flat homology semibiplane embedded in the affine part of the classical flat Möbius plane

As an example, choose p to be the infinite point of the affine part of the classical flat Möbius plane and q the origin of $\mathbf{R}^2$. Then our semibiplane has as points the Euclidean lines not containing the origin and as blocks the Euclidean circles through the origin; see Figure 6.11.

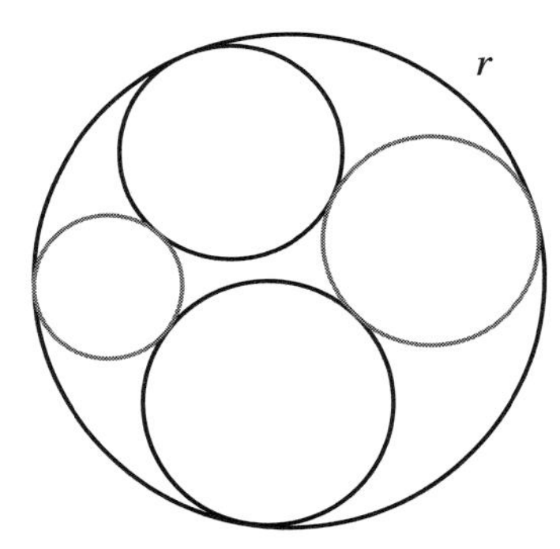

Fig. 6.12. Two nonparallel blocks that intersect in two nonparallel points of a flat homology semibiplane embedded in the classical flat Möbius plane

Case where p and q are two cycles that only differ in their orientation: Again we do not have to worry about orientation of the cycles but, in contrast to the previous case, we can describe the full geometry $\mathcal{G}(p,q)$ in terms of circles of $\mathcal{M}$. Let r be the circle that underlies both p and q.

points: circles that touch r properly
blocks: circles that touch r properly

Incidence is touching at points off r. Note that the point and block sets of this model of $\mathcal{G}(p,q)$ coincide. If S is the point set of the Möbius plane, then the two semibiplanes that make up $\mathcal{G}(p,q)$ are the restrictions of $\mathcal{G}(p,q)$ to the two topological disks that make up $S\backslash r$; see Figure 6.12.

Case where p and q are two cycles as in Figure 6.13: Here one of the flat homology semibiplanes in $\mathcal{G}(p,q)$ has a description in terms of circles of $\mathcal{M}$ and lives on the closed annulus Z bounded by p and q. The objects of this semibiplane are as follows.

points: circles contained in Z that touch p properly but do not touch q

blocks: circles contained in Z that touch q properly, but do not touch p

Incidence is touching within the annulus Z; see Figure 6.13.

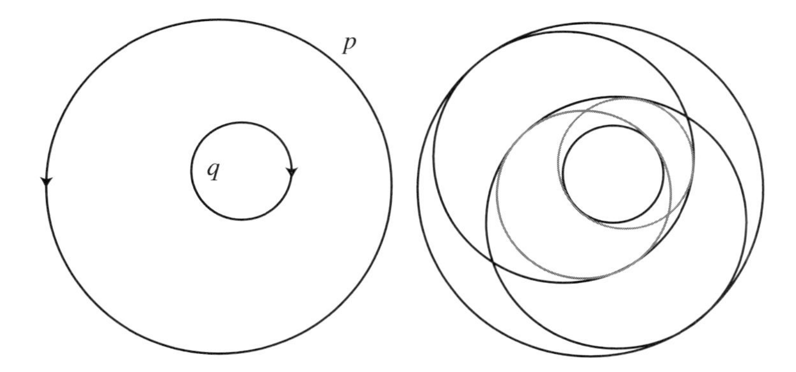

Fig. 6.13. Two nonparallel points and their connecting (grey) blocks of a flat homology semibiplane embedded in the classical flat Möbius plane

6.6.5 Split Semibiplanes

There is one more essentially different construction of semibiplanes in A3GQs. The resulting semibiplanes are not homology semibiplanes.

Let $\mathcal{G}$ be an A3GQ with point set P, let p, q, r, and s be distinct collinear points with connecting line L, and let $P_p^L = p^\perp \setminus L$. Furthermore, let

$$\mathcal{G}(p,q,r,s) = (P_p^L \cup P_q^L, P_r^L \cup P_s^L, \perp).$$

Note that a point of this geometry is incident with a block if they correspond to collinear points of the quadrangle.

THEOREM 6.6.2 (A3GQ to Semibiplane II) *Let $\mathcal{G}$ be an A3GQ and let $p, q, r,$ and s be four distinct collinear points with connecting line L. Then $\mathcal{G}(p, q, r, s)$ is a semibiplane if and only if p and q are contained in different connected components of $L \setminus \{r, s\}$. If $\mathcal{G}(p, q, r, s)$ is a semibiplane, then it is divisible.*

For a proof of this result see Polster–Schroth [1997b]. Alternatively, try to prove it yourself using the solution of the Apollonius problem for flat Laguerre planes; see Theorem 6.8.1.

For $x \in \{p, q\}$ and $y \in \{r, s\}$ the biaffine plane $\mathcal{G}(x, y) = (P_x^L, P_y^L, \perp)$ is a subgeometry of $\mathcal{G}(p, q, r, s)$. Given $x \in \{p, q\}$, the point sets of $\mathcal{G}(x, r)$ and $\mathcal{G}(x, s)$ coincide and points are parallel in $\mathcal{G}(x, r)$ if and only if they are parallel in $\mathcal{G}(x, s)$. The dual statement is true for blocks. Note also that the four sets P_p^L, P_q^L, P_r^L, and P_s^L are pairwise disjoint. This means that $\mathcal{G}(p, q, r, s)$ is the 'disjoint union' of the two biaffine planes $\mathcal{G}(v, w)$ and $\mathcal{G}(x, y)$ whenever $\{v, x\} = \{p, q\}$ and $\{w, y\} = \{r, s\}$ and that it can be represented as such a union in two different ways. This is the reason why the semibiplanes of the form $\mathcal{G}(p, q, r, s)$ are referred to as *split semibiplanes.*

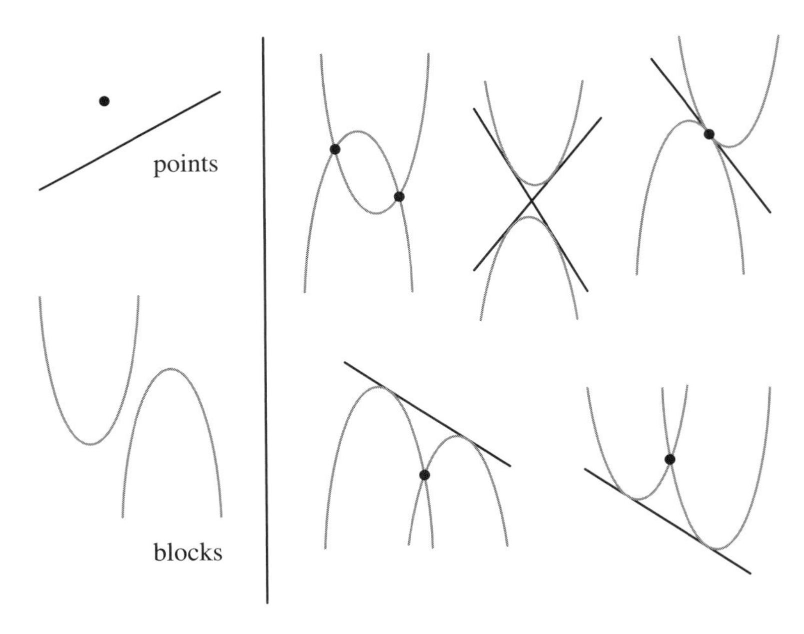

Fig. 6.14. Five ways for two nonparallel (grey) blocks to intersect in two nonparallel points of a split semibiplane

Let us have a look at an example of a split semibiplane in the affine part of the classical Laguerre plane. We choose L to be the parallel class at infinity $\{\infty\}\times\mathbf{R}$ that has been extended (to an extended parallel class) by the point p. Let $q = (\infty, 0)$ be the infinite point of the Euclidean lines, $r = (\infty, 1)$, and $s = (\infty, -1)$. Then the points of $\mathcal{G}(p, q, r, s)$ are the points of $\mathbf{R}^2$, and the Euclidean lines. The blocks are the images of the Euclidean parabolas $\{(x, x^2) \mid x \in \mathbf{R}\}$ and $\{(x, -x^2) \mid x \in \mathbf{R}\}$ under all possible translations. Figure 6.14 shows the five essentially different ways in which two nonparallel points can be connected by two nonparallel blocks in this semibiplane.

For more detailed information about split semibiplanes see Polster–Schroth [1997b].

6.7 Different Ways to Cut and Paste

Cut-and-paste constructions play an important role in constructing new flat circle planes from known ones. Of course, cut-and-paste constructions of flat Laguerre, Möbius, and Minkowski planes can also be translated into the quadrangle setting.

In the following, we only present three particularly attractive examples of 'separating' sets in A3GQs and describe how these translate into cut-and-paste techniques for producing new A3QGs.

Let $\mathcal{G}_1$ and $\mathcal{G}_2$ be two A3GQs that share the same point set P. This is not asking for too much since, by Theorem 6.2.1, the point sets of any two A3GQs are homeomorphic. Assume that the two quadrangles also share one of the following sets S consisting of points, lines, and involutory automorphisms.

(i) All points in $p^\perp \cup q^\perp$, $p, q \in P$, $p \neq q$, $p \perp q$, plus all the lines through p and/or q.

(ii) All points and lines in $p^\perp$, $p \in P$, plus a Möbius involution that fixes p, plus the fixed points of this involution.

(iii) All points and lines in $p^\perp$, $p \in P$, plus a Minkowski involution that fixes p, plus the fixed points and lines of this involution.

Then $P \setminus S$ has two connected components P_1 and P_2, and every line not in S intersects the set $P \setminus S$ in one or two points. Every line of the first kind is completely contained in one of the sets $P_i \cup (S \cap P)$. The set consisting of all pairs of points of intersection of lines of the second kind with $S \cap P$ is the same for the two quadrangles.

We construct a new point–line geometry $\mathcal{G}_3$ as follows. The point set is P, the restriction of $\mathcal{G}_3$ to $P_i \cup (S \cap P)$, $i = 1, 2$, coincides with that of $\mathcal{G}_i$. Given a line segment s_i in $P_i \cup (S \cap P)$ whose two endpoints are points of S, there is a unique line segment s_{3-i} in $P_{3-i} \cup (S \cap P)$ whose two endpoints are the same two points of S. Then the combined set $s_i \cup s_{3-i}$ is homeomorphic to the 1-sphere. These combined sets together with the lines in S and the lines of $\mathcal{G}_i$ completely contained in $P_i \cup (S \cap P)$ form the line set of $\mathcal{G}_3$. Finally, we equip the line set of $\mathcal{G}_3$ with the natural Hausdorff topology.

THEOREM 6.7.1 (Cut and Paste) *Let $\mathcal{G}_1$ and $\mathcal{G}_2$ be two A3GQs that share the same point set P and set S, as described above. Then the geometry $\mathcal{G}_3$ is also an A3GQ.*

This result follows from Theorem 6.2.1, the fact that the topologies on the point and line spaces of A3GQs are uniquely determined by the topologies on the point and circle sets of their associated flat Laguerre planes, and the respective cut-and-paste result in flat Laguerre planes; see Propositions 5.3.8, 5.3.10, and 5.3.12. See also Polster–Steinke [1997] Section 5.

We need to emphasize that combining two (or more) A3GQs into new A3GQs using separating sets can be performed at three different levels: at the (highest) level of the quadrangles themselves, as described above; at the level of a pair of associated circle planes; or at the level of the affine parts of a pair of associated circle planes of the same type. The fact that a cut-and-paste technique based on some separating set can be applied at one of these levels does not imply that it can also be applied in some form at a higher level; see the respective comments about the two levels of cutting and pasting in flat projective planes and flat circle planes in Subsubsection 2.7.10.2 and Polster–Steinke [1997], respectively.

Note that all three sets considered above have to do with geometries that we considered earlier on. For example, the first set consists of the points p and q and all the points and lines of the derived geometry $G(p, q)$. This set is also very similar to the basic separating set for flat projective planes consisting of two lines and all points on these lines.

6.8 The Apollonius Problem

According to Heath [1963], the Apollonius problem in its original form is the following ancient problem in plane geometry posed and solved by Apollonius of Perga.

Given three distinct objects, each either a point, a Euclidean line, or a Euclidean circle, construct a Euclidean circle that touches all three objects. (For a Euclidean circle to touch a point simply means that the circle contains the point.)

Note that this problem is very closely related to antiregularity in A3GQs constructed from flat circle planes. For example, if we are given a triad of points in an A3GQ associated with a flat Möbius plane $\mathcal{M}$, that is, three nontouching points/cycles in the Möbius plane, then to actually check that there are no or two points in the quadrangle collinear to all three points involves solving the following two problems.

(i) Determine the number and relative positions of all circles in the Möbius plane that touch all three points/cycles.
(ii) Determine for every single one of the touching circles whether one of its associated cycles touches all three points in such a way that orientations in the points of touching coincide.

It turns out that (ii) is trivial once (i) is solved. Let us consider an example. The left diagram in Figure 6.15 shows two nontouching cycles and a point in the classical flat Möbius plane. The diagram in the middle shows that there are four common touching circles. It also shows the position of these touching circles. The diagram on the right shows the two of the eight corresponding cycles that touch the given cycles. This means that the centre of the triad of points in the associated A3GQ that we started with has exactly two elements.

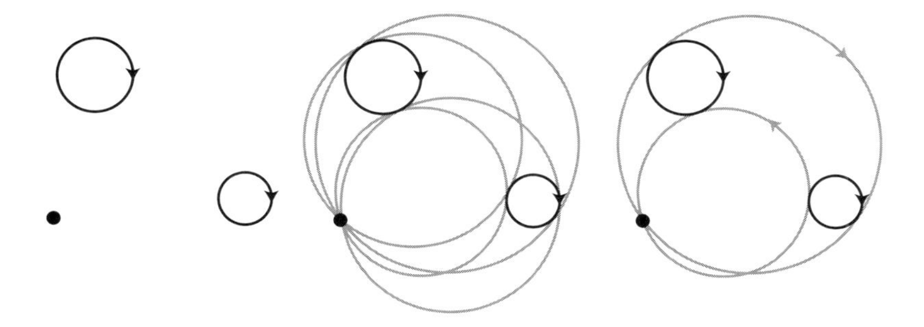

Fig. 6.15. Determining the common touching cycles of one point and two cycles in the classical flat Möbius plane

Let $\mathcal{K}$ be a flat Möbius, Laguerre, or Minkowski plane with point set S and circle set $\mathcal{C}$. Then the Apollonius problem for this plane can be phrased as follows.

Given three distinct elements of $S \cup \mathcal{C}$*, determine the number and relative positions of all circles in* $\mathcal{K}$ *that touch the given objects.*

It turns out that the solution of this problem depends only on what type of circle plane we are dealing with, and the essentially different configurations of the three given objects in the respective point set. Even stronger, all essentially different configurations are already present in the respective classical circle planes, and it therefore suffices to give the solutions of the problem for the classical flat Laguerre, Möbius, and Minkowski planes. See Bruen et al. [1983] for a very readable exposition of the solution of the Apollonius problem in the case of the classical flat Möbius plane.

The only other class of circle planes for which such a strong result has been proved are the 4-dimensional Laguerre and Minkowski planes. This very difficult task has also been accomplished by Schroth in his monograph Schroth [1995a].

The following theorem and tables give the complete solutions of the Apollonius problem for flat Möbius, Laguerre, and Minkowski planes; see Schroth [1995a] Chapter 7. It only lists the essentially different configurations of the three points/circles and the corresponding number of touching circles, not the position of the touching circles with respect to the given three points/circles. The reason for this is twofold. First, to describe the relative positions of the touching circles in every single case is very cumbersome. Second, once we know which of the configurations we are dealing with and how many touching circles to expect, it is not difficult to construct these touching circles in the classical plane and thereby arrive at a complete picture for both the classical and general cases.

THEOREM 6.8.1 (Solution of the Apollonius Problem) *Let* $\mathcal{K}$ *be a flat Laguerre, Möbius, or Minkowski plane with point set* S *and circle set* $\mathcal{C}$*. Then Tables 6.1, 6.2, and 6.3 give the number of circles collinear with three distinct elements of* $S \cup \mathcal{C}$ *for all possible configurations of these three elements in* S*. In every table the entries in the middle column specify the type and relative position of the given three elements, while the entries in the right column give the number of common touching circles.*

i	*two parallel points plus one element of* $S \cup \mathcal{C}$	0
ii	*three pairwise nonparallel points*	1
iii	*two nonparallel points* s_1, s_2*; one circle* K	
1	$s_1 \in K$	1
2	s_1, s_2 in different connected components of $S \setminus K$	0
3	s_1, s_2 in the same connected component of $S \setminus K$	2
iv	*one point* s*; two circles* K_1, K_2	
1	$K_1 \cap K_2 = \{s\}$	∞
2	$K_1 \cap K_2 = \{u\}$ and s not parallel to u	1
3	$K_1 \cap K_2 = \{u\} \neq \{s\}$ and s parallel to u	0
4	$\lvert K_1 \cap K_2\rvert \neq 1$; $s \in K_1 \setminus K_2$	1
5	$\lvert K_1 \cap K_2\rvert = 2$; $s \in K_1 \cap K_2$	0
6	$\lvert K_1 \cap K_2\rvert = 2$; $\lvert K_1 \cap K_2 \cap \Gamma(s) \setminus \{s\}\rvert = 1$	1
7	$\lvert K_1 \cap K_2\rvert = 2$; s in one of the relatively compact connected components of $S \setminus (K_1 \cup K_2)$	0
8	$\lvert K_1 \cap K_2\rvert = 2$; $K_1 \cap K_2 \cap \Gamma(s) = \emptyset$; s not in a relatively compact connected component of $S \setminus (K_1 \cup K_2)$	2
9	$K_1 \cap K_2 = \emptyset$; s not in the relatively compact connected component of $S \setminus (K_1 \cup K_2)$	0
10	$K_1 \cap K_2 = \emptyset$; s in the relatively compact connected component of $S \setminus (K_1 \cup K_2)$	2
v	*three circles* K_1, K_2, K_3	
1	pairwise disjoint	0
2	$K_1 \cap K_2 = \{u\} = K_1 \cap K_3$	∞
3	$K_1 \cap K_2 = \{u\} \neq K_1 \cap K_3$; $u \in K_1 \cap K_3$	0
4	$K_1 \cap K_2 = \{u\} \not\subset K_1 \cap K_3$	1
5	$\lvert K_1 \cap K_2\rvert = 2 = \lvert K_1 \cap K_3\rvert$; $K_2 \cap K_3 = \emptyset$	0
6	$\lvert K_1 \cap K_2\rvert = 2$; $\lvert K_1 \cap K_3\rvert = \lvert K_2 \cap K_3\rvert \neq 1$	2
7	$\lvert K_i \cap K_j\rvert = 2, i \neq j$; $\lvert K_1 \cap K_2 \cap K_3\rvert = 1$	1
8	$\lvert K_i \cap K_j\rvert = 2, i \neq j$; $\lvert K_1 \cap K_2 \cap K_3\rvert = 2$	0

Here $\Gamma(s)$ denotes the set of all points that are parallel to the point s.

Table 6.1. Laguerre planes

i	*three points* s_1, s_2, s_3	1
ii	*two points* s_1, s_2*; one circle* K	
1	$s_1 \in K$	1
2	s_1, s_2 in the same connected component of $S \setminus K$	2
3	s_1, s_2 in different connected components of $S \setminus K$	0
iii	*one point* s*; two circles* K_1, K_2	
1	$K_1 \cap K_2 = \{s\}$	∞
2	$s \in K_1 \setminus K_2$	2
3	$K_1 \cap K_2 = \{x\} \neq \{s\}$; s in the connected component of $S \setminus (K_1 \cup K_2)$ bounded by $K_1 \cup K_2$	3
4	$K_1 \cap K_2 = \{x\} \neq \{s\}$; s not in the connected component of $S \setminus (K_1 \cup K_2)$ bounded by $K_1 \cup K_2$	1
5	$\|K_1 \cap K_2\| = 2$; $s \in K_1 \cap K_2$	0
6	$\|K_1 \cap K_2\| = 2$; $s \notin K_1 \cup K_2$	2
7	$K_1 \cap K_2 = \emptyset$; $s \notin K_i$; s in the connected component of $S \setminus (K_1 \cup K_2)$ bounded by $K_1 \cup K_2$	4
8	$K_1 \cap K_2 = \emptyset$; $s \notin K_i$; s not in the connected component of $S \setminus (K_1 \cup K_2)$ bounded by $K_1 \cup K_2$	0
iv	*three circles* K_1, K_2, K_3	
1	pairwise disjoint; K_i separates K_j and K_l for all $\{i, j, l\} = \{1, 2, 3\}$	8
2	pairwise disjoint; K_1 separates K_2 and K_3	0
3	$\|K_1 \cap K_2 \cap K_3\| = 2$	0
4	$\|K_1 \cap K_2 \cap K_3\| = 1 = \|K_i \cap K_j\|, i \neq j$	∞
5	$\|K_1 \cap K_2 \cap K_3\| = 1 = \|K_1 \cap K_3\|$; $\|K_1 \cap K_2\| = \|K_2 \cap K_3\| = 2$	2
6	$\|K_1 \cap K_2 \cap K_3\| = 1$; $\|K_i \cap K_j\| = 2$, $i \neq j$	4
7	$K_1 \cap K_2 \cap K_3 = \emptyset$; $\|K_i \cap K_j\| = 2$, $i \neq j$	8
8	$K_1 \cap K_2 \cap K_3 = \emptyset$; $\|K_1 \cap K_2\| = \|K_2 \cap K_3\| = 2$; $\|K_1 \cap K_3\| = 1$	6
9	$K_1 \cap K_2 \cap K_3 = \emptyset$; $\|K_1 \cap K_2\| = 2$; $\|K_1 \cap K_3\| = \|K_2 \cap K_3\| = 1$	5
10	$K_1 \cap K_2 \cap K_3 = \emptyset$; $\|K_i \cap K_j\| = 1$, $i \neq j$	5
11	$K_1 \cap K_3 = \emptyset = K_1 \cap K_3$; $\|K_1 \cap K_2\| = 1$; K_2 separates K_1 and K_3	2
12	$K_1 \cap K_3 = \emptyset = K_1 \cap K_3$; $\|K_1 \cap K_2\| = 1$; K_2 does not separate K_1 and K_3	6
13	$K_1 \cap K_3 = \emptyset$; $\|K_1 \cap K_2\| = 2$	4
14	$K_1 \cap K_3 = \emptyset$; $\|K_1 \cap K_2\| = \|K_2 \cap K_3\| = 1$; K_2 does not separate K_1 and K_3	5
15	$K_1 \cap K_3 = \emptyset$; $\|K_1 \cap K_2\| = \|K_2 \cap K_3\| = 1$; K_2 separates K_1 and K_3	3

Table 6.2. Möbius planes

i	*two parallel points plus one element of* $S \cup \mathcal{C}$ or two circles *contained in different halves of* $\mathcal{K}$ *plus one element of* $S \cup \mathcal{C}$	0
ii	*three pairwise nonparallel points*	1
iii	*two nonparallel points* s_1, s_2; *one circle* K	
1	$s_1 \in K$	1
2	$\{\delta(s_1,s_2),\delta(s_2,s_1)\} \cap K = \emptyset$; $\Gamma(s_1) \cup K$ does not separate s_2 and $\delta(\gamma_-(s_1,K),\gamma_+(s_1,K))$.	2
3	$\{\delta(s_1,s_2),\delta(s_2,s_1)\} \cap K = \emptyset$; $\Gamma(s_1) \cup K$ separates s_2 and $\delta(\gamma_-(s_1,K),\gamma_+(s_1,K))$.	0
4	$\delta(s_1,s_2) \in K$; $\delta(s_2,s_1) \notin K$	1
5	$\{\delta(s_1,s_2),\delta(s_2,s_1)\} \subset K$	0
iv	*one point* s; *two circles* K_1, K_2 *contained in the same half of* $\mathcal{K}$	
1	$K_1 \cap K_2 = \{s\}$	∞
2	$K_1 \cap K_2 = \{x\}$; $s \notin \Gamma(x)$	1
3	$K_1 \cap K_2 = \{x\}$; $s \in \Gamma(x) \setminus \{x\}$	0
4	$\|K_1 \cap K_2\| = 2$; $s \in K_1 \cap K_2$	0
5	$\|K_1 \cap K_2\| = 2$; $\|K_1 \cap K_2 \cap \Gamma(s)\| = 2$	0
6	$\|K_1 \cap K_2\| = 2$; $\|K_1 \cap K_2 \cap \Gamma(s) \setminus \{s\}\| = 1$	2
7	$\|K_1 \cap K_2\| = 2$; $\|K_1 \cap K_2 \cap \Gamma(s)\| = 0$; s in the component of $S \setminus (K_1 \cup K_2)$ containing points parallel to $K_1 \cap K_2$	4
8	$\|K_1 \cap K_2\| = 2$; $\|K_1 \cap K_2 \cap \Gamma(s)\| = 0$; $s \notin K_i$; s not in the component of $S \setminus (K_1 \cup K_2)$ containing points parallel to $K_1 \cap K_2$	0
9	$s \in K_1$; $s \notin K_2$	2
10	$K_1 \cap K_2 = \emptyset$; $s \notin K_i$	2
v	*three circles* K_1, K_2, K_3 *contained in the same half of* $\mathcal{K}$	
1	pairwise disjoint	0
2	$\|K_1 \cap K_2 \cap K_3\| = 2$	0
3	$\|K_1 \cap K_2 \cap K_3\| = 1 = \|K_i \cap K_j\|, i \neq j$	∞
4	$\|K_1 \cap K_2 \cap K_3\| = 1 = \|K_1 \cap K_3\|$; $\|K_1 \cap K_3\| = \|K_1 \cap K_3\| = 2$	2
5	$\|K_1 \cap K_2 \cap K_3\| = 1$; $\|K_i \cap K_j\| = 2$, $i \neq j$	4
6	$K_1 \cap K_2 \cap K_3 = \emptyset$; $\|K_i \cap K_j\| = 2$, $i \neq j$; K_1 meets K_3 in the same connected component of $K_3 \setminus \Gamma(s)$ as K_2, where $s \in K_1 \cap K_2$	8
7	$K_1 \cap K_2 \cap K_3 = \emptyset$; $\|K_i \cap K_j\| = 2$, $i \neq j$; K_1 meets K_3 not in the same connected component of $K_3 \setminus \Gamma(s)$ as K_2, where $s \in K_1 \cap K_2$	0
8	$K_1 \cap K_2 \cap K_3 = \emptyset$; $\|K_1 \cap K_2\| = \|K_2 \cap K_3\| = 2$; $\|K_1 \cap K_3\| = 1$; K_1 meets K_3 in the same connected component of $K_3 \setminus \Gamma(s)$ as K_2, where $s \in K_1 \cap K_2$	6

The following abbreviations are used in this table: Let r, s be two points. Then $\Gamma(s)$ is the set of all points that are parallel to s and $\delta(s,t)$ is the point that is $(+)$-parallel to s and $(-)$-parallel to t.

Table 6.3. Minkowski planes (continued on next page)

9	$K_1 \cap K_2 \cap K_3 = \emptyset$; $\lvert K_1 \cap K_2 \rvert = \lvert K_2 \cap K_3 \rvert = 2$; $\lvert K_1 \cap K_3 \rvert = 1$; K_1 meets K_3 not in the same connected component of $K_3 \setminus \Gamma(s)$ as K_2, where $s \in K_1 \cap K_2$	2
10	$K_1 \cap K_2 \cap K_3 = \emptyset$; $\lvert K_1 \cap K_2 \rvert = 2$; $\lvert K_1 \cap K_3 \rvert = \lvert K_2 \cap K_3 \rvert = 1$; K_1 meets K_3 in the same connected component of $K_3 \setminus \Gamma(s)$ as K_2, where $s \in K_1 \cap K_2$	5
11	$K_1 \cap K_2 \cap K_3 = \emptyset$; $\lvert K_1 \cap K_2 \rvert = 2$; $\lvert K_1 \cap K_3 \rvert = \lvert K_2 \cap K_3 \rvert = 1$; K_1 meets K_3 not in the same connected component of $K_3 \setminus \Gamma(s)$ as K_2, where $s \in K_1 \cap K_2$	3
12	$K_1 \cap K_2 \cap K_3 = \emptyset$; $\lvert K_i \cap K_j \rvert = 1$, $i \neq j$	3
13	$K_1 \cap K_2 = \emptyset$; $\lvert K_1 \cap K_2 \rvert = 2$	4
14	$K_1 \cap K_2 = \emptyset$; $\lvert K_1 \cap K_2 \rvert = \lvert K_2 \cap K_3 \rvert = 1$	3

Table 6.3. Minkowski planes (continued from previous page)

7
Tubular Circle Planes

Many of the different types of geometries we are concentrating on in this book have representations as n-unisolvent sets, that is, sets of continuous functions that solve one of the Lagrange interpolation problems. For example, the Euclidean plane corresponds, in the obvious way, to the set of all linear functions over the reals that solves the Lagrange interpolation problem of order 2. Also, the tubular circle planes of rank n correspond to sets of continuous periodic or half-periodic functions that solve the Lagrange interpolation problem of order n.

In this chapter we summarize many important results about interpolating sets of functions and their corresponding geometries following the exposition in Polster [1998d], [1998f], and Polster–Steinke [20XXd]. We find that many of the results that we encountered in the previous chapters have counterparts in this very general setting. However, many more of these counterparts are still waiting to be proved.

The results in this chapter form part of the topological foundation of the theory of approximation and interpolation. There are two properties that make n-unisolvent sets important for classical interpolation and approximation theory.

- An n-unisolvent set on an interval I solves the Lagrange interpolation problem of order n.
- Given an n-unisolvent set F on a closed interval I and a continuous function $g : I \to \mathbf{R}$, there is a unique 'best approximant' to g in F. This means that there is a unique function $f \in F$ for which

  $$\max_{x \in I} |f(x) - g(x)|$$

  is minimal.

For more background information about n-unisolvent sets the reader is referred to the papers and books by Curtis [1959], Kemperman [1969], Morozov [1952], Nemeth [1969], Rice [1964], [1969], Tornheim [1950]. We recommend the books by Karlin and Studden [1966], Schumaker [1981], and Zielke [1973], [1975], [1979] for information about Chebyshev spaces, that is, linear n-unisolvent sets. Also worth a look is the excellent book by Braess [1980] about nonlinear approximation theory.

7.1 Unisolvent Sets of Functions

Let $n \geq 1$ be an integer, let $I \subseteq \mathbf{R}$ be an interval with nonempty interior, and let F be a set of continuous functions $I \to \mathbf{R}$. Then F is called an *n-unisolvent set on I* if, for any $x_1 < x_2 < \cdots < x_n \in I$ and any set of n real numbers $y_1, y_2, \ldots, y_n$, there is a unique $f \in F$ such that

$$f(x_i) = y_i, \; i = 1, \ldots, n.$$

Note that two distinct elements of F can never be equal at more than $n-1$ distinct points. We will also say that f *interpolates* the points $(x_i, y_i) \in I \times \mathbf{R}, i = 1, \ldots, n$. Also, another way of expressing that F is n-unisolvent is to say that it solves the *Lagrange interpolation problem of order n*.

Let I be the closed interval $[a, b]$ and let F be n-unisolvent when restricted to the half-open interval $[a, b)$. Then F is called a *periodic n-unisolvent set* if all its elements are periodic functions on I, that is, $f(a) = f(b)$ for all $f \in F$. Similarly, F is called a *half-periodic n-unisolvent set* if all its elements are half-periodic functions, that is, if $f(a) = -f(b)$ for all $f \in F$. It turns out that periodic n-unisolvent sets exist if and only if n is odd and half-periodic n-unisolvent sets exist if and only if n is even; see Theorem 7.1.3 below. In view of this result and the fact that periodic and half-periodic unisolvent sets behave in a very similar fashion, we will often jointly refer to them as *n-phunisolvent sets*. Depending on n, it will be clear whether we have a periodic or a half-periodic n-unisolvent set in mind. Whenever we talk about *n-(ph)unisolvent* sets, we are referring to all three kinds of unisolvent sets. If we do not specify which interval an n-(ph)unisolvent set lives on, this interval will be the whole of $\mathbf{R}$ or $[-\frac{\pi}{2}, \frac{\pi}{2}]$ if F is n-unisolvent or n-phunisolvent, respectively. In general, when dealing with an n-phunisolvent set on $[a, b]$, we consider a and b as being identified.

The above unique interpolating function f will be denoted by

$$F(x_1, x_2, \ldots, x_n, y_1, y_2, \ldots, y_n).$$

We further write

$$F(x_1, x_2, \ldots, x_n; g) = F(x_1, x_2, \ldots, x_n, g(x_1), g(x_2), \ldots, g(x_n)),$$

where g is a function $I \to \mathbf{R}$. If F is n-phunisolvent on $[a, b]$, then, since a and b are identified, we may require $x_n \neq b$ in both definitions.

Let J be a subinterval of I, and let g be a function defined on I. Then g_J will denote the restriction of g to J and F_J will denote the set of all restrictions of functions in F to J.

7.1.1 Fibrated Circle Planes

Let F be an n-(ph)unisolvent set on the interval I. The circle plane associated with F has point set $P = I \times \mathbf{R}$. Furthermore, if F is periodic, then the left and right boundaries of P get identified by the map

$$\{a\} \times \mathbf{R} \to \{b\} \times \mathbf{R} : (a, y) \mapsto (b, y).$$

Similarly, if F is half-periodic, then the left and right boundaries of P get identified by the map

$$\{a\} \times \mathbf{R} \to \{b\} \times \mathbf{R} : (a, y) \mapsto (b, -y).$$

The circles are the graphs of the functions in F. Parallelism of points defines an equivalence relation whose parallel classes are the verticals contained in the point sets of the geometries. All this means that the geometry associated with an n-phunisolvent set is a *tubular circle plane* of rank n (see Subsection 1.2.2) and the geometry associated with an n-unisolvent set lives on a vertical strip with all circles being homeomorphic to I; we call such a plane a *strip circle plane.* We also introduce *fibrated circle planes of rank n* as a common name for the three different kinds of circle planes associated with the three different kinds of n-(ph)unisolvent sets.

Note that a strip circle plane corresponding to a 2-unisolvent set over an open interval turns into an $\mathbf{R}^2$-plane if we add the verticals to its line set. Similarly, the tubular circle plane that corresponds to a half-periodic 2-unisolvent set turns into a point Möbius strip plane if we add the verticals to its line/circle set.

Two n-(ph)unisolvent sets are called (topologically) isomorphic if their associated circle planes are (topologically) isomorphic. Note also that in an n-phunisolvent set on $[a, b]$ the point which corresponds to a and b is not really distinguished among the points of the interval, that is, we can always 'rotate' the n-unisolvent set to arrive at a topologically isomorphic n-phunisolvent set on $[a,b]$ in which the corresponding point is to be found somewhere in the interior of the interval.

The results in this chapter can all be expressed in both an incidence- and an interpolation-theoretic setting. Since it is no problem to switch back and forth between the two, we will usually state a result only in the setting in which it is most easily expressed and proved.

7.1.2 Models of the Classical Tubular Circle Planes

In the following we give a description of the classical tubular circle plane of rank n, $n > 1$, as the geometry of nontrivial hyperplane sections of a cone in $\mathrm{PG}(n, \mathbf{R})$ over a normal rational curve in $\mathrm{PG}(n-1, \mathbf{R})$. This generalizes the construction of the classical flat Laguerre plane as the geometry of nontrivial plane sections of a quadratic cone in $\mathrm{PG}(3, \mathbf{R})$; see Chapter 5. We also establish an isomorphism between this model of the classical tubular circle plane of rank n and the polynomial geometry $\overline{\mathrm{Poly}}(n, \mathbf{R})$ as we defined it in Subsection 1.1.4. The classical n-unisolvent set gives rise to the polynomial geometry while the classical n-phunisolvent set corresponds to the geometry of nontrivial hyperplane sections.

The classical tubular circle plane of rank 1 has a vertical cylinder in $\mathbf{R}^3$ as its point set and all the horizontal plane sections of this cylinder as its circle set. Clearly, this geometry is isomorphic to $\overline{\mathrm{Poly}}(1, \mathbf{R})$.

7.1.2.1 The Geometry of Nontrivial Hyperplane Sections

Let c_n be the normal rational curve in $\mathrm{PG}(n-1, \mathbf{R})$, $n \geq 2$ given by

$$\begin{aligned} c_n &= \{(s^{n-1} : s^{n-2}t : \cdots : st^{n-2} : t^{n-1}) \in \mathrm{PG}(n-1, \mathbf{R}) \mid \\ &\qquad s, t \in \mathbf{R}, (s,t) \neq (0,0)\}. \end{aligned}$$

The curve c_n has the property that any n mutually distinct points on it span the entire projective space $\mathrm{PG}(n-1, \mathbf{R})$. We embed $\mathrm{PG}(n-1, \mathbf{R})$ in $\mathrm{PG}(n, \mathbf{R})$ as the hyperplane $z_n = 0$ and form the cone C_n over c_n with vertex $v = (0 : \cdots : 0 : 1) \in \mathrm{PG}(n, \mathbf{R})$, that is,

$$\begin{aligned} C_n &= \{(s^{n-1} : s^{n-2}t : \cdots : st^{n-2} : t^{n-1} : z) \in \mathrm{PG}(n, \mathbf{R}) \mid \\ &\qquad s, t, z \in \mathbf{R}, (s,t) \neq (0,0)\}. \end{aligned}$$

A hyperplane intersects C_n in either v, at most n lines through v, or a rational curve. The latter occurs if and only if the hyperplane does not pass through v. We call the intersection of a hyperplane with the cone *nontrivial* if it intersects the cone in such a curve. Then a model of the classical tubular circle plane of rank n has as point set the set $C_n \setminus \{v\}$, as circles the nontrivial hyperplane intersections of the cone, and as parallel classes the lines on C_n through v that have been punctured at v.

It readily follows that every point on c_n can be obtained for $s = \cos x$ and $t = \sin x$, where $x \in [-\frac{\pi}{2}, \frac{\pi}{2}]$. Note that $x = \frac{\pi}{2}$ and $x = -\frac{\pi}{2}$ yield the same point $(0 : \cdots : 0 : 1) \in c_n$. We can therefore represent $C_n \setminus \{v\}$ on the strip $S = [-\frac{\pi}{2}, \frac{\pi}{2}] \times \mathbf{R}$ where a boundary point $(-\frac{\pi}{2}, y)$ becomes identified with $(\frac{\pi}{2}, (-1)^{n-1}y)$. The map from S to C_n given by

$$(x, y) \mapsto (\cos^{n-1} x : \cos^{n-2} x \sin x : \cdots : \sin^{n-1} x : y)$$

defines a bijection.

The intersection of C_n with a hyperplane not passing through v given by $z_n = a_0 z_0 + a_1 z_1 + \cdots + a_{n-1} z_{n-1}$ then corresponds on S to the curve that is defined by the equation

$$y = a_0 \cos^{n-1} x + a_1 \cos^{n-2} x \sin x + \cdots + a_{n-1} \sin^{n-1} x.$$

Hence circles are represented by linear combinations of the functions

$$\cos^{n-1} x, \cos^{n-2} x \sin x, \ldots, \sin^{n-1} x.$$

Note that for $n = 2$ the normal rational curve c_2 is just a projective line and the cone C_2 becomes the entire projective plane $\mathrm{PG}(2, \mathbf{R})$. Clearly, $C_2 \setminus \{v\}$ is a Möbius strip and the resulting geometry is essentially the classical point Möbius strip plane minus the verticals; see Subsection 2.1.3. For $n = 3$ the normal rational curve c_3 describes a conic in $\mathrm{PG}(2, \mathbf{R})$ and we obtain a model of the classical flat Laguerre plane as in Subsection 5.1.1.

7.1.2.2 The Polynomial Geometry

The following map defines an isomorphism between the geometry of nontrivial hyperplane sections described above and the geometry $\overline{\mathrm{Poly}}(n, \mathbf{R})$ as we defined it in Subsection 1.1.4.

$$\begin{aligned} \left(-\frac{\pi}{2}, \frac{\pi}{2}\right] \times \mathbf{R} \quad &\to \quad (\mathbf{R} \cup \{\infty\}) \times \mathbf{R} : \\ (x, y) \quad &\mapsto \quad \begin{cases} \left(\tan x, \dfrac{y}{\cos^{n-1} x}\right) & \text{for } x \in (-\frac{\pi}{2}, \frac{\pi}{2}), \\ (\infty, y) & \text{for } x = \frac{\pi}{2}. \end{cases} \end{aligned}$$

Here the circle that corresponds to the function $[-\frac{\pi}{2}, \frac{\pi}{2}] \mapsto \mathbf{R}$ given by

$$x \to a_0 \cos^{n-1} x + a_1 \cos^{n-2} x \sin x + \cdots + a_{n-1} \sin^{n-1} x$$

gets mapped to the circle in $\overline{\mathrm{Poly}}(n, \mathbf{R})$ that corresponds to the function

$$\mathbf{R} \to \mathbf{R} : t \mapsto a_0 + a_1 t + \cdots + a_{n-1} t^{n-1}.$$

This means that $\mathrm{Poly}(n, \mathbf{R})$ is the 'affine part' of the classical tubular circle plane of rank n in the same sense as the geometry of nonvertical Euclidean lines and Euclidean parabolas with vertical symmetry axes is the affine part of the classical flat Laguerre plane. As in the case of the classical flat Laguerre plane this correspondence can also be established via a 'stereographic projection' whenever $n \geq 3$.

Let Z be the $(n-3)$-dimensional subspace

$$\begin{aligned} Z &= \{(0 : 0 : z_2 : \cdots : z_{n-1} : 0) \in \mathrm{PG}(n, \mathbf{R}) \mid \\ &\qquad z_2, \ldots, z_{n-1} \in \mathbf{R}, (z_2, \ldots, z_{n-1}) \neq (0, \ldots, 0)\}. \end{aligned}$$

For $n \geq 3$ this subspace of $\mathrm{PG}(n, \mathbf{R})$ intersects the cone C_n only in the point $p = (0 : \cdots : 0 : 1 : 0) \in c_n$. We now project C_n from Z onto the plane

$$\begin{aligned} E &= \{(z_0 : z_1 : 0 : \cdots : 0 : z_n) \in \mathrm{PG}(n, \mathbf{R}) \mid \\ &\qquad z_0, z_1, z_n \in \mathbf{R}, (z_0, z_1, z_n) \neq (0, 0, 0)\}. \end{aligned}$$

We identify E with $\mathrm{PG}(2, \mathbf{R})$ via the natural map

$$(z_0 : z_1 : 0 : \cdots : 0 : z_n) \mapsto (z_0 : z_1 : z_n).$$

Under the above 'stereographic' projection from Z points on the line L in C_n through v and p have the same image, but all other points of the affine part $C_n \setminus L$ have exactly one image in E. Furthermore, all points of the plane E occur as images of points in $C_n \setminus L$ except for the points on the projective line $z_0 = 0$. We thus have, in fact, a bijection $C_n \setminus L \to \mathbf{R}^2$ given by

$$(s^{n-1} : s^{n-2} t : \cdots : s t^{n-2} : t^{n-1} : z) \mapsto \left(\frac{t}{s}, \frac{z}{s^n}\right).$$

Since for points of $C_n \setminus L$ we can assume that $s = 1$, the above map becomes $(1, t, \ldots, t^{n-1}, z) \mapsto (t, z)$. Clearly, a line of C_n is taken to a vertical of $\mathbf{R}^2$. Furthermore, the intersection of C_n with the hyperplane

$$H = \{(z_0 : z_1 : \cdots : z_n) \in \mathrm{PG}(n, \mathbf{R}) \mid a_0 z_0 + a_1 z_1 + \cdots + a_n z_n = 0\}$$

not passing through the vertex (that is, $a_n \neq 0$) is taken to

$$\left\{\left(t, -\frac{1}{a_n}(a_0 + a_1 t + \cdots + a_{n-1} t^{n-1})\right) \,\middle|\, t \in \mathbf{R}\right\}.$$

This is a circle of $\mathrm{Poly}(n, \mathbf{R})$ and, clearly, all such circles arise in this manner. Since the group of collineations of $\mathrm{PG}(n, \mathbf{R})$ that globally fix the cone C_n acts transitively on the points not the vertex, this shows again that the geometry $\mathrm{Poly}(n, \mathbf{R})$ 'is' the affine part of the classical tubular circle plane of rank n.

7.1.2.3 Examples of (Ph)unisolvent Sets

The two models of the classical tubular circle plane of rank n correspond to first examples of n-(ph)unisolvent sets. The geometry $\mathrm{Poly}(n, \mathbf{R})$ corresponds to the set of all polynomials of degree at most $n-1$

$$\mathrm{span}\{1, x, x^2, \ldots, x^{n-1}\}.$$

This set is an n-unisolvent set on any interval.

The geometry of nontrivial hyperplane sections corresponds to the n-phunisolvent set

$$\mathrm{span}\{\cos^{n-1} x, \cos^{n-2} x \sin x, \ldots, \sin^{n-1} x\}.$$

These first examples of n-(ph)unisolvent sets are n-dimensional real vector spaces. Any n-dimensional real vector space which is an n-(ph)unisolvent set is called a *Chebyshev space* and every basis for such a space a *Chebyshev system*. Chebyshev spaces give rise to fibrated circle planes that are the counterpart of the ovoidal flat Laguerre planes in the following sense. Given a Chebyshev system

$$\{f_0, f_1, f_2, \ldots, f_{n-1}\}$$

on the interval I, we can replace the normal rational curve in the model of nontrivial hyperplane sections by the curve

$$\{(f_0(t) : f_1(t) : f_2(t) : \cdots : f_{n-1}(t)) \in \mathrm{PG}(n-1, \mathbf{R}) \mid t \in I\}$$

to arrive at a model of the fibrated circle plane that corresponds to the Chebyshev system.

7.1.3 Basic Properties

In the following we prove a number of basic results about n-(ph)unisolvent sets. We start with three obvious methods for constructing n-unisolvent sets. In incidence-geometric terms these methods translate into various ways in which fibrated circle planes occur as subgeometries of other fibrated circle planes. Also, these constructions can be considered as extensions of the derived geometry construction.

THEOREM 7.1.1 (Basic Constructions) *Let F be an n-unisolvent set on the interval I. Then the following hold.*

(i) *If J is a subinterval of I, then F_J is an n-unisolvent set on J.*

(ii) *Let $(x_i, y_i) \in I \times \mathbf{R}$, $i = 0, 1, \ldots, m$, $m \leq n-1$, be pairwise nonparallel points. Let J be a subinterval of I that does not contain any of the x_is and let $G = \{f \mid f \in F, f(x_i) = y_i, i = 1, 2, \ldots, m\}$. Then G_J is an $(n-m)$-unisolvent set on J.*

Let F be an n-phunisolvent set on $[a, b]$ and let $y_0 \in \mathbf{R}$. Moreover, let $G = \{f \mid f \in F, f(b) = y_0\}$. Then $G_{(a,b)}$ is an $(n-1)$-unisolvent set on the interval (a, b).

We say that a continuous function $f : I \to \mathbf{R}$ defined on an interval I *changes sign* at one of its zeros s if either

- the zero s is one of the boundary points of I and there is a neighbourhood $U \subseteq I$ of s such that $f(x) \neq 0$ for $x \in U \setminus \{s\}$, or
- the zero s is contained in the interior of I and there is a neighbourhood $U \subseteq I$ of s such that $f(t)f(u) < 0$ for $t, u \in U$, $t < s < u$.

Let f be (half)periodic on $[a, b]$. Then a is a zero of f if and only if b is. If f has a finite number of zeros including a, to change sign at a means that f has opposite sign or the same sign 'close' to a and b, depending on whether f is periodic or halfperiodic. As an element of a phunisolvent set on $[a, b]$ we count a pair of zeros of f at a and b as one zero. Zeros at which a function changes sign are called *nodal* zeros.

THEOREM 7.1.2 (Intersection) *Let F be an n-(ph)unisolvent set on the interval I and let $f, g \in F$ be distinct. If f and g are equal in $n-1$ distinct points of I, then the function $f - g$ has exactly $n-1$ zeros and changes sign at every single one of these zeros.*

Of course, this just means that if two circles in a fibrated circle plane of rank n intersect in $n-1$ points, then they intersect only in these points and they intersect transversally in every single one of the points. We have come across counterparts of this result in previous chapters.

Proof of theorem. The first part of this result is an immediate consequence of the definition of unisolvent sets of functions and it therefore suffices to show that the function $f - g$ changes sign at its $n-1$ zeros.

We first prove this for n-unisolvent sets defined on an interval I. Let t be one of the zeros. We may assume that t is contained in the interior of I.

Let F be 2-unisolvent. Then t is the only zero of the function $f-g$. Assume that $f-g$ does not change sign at t. Then, without loss of generality, we may assume that $f(x) > g(x)$ for $x \in I \setminus \{t\}$. Choose two points $s, u \in I$ such that $s < t < u$. Then $h = F(s, u, f(s), g(u))$ is a function in F different from both f and g. However, it is easy to see that either $f-h$ or $g-h$ has at least two zeros. This is a contradiction, which proves our result for the case $n = 2$.

Assume that the theorem is true for $(n-1)$-unisolvent sets on I, $n > 2$, and let F be n-unisolvent. Then $f-g$ has $n-1$ zeros. Let J be a subinterval of I that contains t in its interior and contains all the other zeros excepting one; let us call it s. Let G be the restriction of all functions in F which interpolate the point $(s, f(s))$ to the subinterval J. By Theorem 7.1.1, G is an $(n-1)$-unisolvent set that contains the restrictions of f and g to J. Hence, $f-g$ changes sign at t.

The proof for n-phunisolvent sets on $[a, b]$ is based on the fact that we can argue over open subsets of this interval. Whenever in an argument we are dealing with a point on the boundary, we can shift our point of view by a 'rotation' to an n-phunisolvent set that is topologically isomorphic to the set we started out with and in which these boundary points are moved into the interior of the interval. □

THEOREM 7.1.3 (Existence) *Periodic n-unisolvent sets exist if and only if n is odd. Half-periodic n-unisolvent sets exist if and only if n is even.*

See also Curtis [1959] Corollary on p. 1016.

Proof of theorem. We know that periodic n-unisolvent sets exist for all odd n; see Subsection 7.1.2.3. Let F be a periodic n-unisolvent set on $[a, b]$. Choose two distinct functions f and g such that $f-g$ has $n-1$ zeros in (a, b). Since both functions are periodic, this difference takes on the same value at the interval ends. On the other hand, by the previous result, if n is even, this is not possible. This proves the first part of the theorem. The even case follows similarly. □

COROLLARY 7.1.4 (Maximal Orthogonal Arrays) *Every tubular circle plane of rank $n \geq 2$ is a maximal orthogonal array of rank n.*

Note that a tubular circle plane of rank 1 is not maximal; see Subsection 1.1.4 for a definition of the term maximal.

Proof of corollary. Let $\mathcal{C}$ be a tubular circle plane of odd rank $n > 2$. Assume that there exists an extension of $\mathcal{C}$ by a parallel class to a larger orthogonal array of rank n. Then the set of all circles in $\mathcal{C}$ that get extended by a certain fixed point in this parallel class is a periodic $(n-1)$-unisolvent set. This contradicts Theorem 7.1.3. The even case can be dealt with in a similar manner. $\square$

THEOREM 7.1.5 (Continuity) *Let F be an n-(ph)unisolvent set on the interval I and let $D \subset \mathbf{R}^{2n}$ be the set*

$$\{(x_1, x_2, \ldots, x_n, y_1, y_2, \ldots, y_n) \mid x_1 < x_2 < \cdots < x_n, x_i \in I, y_i \in \mathbf{R}\}.$$

Then the function $D \times I \to \mathbf{R}$ defined by

$$(x_1, x_2, \ldots, x_n, y_1, y_2, \ldots, y_n, x) \mapsto F(x_1, x_2, \ldots, x_n, y_1, y_2, \ldots, y_n)(x)$$

is continuous.

This result corresponds to the continuity results for the geometries on surfaces that we considered before. For detailed proofs of the n-unisolvent part of this result see Tornheim [1950] Theorem 5 and for the periodic n-unisolvent part Curtis [1959] Theorem 1. The half-periodic n-unisolvent part can be dealt with like the periodic n-unisolvent part.

Let F be an n-(ph)unisolvent set on I. Fix n points $t_1 < t_2 < \cdots < t_n$ in the interval I and identify the elements of F with the points of $\mathbf{R}^n$ via the function

$$F \to \mathbf{R}^n : f \mapsto (f(t_1), f(t_2), \ldots, f(t_n)).$$

The corresponding topology on F is its *natural topology* as it is the finest topology on F with respect to which the map that assigns n pairwise nonparallel points in $I \times \mathbf{R}$ their interpolating function in F is continuous. In particular, the natural topology is independent of the choice of the points t_i.

THEOREM 7.1.6 (Spaces of Functions and Continuity) *Let F be an n-(ph)unisolvent set on an interval I and let D be defined as in Theorem 7.1.5. Then F, equipped with its natural topology, is a topological space homeomorphic to $\mathbf{R}^n$. The set of functions interpolating $k < n$ mutually nonparallel points in $I \times \mathbf{R}$ is homeomorphic to $\mathbf{R}^{n-k}$.*

Furthermore, the interpolating map

$$D \to F : (x_1, \ldots, x_n, y_1, \ldots, y_n) \mapsto F(x_1, \ldots, x_n, y_1, \ldots, y_n)$$

is continuous.

Of course, this translates into the fact that, equipped with its natural topology, the circle space of a fibrated circle plane of rank n is homeomorphic to $\mathbf{R}^n$ and that the operation of joining n pairwise nonparallel points is continuous. Furthermore, the flag space of a tubular circle plane can be described as follows.

PROPOSITION 7.1.7 (Flag Space) *The flag space $\mathcal{F}$ of a tubular circle plane of rank n is a connected manifold of dimension $n+1$. More precisely, $\mathcal{F}$ is homeomorphic to $\mathbf{S}^1 \times \mathbf{R}^n$.*

THEOREM 7.1.8 (More Constructions) *Let F be an n-unisolvent set, $n \geq 2$, on the interval $[a,b]$. Then $F' = \{f \in F \mid f(a) = (-1)^n f(b)\}$ is a (half)periodic $(n-1)$-unisolvent set on $[a,b]$ for (odd) even n.*

Proof. Let n be even. Let $(x_i, y_i) \in [a,b) \times \mathbf{R}$, $i = 1, 2, \ldots, n-1$, be pairwise nonparallel points. We show that there exists a uniquely determined function $f \in F'$ that interpolates these points. If the abscissa of one of the points is a, then this is clearly true. If this is not the case, let G be the set of all functions in F that interpolate these $n-1$ points. Let $f, g \in G, f \neq g$. Then, by definition, the function $f - g$ has $n-1$ distinct zeros. By Theorem 7.1.2, these are all the zeros of $f - g$ and $f - g$ changes sign exactly $n-1$ times. Since $n-1$ is odd, we conclude that $f(a) < g(a)$, $f(a) = g(a)$, or $f(a) > g(a)$ implies that $f(b) > g(b)$, $f(b) = g(b)$, or $f(b) < g(b)$, respectively. Let $f_t, t \in \mathbf{R}$, be the uniquely determined function in G with $f_t(a) = t$. Define a map $\lambda : \mathbf{R} \to \mathbf{R} : t \mapsto f_t(b)$. From Theorem 7.1.5 it is clear that this map is strictly decreasing and bijective, that is, it is a strictly decreasing homeomorphism. Every such homeomorphism has exactly one fixed point $t_0 \in \mathbf{R}$. Hence $f_{t_0}(b) = t_0 = f_{t_0}(a)$ and f_{t_0} is the function we have been looking for.

The case where n is odd is dealt with in a similar fashion. □

Note that if F in the above theorem is a Chebyshev space, then F' is also a Chebyshev space.

Given a periodic or half-periodic function on an interval $[a,b]$, we define the *open neighbourhoods* of the identified point $a = b$ to be the complements of proper closed subintervals of (a,b). A set of (periodic or half-periodic) continuous functions F on an interval I is called *locally n-(ph)unisolvent* if for every point $p \in I$ there exists an open neighbourhood J of p such that the restriction of F to J is n-unisolvent.

THEOREM 7.1.9 (Local Implies Global) *Let F be a set of periodic or half-periodic continuous functions or a set of continuous functions on an open interval. Then F is n-(ph)unisolvent if and only if it is locally n-unisolvent and the difference of two distinct functions in the set has at most $n-1$ zeros.*

For a proof of this result for n-unisolvent sets on open intervals see Hartman [1971] Theorem II 1.1. The respective result for periodic and half-periodic n-unisolvent sets is an easy corollary of Hartman's result. Also see the corresponding result for flat projective planes Theorem 2.2.8.

7.2 Nested (Ph)unisolvent Sets and Their Circle Planes

Certain n-(ph)unisolvent sets and their associated circle planes have a richer local structure than suggested by the Axiom of Joining. These special mathematical objects have been investigated in both incidence geometry and the theory of interpolation in the guise of flat Laguerre planes (only rank 3) and as sets of functions which solve the Hermite interpolation problem (all ranks), respectively.

We first have a closer look at these sets of functions and geometries before we introduce 'nested circle planes' that are the natural 'topological' generalization of both concepts.

7.2.1 Unrestricted (Ph)unisolvent Sets

A SIC (set of initial conditions) S of order $m \in \mathbf{N}$ on an interval I is an ordered triple (X, Λ, Y), where $X = \{x_1, x_2, \ldots, x_k\}$ is an ordered set of distinct points in I, $\Lambda = \{\lambda_1, \lambda_2, \ldots, \lambda_k\}$ is an ordered set of positive integers such that $\lambda_1 + \lambda_2 + \cdots + \lambda_k = m$, and $Y = \{Y_1, Y_2, \ldots, Y_k\}$ is an ordered set of ordered sets $Y_i = \{y_i^{(0)}, y_i^{(1)}, \ldots, y_i^{(\lambda_i - 1)}\}$, $i = 1, 2, \ldots, k$, of real numbers. We abbreviate all this by writing

$$S = \{x_1, \ldots, x_k \mid \lambda_1, \ldots, \lambda_k \mid y_1^{(0)}, \ldots, y_k^{(\lambda_k - 1)}\}.$$

An $m-1$ times continuously differentiable function $f : I \to \mathbf{R}$ *satisfies* S if

$$f^{(j)}(x_i) = y_i^{(j)},$$

for all $i = 1, 2, \ldots, k, j = 0, 1, \ldots, \lambda_i - 1$. Here $f^{(0)} = f$ and $f^{(j)}$, $j \geq 1$ denotes the jth derivative of f.

We call a set F of $n-1$ times continuously differentiable functions on I an *unrestricted n-unisolvent set* if every SIC of order n on I is satisfied by exactly one $f \in F$. A (half-)periodic function f on an interval $[a,b]$ is n times continuously differentiable if it is so at every point of the open interval (a,b) and if $f^{(i)}(a) = (-1)^{n+1} f^{(i)}(b)$ for $i = 0, 1, \ldots, n-1$ where $f^{(i)}(a)$ and $f^{(i)}(b)$ are the one-sided ith derivatives of f at a and b, respectively. A set F of $n-1$ times continuously differentiable (half-)periodic functions on $[a,b]$ is called a *(half-)periodic unrestricted n-unisolvent set* if the restriction of F to the half-open interval $[a,b)$ is an unrestricted n-unisolvent set.

Note that an unrestricted n-(ph)unisolvent set is automatically an n-(ph)unisolvent set since a function satisfying a SIC of the form

$$\{x_1, \ldots, x_n \mid 1, 1, \ldots, 1 \mid y_1^{(0)}, y_2^{(0)}, \ldots, y_n^{(0)}\}$$

interpolates the n points

$$(x_1, y_1^{(0)}), (x_2, y_2^{(0)}), \ldots, (x_n, y_n^{(0)}).$$

For F to be an unrestricted n-(ph)unisolvent set just means that it solves the *Hermite interpolation problem of order n*; see Lorentz–Jetter–Riemenschneider [1983].

The n-(ph)unisolvent sets in Subsubsection 7.1.2.3 are unrestricted n-(ph)unisolvent. The Chebyshev systems that give rise to unrestricted n-(ph)unisolvent sets of functions are called *extended Chebyshev sets* and their bases *extended Chebyshev systems.*

A set G of $n-1$ times continuously differentiable functions on the interval I is said to have *the property of unique n initial values* if all SICs of the form

$$\{x_1 \mid n \mid y_1^{(0)}, y_1^{(1)}, \ldots, y_1^{(n-1)}\}$$

on I are satisfied by exactly one $f \in G$.

THEOREM 7.2.1 (Unisolvent + Initial Values = Unrestricted) *Let F be a set of $n-1$ times continuously differentiable functions on an open interval or a set of $n-1$ times continuously differentiable periodic or half-periodic functions. Then F is an unrestricted n-(ph)unisolvent set if and only if it is an n-(ph)unisolvent set and it has the property of unique n initial values.*

In the case that F is unisolvent on an open interval this important result is due to Hartman [1958]. The second part of this result follows immediately from Hartman's result; see also Polster [1998d] Corollary 4.4.

7.2.2 Integrating Unisolvent Sets

Let F be an n-unisolvent set on I. We denote by $S(F)$ the set of all functions

$$I \to \mathbf{R} : x \mapsto \int_a^x f(t)dt + b,$$

where $a \in I, b \in \mathbf{R}, f \in F$.

We call $S(F)$ the *integral* of F. Note that if F is the set of all polynomials of degree at most $n-1$, then $S(F)$ is the set of all polynomials of degree at most n.

THEOREM 7.2.2 (Integral I) *Let F be an n-unisolvent set on the open or half-open interval I. Then $S(F)$ is an $(n+1)$-unisolvent set on I.*

For a proof of this result see Polster [1998c] Theorem 1 and Polster [1995c] Proposition 2.1. Clearly, this result also remains true if we are dealing with an n-unisolvent set F on a closed interval I as long as there are an n-unisolvent set $\overline{F}$ and an open or half-open interval $\overline{I}$ such that $I \subset \overline{I}$ and F is the restriction of $\overline{F}$ to I. See also Subsections 2.7.9 and 5.3.4 for important applications of this result.

THEOREM 7.2.3 (Integral II) *Let F be an unrestricted n-unisolvent set on the open interval I. Then $S(F)$ is an unrestricted $(n+1)$-unisolvent set on I.*

For a proof of this result see Polster [1998c] Theorem 2. As above we note that this result also stays true if we are dealing with an unrestricted n-unisolvent set F on a half-open or closed interval I as long as there are an unrestricted n-univolvent set $\overline{F}$ and an open interval $\overline{I}$ such that $I \subset \overline{I}$ and F is the restriction of $\overline{F}$ to I.

7.2.3 Nested Tubular Circle Planes

Remember that the derived geometry of a fibrated circle plane $\mathcal{C}$ at a point p consists of all the points of $\mathcal{C}$ not parallel to p and all circles of $\mathcal{C}$ through p that have been punctured at p. Furthermore its derived plane at p is the derived geometry to whose circle set all parallel classes not through p have been added. Now it is clear that the derived geometry of a cylindrical circle plane $\mathcal{C}$ (= tubular circle plane of rank 3) at a point is a strip circle plane of rank 2 and the derived plane at a point is an $\mathbf{R}^2$-plane. Remember that $\mathcal{C}$ is a flat Laguerre plane if and only if the derived plane at every point is a flat affine plane. This implies that any derived geometry of a flat Laguerre plane can be extended, in an essentially unique way, by a 'parallel class at infinity' such that nonvertical parallel lines in the corresponding affine plane get extended by a common point at infinity and such that the resulting geometry is topologically isomorphic to a tubular circle plane of rank 2. This means that associated with any point of a flat Laguerre plane is a tubular circle plane of rank 2; see Figure 7.1.

Now, let $\mathcal{C}$ be a tubular circle plane of rank 2, that is, a point Möbius strip plane from which the verticals have been removed. Then the derived geometry of $\mathcal{C}$ at a point is a strip circle plane of rank 1 on an open interval which can also be extended by a 'parallel class at infinity' such that the resulting geometry is a tubular circle plane of rank 1.

These considerations show that both flat Laguerre planes and rank 2 tubular circle planes are *nested structures*. In the following we will define what it means for a tubular circle plane of rank n to be 'nested'.

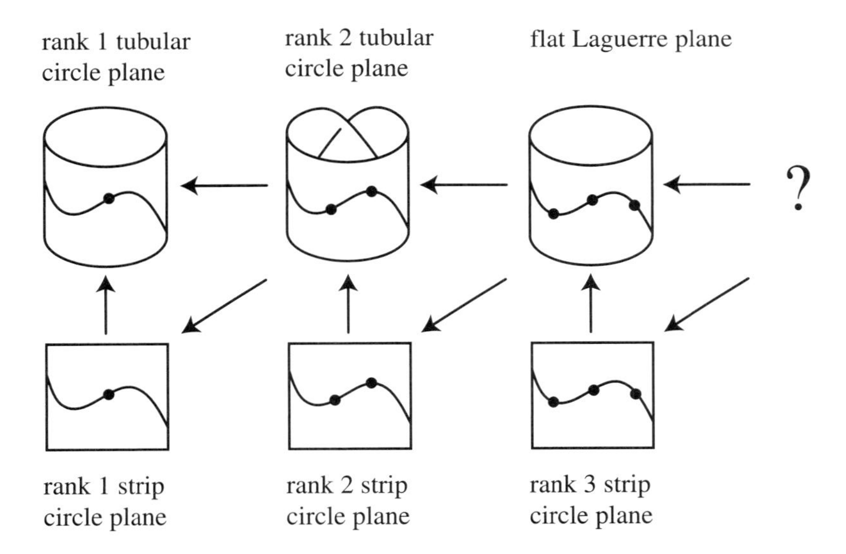

Fig. 7.1. Nested tubular circle planes

Let $\mathcal{C}$ be a tubular circle plane of rank n and let p be one of its points. Let $\mathcal{C}_p$ be the derived geometry of $\mathcal{C}$ at p and let $\mathcal{C}^p$ be the geometry whose point set is the same as that of $\mathcal{C}_p$ and whose lines are the circles of $\mathcal{C}$ that have been punctured at the vertical through p. We call $\mathcal{C}^p$ the *restricted geometry* of $\mathcal{C}$ at p. Note that in the case that $\mathcal{C}$ is a cylindrical circle plane, the restricted geometry is essentially one of the affine parts of this cylindrical circle plane.

THEOREM 7.2.4 (Unique Extension) *Let $\mathcal{C}_1$ and $\mathcal{C}_2$ be two tubular circle planes whose restricted geometries at the two points $p_1 \in \mathcal{C}_1$ and $p_2 \in \mathcal{C}_2$ are topologically isomorphic. Then the tubular circle planes themselves are topologically isomorphic. More precisely, every continuous isomorphism from the restricted geometry $\mathcal{C}_1^{p_1}$ to the restricted geometry $\mathcal{C}_2^{p_2}$ extends to a continuous isomorphism from $\mathcal{C}_1$ to $\mathcal{C}_2$.*

This result implies that if a strip circle plane can be extended to a tubular circle plane (by attaching one parallel class), then this can be done in an essentially unique way. For a proof of this result see Polster [1998d] Proposition 3.3.

A tubular circle plane $\mathcal{D}$ of rank $n-1$ is called a *derived tubular circle plane* of $\mathcal{C}$ at p if there exists a point $q \in \mathcal{D}$ such that $\mathcal{C}_p$ is topologically isomorphic to $\mathcal{D}^q$, that is, there is a homeomorphism of the point set of $\mathcal{C}_p$ to the point set of $\mathcal{D}^q$ that induces an isomorphism of the two geometries. If a derived tubular circle plane exists, then Theorem 7.2.4 guarantees that it is uniquely determined up to topological isomorphism. It therefore makes sense to speak of *the* derived tubular circle plane at a point.

THEOREM 7.2.5 (Derived Classical Plane Is Classical) *The derived tubular circle planes at all points of the classical tubular circle plane of rank $n > 1$ exist, are isomorphic, and are themselves classical.*

Proof. The derived geometry of the tubular circle plane $\overline{\mathrm{Poly}}(n, \mathbf{R})$ at the point $(\infty, 0)$ is $\mathrm{Poly}(n-1, \mathbf{R})$. Hence the derived tubular circle plane at this point exists and is classical. By Theorem 7.2.14 the automorphism group of $\overline{\mathrm{Poly}}(n, \mathbf{R})$ acts transitively on the point set. Hence the derived tubular circle planes at all points exist and are isomorphic. □

Nested Tubular Circle Planes

A tubular circle plane $\mathcal{C}$ of rank n is *nested* either if $n = 1$ or if the derived tubular circle planes at all its points exist and are nested themselves.

With this recursive definition we can summarize everything that we said at the beginning of this subsection.

THEOREM 7.2.6 (Nested Lower-Rank Circle Planes) *Every tubular circle plane of rank 2 is nested. A tubular circle plane of rank 3 is nested if and only if it is a flat Laguerre plane.*

Note that the 'only if' part follows from the fact that all derived planes of a cylindrical circle plane are flat affine planes only if the cylindrical circle plane is a flat Laguerre plane.

THEOREM 7.2.7 (Unrestricted Implies Nested) *Tubular circle planes that correspond to unrestricted n-phunisolvent sets are nested.*

Proof. We start with an unrestricted n-unisolvent set F on the open interval $I = (a, b)$ and let O be the strip circle plane associated with it. With every SIC $S = \{x_1, \cdots, x_k \mid \lambda_1, \ldots, \lambda_k \mid y_1^{(0)}, \ldots, y_k^{(\lambda_k - 1)}\}$ on I of order $m \leq n - 1$ we will associate a strip circle plane O_S of rank $n - m$ that has the same point set as O. Let us assume that $x_1 < x_2 < \cdots < x_k$.

[In the course of this construction it helps to keep one specific example in mind. Comments about this specific example will always be inserted in square brackets. In our example

$$n = 5, I = (0, 3), k = 2,$$
$$S = \{x_1 = 1, x_2 = 2 \mid \lambda_1 = 1, \lambda_2 = 2 \mid y_1^{(0)} = 1, y_2^{(0)} = 1, y_2^{(1)} = 0\}.$$

So in this case $m = \lambda_1 + \lambda_2 = 3$ and O_S will be a strip circle plane of rank 2.]

Let

$$F_S = \{f \in F \mid f^{(j)}(x_i) = y_i^{(j)}, i = 1, 2, \ldots, k, j = 0, 1, \ldots, \lambda_i - 1\},$$

that is, F_S is the set of all $f \in F$ that satisfy S. Let $I' = I \setminus \{x_1, \ldots, x_k\}$ and let F'_S be the restriction of F_S to I'. All graphs of functions in F_S pass through the k points $(x_1, y_1^{(0)}), (x_2, y_2^{(0)}), \ldots, (x_k, y_k^{(0)})$. Now we choose $\varepsilon_i > 0$, $i = 1, 2, \ldots, k$, such that the k intervals $I_i = [x_i - \varepsilon_i, x_i + \varepsilon_i]$ are disjoint and all contained in I.

[To illustrate the following transformations of the set F_S we concentrate on two typical functions $f, g \in F_S$, $f \neq g$. We choose f and g such that $f^{(1)}(1) = f^{(2)}(2) = 1 \neq -1 = g^{(1)}(1) = g^{(2)}(2)$. Then $f - g$ has a nodal zero at 1 and a nonnodal zero at 2; see Figure 7.2 (left).]

It is possible to construct a homeomorphism γ of $I' \times \mathbf{R}$ that leaves

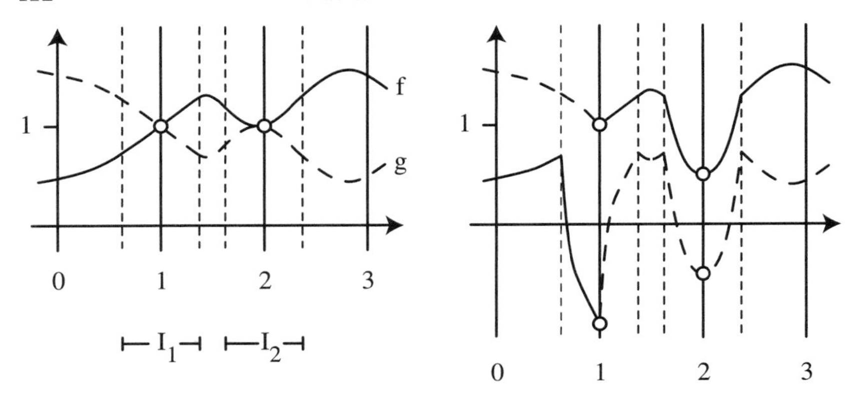

Fig. 7.2.

all verticals in this set globally fixed and if a vertical is not contained in one of the strips $I_i \times \mathbf{R}$, γ leaves this vertical pointwise fixed. Note that in between two adjacent strips there always exists such a vertical. This means that γ is orientation-preserving when restricted to the connected components of $I' \times \mathbf{R}$. In addition, we will choose γ such that the images under γ of the restrictions to $I' \times \mathbf{R}$ of the graphs of the functions in F_S are aligned in a special way. Aligned in this special way these graphs can then be easily extended and combined into the circles of the strip circle plane O_S that we are trying to construct.

For arbitrary δ_i with $0 < \delta_i < \varepsilon_i$, $i = 1, 2, \ldots, k$, the homeomorphism γ can be constructed in such a way that restricted to the strip

$$((x_i - \delta_i, x_i) \cup (x_i, x_i + \delta_i)) \times \mathbf{R}$$

it looks as follows.

$$\gamma(x, y) = \left(x, (\operatorname{sgn}(x - x_i))^{\lambda_i} \frac{y - \sum_{j=0}^{\lambda_i - 1} y_i^{(j)} \frac{(x-x_i)^j}{j!}}{\frac{(x-x_i)^{\lambda_i}}{\lambda_i!}} \right).$$

Note that the factor $(\operatorname{sgn}(x - x_i))^{\lambda_i}$ makes sure that this map is really orientation-preserving. Here sgn is the function $\mathbf{R} \to \mathbf{R}$ that maps the number $x \in \mathbf{R}$ to -1, 0, 1 if $x < 0$, $x = 0$, $x > 0$, respectively. Furthermore, for $f \in F'_S$ we find

$$\lim_{x \to x_i \pm 0} \gamma(x, f(x)) = (x_i, (\pm 1)^{\lambda_i} f^{(\lambda_i)}(x_i)).$$

[Let us have a look at what is happening in our example; see Figure 7.2 (right). We note that if we flip both the strip $(1, 2) \times \mathbf{R}$ and

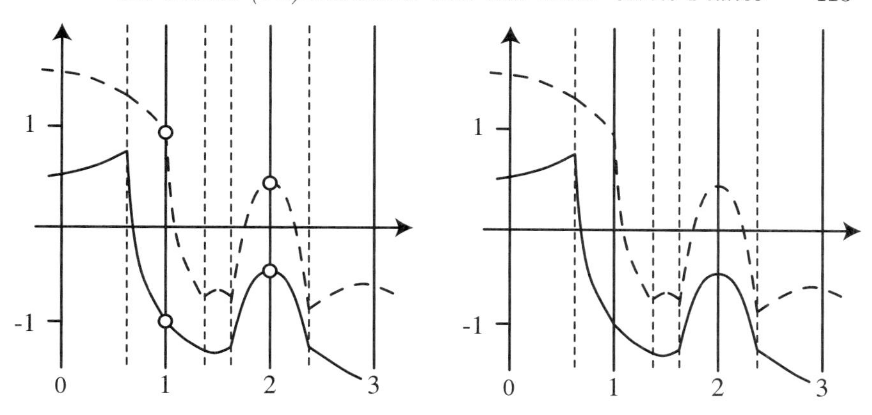

Fig. 7.3.

the strip $(2,3) \times \mathbf{R}$ about the x-axis, both the dotted line and the solid line can be completed to continuous curves by adding two points; see Figure 7.3.]

As in our example, the images of all graphs of functions in F_S under the map γ are still discontinuous. But by starting from the left and working towards the right, we can match up the ends on both sides of the verticals $\{x_i\} \times \mathbf{R}$ by flipping connected components of $I' \times \mathbf{R}$ about the x-axis whenever necessary. This flipping operation amounts to an involutory homeomorphism γ' of $I' \times \mathbf{R}$. Let $\mathrm{sig}_i = \prod_{j=1}^{i}(-1)^{\lambda_j}$ for $i = 1, 2, \ldots, k$ and let $x_{k+1} = b$. Then γ' looks as follows.

$$\gamma' : I' \times \mathbf{R} \to I' \times \mathbf{R} : (x, y) \mapsto \begin{cases} (x, y) & \text{for} \quad a < x < x_1, \\ (x, \mathrm{sig}_i y) & \text{for} \quad x_i < x < x_{i+1}. \end{cases}$$

The image of a graph of a function in F'_S under $\gamma'\gamma$ can be completed to a continuous curve by adding to it all its boundary points in $I \times \mathbf{R}$; see Figure 7.3 (right). Let L_S be the set of all curves that have been completed in this manner and let O_S be the strip circle plane with point set $I \times \mathbf{R}$ and line set L_S.

To show that O_S is a strip circle plane of rank $n - m$ it is necessary to prove that the set E of continuous functions $I \to \mathbf{R}$ that corresponds to O_S is $(n-m)$-unisolvent. Given a set of $n - m$ pairwise nonparallel points in $I \times \mathbf{R}$ the problem of finding a function in E that interpolates these points can be translated in a straightforward manner into the problem of finding a function in our original unrestricted n-unisolvent set F that satisfies a certain SIC of order n that 'extends' our original

SIC S. Since there is a unique solution to the latter problem, the same is true for the first problem, which shows that O_S is indeed $(n-m)$-unisolvent. For details see Polster [1998d] Proposition 4.7.

Let us go back, for the moment, to general nested tubular circle planes of rank n. Deriving at a point p, that is, constructing the derived tubular circle plane at p, means concentrating on the circles through the point p, removing the parallel class p is contained in, 'twisting' the strip $(-\frac{\pi}{2}, \frac{\pi}{2}) \times \mathbf{R}$, that is, flipping the strip on just one side, in order to move from cylinder to Möbius strip, or the other way around, and finally gluing in another parallel class to fit the left and right sides of the strip together. It is convenient to 'identify' the parallel class that gets removed with the parallel class that gets added. In this way it makes sense to say: 'We derive at m points of the same parallel class', that is, we choose a point and derive, choose another point and derive again, and so on, m times. All this sounds very much like what we just did.

So let F be an unrestricted n-phunisolvent set and let O be the tubular circle planes of rank n associated with it. Let

$$S = \{x_1, \ldots, x_k \mid \lambda_1, \ldots, \lambda_k \mid y_1^{(0)}, \ldots, y_k^{(\lambda_k - 1)}\}$$

be a SIC of order $m \leq n-1$ on $[-\frac{\pi}{2}, \frac{\pi}{2})$. Then we can construct a tubular circle plane O_S of rank $n-m$ in exactly the same manner as before when F was an unrestricted n-unisolvent set. The only difference in our picture is that the left and right sides of the interval $I = [-\frac{\pi}{2}, \frac{\pi}{2}]$ get identified and therefore, when we are busy flipping the strips, we may actually have to introduce a twist at the last flip in order to be able to glue respective sides together (thereby forming a Möbius strip). Note that odd λ_is introduce a flip (or un-flip) and every even λ_i introduces no flip at all. Furthermore, if $n - m = n - (\lambda_1 + \lambda_2 + \cdots + \lambda_k)$ is even, we end up with a Möbius strip and if $n - m$ is odd, with a cylinder.

Now, clearly, if O is nested, then for $p = (x_1, y_1) \in [-\frac{\pi}{2}, \frac{\pi}{2}) \times \mathbf{R}$ and $S = \{x_1 \mid \lambda_1 = 1 \mid y_1^{(0)} = y_1\}$, O_S is the derived tubular circle plane at the point p. We already know that O_S is a tubular circle plane of rank $n-1$. We just have to check that it is nested. The possible candidates for all the derived tubular circle planes of rank $n-2$ of this tubular circle plane are the tubular circle planes of rank $n-2$ that correspond to SICs of order 2 on $[-\frac{\pi}{2}, \frac{\pi}{2})$ that extend S. Again it remains to check that all these rank $n-2$ tubular circle planes are nested. Let S' be a SIC of order 2 that extends S. Then the possible candidates for all the derived (nested) tubular circle planes of rank $n-3$ of the tubular circle planes of rank $n-2$ associated with S' are the tubular circle planes of rank $n-3$ that correspond to SICs of order 3 on $[-\frac{\pi}{2}, \frac{\pi}{2})$ that extend S'. Again it remains to check that all these tubular circle planes are nested. We can continue to argue in this manner until we are left with tubular circle planes of rank 1, which are automatically nested. This, of course, shows that all the tubular circle planes of rank $n-m$ that we encountered on the way are nested themselves. □

COROLLARY 7.2.8 (Laguerre Planes) *The tubular circle plane that corresponds to an unrestricted* 3*-phunisolvent set is a flat Laguerre plane.*

It is also possible to define *nested strip circle planes of rank* n. This definition is messy but basically the same as that for nested tubular circle planes. With this definition it is possible to extend Theorem 7.2.7 to show that, in general, fibrated circle planes that correspond to unrestricted n-(ph)unisolvent sets are nested; see Polster [1998d].

We have already observed that all fibrated circle planes of rank 1 or 2 are nested. On the other hand, it is easy to construct nonnested higher-rank fibrated circle planes using Theorem 7.1.8.

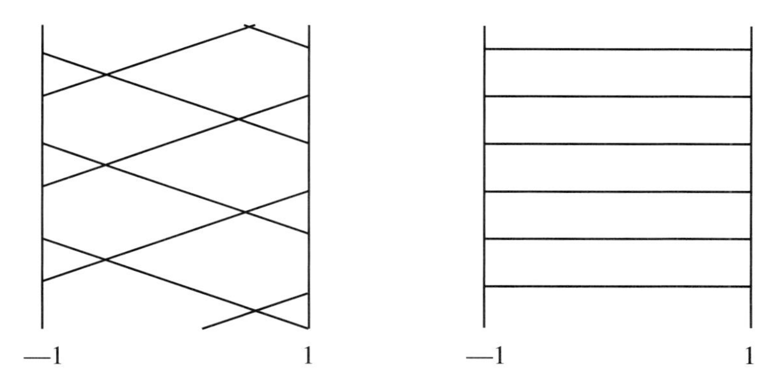

Fig. 7.4.

THEOREM 7.2.9 (Nonnested Planes Exist) *There exist tubular circle planes of rank* n, $n \geq 3$, *that are not nested.*

Proof. In Theorem 7.1.8 let F be the set of polynomials P_n of degree at most $n-1$ for some odd $n > 3$, and let $I = [-1, 1]$. We show that the tubular circle plane that corresponds to the half-periodic $(n-1)$-phunisolvent set F' is not nested. For this we have a closer look at the set G of all polynomials in F' that interpolate the point $(1,0)$. Clearly, $G = (x-1)(x+1)P_{n-2}$. This means that the restriction $G_{(-1,1)}$ is topologically isomorphic to the restriction of the set P_{n-2} to the interval $(-1,1)$. The corresponding strip circle plane $\mathcal{S}$ of rank $n-2$ is not topologically isomorphic to the restriction at a point of a tubular circle plane of rank $n-2$ (living on the cylinder). To see this, consider the left diagram in Figure 7.4.

If there were an extension of $\mathcal{S}$ to a tubular circle plane, every pair of intersecting line segments in the diagram would have to be extended by the same point. This diagram is supposed to be continued indefinitely above and below. Because of the way line segments in the infinitely extended diagram intersect, all of them would be extended by the same point. In fact, every single element in $\mathcal{S}$ would be extended by the same point. This is a contradiction.

The even case is dealt with in a similar manner using the diagram on the right. □

7.2.4 Automorphisms of Nested Tubular Circle Planes

As a consequence of the iterated derivation process in a nested tubular circle plane $\mathcal{N} = (P, \mathcal{C})$ of rank $n > 1$, we can define the *multiplicity* with which two circles intersect at a point. Let C and D be distinct circles of $\mathcal{N}$ that have the point p in common. We form the derived tubular circle plane $\mathcal{N}'$ at p. Both planes share the set P_p of all points not parallel to p. The circles C and D give rise to circles C' and D' such that $C \cap P_p = C' \cap P_p$ and likewise $D \cap P_p = D' \cap P_p$. Let p' be the point on C' the set $C \cap P_p$ gets extended with in the derived plane $\mathcal{N}'$, that is, $\{p'\} = C' \setminus P_p$. If $p' \in D'$, we derive at p' to obtain another tubular circle plane $\mathcal{N}''$ and circles C'' and D'' and a point p'', where $\{p''\} = C'' \setminus P_p$. We can continue in this way until, say after k steps, we obtain a tubular circle plane $\mathcal{N}^{(k)}$ and circles $C^{(k)}$ and $D^{(k)}$ such that $C^{(k)} \setminus P_p$ and $D^{(k)} \setminus P_p$ are different points. Since circles in a tubular circle plane of rank 1 are disjoint, we conclude that $k < n$. Note that $C \cap P_p = C^{(m)} \cap P_p$ and likewise $D \cap P_p = D^{(m)} \cap P_p$ for $1 \leq m \leq k$. We then say that the circles C and D intersect in p with multiplicity k. We say that C intersects itself at p with multiplicity n.

Assume that the rank of $\mathcal{N}$ is greater than 2 and fix a point $p \in P$ and a circle $C \in \mathcal{C}$ through p. Let $\mathcal{C}_{p,C}$ be the collection of all circles in $\mathcal{C}$ that intersect C in p with multiplicity at least $n-2$. Note that we can obtain the set $\mathcal{C}_{p,C}$ by deriving $\mathcal{N}$ successively $n-2$ times at the 'infinite point' of C. For example, in the derived tubular circle plane $\mathcal{N}'$ at p the restriction $C \setminus |p|$ is extended by an infinite point p' to give a circle C' in $\mathcal{N}'$. Then one derives $\mathcal{N}'$ at p' and so on. That means that $\mathcal{C}_{p,C}$ when restricted to $P \setminus |p| \simeq \mathbf{R}^2$ plus the parallel classes contained in $P \setminus |p|$ form the line set of an $\mathbf{R}^2$-plane $\mathcal{R}$. More precisely, because $\mathcal{C}_{p,C}$ essentially is the circle set of a tubular circle plane of rank 2, or equivalently, a point Möbius strip plane, $\mathcal{R}$ is a flat affine plane.

LEMMA 7.2.10 (Affine Planes from Tubular Circle Planes) *Let $\mathcal{N} = (P, \mathcal{C})$ be a nested tubular circle plane of rank $n > 2$. For a flag (p, C) of $\mathcal{N}$ let $\mathcal{C}_{p,C}$ be the collection of all circles in $\mathcal{C}$ that intersect C in p with multiplicity at least $n - 2$. Then $\mathcal{C}_{p,C}$ is the set of nonvertical lines of a flat affine plane on $P \setminus |p|$ when circles in $\mathcal{C}_{p,C}$ are punctured at p.*

For all the circle planes that we encountered in the previous chapters we know that isomorphisms are continuous. The same is true for nested tubular circle planes whose rank is greater than 1. To see this first note that isomorphisms between tubular circle planes of rank 2 are also isomorphisms of the corresponding point Möbius strip planes and therefore continuous; see Theorem 2.4.2. Note also that any tubular circle plane of rank 1 admits automorphisms that are not continuous. For example, given any permutation of the circles in such a plane, there is an automorphism that induces this permutation. Of course, any such plane also admits a large group of continuous automorphisms.

Now let $\mathcal{N} = (P, \mathcal{C})$ and $\mathcal{N}' = (P', \mathcal{C}')$ be two nested tubular circle planes of the same rank $n > 2$ and let $\gamma : P \to P'$ be an isomorphism from $\mathcal{N}$ to $\mathcal{N}'$. We fix a flag (p, C) of $\mathcal{N}$. By Lemma 7.2.10 we can associate a flat affine plane $\mathcal{R}$ with (p, C). The isomorphism γ induces a bijection from $P \setminus |p|$ onto $P' \setminus |\gamma(p)|$. In fact, γ induces an isomorphism from $\mathcal{R}$ onto a flat affine plane living on $P' \setminus |\gamma(p)|$. But such an isomorphism is continuous by Theorem 2.4.2. Changing the point p we see that γ must be continuous everywhere.

THEOREM 7.2.11 (Isomorphisms Are Continuous) *An isomorphism between nested tubular circle planes of the same rank $n > 1$ is continuous. In particular, every automorphism of such a nested tubular circle plane is continuous.*

Using Lemma 7.2.10 and Corollaries 2.4.3 and 2.4.4 one readily obtains the following result.

COROLLARY 7.2.12 (Stabilizers) *An isomorphism between two nested tubular circle planes of the same rank $n > 1$ is determined by a circle C, two distinct points p_1, p_2 on C, a third point p_3 off C parallel to neither p_1 nor p_2, and the images of these four objects under the isomorphism.*

In particular, an automorphism of such a nested tubular circle plane that fixes the circle C and the points p_1, p_2, and p_3 chosen as above is the identity.

Just note that for $n > 2$ in the projective extension of the flat affine plane associated with the flag (p_1, C) the points p_2 and p_3 and the infinite points of the verticals and of C form a nondegenerate quadrangle.

The collection of all automorphisms of a nested tubular circle plane $\mathcal{N}$ forms a group Γ with respect to composition. If the rank n of $\mathcal{N}$ is at least 2, then Γ is an effective topological transformation group of the point set (the cylinder or the Möbius strip for n odd or n even, respectively), when Γ carries the compact-open topology (or, equivalently, the topology of uniform convergence on compact sets). For Γ to be a Lie group we essentially need that Γ is locally compact; compare Theorem A2.3.5. This condition can be verified for nested tubular circle planes; see Polster–Steinke [20XXd].

THEOREM 7.2.13 (Automorphism Group) *The collection Γ of all automorphisms of a nested tubular circle plane of rank $n > 1$ is a Lie group with respect to the compact-open topology of dimension at most $n + 4$.*

The statement on the dimension of the automorphism group is an immediate consequence of the dimension formula A2.3.6 and Corollary 7.2.12. A flag (p, C) has an orbit of dimension at most $n + 1$; see Proposition 7.1.7. The orbit of a second point on C is at most 1-dimensional under the stabilizer $\Gamma_{p,C}$ of (p, C) and a third point has an orbit of dimension at most 2.

The maximum group dimensions are attained in the classical tubular circle planes. In fact, the classical tubular circle planes are characterized by this dimension; see Theorem 7.2.18.

THEOREM 7.2.14 (Automorphism Group of Classical Planes) *The automorphism group of the classical tubular circle plane $\overline{\text{Poly}}(n, \mathbf{R})$ of rank $n > 1$ with point set $(\mathbf{R} \cup \{\infty\}) \times \mathbf{R}$ is the $(n + 4)$-dimensional*

Lie group consisting of the automorphisms

$$(x, y) \to \left(\frac{ax+b}{cx+d}, \frac{ry + \sum_{i=0}^{n-1} a_i x^i}{(cx+d)^{n-1}} \right),$$

where $r \in \mathbf{R} \setminus \{0\}$, $a, b, c, d \in \mathbf{R}$, $ad - bc \neq 0$, *and* $a_i \in \mathbf{R}$, $i = 0, 1, 2, \ldots, n-1$.

This group is flag-transitive and thus acts transitively on both the point and circle sets of this plane.

For a proof of this result see Luksch [1987].

The following useful result for the classification of nested tubular circle planes with respect to their group dimensions is an immediate consequence of Theorems 7.2.4 and 7.2.11.

THEOREM 7.2.15 (Induced Automorphism) *Let $\mathcal{N}$ be a nested tubular circle plane of rank $n > 1$ and let γ be an automorphism of $\mathcal{N}$ that fixes a point p. Then γ induces an automorphism of the derived tubular circle plane $\mathcal{N}_p$ at p.*

The automorphism group Γ of a nested tubular circle plane $\mathcal{N}$ has a distinguished subgroup, the *kernel* T of the action of Γ on the set of parallel classes, that is, T consists of all automorphisms that fix each parallel class globally. This collection of automorphisms is a closed normal subgroup of the automorphism group of the circle plane.

To calculate the dimension of the kernel one again uses the dimension formula A2.3.6 and Corollary 7.2.12. Note that now each point has an orbit of dimension at most 1 and that, if a circle is fixed, then it is fixed pointwise.

THEOREM 7.2.16 (Dimension of Kernel) *The dimension of the kernel of a nested tubular circle plane of rank $n > 1$ is at most $n + 1$.*

The maximum dimension is attained in the tubular circle planes over *nested Chebyshev systems*, that is, Chebyshev systems that give rise to nested tubular circle planes. Indeed, if $\mathcal{S} = \{f_0, f_1, \ldots, f_{n-1}\}$ is such a nested Chebyshev system of rank n, then circles of the associated tubular circle plane $\mathcal{N}(\mathcal{S})$ are the graphs of functions

$$\left[-\frac{\pi}{2}, \frac{\pi}{2}\right] \to \mathbf{R} : x \mapsto a_0 f_0(x) + a_1 f_1(x) + \cdots + a_{n-1} f_{n-1}(x),$$

where $a_0, a_1, \ldots, a_{n-1} \in \mathbf{R}$, and the transformations

$$(x, y) \mapsto (x, ry + c_0 f_0(x) + c_1 f_1(x) + \cdots + c_{n-1} f_{n-1}(x)),$$

where $r, c_0, c_1, \ldots, c_{n-1} \in \mathbf{R}$, $r \neq 0$, are automorphisms in the kernel T of $\mathcal{N}(\mathcal{S})$. Hence T is $(n+1)$-dimensional. In fact, the converse is also true; see Polster–Steinke [20XXd].

THEOREM 7.2.17 ($(n+1)$-Dim. Kernel Implies Chebyshev) *Let $\mathcal{N}$ be a nested tubular circle plane of rank $n > 1$ and assume that the kernel of $\mathcal{N}$ is $(n+1)$-dimensional. Then $\mathcal{N}$ is isomorphic to a tubular circle plane $\mathcal{N}(\mathcal{S})$ over a nested Chebyshev system $\mathcal{S}$.*

The above theorem follows by induction on the rank n. The derived tubular circle plane $\mathcal{N}_p$ of $\mathcal{N}$ at a point p is a nested tubular circle plane of rank $n-1$ and the stabilizer T_p of p induces an n-dimensional group in the kernel of $\mathcal{N}_p$. Also note that $n = 2$ corresponds to the classical point Möbius strip plane and $n = 3$ corresponds to an ovoidal flat Laguerre plane; compare Theorem 5.4.8. Both planes can be described in terms of nested Chebyshev systems. For example, the rank 2 plane corresponds to the Chebyshev system $\{\cos x, \sin x\}$; see Subsubsection 7.1.2.1.

We can now show that the maximum group dimensions are attained only in the classical tubular circle planes; compare Theorem 7.2.14.

THEOREM 7.2.18 (Large Group Implies Classical) *Let $\mathcal{N}$ be a nested tubular circle plane of rank $n > 1$ and let Γ be the automorphism group of $\mathcal{N}$. Assume that there is a parallel class π such that the stabilizer Γ_π of π is (at least) $(n+3)$-dimensional. Then $\mathcal{N}$ is classical. In particular, if Γ is $(n+4)$-dimensional, then $\mathcal{N}$ is classical.*

Note that the derived tubular circle plane $\mathcal{N}_p$ of $\mathcal{N}$ at a point $p \in \pi$ is a nested tubular circle plane of rank $n-1$ and the stabilizer Γ_p of p induces an $(n+2)$-dimensional group of automorphisms of $\mathcal{N}_p$. Furthermore, the kernel of such a plane $\mathcal{N}$ must be $(n+1)$-dimensional and transitive on π. Also note that $n = 2$ and $n = 3$ yield the classical point Möbius strip plane and the classical flat Laguerre plane, respectively; compare Theorems 2.6.5 and 5.4.12. On can therefore use induction on the rank n to prove this result.

7.3 Convexity and Cut-and-Paste Constructions

We have already encountered a number of construction principles that allow us to derive new (ph)unisolvent sets from other (ph)unisolvent sets. In this section we introduce a notion of convexity in (ph)unisolvent sets that extends the notion of convexity that we dealt with in Section 2.2. Based on this notion are a number of cut-and-paste constructions that allow us to combine different n-(ph)unisolvent sets into new n-(ph)unisolvent sets. Again, these cut-and-paste constructions extend and generalize the cut-and-paste constructions that we encountered in previous chapters. Apart from Theorem 7.3.7 this section is a summary of Polster [1998f]. For more background information see, in addition to the references mentioned at the beginning of this chapter, the articles by Moldovan [1959], [1964], [1965], [1966], and Popoviciu [1978], [1979].

7.3.1 Convexity

In the following let F be an n-unisolvent set on the open interval I. A function $g : I \to \mathbf{R}$ is said to be *convex (concave) relative to* F if

$$(-1)^{n-i}(g(x) - F(x_1, x_2, \ldots, x_n; g)(x)) \geq 0 \ (\leq 0)$$

for $x_i < x < x_{i+1}$, $i = 0, 1, \ldots, n$, whenever the $n + 2$ points $x_i \in I$ are chosen such that $x_0 < x_1 < \cdots < x_{n+1}$. The function g is *strictly convex (concave) relative to* F if all the above inequalities are strict.

Clearly, if g is (strictly) convex or concave relative to F and J is an open subinterval of I, then g_J is (strictly) convex or concave relative to F_J, which, of course, is an n-unisolvent set on J.

If $n = 2$ and the 2-unisolvent system under consideration is the set of all linear functions, a continuous function is (strictly) convex in the above sense if and only if it is (strictly) convex in the usual sense.

THEOREM 7.3.1 (Transitivity of Convexity) *Let H_1 and H_2 be two n-unisolvent sets on the open interval I.*

If all elements of H_1 are convex with respect to H_2 and $f : I \to \mathbf{R}$ is a continuous function that is (strictly) convex with respect to H_1, then f is (strictly) convex with respect to H_2.

If all elements of H_1 are concave with respect to H_2 and $f : I \to \mathbf{R}$ is a continuous function that is (strictly) concave with respect to H_1, then f is (strictly) concave with respect to H_2.

This result follows immediately from our definition of convexity; see also Polster [1998f] Proposition 1.

If H is an $(n-1)$-unisolvent subset of F, let $F^+(H)$ and $F^-(H)$ denote the sets of all strictly convex and concave functions, respectively, relative to H in $F \setminus H$.

COROLLARY 7.3.2 *Let F be an n-unisolvent set, $n \geq 2$, on the open interval I and let H be an $(n-1)$-unisolvent subset of F. Then F is the disjoint union of the sets H, $F^+(H)$, and $F^-(H)$.*

Proof. Let g be one of the functions in $F \backslash H$. As a consequence of Theorem 7.1.5, considered as a function in the n variables $x_1, x_2, \ldots, x_{n-1}, x$

$$g(x) - H(x_1, x_2, \ldots, x_{n-1}; g)(x)$$

is continuous. Furthermore, by Theorem 7.1.2, as a function of x alone, this function has exactly $n-1$ zeros (the x_is) and changes sign at every single one of these zeros. As we continuously vary $x_1, x_2, \ldots, x_{n-1}, x_n$ in such a way that at all times $x_1 < x_2 < \cdots < x_{n-1} < x_n$, the sign of

$$g(x_n) - H(x_1, x_2, \ldots, x_{n-1}; g)(x_n)$$

does not change. Hence, as a consequence of Theorem 7.1.2, if this sign is positive or negative, then g is strictly convex or concave with respect to H, respectively. □

7.3.2 Different Ways to Cut and Paste

Remember that, given n distinct fixed points $t_1 < t_2 < \cdots < t_n$ in the open interval I, we can identify an n-unisolvent set F on I with $\mathbf{R}^n$ by mapping the function $f \in F$ to the point $(f(t_1), f(t_2), \ldots, f(t_n)) \in \mathbf{R}^n$; see Theorem 7.1.6. Under this identification an $(n-1)$-unisolvent subset H of F corresponds to a closed subset of $\mathbf{R}^n$ homeomorphic to $\mathbf{R}^{n-1}$ that separates $\mathbf{R}^n$ into two open components. These two open components correspond to $F^+(H)$ and $F^-(H)$. On top of this topological separating property, H also separates F in a geometrical way.

THEOREM 7.3.3 (Cut and Paste I) *Let F_1 and F_2 be n-unisolvent sets, $n \geq 2$, on an open interval and let H be a common $(n-1)$-unisolvent subset of F_1 and F_2. Then $F_3 = H \cup F_1^+(H) \cup F_2^-(H)$ is an n-unisolvent set. If F_1, F_2, and H are unrestricted, then F_3 is unrestricted as well.*

As a consequence of Theorem 7.1.3, n-phunisolvent sets do not have any $(n-1)$-phunisolvent subsets. This means that this theorem does not have a 'phunisolvent counterpart'.

For the proof of the second part of this result, we require the following lemma.

LEMMA 7.3.4 *Let F be an unrestricted n-unisolvent set, $n \geq 2$, on the open interval I and let H be an unrestricted $(n-1)$-unisolvent subset of F. Furthermore, let K be the set of all functions in F that satisfy a given SIC $\{x_1 \mid n-1 \mid y_1^{(0)}, y_1^{(1)}, \ldots, y_1^{(n-2)}\}$ of order $n-1$ on I. Then K has the following properties.*

(i) *Every point $(s,t) \in I \times \mathbf{R}$, $s \neq x_1$ is interpolated by exactly one element in K.*
(ii) *There is exactly one element $h \in H$ that is contained in K.*
(iii) *A function $f \in K \setminus \{h\}$ is contained in $F^+(H)$ ($F^-(H)$) if and only if $f(x) - h(x)$ is positive (negative) for any and therefore all $x > x_1$.*
(iv) *If $f \in F^+(H) \cap K$ and $x < x_1$, then $f(x) - h(x)$ is positive (negative) if n is odd (even).*
(v) *If $f \in F^-(H) \cap K$ and $x < x_1$, then $f(x) - h(x)$ is negative (positive) if n is odd (even).*

Proof. Parts (i) and (ii) follow immediately from the definition of unrestricted unisolvence. Furthermore, (iii), (iv), and (v) are corollaries of Mathsen [1982] Corollary 1 and the remark following that corollary. □

Proof of Theorem 7.3.3. It follows from Theorem 7.1.2 that for any choice of distinct points $x_1, x_2, \ldots, x_n \in I$ with $x_1 < x_2 < \cdots < x_n$ and real numbers $y_1, y_2, \ldots, y_n$ the sign of

$$y_n - H(x_1, x_2, \ldots, x_{n-1}, y_1, y_2, \ldots, y_{n-1})(x_n)$$

alone determines whether $F_i(x_1, x_2, \ldots, x_n, y_1, y_2, \ldots, y_n)$ is strictly convex with respect to H, belongs to H, or is strictly concave with respect to H.

Theorem 7.3.3 and Theorem 7.2.1 show that for the second part of this result we have to prove that the following holds. Given any SIC

$$S = \{x_1 \mid n \mid y_1^{(0)}, y_1^{(1)}, \ldots, y_1^{(n-1)}\}$$

of order n, and any unrestricted n-unisolvent set F that contains H, looking at H alone suffices to determine whether the uniquely determined function $f \in F$ satisfying S will be contained in $F^+(H)$, H, or $F^-(H)$.

Let $h \in H$ be the uniquely determined (by Lemma 7.3.4) function that satisfies the SIC

$$\{x_1 \mid n-1 \mid y_1^{(0)}, y_1^{(1)}, \ldots, y_1^{(n-2)}\}$$

of order $n-1$. Both $y_1^{(n-1)} - h^{(n-1)}(x_1)$ and $f(x) - h(x)$, $x > x_1$ have the same sign. Hence, by Lemma 7.3.4, f will belong to $F^+(H)$, H, or $F^-(H)$ if and only if $y_1^{(n-1)} - h^{(n-1)}(x_1)$ is positive, zero, or negative, respectively. □

Example. Let $F_i = \text{span}\{1, x, x^2, \ldots, x^{n-2}, f_i(x)\}$, $i = 1, 2$, such that the function $f_i : \mathbf{R} \to \mathbf{R}$ is $n-1$ times continuously differentiable and its $(n-1)$st derivative is a positive function. Then F_i is an unrestricted n-unisolvent set on $\mathbf{R}$ and $H = \text{span}\{1, x, x^2, \ldots, x^{n-2}\}$ is an unrestricted $(n-1)$-unisolvent set contained in both F_1 and F_2. Using Theorem 7.3.3, we can combine these two sets into an n-unisolvent set that, in general, is no longer linear.

Because of the last theorem, given an n-unisolvent set F, we are interested in constructing $(n-1)$-unisolvent subsets of F as well as other subsets that separate F in a similar way. A *separating set* such as this can then be used to construct a new n-unisolvent set from two given n-unisolvent sets that share this subset, as demonstrated above.

Let F be an n-(ph)unisolvent set on the open interval I. Furthermore, let $p = (s, t)$ be a point in $I \times \mathbf{R}$. Let $F(p)$, $F^+(p)$, $F^-(p)$ denote the set of all functions $f \in F$ such that $t = f(s)$, $t < f(s)$, and $t > f(s)$, respectively. The following obvious result gives one way of finding $(n-1)$-unisolvent subsets of n-unisolvent sets.

LEMMA 7.3.5 (Unisolvent Subsets of Unisolvent Sets) *Let F be an n-unisolvent set on I, $n \geq 2$, let J be a proper open subinterval of I, and let $p = (s, t) \in (I \setminus J) \times \mathbf{R}$. Then $F(p)_J$ is an $(n-1)$-unisolvent subset of F_J. Furthermore, if the value of s is greater than all elements of J, then*

$$F_J^+(F(p)_J) = F^+(p)_J,$$
$$F_J^-(F(p)_J) = F^-(p)_J.$$

Similar equations hold in the case that s is less than all elements of I.

This means that if we want to construct an $(n-1)$-unisolvent subset of an n-unisolvent set on the interval J, we can first try to embed it in a larger n-unisolvent set on an interval I, and then concentrate on the functions that interpolate a point $p \in (I \setminus J) \times \mathbf{R}$.

It is not known whether all n-unisolvent sets on open intervals can be embedded into larger n-unisolvent sets in this way. If G is an n-phunisolvent set, $n \geq 2$, on $[a, b]$, then the restriction of G to $[a, b)$

is an example of an n-unisolvent set on a half-open interval that cannot be embedded into a larger n-unisolvent set. This is an immediate consequence of the fact that n-phunisolvent sets, $n \geq 2$, are maximal orthogonal arrays; see Corollary 7.1.4.

Now, instead of concentrating on the functions that interpolate an 'exterior' point $p \in (I \setminus J) \times \mathbf{R}$, let us focus on the functions that interpolate an 'interior' point $p \in I \times \mathbf{R}$.

THEOREM 7.3.6 (Cut and Paste II) *Let F_1 and F_2 be n-(ph)unisolvent sets, $n \geq 2$, on the interval I, and let $p = (s,t) \in I \times \mathbf{R}$. Furthermore, let $F_1(p) = F_2(p)$. Then $F_3 = F_1(p) \cup F_1^+(p) \cup F_2^-(p)$ is an n-(ph)unisolvent set. If both F_1 and F_2 are unrestricted, then F_3 is unrestricted, too.*

Note that, unlike Theorem 7.3.3, this result also applies to n-phunisolvent sets.

Proof of theorem. The first part is a corollary of a more general result on separating sets; see Theorem 7.3.7. See also Polster [1998f] Theorem 2.

For the second part we have to prove that the following holds. Given any SIC

$$S = \{x_1 \mid n \mid y_1^{(0)}, y_1^{(1)}, \ldots, y_1^{(n-1)}\}$$

of order n, and any unrestricted n-unisolvent set F that contains $F(p)$, looking at $F(p)$ alone suffices to determine whether the uniquely determined (by Lemma 7.3.4) function $f \in F$ satisfying S will be contained in $F^+(p)$, $F(p)$, or in $F^-(p)$.

If $s = x_1$ and $t > y_1^{(0)}$, $t = y_1^{(0)}$, or $t < y_1^{(0)}$, then it is easy to check that $f \in F^+(p)$, $f \in F(p)$, or $f \in F^-(p)$, respectively. Let $x_1 < s$ and let h be the uniquely determined function in $F(p)$ that satisfies the SIC $\bar{S} = \{x_1 \mid n \mid y_1^{(0)}, y_1^{(1)}, \ldots, y_1^{(n-2)}\}$ of order $n-1$. Then f will be contained in $F^+(p)$, $F(p)$, or $F^-(p)$ depending on whether the value $y_1^{(n-1)} - h^{(n-1)}(x_1)$ is positive, zero or negative. For $s < x_1$ a similar argument applies. □

In the following it is again important to remember that n-(ph)unisolvent sets of functions provided with their natural topology (see p. 404) are topological spaces homeomorphic to $\mathbf{R}^n$.

All separating sets that we looked at so far have the property that, considered as subsets of such an $\mathbf{R}^n$, they separate this space into two path-connected components. The following result states that this is basically all that is necessary for a subset of an n-(ph)unisolvent set to be a separating set.

THEOREM 7.3.7 (General Cut and Paste) *Let F_1 and F_2 be two n-(ph)unisolvent sets, $n \geq 2$, on I and let $S \subset F_1 \cap F_2$. Suppose that $F_1 \setminus S$ and $F_2 \setminus S$ each have two nonempty path-connected components F_1^+, F_1^- and F_2^+, F_2^-, respectively. Furthermore, the labelling of the components F_1^+ and F_2^+ agrees, that is, there are n pairwise nonparallel points in $I \times \mathbf{R}$ such that the functions in F_1 and F_2 interpolating these points are contained in F_1^+ and F_2^+. Then $F_3 = S \cup F_1^+ \cup F_2^-$ is an n-(ph)unisolvent set.*

Proof. We only prove this for n-unisolvent sets over an interval I. The proof for n-phunisolvent sets runs along the same lines.

Let $x_1 < x_2 < \cdots < x_n \in I$ and let $y_1, y_2, \ldots, y_n \in \mathbf{R}$ be such that the function $f_i = F_i(x_1, x_2, \ldots, x_n, y_1, y_2, \ldots, y_n)$, $i = 1, 2$, belongs to $F_i \setminus S$ and $f_1 \in F_1^+$. Furthermore, let $u_1 < u_2 < \cdots < u_n \in I$, let $v_1, v_2, \ldots, v_n \in \mathbf{R}$, and let $g_i = F_1(u_1, u_2, \ldots, u_n, v_1, v_2, \ldots, v_n)$ belong to F_i^+. This means that g_1 and g_2 are 'labelling' functions. Since F_1^+ is path-connected, there is a path in F_1^+ from g_1 to f_1, that is, there is a continuous map $\gamma : [0, 1] \to F_1^+$ such that $\gamma(0) = g_1$ and $\gamma(1) = f_1$.

We now let $w_i(t) = \gamma(t)[u_i]$ and $z_i(t) = u_i(1-t)+x_i t$ for $i = 1, 2, \ldots, n$, and $0 \leq t \leq 1$. We define the function $h_i^t : I \to \mathbf{R}$, $i = 1, 2$, to be equal to

$$F_i(u_1, u_2, \ldots, u_n, w_1(2t), w_2(2t), \ldots, w_n(2t))$$

for $0 \leq t \leq \frac{1}{2}$ and

$$\begin{aligned} F_i(&z_1(2t-1), z_2(2t-1), \ldots, z_n(2t-1), \\ &f_1(z_1(2t-1)), f_1(z_2(2t-1)), \ldots, f_1(z_n(2t-1))) \end{aligned}$$

for $\frac{1}{2} \leq t \leq 1$. Then the map $\gamma_i : [0, 1] \to F_i : t \mapsto h_i^t$ is continuous. In fact, γ_1 defines essentially the same path as γ, that is, a path connecting g_1 and f_1 that is completely contained in F_1^+ or, equivalently, $h_1^t \in F_1^+$ for all $t \in [0, 1]$. Clearly, $h_2^0 = g_2$ is a function in F_2^+. In fact, $h_2^t \in F_2^+$ for all $t \in [0, 1]$ as well. This follows from the connectedness of the interval $[0, 1]$ and the fact that the sets $\{t \in [0, 1] \mid h_2^t \in F_2^+\}$ and $\{t \in [0, 1] \mid h_2^t \in F_2^-\}$ form a partition of $[0, 1]$. The set $\{t \in [0, 1] \mid h_2^t \in S\}$ is empty because, by construction, $\{t \in [0, 1] \mid h_1^t \in S\}$ is empty. Since the first of the partitioning sets is nonempty, we must have that $h_2^t \in F_2^+$ for all $t \in [0,1]$. By exchanging the rôles of F_1 and F_2, we see that n pairwise nonparallel points can be interpolated in F_1 by a function in F_1^+ if and only if they can be interpolated in F_2 by a function in F_2^+.

It now readily follows that any n pairwise nonparallel points in $I \times \mathbf{R}$ are interpolated by a unique function in $F_3 = S \cup F_1^+ \cup F_2^-$. □

7.4 Open Problems

The theory of n-(ph)unisolvent sets of functions, $n > 3$, and their corresponding geometries is nowhere near as developed as the theories of the lower-rank circle planes as described in the previous chapters. As we have seen, many results about the lower-rank circle planes have counterparts for the higher-rank geometries. Many more such counterparts are waiting to be proved. In particuar, the group-dimension classifications of higher-rank tubular circle planes should prove a rich source of problems for future research. In the following we only list some problems that aim at expanding the results mentioned in this chapter.

PROBLEM 7.4.1 *Which of our results about unrestricted n-(ph)unisolvent sets can be extended to results about nested n-(ph)unisolvent sets?*

For example, Theorem 7.2.3 suggests checking whether the integral of a nested n-unisolvent set is a nested $(n+1)$-unisolvent set. Theorem 5.3.13 implies that this is the case for $n = 2$.

There are many more shades of unisolvence and corresponding notions of convexity in between n-unisolvence and unrestricted n-unisolvence; see, for example, the papers by Mathsen [1969], [1972], [1982] and Umamaheswaram [1978a], [1978b]. For many of these it should be possible to generalize the results in this chapter. Are any of these other shades of unisolvence of any importance for incidence geometry?

PROBLEM 7.4.2 *Apart from the tubular circle planes, are there any other fibrated circle planes of rank $n \geq 2$ that are maximal orthogonal arrays of rank n?*

If you think about this problem for a second, you will probably come up with the following examples. Let F be an n-phunisolvent set on $[a,b]$. Then the strip circle plane associated with $F_{[a,b)}$ is a maximal orthogonal array of rank n. Of course, this does not really tell us anything new. Are there any other examples? If you are interested in this question, first check out some of the literature that deals with extending Chebyshev systems to larger Chebyshev systems; see Zalik–Zwick [1989] and the references given there. Compare also the results about maximal stable planes listed in Subsection 2.8.4.

PROBLEM 7.4.3 *Are $(n-1)$-unisolvent subsets of unrestricted or nested n-unisolvent sets automatically unrestricted or nested.*

This problem is inspired by Theorem 7.3.3. Note that the 2-unisolvent set $\text{span}\{1, x^3\}$ is not unrestricted and is a subset of the unrestricted 4-unisolvent set $\text{span}\{1, x, x^2, x^3\}$.

Appendix 1
Tools and Techniques from Topology and Analysis

The geometries encountered in this book all live on point and line sets that are well-behaved topological spaces. In most cases these spaces can be shown to be metric spaces or even manifolds. In this section we list some basic results about these special kinds of topological spaces for easy reference. For more information about topological and metric spaces we refer to Dugundji [1966] and Schubert [1968].

A1.1 Metric Spaces

A *metric space* (X, d) consists of a nonempty set X together with a function $d : X \times X \to \mathbf{R}$, called a *metric*, satisfying the following axioms.

Axioms for Metric Spaces

(M1) $d(x, y) \geq 0$ for all $x, y \in X$.

(M2) $d(x, y) = 0$ if and only if $x = y$.

(M3) $d(x, y) = d(y, x)$ for all $x, y \in X$.

(M4) $d(x, z) \leq d(x, y) + d(y, z)$ for all $x, y, z \in X$ (*triangle inequality*).

The Euclidean space $\mathbf{R}^n$ is a metric space with respect to the *Euclidean metric* given by

$$d(x, y) = \left(\sum_{j=1}^{n} (x_j - y_j)^2 \right)^{\frac{1}{2}}.$$

We always assume that $\mathbf{R}^n$ is equipped with this metric. We further carry over this metric to subspaces of $\mathbf{R}^n$. More generally, if (X, d) is a metric space and Y is a nonempty subset of X, then by restricting the metric of X to Y we obtain a metric space $(Y, d|_{Y\times Y})$, called a *subspace of* X with respect to the *induced metric*. We always use the metric induced by the Euclidean metric of $\mathbf{R}^n$ when dealing with subsets of Euclidean spaces such as the sphere $\mathbf{S}^2$, the cylinder $\mathbf{S}^1 \times \mathbf{R}$, and the torus $\mathbf{S}^1 \times \mathbf{S}^1$ in $\mathbf{R}^3$.

A *neighbourhood* of a point p in a metric space (X, d) is a subset U of X that contains an open ball centred at p, that is, there exists $r > 0$ such that the *open ball* $B_r(p) = \{x \in X \mid d(x,p) < r\}$ *of radius* r *and centre* p is entirely contained in U. A subset V of X is called *open* if it is a neighbourhood of each of its points. A subset W of X is called *closed*, if $X \setminus W$ is open. Note that the empty set $\emptyset$ and the entire space X are open sets and also closed sets of X.

A subset A of a metric space X is closed if and only if A contains all its accumulation points, that is, if (a_n) is a sequence of points in A that converges to $a \in X$, then $a \in A$.

For each subset Y of a metric space X there are a smallest closed subset $\overline{Y}$ of X containing Y, called the *closure of* Y, and a largest open set Y° contained in Y, called the *interior of* Y. The *boundary* ∂Y *of* Y then is $\partial Y = \overline{Y} \setminus Y^\circ$. In terms of sequences, the boundary of Y is the collection of all points that occur as accumulation points of sequences in X each element of which belongs to Y as well as sequences in X no element of which belongs to Y.

Let (X, d) be a metric space with *bounded metric* d, that is, there is a positive number b such that $d(x, y) \leq b$ for all $x, y \in X$. For nonempty closed subsets A and B of X let

$$d'(A, B) = \sup_{a\in A, b\in B}\{\inf_{u\in A} d(u, b), \inf_{v\in B} d(a, v)\}.$$

Then d' is a metric on the collection of nonempty closed subsets of X, called the *Hausdorff metric*.

We can use this construction to obtain a metric on the real projective plane, which is not a subset of $\mathbf{R}^3$. Points of the real projective plane are obtained from the sphere by identifying antipodal points on $\mathbf{S}^2$, that is, each point of this topological space can be identified with the set $\{x, -x\} \subset \mathbf{S}^2$ for some $x \in \mathbf{S}^2$. Since the sets $\{x, -x\}$ are closed subsets of $\mathbf{S}^2$, the Hausdorff metric

$$d'(\{x, -x\}, \{y, -y\}) = \min\{d(x, y), d(x, -y)\},$$

where d is the Euclidean metric on $\mathbf{S}^2$, yields a metric on the real projective plane.

The assumption of a bounded metric is no real restriction since we can always replace a metric by a bounded metric without changing the open sets.

LEMMA A1.1.1 (Bounded Metric) *Let* (X,d) *be a metric space and let*

$$d' : X \times X \to \mathbf{R} : (x,y) \mapsto \frac{d(x,y)}{1+d(x,y)}.$$

Then (X,d') *is a metric space that has the same open subsets as* (X,d)*. (That is,* (X,d) *and* (X,d') *are homeomorphic; for a definition of 'homeomorphic' see the end of this section.)*

When dealing with topological geometries we are dealing not only with topologies on the point sets but also with topologies on the line/circle sets. For the topological geometries we are interested in these topologies coincide with the topologies induced by Hausdorff metrics on the sets of lines/circles. Note that in our geometries lines and circles are closed subsets of the respective point sets.

In the following we list some of the most important properties metric spaces can have. In fact, many of these properties do not use the metric but only refer to the *topology* of the metric space, that is, the collection of open subsets. Most of these properties can therefore be defined for the larger class of *topological spaces*, that is, sets that have a topology specified. We do not want to give a formal definition of topological space, because in the spaces we encounter in this book we always have an underlying metric and also the most useful topological spaces require certain separation axioms. For more information on topological spaces we refer to Dugundji [1966] or Schubert [1968]. We just mention that a *nonindiscrete topological space* has at least one proper nonempty open subset and that there are various ways to describe the open subsets in a topological space.

For example, instead of all open subsets one can just specify a *basis* for the topology, that is, a family of open sets such that every open set is the union of a certain number of sets belonging to this family, or a *subbasis* for the topology, that is, a family of open sets such that the collection of finite intersections of its elements is a basis.

One can also use the collection of neighbourhoods of points to define the topology. Clearly, every set that contains a neighbourhood of a point p is again a neighbourhood of p. A *neighbourhood basis* of p then is a family of neighbourhoods of p such that every neighbourhood of p contains one of its elements.

A metric space X is called *connected* if there does not exist a pair of nonempty open sets A, B such that $X = A \cup B$ and $A \cap B = \emptyset$. Equivalently, X is connected if and only if X and $\emptyset$ are the only subsets of X that are both open and closed. A subset A of X is connected if and only if A is a connected metric space relative to the induced metric. A metric space X is called *locally connected* if for each point $x \in X$ and each neighbourhood U of x there is a connected neighbourhood V of x that is entirely contained in U.

LEMMA A1.1.2 (Connected Subsets) *If C is a connected subset in a metric space X, then the closure $\overline{C}$ of C is also connected.*

A maximally connected subset C of a metric space X, that is, one such that every connected subset D of X containing C equals C, is called a *connected component of* X. The 1-sphere $\mathbf{S}^1$ is connected and locally connected and so are $\mathbf{R}$ and the closed interval $[0, 1]$. However, $\mathbf{R} \setminus \{0\}$ is not connected (but still locally connected). Its connected components are $\mathbf{R}^+$ and $\mathbf{R}^-$.

By the above lemma, each maximally connected subset must be closed.

PROPOSITION A1.1.3 (Connected Components) *The connected components of a metric space X form a partition of X into closed subsets of X.*

A subset S of a metric space X is called a *separating set of* X if $X \setminus S$ is not connected.

A metric space (X, d) is called *separable* if X contains an at most countable subset D such that every nonempty open subset contains a point of D, that is, D is *dense* in X. For example, $\mathbf{R}^n$ with the Euclidean metric is separable; $D = \mathbf{Q}^n$ is a countable dense subset of $\mathbf{R}^n$.

Using a different characterization of separable metric spaces in terms of a basis for the topology one obtains the following.

PROPOSITION A1.1.4 (Subspaces of Separable Spaces) *Every subspace of a separable metric space is separable.*

A set of subsets of X is a *covering of* X if X is the union of these subsets. A metric space X is called *compact* if for every covering C

of X consisting of open sets there exists a finite subset $F \subseteq C$ that is a covering of X. Using sequences, a metric space X is compact if and only if every sequence in X has an accumulation point in X.

A subset A of X is compact if and only if A is a compact metric space relative to the induced metric. The compact subsets of $\mathbf{R}^n$ with the Euclidean metric are precisely the closed and bounded subsets.

A metric space X is called *locally compact* if and only if every point in X has a compact neighbourhood. The Euclidean n-space $\mathbf{R}^n$ with the Euclidean metric is locally compact. The closed balls $\overline{B}_r(p)$, $r \in \mathbf{R}^+$, are compact neighbourhoods of a point $p \in \mathbf{R}^n$.

PROPOSITION A1.1.5 (Compact Metric Spaces) *Every compact metric space is separable.*

Let (X_1, d_1) and (X_2, d_2) be metric spaces. There are many ways to introduce a metric on the product set $X = X_1 \times X_2$. One such way is to define

$$X \times X \to \mathbf{R} : ((x_1, x_2), (y_1, y_2)) \mapsto d_1(x_1, y_1) + d_2(x_2, y_2).$$

Then X equipped with this metric is called the *direct product of* (X_1, d_1) *and* (X_2, d_2). Note that the direct product $\mathbf{R} \times \mathbf{R}$ is a metric space with a metric different from the Euclidean metric of $\mathbf{R}^2$. However, the two spaces with the respective metrics are homeomorphic (see the definition below), that is, both metrics describe the same open sets.

Let (X_1, d_1) and (X_2, d_2) be metric spaces. A map $f : X_1 \to X_2$ is called *continuous at a point* $p_1 \in X_1$ if and only if $f^{-1}(U_2)$ is a neighbourhood of p_1 for every neighbourhood U_2 of $f(p_1)$. The map f is called *continuous* if the inverse image $f^{-1}(U_2)$ of any open subset U_2 of X_2 is open in X_1. It readily follows that a map between metric spaces is continuous if and only if it is continuous at every point and that the composition of continuous maps is continuous.

A metric space X is called *pathwise connected* or *path-connected* if any two of its points can be joined by a path, that is, if $x_0, x_1 \in X$, then there is a continuous map $f : [0, 1] \to X$ such that $f(0) = x_0$ and $f(1) = x_1$. Since the unit interval $[0, 1]$ is connected, it readily follows that a pathwise connected space is connected; the converse is not true however. Both $\mathbf{R}^n$ and $\mathbf{S}^n$ for $n \geq 1$ are pathwise connected.

A pathwise connected metric space X is called *simply connected* if for each point $x_0 \in X$ and each continuous map $f : [0, 1] \to X$ such that $f(0) = f(1) = x_0$ there is a continuous map F from the unit square $[0,1] \times [0,1]$ into X such that $F(s,0) = f(s)$ and such that, furthermore, $F(0,t) = F(1,t) = F(s,1) = x_0$ for all $s,t \in [0,1]$. This just means that every loop at x_0 can be continuously contracted to the point x_0. Examples of simply connected spaces are $\mathbf{R}^n$, $n \geq 1$, and $\mathbf{S}^m$ for $m \geq 2$. However, $\mathbf{S}^1$ is not simply connected.

A bijection $f : X_1 \to X_2$ is called a *homeomorphism* if f and f^{-1} are continuous. If such a map exists two spaces X_1 and X_2 are called *homeomorphic*.

Note that homeomorphic spaces have the same topological properties. More generally, many topological properties are preserved under continuous maps. For example, the continuous image of a compact or connected space is compact or connected, respectively.

Using the canonical projection from $\mathbf{S}^2$ onto the real projective plane that identifies antipodal points, many topological properties of $\mathbf{S}^2$ can be carried over to the real projective plane. In particular, it is easy to see that this space is compact, connected, locally connected, and pathwise connected. However, it is not simply connected; any loop that comes from a great circle on the sphere $\mathbf{S}^2$ cannot be continuously contracted to a point.

A1.2 Manifolds

We are mostly dealing with topological manifolds, but occasionally we refer to differentiable manifolds as well, and, of course, the automorphism groups of many of the geometries in this book are Lie groups and, as such, fall into this category. Therefore we include the wider definition of smooth (differentiable) manifolds. For more information about manifolds we refer to Warner [1983].

A separable metric space M is called an *m-dimensional (topological) manifold* or simply an *m-manifold* if there exist collections of sets $(U_j)_{j\in J}$, $(V_j)_{j\in J}$, and maps $(\varphi_j)_{j\in J}$ such that the following hold.

- Each $U_j \subseteq M$ is open in M and $\bigcup_{j\in J} U_j = M$.
- Each $V_j \subseteq \mathbf{R}^m$ is open in $\mathbf{R}^m$.
- Each $\varphi_j : U_j \to V_j$ is a homeomorphism.

A topological manifold M is called a *smooth m-dimensional (differentiable) manifold* where the collections $(U_j)_{j\in J}$, $(V_j)_{j\in J}$, $(\varphi_j)_{j\in J}$ of sets and maps have the following additional property.

- The map $\varphi_j\varphi_k^{-1} : \varphi_k(U_j \cap U_k) \to \varphi_j(U_j \cap U_k)$ is infinitely differentiable for all $j, k \in J$ whenever $U_j \cap U_k$ is nonempty.

Each triple $(U_j, \varphi_j, V_j)_{j\in J}$ is called a *coordinate system of* M. The collection $(U_j, \varphi_j, V_j)_{j\in J}$ of coordinate systems forms an *atlas* of M. The metric and an atlas define the differentiable structure of the manifold. Each map φ_j is a *coordinate map of* M and its inverse $\varphi_k^{-1} : V_k \to U_k$ is called a *local parametrization.*

The concept of a manifold is a local notion. Every nonempty open subset of a manifold is a manifold. The separability of a manifold implies that at most countably many coordinate maps are needed in order to describe the manifold completely.

It is easy to verify that the 2-sphere $\mathbf{S}^2$, the cylinder $\mathbf{S}^1 \times \mathbf{R}$, and the torus $\mathbf{S}^1 \times \mathbf{S}^1$ are 2-dimensional smooth manifolds. Coordinate maps can be obtained by stereographic projection from points of these sets onto a plane not passing through the point from which we project by restricting to suitable subsets. Restricting coordinate maps of $\mathbf{S}^2$ to open subsets of hemispheres we further see that the real projective plane is also a 2-dimensional smooth manifold.

These manifolds and many of the classical Lie groups can also be described in a different way. To this end let $U \subseteq \mathbf{R}^n$ be a nonempty open set and let $f : U \to \mathbf{R}^k$, $k \leq n$, be a differentiable map. A point $a \in f(U) \subseteq \mathbf{R}^k$ is called a *regular value* of f if the derivative $D_x f$ of f at x has maximum rank (that is, rank k, or equivalently, the linear map $D_x f$ is surjective) for all $x \in f^{-1}(a)$.

PROPOSITION A1.2.1 (Regularity Theorem) *Let $U \subseteq \mathbf{R}^n$ be a nonempty open set, let $f : U \to \mathbf{R}^k$, $k \leq n$, be a smooth map, and let $a \in \mathbf{R}^k$ be a regular value of f. Then the set $M = f^{-1}(a) \subseteq \mathbf{R}^n$ is an $(n-k)$-dimensional smooth manifold.*

For example, the sphere $\mathbf{S}^2$, the cylinder $\mathbf{S}^1 \times \mathbf{R}$, and the torus $\mathbf{S}^1 \times \mathbf{S}^1$ can be obtained in this way from the maps $f : \mathbf{R}^3 \to \mathbf{R}$, $g : \mathbf{R}^3 \to \mathbf{R}$, and $h : \mathbf{R}^4 \to \mathbf{R}^2$ given by $f(x, y, z) = x^2 + y^2 + z^2$, $g(x, y, z) = x^2 + y^2$, and $h(w, x, y, z) = (w^2 + x^2, y^2 + z^2)$, respectively, such that $\mathbf{S}^2 = f^{-1}(1)$, $\mathbf{S}^1 \times \mathbf{R} = g^{-1}(1)$, and $\mathbf{S}^1 \times \mathbf{S}^1 = h^{-1}(1, 1)$.

Let M_1 and M_2 be smooth manifolds with respective dimensions m_1 and m_2 and let $f : M_1 \to M_2$ be a continuous map. Then f is said to be smooth if for all coordinate maps $\varphi_1 : U_1 \to V_1$ and $\varphi_2 : U_2 \to V_2$

of M_1 and M_2, respectively, such that $U_1' = U_1 \cap f^{-1}(U_2) \neq \emptyset$, the map

$$\varphi_2 \circ f \circ \varphi_1^{-1} : \varphi_1(U_1') \to \mathbf{R}^{m_2}$$

between open subsets of Euclidean spaces is smooth. A *(smooth) diffeomorphism* from M_1 to M_2 is a bijection $f : M_1 \to M_2$ such that f and f^{-1} are smooth.

For submanifolds of Euclidean spaces differentiability of maps between them can be recognized more easily. If $M_1 \subseteq \mathbf{R}^{n_1}$ and $M_2 \subseteq \mathbf{R}^{n_2}$ are smooth submanifolds of dimensions m_1 and m_2, respectively, and the function $f : M_1 \to M_2$ is continuous, then f is smooth if and only if f can be locally extended to a smooth map between open subsets of $\mathbf{R}^{n_1}$ and $\mathbf{R}^{n_2}$.

Manifolds constructed by using the Regularity Theorem are subsets of some Euclidean space. In fact, there is no loss of generality in considering only submanifolds of Euclidean spaces.

THEOREM A1.2.2 (Whitney Embedding Theorem) *For every smooth manifold M of dimension m there is a map $f : M \to \mathbf{R}^{2m+1}$ such that $f : M \to f(M)$ is a diffeomorphism and $f(M)$ is closed in $\mathbf{R}^{2m+1}$ (so that $f(M)$ is a smooth submanifold of $\mathbf{R}^{2m+1}$).*

The 2-dimensional connected manifolds are referred to as *surfaces*. A surface locally looks like $\mathbf{R}^2$, but globally it may be very different. The compact surfaces are classified; see Seifert–Threlfall [1980] Chapter 6. They come in two different infinite families. One family, comprising the so-called *orientable* compact surfaces, is based on the 2-sphere $\mathbf{S}^2$; the other family, comprising the so-called *nonorientable* compact surfaces, is based on the real projective plane. One then successively adds more and more 'handles' to these two basic surfaces to obtain the other compact surfaces. The number of handles then uniquely determines the respective surface. The compact surfaces we are dealing with as point sets of geometries are the 2-sphere, the torus (= 2-sphere with one handle) and the real projective plane. The other surfaces in these two families do not seem to admit nice geometries.

In the previous section A1.1 we introduced the notions of connected and pathwise connected spaces. For manifolds these two classes can be shown to coincide.

PROPOSITION A1.2.3 (Connected = Pathwise Connected) *A topological manifold is connected if and only if it is pathwise connected.*

We conclude this section by having a look at embeddings of 1-spheres into 2-spheres. The following result is a special case of the Jordan–Brouwer Separation Theorem; see Spanier [1966] Theorem 4.8.15.

THEOREM A1.2.4 (Jordan–Schoenflies Separation Theorem) *A 1-sphere $\mathbf{S}^1$ embedded in a 2-sphere $\mathbf{S}^2$ separates $\mathbf{S}^2$ into two components of which it is the common boundary. There is a homeomorphism of $\mathbf{S}^2$ onto itself that takes $\mathbf{S}^1$ onto the equator.*

Note that for the real projective plane the corresponding result is not true. Here are two different embeddings of $\mathbf{S}^1$ in this surface. One embedding occurs as a line L in the classical flat projective plane; here the complement of $L \cong \mathbf{S}^1$ is homeomorphic to $\mathbf{R}^2$ and thus connected. A different embedding occurs as a conic C in the classical flat projective plane; here the complement of $C \cong \mathbf{S}^1$ has two connected components, one homeomorphic to $\mathbf{R}^2$ and the other homeomorphic to a Möbius strip.

A1.3 Topological Dimension

A dimension function assigns each space in a certain class of topological spaces a 'number', its *dimension*, and measures in some sense how 'big' the space is. There are several notions of dimension for topological spaces. In particular, the inductive dimensions 'ind' and 'Ind' and the covering dimension 'dim' are widely used. These dimension functions agree on nice spaces such as separable metric spaces or locally compact groups. The dimension theory for the former class of spaces was developed in Hurewicz–Wallman [1948].

Since we are not interested in general spaces, we just mention the definition of the *small inductive dimension* of a separable metric space because this is one of the most easily visualizable dimension functions. We set $\operatorname{ind}(X) = -1$ for a separable metric space X if and only if $X = \emptyset$. We then define $\operatorname{ind}(X)$ inductively as follows. If n is a non-negative integer and X a separable metric space, then we say that $\operatorname{ind}(X) \leq n$ if and only if every point has arbitrarily small open neighbourhoods U whose boundaries $\partial U = \overline{U} \backslash U$ have inductive dimension $\operatorname{ind}(\partial U) \leq n-1$.

A separable metric space is 0-dimensional if and only if it is totally disconnected, that is, no connected subset contains more than one point.

Clearly, every subset of a separable metric space X has dimension at most the dimension of X.

PROPOSITION A1.3.1 (Sum Theorem) *Let M be a topological manifold of dimension n and suppose that M is the union of countably many closed subsets A_i of M. Then there is at least one i such that A_i has dimension n.*

It readily follows by induction that $\mathrm{ind}(\mathbf{R}^n) \leq n$. In fact, one has equality; see Hurewicz–Wallman [1948] Theorem IV.1. This carries over to manifolds and to open subsets therein. Conversely, Hurewicz–Wallman [1948] Theorem IV.3 shows the following.

THEOREM A1.3.2 (Subsets of Maximal Dimension) *Let M be a topological manifold of dimension n. If $U \subseteq M$ is a subset of dimension n, then U contains a nonempty open subset of M.*

Separating sets in connected manifold cannot be too small. Hurewicz–Wallman [1948] Theorem IV.4 shows the following.

PROPOSITION A1.3.3 (Separating Sets) *A separating set in a topological n-dimensional connected manifold is at least $(n-1)$-dimensional.*

An application of Theorem A1.3.2 is the following.

THEOREM A1.3.4 (Brouwer Theorem: Invariance of Domain) *If U and V are homeomorphic subsets of a topological manifold M of finite dimension and U is open in M, then V is open in M.*

We conclude this section with a topological version of Whitney's embedding theorem A1.2.2; see Hurewicz–Wallman [1948] Theorem V.3.

THEOREM A1.3.5 (Topological Embedding Theorem) *A separable metric space X of finite dimension at most n is homeomorphic to a subset of $\mathbf{R}^{2n+1}$.*

A1.4 Continuous Maps and Fixed Points

When verifying the geometric axioms in the geometries we are dealing with in this book, one often reduces this task to one where one has to determine the number of fixed points of certain functions. There are a number of fixed-point theorems to help in this situation. Usually these general theorems are obtained using methods from algebraic topology and, in particular, homology groups of spheres. However, some of the most useful ones require only elementary calculus.

We begin with the 1-dimensional case. An orientation-preserving (orientation-reversing) permutation of $\mathbf{R}$ is just a strictly increasing (strictly decreasing) permutation $f : \mathbf{R} \to \mathbf{R}$, that is, $f(x) < f(y)$ $(f(x) > f(y))$ for all $x, y \in \mathbf{R}$, $x < y$. Note that an orientation-preserving (orientation-reversing) permutation of $\mathbf{R}$ must be continuous and thus a homeomorphism from $\mathbf{R}$ onto itself. If f is even differentiable, one can determine whether or not f is orientation-preserving by using its derivative f'. In this case f is orientation-preserving if and only if $f'(x) \geq 0$ for all $x \in \mathbf{R}$ and f is injective and unbounded.

The notion of orientation-preserving (orientation-reversing) permutations of $\mathbf{R}$ extends to one for homeomorphisms of $\mathbf{S}^1$. Given a homeomorphism $h : \mathbf{S}^1 \to \mathbf{S}^1$ we can always find a rotation ρ of $\mathbf{S}^1$ such that $\rho \circ h$ fixes a particular point p of $\mathbf{S}^1$. Hence $\rho \circ h$ gives rise to a homeomorphism of $\mathbf{R}$ by restricting $\rho \circ h$ to $\mathbf{S}^1 \setminus \{p\}$ and identifying $\mathbf{S}^1 \setminus \{p\}$ with $\mathbf{R}$. We then say that h is orientation-preserving (orientation-reversing) if and only if the above restriction is an orientation-preserving (orientation-reversing) of $\mathbf{R}$. Intuitively, we can move around on a circle in two ways; this gives rise to two possible orientations of $\mathbf{S}^1$.

Fig. A1.1. The images of two points determine whether a homeomorphism $\mathbf{R} \to \mathbf{R}$ preserves orientation

PROPOSITION A1.4.1 (Homeomorphisms of R and $\mathbf{S}^1$) *For homeomorphisms of* $\mathbf{R}$ *and* $\mathbf{S}^1$ *the following hold.*

(i) *Let $g : \mathbf{R} \to \mathbf{R}$ be a homeomorphism. Then the images of any two distinct points a and b determine whether g is orientation-preserving or orientation-reversing; see Figure A1.1.*
(ii) *If g is orientation-reversing, then it has exactly one fixed point.*
(iii) *Let $h : \mathbf{S}^1 \to \mathbf{S}^1$ be a homeomorphism. Then the images of any three distinct points determine whether h is orientation-preserving or orientation-reversing.*
(iv) *If the homeomorphism h is orientation-reversing, then it has exactly two fixed points.*
(v) *If h has exactly one or at least three fixed points, then it is orientation-preserving.*

For a proof of this result see Polster [1998a] Lemma 2.1.1.

Involutions play a prominent role in the investigation of the automorphism groups of geometries. A (continuous) involution of $\mathbf{S}^1$ is called *elliptic* if it has no fixed points. It is called *hyperbolic* if it has exactly two fixed points.

PROPOSITION A1.4.2 (Involutions of R and $\mathbf{S}^1$) *For involutions of* $\mathbf{R}$ *and* $\mathbf{S}^1$ *the following hold.*

(i) *Every involutory homeomorphism* $\mathbf{R} \to \mathbf{R}$ *has exactly one fixed point and is topologically equivalent to a reflection of* $\mathbf{R}$ *about one of its points.*

(ii) *Every involutory homeomorphism* $h : \mathbf{S}^1 \to \mathbf{S}^1$ *is topologically equivalent to a reflection of* $\mathbf{S}^1$ *about its centre or a line through the centre. In the first case it is an elliptic involution and orientation-preserving. In the second case it is a hyperbolic involution and orientation-reversing.*

(iii) *Let* p *and* q *be two distinct points in* $\mathbf{S}^1$ *that do not get exchanged by* h. *If* p *or* q *is fixed, then* h *is hyperbolic. Otherwise, the set* $\mathbf{S}^1 \setminus \{p, h(p)\}$ *has two connected components. The involution is elliptic (hyperbolic) if and only if the two points* q *and* $h(q)$ *are contained in different components (the same component).*

For higher-dimensional manifolds orientation-preserving and orientation-reversing homeomorphisms are more difficult to define. We just mention that for a differentiable homeomorphism f one can use its derivative f'. Then f is orientation-preserving if and only if the determinant $|f'|$ of f' is always nonnegative.

We finally list some results that do not require this notion. A particularly nice result that always guarantees a fixed point is the following; see Spanier [1966] Theorem 4.7.5.

THEOREM A1.4.3 (Brouwer Fixed-Point Theorem) *Every continuous map from the closed* n*-ball*

$$\overline{B}_1^n(0) = \{x \in \mathbf{R}^n \mid d(x, 0) \leq 1\},$$

where $n \geq 0$, *to itself has a fixed point.*

The above theorem does not apply to spheres. For example, the antipodal map $x \mapsto -x$ is a fixed-point-free homeomorphism. However, for $\mathbf{S}^2$ we obtain the following.

THEOREM A1.4.4 (Homeomorphisms of $\mathbf{S}^2$) *Every homeomorphism from $\mathbf{S}^2$ to itself has a fixed point or exchanges a pair of antipodal points. Every homeomorphism from the real projective plane to itself has a fixed point.*

Finally, involutions of $\mathbf{S}^2$ can be classified as follows; see Brouwer [1919].

THEOREM A1.4.5 (Involutions of $\mathbf{S}^2$) *Every involutory homeomorphism of $\mathbf{S}^2$ to itself is topologically equivalent to a reflection of $\mathbf{S}^2$ about its centre, a line through the centre, or an equatorial plane.*

We conclude this section with a summary of 'nice' homeomorphisms of $\mathbf{R}$ or $\mathbf{S}^1$ that we frequently encounter in the construction of our topological geometries and in the classification of these geometries with large group dimensions. These homeomorphisms usually satisfy certain functional equations and are obtained as the general solutions of these functional equations. To begin with an *additive* function of $\mathbf{R}$ is a map $f : \mathbf{R} \to \mathbf{R}$ such that

$$f(x + y) = f(x) + f(y)$$

for all $x, y \in \mathbf{R}$. There are many additive functions of $\mathbf{R}$; most of them are 'wild'. However, if we restrict ourselves to additive functions that are homeomorphisms, only a few can occur. Clearly, the restriction of such an additive function f to the rational numbers $\mathbf{Q}$ is of the form $f(x) = sx$ for some $s \in \mathbf{R}\setminus\{0\}$ and all $x \in \mathbf{Q}$. Since $\mathbf{Q}$ is dense in $\mathbf{R}$, the continuity of f then implies that $f(x) = sx$ for all $x \in \mathbf{R}$. We call a function of this form a *multiplication* and usually denote it by μ_s. In generalization, a *semi-linear* function is one that is made up of two multiplications, that is, it is of the form

$$\mathbf{R} \to \mathbf{R} : x \mapsto \begin{cases} rx & \text{for } x \geq 0, \\ sx & \text{for } x < 0, \end{cases}$$

where $r, s \in \mathbf{R}^+$. Multiplications are obtained for $s = r$.

An *affine* function of $\mathbf{R}$ is a map of the form

$$\mathbf{R} \to \mathbf{R} : x \mapsto ax + b,$$

where $a, b \in \mathbf{R}$, $a \neq 0$. A *fractional linear map* of $\mathbf{S}^1 \simeq \mathbf{R} \cup \{\infty\}$ is a map of the form

$$x \mapsto \frac{ax + b}{cx + d},$$

where $a, b, c, d \in \mathbf{R}$, $ad - bc \neq 0$; compare Subsubsection 2.9.2.1. Of course, the above expression only makes sense as long as $cx + d \neq 0$ and $x \in \mathbf{R}$. The exceptional values for x are determined as follows. If $c = 0$, we just obtain an affine map and we let the special point ∞ be taken to ∞. For $c \neq 0$ the point ∞ is taken to $\frac{a}{c}$ and $-\frac{d}{c}$ is taken to ∞. We always make these conventions when dealing with fractional linear maps. In fact, fractional linear maps are homeomorphisms of $\mathbf{S}^1$, the inverse of the above map is given by

$$x \mapsto \frac{dx - b}{-cx + a}.$$

A homeomorphism $h : \mathbf{R} \to \mathbf{R}$ is called *semi-multiplicative* if

$$h(xy) = h(x)h(y)$$

for all $x, y \in \mathbf{R}$, $x > 0$. The homeomorphism is called *multiplicative* if the above identity is satisfied for all $x, y \in \mathbf{R}$. We further extend the notion of (semi-)multiplicative homeomorphisms to homeomorphisms of the circle $\mathbf{S}^1 \simeq \mathbf{R} \cup \{\infty\}$ by requiring that ∞ is fixed.

Clearly, $h(0) = 0$ and $h(1) = 1$ for every semi-multiplicative homeomorphism h. Furthermore, h must be orientation-preserving. In particular, $h(x) > 0$ for $x > 0$. Taking the natural logarithm of h for positive x one obtains a function that satisfies 'half' the identity of an additive function. It then readily follows that this combined function looks like the restriction of a multiplication to $\mathbf{R}^+$ and, consequently, that a semi-multiplicative homeomorphism is of the form

$$f_{r,s}(x) = \begin{cases} x^r & \text{for } x \geq 0, \\ -s|x|^r & \text{for } x < 0, \\ \infty & \text{for } x = \infty, \end{cases}$$

where $r, s \in \mathbf{R}^+$. The multiplicative homeomorphisms are obtained precisely for $s = 1$.

We say that a homeomorphism h of $\mathbf{S}^1$ to itself is *inversely semi-multiplicative* if $h(\infty) = 0$, $h(0) = \infty$, and $1/h$ is a semi-multiplicative homeomorphism. Clearly, an inversely semi-multiplicative homeomorphism h of $\mathbf{S}^1$ has the form

$$h(x) = \begin{cases} x^r & \text{for } x > 0, \\ -s|x|^r & \text{for } x < 0, \\ \infty & \text{for } x = 0, \\ 0 & \text{for } x = \infty, \end{cases}$$

where $r \in \mathbf{R}^-$, $s \in \mathbf{R}^+$.

Finally, a *skew parabola function* f on $\mathbf{R}$ almost looks like a semi-multiplicative homeomorphism. It satisfies the same functional equation but is no longer a homeomorphism. More precisely, f is of the form

$$f(x) = \begin{cases} x^r & \text{for } x \geq 0, \\ s|x|^r & \text{for } x < 0, \end{cases}$$

where $r, s \in \mathbf{R}^+$, $r > 1$. The restriction $r > 1$ implies that a skew parabola function is differentiable with derivative given by

$$x \mapsto \begin{cases} rx^{r-1} & \text{for } x \geq 0, \\ -rs|x|^{r-1} & \text{for } x < 0, \end{cases}$$

that is, the derivative is a positive multiple of a semi-multiplicative homeomorphism. In particular, a skew parabola function is strictly convex.

Appendix 2
Lie Transformation Groups

When dealing with the automorphism groups of geometries on surfaces as considered in this book we usually obtain Lie groups. Furthermore, these Lie groups are of dimension at most 8 except in the case of the tubular circle planes of rank greater than 3. In this appendix we compile some useful results on lower-dimensional Lie groups. For general information on Lie groups we refer to Freudenthal–de Vries [1969], Hochschild [1965], and Varadarajan [1974].

We begin with topological groups, which are the basic objects underlying Lie groups, and list some of their properties.

A2.1 Topological Groups

A *topological group* G is a group G equipped with a Hausdorff topology such that the two basic group operations $G \times G \to G : (x, y) \mapsto xy$ and $G \to G : x \mapsto x^{-1}$ are continuous. Most of the groups we encounter in this book operate on some manifold. It then follows that such a group can be equipped with a metric and that it is separable; see Section A2.3. So, if you are unfamiliar with the more general setting of topological spaces, you can always assume you are dealing with a metric space.

A topological group G is called (locally) compact, (locally) connected, finite-dimensional, etc., if the underlying topological space G has the corresponding property.

For each element a in a topological group G *left translation by* a, that is, the map $\lambda_a : G \to G : x \mapsto ax$, is a homeomorphism of G. This implies that G looks the same at each point.

For a topological group G we denote by G^1 the connected component that contains the identity element.

PROPOSITION A2.1.1 (Connected Component) *The connected component G^1 of a topological group G is a closed normal subgroup of G. If G is connected, then G is generated, as an abstract group, by each neighbourhood of the identity element.*

Therefore neighbourhoods of the identity element contain most of the information about a connected topological group.

PROPOSITION A2.1.2 (Factor Groups) *If H is a closed normal subgroup of a locally compact topological group G, then the quotient or factor group G/H can be equipped with a Hausdorff topology such that G/H becomes a locally compact topological group.*

There are many more results that can be stated in the more general context of locally compact topological groups. However, in this book we only need more detailed information about topological groups that are Lie groups.

A2.2 Lie Groups

The collection $\mathrm{M}_n(\mathbf{R})$ of all $n \times n$ matrices with real entries can be viewed as the Euclidean space $\mathbf{R}^{n^2}$. Groups of matrices therefore inherit a differentiable structure from $\mathrm{M}_n(\mathbf{R})$. Furthermore, matrix multiplication is differentiable with respect to this differentiable structure. One can similarly equip the collection $\mathrm{M}_n(\mathbf{C})$ of all $n \times n$ matrices with complex entries with a differentiable structure by canonically embedding $\mathrm{M}_n(\mathbf{C})$ into $\mathrm{M}_{2n}(\mathbf{R})$ by replacing each complex entry $a + bi$ by the matrix

$$\begin{pmatrix} a & -b \\ b & a \end{pmatrix}.$$

A *(real) Lie group G* is a group G that is equipped with a differentiable smooth manifold structure such that the two basic group operations $G \times G \to G : (x, y) \mapsto xy$ and $G \to G : x \mapsto x^{-1}$ are smooth.

The prototype examples of Lie groups are the so-called *general linear groups* $\mathrm{GL}_n(\mathbf{R})$ and $\mathrm{GL}_n(\mathbf{C})$ as well as their closed subgroups the *special linear groups* $\mathrm{SL}_n(\mathbf{C})$ and $\mathrm{SL}_n(\mathbf{C})$, the *special orthogonal groups* $\mathrm{SO}_n(\mathbf{R})$ and $\mathrm{SO}_n(\mathbf{C})$, and the *special unitary groups* $\mathrm{SU}_n(\mathbf{C})$ and $\mathrm{SU}_n(\mathbf{C}, k)$. These groups are defined as follows.

$$\begin{array}{lcl}
\mathrm{GL}_n(\mathbf{R}) & = & \{A \in \mathrm{M}_n(\mathbf{R}) \mid |A| \neq 0\}, \\
\mathrm{GL}_n(\mathbf{C}) & = & \{A \in \mathrm{M}_n(\mathbf{C}) \mid |A| \neq 0\}, \\
\mathrm{SL}_n(\mathbf{R}) & = & \{A \in \mathrm{M}_n(\mathbf{R}) \mid |A| = 1\}, \\
\mathrm{SL}_n(\mathbf{C}) & = & \{A \in \mathrm{M}_n(\mathbf{C}) \mid |A| = 1\}, \\
\mathrm{SO}_n(\mathbf{R}) & = & \{A \in \mathrm{SL}_n(\mathbf{R}) \mid A^t A = I_n\}, \\
\mathrm{SO}_n(\mathbf{C}) & = & \{A \in \mathrm{SL}_n(\mathbf{C}) \mid A^t A = I_n\}, \\
\mathrm{SU}_n(\mathbf{C}) & = & \{A \in \mathrm{SL}_n(\mathbf{C}) \mid \overline{A}^t A = I_n\}, \\
\mathrm{SU}_n(\mathbf{C}, k) & = & \{A \in \mathrm{SL}_n(\mathbf{C}) \mid \overline{A}^t B_{n,k} A = B_{n,k}\},
\end{array}$$

where $|A|$ denotes the determinant of A, I_n is the $n \times n$ identity matrix and $B_{n,k}$ is an $n \times n$ diagonal matrix with the first $n - k$ entries on the diagonal being 1s followed by k entries of (-1)s. The underlying sets are smooth manifolds since they are open subsets of a Euclidean space or by the Regularity Theorem A1.2.1. Note that the determinant function and the maps $A \mapsto A^t$ and $A \mapsto \overline{A}$ are smooth. Furthermore, the group $\mathrm{GL}_n(\mathbf{R})$ is canonically a subgroup of $\mathrm{GL}_n(\mathbf{C})$ and $\mathrm{GL}_n(\mathbf{C})$ is canonically a subgroup of $\mathrm{GL}_{2n}(\mathbf{R})$.

The centre C of each of the above groups G is a normal subgroup consisting of all scalar matrices in G and the factor group G/C can be canonically made into a Lie group. One correspondingly obtains the *projective linear groups* $\mathrm{PGL}_n(\mathbf{R})$ and $\mathrm{PSL}_n(\mathbf{C})$ as well as their closed subgroups the *projective special linear groups* $\mathrm{PSL}_n(\mathbf{R})$ and $\mathrm{PSL}_n(\mathbf{C})$, and the *projective special unitary groups* $\mathrm{PSU}_n(\mathbf{C})$ and $\mathrm{PSU}_n(\mathbf{C}, k)$. These groups are defined as follows.

$$\begin{array}{lcl}
\mathrm{PGL}_n(\mathbf{R}) & = & \mathrm{GL}_n(\mathbf{R})/\{rI_n \mid r \in \mathbf{R}, r \neq 0\}, \\
\mathrm{PGL}_n(\mathbf{C}) & = & \mathrm{GL}_n(\mathbf{C})/\{cI_n \mid c \in \mathbf{C}, c \neq 0\}, \\
\mathrm{PSL}_n(\mathbf{R}) & = & \mathrm{SL}_n(\mathbf{R})/\{rI_n \mid r = \pm 1, r^n = 1\}, \\
\mathrm{PSL}_n(\mathbf{C}) & = & \mathrm{SL}_n(\mathbf{C})/\{cI_n \mid c \in \mathbf{C}, c^n = 1\}, \\
\mathrm{PSU}_n(\mathbf{C}) & = & \mathrm{SU}_n(\mathbf{C})/\{cI_n \mid c \in \mathbf{C}, |c| = c^n = 1\}, \\
\mathrm{PSU}_n(\mathbf{C}, k) & = & \mathrm{SU}_n(\mathbf{C}, k)/\{cI_n \mid c \in \mathbf{C}, |c| = c^n = 1\}.
\end{array}$$

Note that $\mathrm{PGL}_n(\mathbf{C}) = \mathrm{PSL}_n(\mathbf{C})$ but $\mathrm{PSL}_n(\mathbf{R})$ is only the connected component of $\mathrm{PGL}_n(\mathbf{R})$ that contains the identity.

The assumptions in the definition of a Lie group that the manifold is smooth and that the group operations are smooth are overly restrictive. In fact, for most purposes one can make do with manifolds for which all coordinate changes $\varphi_j \varphi_k^{-1}$ and the group operations are twice continuously differentiable; see Freudenthal–de Vries [1969]. One can drop even differentiability conditions altogether. The famous Hilbert Fifth Problem asks for a characterization of Lie groups without differentiability assumptions; see Montgomery–Zippin [1955] Section 4.10.

THEOREM A2.2.1 (Solution to Hilbert's Fifth Problem) *Let G be a locally compact topological group. If G is locally connected and of finite dimension, then G is a Lie group.*

For each $a \in G$ the left translation $\lambda_a : G \to G : x \mapsto ax$ is a (smooth) diffeomorphism of G.

Usually, when dealing with a Lie group that is an automorphism group of a geometry, we use not only the entire automorphism group but also certain subgroups. Subgroups of Lie groups are again Lie groups. However, one has to be careful about the differentiable manifold structure. For example, the group

$$G_c = \left\{ \begin{pmatrix} e^{2\pi it} & 0 \\ 0 & e^{2\pi ict} \end{pmatrix} \;\middle|\; t \in \mathbf{R} \right\},$$

where $c \in \mathbf{R}$, is a subgroup of $\mathrm{GL}_2(\mathbf{C})$. It is a manifold with respect to the differentiable structure obtained by carrying over the differentiable structure of $\mathbf{R}$ via the map

$$t \mapsto \begin{pmatrix} e^{2\pi it} & 0 \\ 0 & e^{2\pi ict} \end{pmatrix}.$$

If c is a rational number, then G_c is a closed (in the topological sense as a subset of $\mathrm{GL}_2(\mathbf{C})$) subgroup of $\mathrm{GL}_2(\mathbf{C})$ and the differentiable structure induced from $\mathrm{GL}_2(\mathbf{C})$ and the one obtained from $\mathbf{R}$ agree. However, if c is not rational, then the differentiable structure obtained from $\mathbf{R}$ is finer than the one induced from $\mathrm{GL}_2(\mathbf{C})$. In this case, G_c winds around the torus

$$\left\{ \begin{pmatrix} e^{2\pi is} & 0 \\ 0 & e^{2\pi it} \end{pmatrix} \;\middle|\; s, t \in \mathbf{R} \right\} \cong \mathrm{SO}_2(\mathbf{R}) \times \mathrm{SO}_2(\mathbf{R})$$

and forms a dense subgroup.

Let G be a Lie group. A *Lie subgroup H of G* is a subgroup $H \leq G$ of G that is a manifold with respect to a differential structure that is possibly finer than the differential structure inherited from G. Then H is a Lie group with respect to this finer differentiable structure, the so-called *Lie topology* of H. However, if H is a closed subgroup of G, then H is a Lie group with respect to the differentiable structure inherited from G. Therefore one usually only looks at closed subgroups of a Lie group.

PROPOSITION A2.2.2 (Closed Subgroups) *Let H be a closed subgroup of a connected Lie group G. Then the following hold.*

(i) *If the coset space G/H is simply connected, then H is connected.*
(ii) *If* $\dim H = \dim G$, *then* $H = G$.
(iii) *If H, in addition, is a normal subgroup of G, then the factor group G/H can be equipped with a manifold structure so that G/H becomes a Lie group. Furthermore,*

$$\dim G = \dim(G/H) + \dim H.$$

In fact, part (iii) in the above theorem can be reversed. If G is a topological group and H is a closed normal subgroup of G such that both H and G/H are Lie groups, then G is also a Lie group.

Let G_1 and G_2 be two Lie groups and let U_1 and U_2 be open neighbourhoods of the identities in G_1 and G_2 respectively. Then a diffeomorphism $\varphi : U_1 \to U_2$ is called a *local isomorphism* from G_1 to G_2, if for all $x, y, z \in U_1$

$$xy = z \text{ if and only if } \varphi(x)\varphi(y) = \varphi(z).$$

If $U_1 = G_1$ and $U_2 = G_2$, then φ is called an *isomorphism* from G_1 to G_2.

Two Lie groups are called *(locally) isomorphic* if there exists a (local) isomorphism between them. The relation of being (locally) isomorphic is an equivalence relation on the collection of all Lie groups. Isomorphic Lie groups are locally isomorphic. The converse is not true. For example, the groups $\mathrm{SL}_2(\mathbf{R})$ and $\mathrm{PSL}_2(\mathbf{R})$ are Lie groups. The canonical homomorphism $\pi : \mathrm{SL}_2(\mathbf{R}) \to \mathrm{PSL}_2(\mathbf{R})$ is a local isomorphism, but the two groups are not isomorphic. To see this just note that the centre of $\mathrm{SL}_2(\mathbf{R})$ is a cyclic group of order 2 generated by $-I_2$ whereas $\mathrm{PSL}_2(\mathbf{R})$ has trivial centre.

A *(linear) representation of a Lie group G on* $\mathbf{R}^m$ is a continuous homomorphism $\varphi : G \to \mathrm{GL}_m(\mathbf{R})$.

THEOREM A2.2.3 (Lie Groups As Linear Groups) *Every Lie group is locally isomorphic to a Lie subgroup of* $\mathrm{GL}_m(\mathbf{R})$ *for some positive integer m. Every subgroup of* $\mathrm{GL}_n(\mathbf{R})$ *is a Lie subgroup of* $\mathrm{GL}_n(\mathbf{R})$.

An important tool in the investigation of the structure of Lie groups and their representations is the Lie algebra associated with a Lie group. We do not want to formally introduce Lie algebras because the Lie groups that occur as automorphism groups of the geometries we are looking at in more detail have small dimensions and their structures are well documented. However, when dealing with higher dimensions Lie algebras provide a convenient and frequently used path to the investigation of the associated Lie groups. The Lie algebra of a Lie group G uniquely determines the local structure of G, that is, two Lie groups are locally isomorphic if and only if their Lie algebras are isomorphic. Furthermore, continuous homomorphisms between Lie groups give rise to homomorphisms between their Lie algebras and conversely a homomorphism between Lie algebras can be lifted to a continuous local homomorphism between the associated Lie groups. This close relationship extends to representations of Lie groups and Lie algebras.

For example, this correspondence between Lie groups and their Lie algebras can be applied to obtain the following result.

THEOREM A2.2.4 (Low-Dimensional Connected Lie Groups) *A connected Lie group of dimension at most 2 is isomorphic to one of the groups* $\mathbf{R}$, $\mathrm{SO}_2(\mathbf{R})$, $\mathbf{R}^2$, L_2, $\mathbf{R} \times \mathrm{SO}_2(\mathbf{R})$, *and* $\mathrm{SO}_2(\mathbf{R}) \times \mathrm{SO}_2(\mathbf{R})$, *where* L_2 *is the connected component of the affine group on* $\mathbf{R}$, *that is,* $\mathrm{L}_2 = \{x \mapsto ax + b \mid a, b \in \mathbf{R}, a > 0\}$.

Note that the two 1-dimensional Lie groups $\mathbf{R}$ and $\mathrm{SO}_2(\mathbf{R})$ have the same Lie algebra so that the two groups are locally isomorphic; the map

$$\varphi : \mathbf{R} \to \mathrm{SO}_2(\mathbf{R}) : t \mapsto \begin{pmatrix} \cos t & -\sin t \\ \sin t & \cos t \end{pmatrix}$$

is a local isomorphism. Similarly, the Lie groups $\mathbf{R}^2$, $\mathbf{R} \times \mathrm{SO}_2(\mathbf{R})$, and $\mathrm{SO}_2(\mathbf{R}) \times \mathrm{SO}_2(\mathbf{R})$ are locally isomorphic and share the same Lie algebra. So we usually have to take extra care and effort to go from the Lie algebra to the group.

A homomorphism $h : G \to H$ between Lie groups G and H is a *covering map* if h is surjective and has discrete kernel. If such a homomorphism h exists, then G is called a *covering group* of H. In particular, G and H are locally isomorphic. For example, the homomorphism φ from above is a covering map of $\mathrm{SO}_2(\mathbf{R})$ and $\mathbf{R}$ is a covering group of $\mathrm{SO}_2(\mathbf{R})$. A simply connected covering group is called a *universal covering group.*

THEOREM A2.2.5 (Universal Covering Groups) *Every connected Lie group* G *has a universal covering group* $\overline{G}$ *and* $\overline{G}$ *is unique up to isomorphism. Moreover, the kernel* K *of a covering map* $h : \overline{G} \to G$ *is a discrete central subgroup of* $\overline{G}$ *and is isomorphic to the fundamental group of* G.

See Hochschild [1965] Section IV.

The simply connected covering groups of the groups $\mathrm{SO}_2(\mathbf{R})$, $\mathrm{SO}_3(\mathbf{R})$, and $\mathrm{PSL}_2(\mathbf{C})$ are isomorphic to $\mathbf{R}$, $\mathrm{SU}_2(\mathbf{C})$, and $\mathrm{SL}_2(\mathbf{C})$, respectively. The group $\mathrm{SL}_2(\mathbf{R})$ is a covering group of $\mathrm{PSL}_2(\mathbf{R})$ but it is not simply connected. In fact, the simply connected covering group $\overline{\mathrm{PSL}_2}(\mathbf{R})$ of $\mathrm{PSL}_2(\mathbf{R})$ has no representation in terms of matrices and cannot easily be written down. See Subsection 2.7.3 for a geometry that admits this group as a group of automorphisms.

Every Lie group that is locally isomorphic to some Lie group G can be obtained from the simply connected covering $\overline{G}$ as a factor group with respect to some subgroup in the centre of $\overline{G}$. For example, $\overline{\mathrm{PSL}_2}(\mathbf{R})$ has an infinitely cyclic centre C. The factor group of $\overline{\mathrm{PSL}_2}(\mathbf{R})$ with respect to a subgroup of C yields a Lie group locally isomorphic to $\mathrm{SL}_2(\mathbf{R})$. In fact, every connected Lie group that is locally isomorphic to $\mathrm{SL}_2(\mathbf{R})$ is isomorphic to precisely one of the groups $\mathrm{PSL}_2^{(k)}(\mathbf{R}) \simeq \overline{\mathrm{PSL}_2}(\mathbf{R})/C_k$ for $k = 1, 2, \ldots, \infty$, where C_k is the cyclic subgroup of $C = \langle c \rangle$ generated by kc, for $k \neq \infty$ and $C_\infty = \{1\}$ is the trivial subgroup. In particular, $\mathrm{PSL}_2^{(k)}(\mathbf{R})$ has cyclic centre of order k and $\mathrm{PSL}_2^{(\infty)}(\mathbf{R}) = \overline{\mathrm{PSL}_2}(\mathbf{R})$.

A noncommutative connected Lie group is called *almost simple* if it has no nontrivial connected proper closed normal subgroup. An almost simple Lie group may not be simple in the abstract group-theoretic sense. However, any nontrivial normal subgroup in such a Lie group is contained in its centre and is discrete.

Almost simple Lie groups are all classified. For small dimensions one obtains the following groups.

THEOREM A2.2.6 (Almost Simple Groups) *Let G be an almost simple connected Lie group of dimension $n \leq 8$. Then $n \in \{3, 6, 8\}$ and G is locally isomorphic to precisely one of the following simple groups.*

$$n = 3 : \mathrm{SO}_3(\mathbf{R}),\ \mathrm{PSL}_2(\mathbf{R})$$
$$n = 6 : \mathrm{PSL}_2(\mathbf{C})$$
$$n = 8 : \mathrm{SL}_3(\mathbf{R}),\ \mathrm{PSU}_3(\mathbf{C}),\ \mathrm{PSU}_3(\mathbf{C}, 1)$$

Using the simply connected covering groups of the above groups one obtains all almost simple connected Lie group of dimension at most 8. For example, the almost simple connected Lie groups in dimensions 3 and 6 are the groups $\mathrm{SO}_3(\mathbf{R})$, $\mathrm{PSU}_2(\mathbf{C})$ (this group is locally isomorphic to $\mathrm{SO}_3(\mathbf{R})$), $\mathrm{PSL}_2^{(k)}(\mathbf{R})$ for $k = 1, 2, \ldots, \infty$ (this group is locally isomorphic to $\mathrm{PSL}_2(\mathbf{R})$), and $\mathrm{PSL}_2(\mathbf{C})$, $\mathrm{SL}_2(\mathbf{C})$ (this group is locally isomorphic to $\mathrm{PSL}_2(\mathbf{C})$).

Almost simple Lie groups are fundamental building blocks of Lie groups in general. The locally direct product of almost simple Lie groups gives rise to the so-called *semi-simple* Lie groups.

The other fundamental building blocks are the *solvable* Lie groups. Every Lie group of dimension at most 2 is solvable. The *radical* of a Lie group G is the largest connected solvable normal subgroup of G. Every Lie group can be built up from semi-simple and solvable Lie groups; see Hochschild [1965] Sections XI and XVIII and Varadarajan [1974].

THEOREM A2.2.7 (Levi's Theorem) *Let G be a connected Lie group. Then G has a unique radical R and R is closed in G.*

There is a maximal semi-simple Lie subgroup S of G, called a Levi complement of R, and $G = SR$. Furthermore, any two Levi complements are conjugate by an element in R.

If G is simply connected, then S is closed in G and G is the semi-direct product of S and R.

Compact subgroups play a prominent role in the theory of Lie groups and as automorphism groups of our geometries. A compact subgroup of a Lie group G is called *maximal compact* if it is not properly contained in any compact subgroup of G.

THEOREM A2.2.8 (Mal'cev–Iwasawa Theorem) *Let G be a connected Lie group. Then there exists a maximal compact subgroup C of G, and each maximal compact subgroup is connected and conjugate to C. Moreover, each compact subgroup of G is contained in a maximal one.*

As a topological space the connected group G is homeomorphic to the space $C \times \mathbf{R}^m$ for some integer $m \geq 0$.

See Iwasawa [1949].

The maximal compact subgroups of the simple Lie groups listed in Theorem A2.2.6 are isomorphic to $\mathrm{SO}_3(\mathbf{R})$, $\mathrm{SO}_2(\mathbf{R})$, $\mathrm{SO}_3(\mathbf{R})$, $\mathrm{SO}_3(\mathbf{R})$, $\mathrm{PSU}_3(\mathbf{C})$, and $\mathrm{SO}_3(\mathbf{R}) \times \mathrm{SO}_2(\mathbf{R})$, respectively.

A2.3 Transformation Groups

Not only are automorphism groups of geometries abstract groups, their elements also act as transformations on the point sets and line/circle sets.

A *(topological) transformation group* (G, M) is a topological group G together with a topological space M and a homomorphism φ from G into the group of homeomorphisms of M such that the map

$$G \times M \to M : (g, x) \mapsto (\varphi(g))(x)$$

is continuous. We say that G *acts* or *operates* on M and often write gx or $g(x)$ for $(\varphi(g))(x)$ for $g \in G$ and $x \in M$. One frequently omits M and refers to G as a transformation group when it is clear what space it operates on. Two transformation groups (G, M) and (H, N) are said to be *equivalent* if there are a homeomorphism $h : M \to N$ and a topological isomorphism $\gamma : G \to H$ such that $h(gx) = \gamma(g)h(x)$ for all $g \in G$ and $x \in M$.

If G is a topological group and H is a closed subgroup, then G acts on the coset space G/H by left translation $xH \mapsto gxH$ and $(G, G/H)$ is a transformation group.

For each $x \in M$ we can form the *orbit* $G(x) = \{gx \mid g \in G\}$ of x under G and the *stabilizer* $G_x = \{g \in G \mid gx = x\}$ of x. Note that $G(x)$ is a subset of M whereas G_x is a subgroup of G. Moreover, the orbits of G form a partition of M. There is a natural bijection from the coset space G/G_x to the orbit $G(x)$ induced by the map $g \mapsto gx$. We say that G *acts freely* on M if each stabilizer G_x is trivial, that is, consists of the identity only. In this case each orbit is in bijection to the group G.

The intersection of all stabilizers G_x for $x \in M$ is the *kernel* of the action of G on M. If this kernel is trivial, that is, consists of the identity element only, we say that G *acts effectively* on M.

LEMMA A2.3.1 (Stabilizers and Kernels) *Let (G, M) be a transformation group and let $x \in M$. Then G_x is a closed subgroup of G and the kernel is a closed normal subgroup of G.*

If (G, M) is a transformation group and K is the kernel of the action of G on M, then G/K is a transformation group that acts effectively on M.

Let G be a group of homeomorphisms acting on a manifold M. The *compact-open topology* on G is generated by the subbasis consisting of all sets $\{g \in G \mid g(C) \subseteq U\}$, where C is is a compact set and U is an open subset of M.

PROPOSITION A2.3.2 (Groups of Homeomorphisms) *Let G be a group of homeomorphisms acting on a manifold M. If G is equipped with the compact-open topology, then G becomes a separable metric space and (G, M) is a topological transformation group.*

Under the assumptions of the above proposition, the compact-open topology on G can also be described by sequences as follows. A sequence (g_n) converges to $g \in G$ if and only if the sequence $(g_n x_n)$ converges to gx whenever (x_n) is a sequence in M that converges to $x \in M$; see Dugundji [1966] Section XII.

A transformation group G is *transitive* on M if G has only one orbit, that is, $G(x) = M$ for one (and thus any) $x \in M$.

LEMMA A2.3.3 (Transitive Groups) *Let G be a transitive topological transformation group on a separable locally compact metric space M. If G is locally compact, then the canonical map from G/G_x to $G(x) = M$ is a homeomorphism and (G, M) is equivalent to $(G, G/G_x)$.*

Locally compact transformation groups acting on manifolds are often very close to being Lie groups. In fact, the most difficult part in the verification of automorphism groups on surfaces being Lie groups is to show that these groups are locally compact.

THEOREM A2.3.4 (Szenthe's Theorem) *If G is a locally compact, connected, effective and transitive transformation group on a connected manifold, then G is a Lie group.*

See Szenthe [1974] Theorem 4. For surfaces special features imply

THEOREM A2.3.5 (Transformation Group on Surface Is Lie) *If G is a locally compact effective transformation group on a surface, then G is a Lie group.*

If G is a group of homeomorphisms acting on a surface M, then (G, M) is a topological transformation group when G is equipped with the compact-open topology by Proposition A2.3.2. Usually G is effective on M so that we only have to show that G is locally compact in order to verify that G is a Lie group. Since G is a topological group, G is locally compact if and only if the identity has a compact neighbourhood.

For a *Lie transformation group* (G, M) we require that G is a Lie group, that M is a smooth manifold and that the action of G on M is smooth.

Orbit and stabilizer of a point determine each other. In particular, their dimensions and the dimension of the transformation group are linked by the following formula; see Halder [1971].

THEOREM A2.3.6 (Dimension Formula) *If the Lie group G acts on a manifold M, then*

$$\dim G = \dim G_p + \dim G(p),$$

where G_p and $G(p)$ are the stabilizer and orbit, respectively, of the point $p \in M$.

The dimension formula is valid for a larger class of topological transformation groups; see Salzmann et al. [1995] 96.10.

A frequent and typical application of Theorem A1.3.2 and the dimension formula is the following; compare Salzmann et al. [1995] 96.11.

COROLLARY A2.3.7 (Actions on Manifolds) *Let G be a Lie transformation group acting on an n-dimensional manifold M.*

(i) *If $p \in M$ is a point whose orbit $G(p)$ has dimension n, then $G(p)$ is open in M. If, in addition, G is compact and M is connected, then $G(p) = M$ and G is transitive on M.*

(ii) *If G is transitive on M and M is connected, then G^1 is transitive.*

As we have already seen, transitive and effective transformation groups are a distinguished class of transformation groups. We describe the classification of transitive and effective transformation groups on 1-manifolds; see Brouwer [1909] or Hofmann–Mostert [1968].

THEOREM A2.3.8 (Brouwer's Theorem) *Let G be a locally compact, connected, effective and transitive transformation group on a connected 1-dimensional manifold M. Then G has dimension at most 3.*

(i) *If $M \cong \mathbf{S}^1$, then G is isomorphic and acts equivalently to the rotation group $\mathrm{SO}_2(\mathbf{R})$ or a finite covering group $\mathrm{PSL}_2^{(k)}(\mathbf{R})$, $1 \leq k < \infty$, of the projective group $\mathrm{PSL}_2(\mathbf{R})$.*

(ii) *If $M \cong \mathbf{R}$, then G is isomorphic and acts equivalently to $\mathbf{R}$, the connected component $\mathrm{L}_2(\mathbf{R})$ of the affine group of $\mathbf{R}$, or the simply connected covering group of the projective group $\mathrm{PSL}_2(\mathbf{R})$.*

Immediate and useful consequences of Brouwer's Theorem are the following two corollaries.

COROLLARY A2.3.9 (Actions on R and $\mathbf{S}^1$ I) *Let G be a locally compact, connected transformation group on a connected 1-dimensional manifold M. If G is at least 4-dimensional, then G cannot be both effective and transitive on M.*

If G is effective and at least 4-dimensional, then G fixes a point of M. Furthermore, if $M \simeq \mathbf{S}^1$, then G fixes at least two points of M.

COROLLARY A2.3.10 (Actions on R and $\mathbf{S}^1$ II) *Let $G \cong \mathrm{SO}_2(\mathbf{R})$ be a transformation group acting on a connected* 1*-manifold.*

(i) *If $M \cong \mathbf{R}$, then G acts trivially on $\mathbf{R}$.*

(ii) *If $M \cong \mathbf{S}^1$, then G acts either trivially or transitively on $\mathbf{S}^1$.*

All transitive and effective transformation groups on $\mathbf{S}^2$ are also completely classified; see Mostow [1950].

THEOREM A2.3.11 (Transitive Actions on $\mathbf{S}^2$) *A locally compact, connected transitive and effective transformation group G of the 2-sphere $\mathbf{S}^2$ is isomorphic and acts equivalently to $\mathrm{PSL}_3(\mathbf{R})$, $\mathrm{PSL}_2(\mathbf{C})$, or $\mathrm{SO}_3(\mathbf{R})$ in their respective standard actions on $\mathbf{S}^2$.*

In particular, if G acts 2-transitively and effectively on $\mathbf{S}^2$, then G is isomorphic and acts equivalently to $\mathrm{PSL}_2(\mathbf{C})$ in its standard action on $\mathbf{S}^2$ as the group of fractional linear transformations.

There are many Lie transformation groups that are transitive on $\mathbf{R}^2$. For 2-transitive groups however there are only a few possibilities; see Tits [1952], [1956].

THEOREM A2.3.12 (2-Transitive Actions on $\mathbf{R}^2$) *Let G be a locally compact, connected* 2*-transitive and effective transformation group of $\mathbf{R}^2$. Then G is isomorphic and acts equivalently to*

- $\mathrm{L}_2(\mathbf{C}) = \{z \mapsto az + b \mid a, b \in \mathbf{C}, a \neq 0\}$,
- $\{x \mapsto Ax + t \mid A \in \mathrm{SL}_2(\mathbf{R}), t \in \mathbf{R}^2\}$, *the semi-direct product of the groups $\mathbf{R}^2$ and $\mathrm{SL}_2(\mathbf{R})$, or*
- $\{x \mapsto Ax + t \mid A \in \mathrm{GL}_2(\mathbf{R})^1, t \in \mathbf{R}^2\}$, *the semi-direct product of the groups $\mathbf{R}^2$ and $\mathrm{GL}_2(\mathbf{C})^1$,*

in their respective standard actions on $\mathbf{R}^2$.

For the dimension of orbits of compact Lie groups see Montgomery–Zippin [1955] 6.3 Theorem 2.

THEOREM A2.3.13 (Orbits of Compact Groups) *Let G be a compact connected effective Lie transformation group on a connected manifold M of dimension n. Then*

(i) $\dim G \leq n(n+1)/2$ *if G is transitive on M;*
(ii) $\dim G \leq n(n-1)/2$ *if G is not transitive on M.*

In particular, a compact connected effective Lie transformation group on a surface can have dimension at most 3.

As we already mentioned in Section A2.2, $\mathbf{R}$ can wind around the torus $\mathbf{S}^1 \times \mathbf{S}^1$ infinitely often without ever closing back in on itself, that is, the transformation group $\mathbf{R}$ can have a dense orbit on $\mathbf{S}^1 \times \mathbf{S}^1$. For the 2-sphere and the real projective plane this cannot occur; see Halder [1973].

THEOREM A2.3.14 (Halder's Theorem) *Let G be a locally compact connected transformation group of $\mathbf{S}^2$ or the real projective plane. Then G has a closed orbit. Every closed G-orbit in $\mathbf{S}^2$ is a manifold so that G fixes a point, has an orbit that is homeomorphic to $\mathbf{S}^1$, or acts transitively on $\mathbf{S}^2$.*

The group $\mathrm{SO}_2(\mathbf{R})$ can act on $\mathbf{S}^2$ only trivially or equivalently to the standard rotation group; see Montgomery–Zippin [1955] p. 260. This implies the following for the orbits of such a group.

THEOREM A2.3.15 ($\mathrm{SO}_2(\mathbf{R})$ Acting on $\mathbf{S}^2$) *Let $G \cong \mathrm{SO}_2(\mathbf{R})$ be a transformation group acting effectively on $\mathbf{S}^2$. Then G acts equivalently to the rotation group.*

In particular, the group G fixes precisely two points a and b of $\mathbf{S}^2$ and operates on every orbit $G(c)$, $c \neq a, b$, effectively and sharply transitively. Furthermore, the orbits $G(c)$ separate the points a and b.

We conclude this section with the classification of the closed subgroups of the group $\mathrm{PSL}_2(\mathbf{C})$ of dimension at least 4; see Coolidge [1916] Section VII.3.

THEOREM A2.3.16 (Subgroups of $\mathrm{PGL}_2(\mathbf{C})$ Acting on $\mathbf{S}^2$) *Let G be the 6-dimensional connected Lie group $\mathrm{PGL}_2(\mathbf{C}) = \mathrm{PSL}_2(\mathbf{C})$ acting on $\mathbf{S}^2$ in the standard way as the group of fractional linear maps.*

The group G is the only 6-dimensional subgroup of G and does not contain any 5-dimensional closed connected subgroup.

A 4-dimensional closed connected subgroup of G is conjugate to the *group* $\mathrm{L}_2(\mathbf{C}) = \{z \mapsto az + b \mid a, b \in \mathbf{C}, a \neq 0\}$, *the stabilizer of* ∞ *under the standard action.*

A connected 3-dimensional subgroup of G *is conjugate to one of the three groups* $\mathrm{SO}_3(\mathbf{R})$, $\mathrm{PSL}_2(\mathbf{R})$, *and* $\{z \mapsto e^{ict}z + b \mid b \in \mathbf{C}, t \in \mathbf{R}\}$ *for some* $c \in \mathbf{R}$, $c \neq 0$.

Bibliography

Anisov, S. S.

[1992] The collineation group of Hilbert's example of a projective plane (Russian), *Uspehi Mat. Nauk* **47**, 147–148.

Arens, R.

[1946] Topologies for homeomorphism groups, *Amer. J. Math.* **68**, 593–610.

Artzy, R. and Groh, H.

[1986] Laguerre and Minkowski planes produced by dilatations, *J. Geom.* **26**, 1–20.

Barlotti, A.

[1957] Le possibili configurazioni del sistema delle coppie puntoretta (A, a) per cui un piano grafico risulta $A - a$ transitivo. *Boll. Un. Mat. Ital.* **12**, 212–226.

Bedürftig, T.

[1974a] Polaritäten ebener projektiver Ebenen, *J. Geom.* **5**, 39–66.

[1974b] Ebene projektive Ebenen über Neokörpern, *Geom. Dedicata* **3**, 21–34.

Benz, W.

[1973] *Vorlesungen über Geometrie der Algebren*, Springer, Berlin.

Betten, D.

[1968] Topologische Geometrien auf dem Möbiusband, *Math. Z.* **107**, 363–379.

[1971] 2-dimensionale differenzierbare projektive Ebenen, *Arch. Math.* **22**, 304–309.

[1972] Projektive Darstellung der Moulton-Ebenen, *J. Geom.* **2**, 107–114.

[1979] Die Projektivitätengruppe der Moulton-Ebenen, *J. Geom.* **13**, 197–209.

Betten, D. and Ostmann, A.

[1978] Wirkungen und Geometrien der Gruppe $L_2 \times \mathbf{R}$, *Geom. Dedicata* **7**, 141–162.

Betten, D. and Wagner, A.

[1982] Eine stückweise projektive topologische Gruppe im Zusammenhang mit den Moulton-Ebenen, *Arch. Math.* **38**, 280–285.

Betten, D. and Weigand, C.

[1985] Groups of projectivities of topological planes, *Math. Rep. Acad. Sci. Canada* **7**, 73–78.

Biliotti, M. and Johnson, N. L.

[1999] Bilinear flocks of quadratic cones, *J. Geom.* **64**, 16–50.

Björner, A., Las Vergnas, M., Sturmfels, B., White, N., and Ziegler, G.

[1993] *Oriented Matroids*, Encyclopedia of Mathematics and its Applications 46, Cambridge University Press, Cambridge.

Bödi, R.

[1997] Smooth stable planes, *Results Math.* **31**, 300–321.

[1998a] Stabilizers of collineation groups of smooth stable planes, *Indag. Math.* **9**, 477–490.

[1998b] Collineations of smooth stable planes, *Forum Math.* **10**, 751–773.

[1998c] Solvable collineation groups of smooth projective planes, *Beiträge Algebra Geom.* **39**, 121–133.

Bödi, R. and Immervoll, S.

[2000] Implicit characterizations of smooth incidence geometries, *Geom. Dedicata* **83**, 63–76.

Bödi, R., Immervoll, S., and Löwe, H.

[2000] Smooth stable planes and the moduli spaces of locally compact translation planes, *Monatsh. Math.* **129**, 303–319.

Bokowski, J.

[1993] Oriented matroids, in: Handbook of Convex Geometry (eds. P. Gruber, J. Wills), pp. 555–602, North-Holland, Amsterdam.

Bonisoli, A.

[1989] On resolvable finite Minkowski planes, *J. Geom.* **36**, 1–7.

Braess, D.

[1980] *Nonlinear Appoximation Theory*, Springer, Berlin.

Breitsprecher, S.

[1967a] Uniforme projektive Ebenen, *Math. Z.* **95**, 139–168.

[1967b] Einzigkeit der reellen und der komplexen projektiven Ebene, *Math. Z.* **99**, 429–432.

[1971] Projektive Ebenen die Mannigfaltigkeiten sind, *Math. Z.* **121**, 157–174.

[1972] Zur topologischen Struktur zweidimensionaler projektiver Ebenen, *Geom. Dedicata* **1**, 21–32.

Bröcker, L.

[1971] Locally Desarguesian surfaces, in: Atti del Convegno di Geometria Combinatoria e sue Applicazioni (Proc. Perugia 1970), pp. 103–111, Univ. Perugia.

Brouwer, L. J.

[1909] Die Theorie der endlichen kontinuierlichen Gruppen, unabhängig von den Axiomen von Lie, *Math. Ann.* **67**, 246–267.

[1919] Über die periodischen Transformationen der Kugel, *Math. Ann.* **80**, 39–41.

Bruen, A., Fisher, J. C., and Wilker, J. B.

[1983] Apollonius by inversion, *Math. Mag.* **56**, 97–103.

Buchanan, T., Hähl, H., and Löwen, R.

[1980] Topologische Ovale, *Geom. Dedicata* **9**, 401–424.

Buckel, W.

[1953] Eine Kennzeichnung des Systems aller Kreise mit nicht verschwindendem Radius, *J. Reine Angew. Math.* **191**, 13–30.

Buekenhout, F.

[1966] Étude intrinsèque des ovales, *Rend. Mat.* (5) **25**, 333–393.

[1969] Ensembles quadratiques des espaces projectifs, *Math. Z.* **110**, 306–318.

[1995] *Handbook of Incidence Geometry*, Elsevier, Amsterdam.

Busemann, H.

[1955] *The Geometry of Geodesics*, Academic Press, New York.

Bush, K. A.
[1952] Orthogonal arrays of index unity, *Ann. Math. Statistics* **23**, 426–434.

Cantwell, J.
[1974] Geometric convexity I, *Bull. Inst. Math. Acad. Sinica* **2**, 289–307.
[1978] Geometric convexity. II: Topology, *Bull. Inst. Math. Acad. Sinica* **6**, 303–311.

Cantwell, J. and Kay, D. C.
[1978] Geometric convexity. III: Embedding, *Trans. Amer. Math. Soc.* **246**, 211–230.

Carathéodory, C.
[1937] The most general transformations of plane regions which transform circles into circles, *Bull. Amer. Math. Soc.* **43**, 573–579.

Castro, S.
[1987] *Miquelsche Minkowski-Ebenen in spiegelungsgeometrischer Darstellung und topologische Minkowski-Ebenen*, Dissertation A, Erfurt.

Castro, S. and Wernicke, B.
[1990] Über topologische Minkowski-Ebenen und topologische affine Ebenen, *Beiträge Algebra Geom.* **30**, 61–68.

Colbourn, C. J. and Dinitz, J. H., eds.
[1996] *The CRC Handbook of Combinatorial Designs*, CRC Press Series on Discrete Mathematics and its Applications. CRC Press, Boca Raton, Fl.

Coolidge, J. L.
[1916] *A Treatise on the Circle and the Sphere*, Chelsea Publishing Co., Bronx, N.Y. Reprinted 1971.

Coxeter, H. S. M.
[1968] The problem of Apollonius, *Amer. Math. Monthly* **75**, 5–15.

Curtis, P. C., Jr
[1959] n-parameter families and best approximation, *Pacific J. Math.* **9**, 1013–1027.

Delandtsheer, A.
[1995] Dimensional linear spaces, in: Buekenhout [1995], pp. 193–294.

Dembowski, P.
[1968] *Finite Geometries*, Springer, Berlin. Reprinted 1997.

Dienst, K. J.

[1977] Minkowski-Ebenen mit Spiegelungen, *Monatsh. Math.* **84**, 197–208.

Doignon, J.-P.

[1971] Sur les espaces projectifs topologiques, *Math. Z.* **122**, 57–60.

[1976] Caractérisations d'espaces de Pasch–Peano, *Bull. Acad. R. Belg. Cl. Sc.* (5) **62**, 679–699.

Drandell, M.

[1952] Generalized convex sets in the plane, *Duke Math. J.* **19**, 537–547.

Dugundji, J.

[1966] *Topology*, Allyn and Bacon, Boston, Mass.

Ewald, G.

[1957] Begründung der Geometrie der ebenen Schnitte einer Semiquadrik, *Arch. Math.* **8**, 203–208.

[1960] Beispiel einer Möbiusebene mit nichtisomorphen affinen Unterebenen, *Arch. Math.* **11**, 146–150.

[1967] Aus konvexen Kurven bestehende Möbiusebenen, *Abh. Math. Sem. Univ. Hamburg* **30**, 179–187.

Faina, G.

[1984] The B-ovals of order ≤ 8, *J. Combin. Theory Ser. A* **36**, 307–314.

Forst, M.

[1981] Topologische 4-Gone, *Mitt. Math. Sem. Giessen* **147**, 65–129.

Förtsch, A.

[1982] *Kollineationsgruppen endlich-dimensionaler topologischer Benz-Ebenen*, Dissertation, Erlangen–Nürnberg.

Freudenthal, H.

[1957a] Kompakte projektive Ebenen, *Illinois J. Math.* **1**, 9–13.

[1957b] Zur Geschichte der Grundlagen der Geometrie. Zugleich eine Besprechung der 8. Auflage von Hilberts 'Grundlagen der Geometrie', *Nieuw Arch. Wisk.* (4) **5**, 105–142.

[1965] Lie groups in the foundations of geometry, *Adv. Math.* **1**, 145–190.

Freudenthal, H. and de Vries, H.

[1969] *Linear Lie Groups*, Pure and Applied Mathematics 35, Academic Press, New York–London.

Freudenthal, H. and Strambach, K.

[1975] Schließungssätze und Projektivitäten in der Möbius- und Laguerregeometrie, *Math. Z.* **143**, 213–234.

Gibbons, J. C. and Webb, C.

[1979] Circle-preserving functions of spheres, *Trans. Amer. Math. Soc.* **248**, 67–83.

Goodman, J. E., Pollack, R., Wenger, R., and Zamfirescu, T.

[1994a] Every arrangement extends to a spread, *Combinatorica* **14**, 301–306.

[1994b] Arrangements and topological planes, *Amer. Math. Monthly* **101**, 866–877.

Groh, H.

[1968] Topologische Laguerreebenen I, *Abh. Math. Sem. Univ. Hamburg* **32**, 216–231.

[1969] Characterization of ovoidal Laguerre planes, *Arch. Math.* **20**, 219–224.

[1970] Topologische Laguerreebenen II, *Abh. Math. Sem. Univ. Hamburg* **34**, 11–21.

[1971a] Laguerre planes generated by Moebius planes, in: Proceedings of the Twenty-Fifth Summer Meeting of the Canadian Mathematical Congress 1971, pp. 373–404, Lakehead Univ., Thunder Bay, Ont.

[1971b] On flat Laguerre planes, *J. Geom.* **1**, 18–40.

[1971c] 1-dimensional orbits in flat projective planes, *Math. Z.* **122**, 117–124.

[1972] Flat Moebius planes, *Geom. Dedicata* **1**, 65–84.

[1973] Moebius planes with locally Euclidean circles are flat, *Math. Ann.* **201**, 149–156.

[1974a] Laguerre planes generated by Moebius planes, *Abh. Math. Sem. Univ. Hamburg* **40**, 43–63.

[1974b] Flat Moebius and Laguerre planes, *Abh. Math. Sem. Univ. Hamburg* **40**, 64–76.

[1974c] Ovals and non-ovoidal Laguerre planes, *J. Reine Angew. Math.* **267**, 50–66.

[1975] $\mathbf{R}^2$-planes with 2-dimensional point transitive automorphism group, in: Proc. of the Lattice Theory Conference (Ulm, 1975), pp. 80–91, Univ. Ulm.

[1976] Point homogeneous flat affine planes, *J. Geom.* **8**, 145–162.

[1977] Flat projective planes whose automorphism group contains $\mathbf{R}^2$, in: Beiträge zur geometrischen Algebra (Proc. Duisburg 1976), pp. 129–131, Birkhäuser, Basel.

[1979] $\mathbf{R}^2$-planes with 2-dimensional point transitive automorphism group, *Abh. Math. Sem. Univ. Hamburg* **48**, 171–202.

[1981] Pasting of $\mathbf{R}^2$-planes, *Geom. Dedicata* **11**, 69–98.

[1982a] $\mathbf{R}^2$-planes with point transitive 3-dimensional collineation group, *Indag. Math.* **44**, 173–182.

[1982b] Isomorphism types of arc planes, *Abh. Math. Sem. Univ. Hamburg* **52**, 133–149.

Groh, H. and Heise, W.

[1973] 3-ovals in Moebius planes, *Geom. Dedicata* **1**, 426–433.

Groh, H., Lippert, M. F., and Pohl, H.-J.

[1983] $\mathbf{R}^2$-planes with 3-dimensional automorphism group fixing precisely a line, *J. Geom.* **21**, 66–96.

Grünbaum, B.

[1966] Continuous families of curves, *Canad. J. Math.* **18**, 529–537.

[1972] *Arrangements and Spreads*, Conference Board of the Mathematical Sciences, Regional Conference Series in Mathematics 10, American Mathematical Society, Providence, R.I.

Grundhöfer, T.

[1988] The groups of projectivities of finite projective and affine planes, *Ars Combin.* **25**, 269–275.

Grundhöfer, T. and Löwen, R.

[1995] Linear topological geometries, Buekenhout [1995], pp. 1255–1324.

Halder, H. R.

[1971] Dimension der Bahnen lokalkompakter Gruppen, *Arch. Math.* **22**, 302–303.

[1973] Über Bahnen lokalkompakter Gruppen auf Flächen, *Geom. Dedicata* **2**, 101–109.

[1974] Ebene n-Schnitt-Geometrien und sphärische Strukturen, *Arch. Math.* **25**, 553–560.

Hartman, P.

[1958] Unrestricted n-parameter families, *Rend. Circ. Mat. Palermo* (2) **7**, 123–142.

[1971] On n-parameter families and interpoation problems for nonlinear ordinary differential equations, *Trans. Amer. Math. Soc.* **154**, 201–226.

Hartmann, E.

[1976] Moulton–Laguerre-Ebenen, *Arch. Math.* **27**, 424–435.

[1979] Eine Klasse nicht einbettbarer Laguerre-Ebenen, *J. Geom.* **13**, 49–67.

[1981] Beispiele nicht einbettbarer reeller Minkowski-Ebenen, *Geom. Dedicata* **10**, 155–159.

[1982a] Minkowski-Ebenen mit Transitivitätseigenschaften, *Resultate Math.* **5**, 136–148.

[1982b] Transitivitätssätze für Laguerre-Ebenen, *J. Geom.* **18**, 9–27

[1984] Ovoide und Möbius-Ebenen über konvexen Funktionen, *Geom. Dedicata* **15**, 377–388.

Haupt, O. and Künneth, H.

[1967] *Geometrische Ordnungen*, Springer, Berlin.

Heath, T. L.

[1963] *A Manual of Greek Mathematics*, Dover Publications, N.Y.

Heise, W.

[1969] Zum Begriff der topologischen Möbius-Ebene, *Abh. Math. Sem. Hamburg* **33**, 216–224.

[1971] Bericht über κ-affine Geometrien, *J. Geom.* **1**, 197–224.

Heise, W. and Karzel, H.

[1973] Symmetrische Minkowski-Ebenen, *J. Geom.* **3**, 5–20.

Hering, C.

[1965] Eine Klassifikation der Möbiusebenen, *Math. Z.* **87**, 252–262.

Herzer, A.

[1995] Chain geometries, in: Buekenhout [1995], pp. 781–842.

Higman, D. G. and McLaughlin, J. E.

[1961] Geometric ABA-groups, *Illinois J. Math.* **5**, 382–397.

Hilbert, D.

[1899] *Grundlagen der Geometrie*, Teubner, Leipzig.

[1903] Über die Grundlagen der Geometrie, *Math. Ann.* **56**, 381–422 (appendix IV in Hilbert [1930]).

[1930] *Grundlagen der Geometrie*, 7th ed., Teubner, Leipzig and Berlin; 13th ed., Teubner, Stuttgart, 1987.

Hochschild, G.

[1965] *The Structure of Lie Groups*, Addison-Wesley, Reading, Mass.

Hofmann, K. H. and Mostert, P. S.

[1968] One dimensional coset spaces, *Math. Ann.* **178**, 44–52.

Hubig, M.

[1987] *Zweidimensionale stabile Ebenen mit auflösbarer, nicht zu $L_2 \times \mathbf{R}$ lokalisomorpher, mindestens dreidimensionaler Kollineationsgruppe*, Diplomarbeit, Tübingen.

Hughes, D. R. and Piper, F. C.

[1973] *Projective Planes*, Graduate Texts in Mathematics 6, Springer, New York–Berlin.

Hurewicz, W. and Wallman, H.

[1948] *Dimension Theory*, Princeton University Press, Princeton, N.J.

Iversen, U.

[1970] Zum Begriff der topologischen Laguerre-Ebene, *Abh. Math. Sem. Hamburg* **34**, 227–237.

Iwasawa, K.

[1949] On some types of topological groups, *Ann. Math.* **50**, 507–558.

Jakóbowski, J., Kroll, H.-J., and Matrás, A.

[20XX] Minkowski planes admitting automorphism groups of small type, preprint.

Johnson, D. J.

[1976] The trigonometric Hermite–Birkhoff interpolation problem, *Trans. Amer. Math. Soc.* **212**, 365–374.

Jónsson, W. J.

[1963] Transitivität und Homogenität projektiver Ebenen, *Math. Z.* **80**, 269–292.

Kaerlein, G. F. F.

[1970] *Der Satz von Miquel in der Pseudo-Euklidischen (Minkowskischen) Geometrie*, Dissertation, Bochum.

Kahn, J.

[1980] Locally projective-planar lattices which satisfy the bundle theorem, *Math. Z.* **175**, 219–247.

Karlin, S. and Studden, W. J.

[1966] *Tchebycheff Systems: with Applications in Analysis and Statistics*, Interscience, New York.

Kemperman, J. H. B.

[1969] On the regularity of generalized convex functions, *Trans. Amer. Math. Soc.* **135**, 69–93.

Klein, M.

[1992] A classification of Minkowski planes by transitive groups of homotheties, *J. Geom.* **43**, 116–128.

Klein, M. and Kroll, H.-J.

[1989] A classification of Minkowski planes, *J. Geom.* **36**, 99–109.

[1994] On Minkowski planes with transitive groups of homotheties, *Abh. Math. Sem. Univ. Hamburg* **64**, 303–313.

Kleinewillinghöfer, R.

[1979] *Eine Klassifikation der Laguerre-Ebenen*, Dissertation, Darmstadt.

[1980] Eine Klassifikation der Laguerre-Ebenen nach $\mathcal{L}$-Streckungen und $\mathcal{L}$-Translationen, *Arch. Math.* **34**, 469–480.

Knuth, D. E.

[1992] *Axioms and Hulls*, Lecture Notes in Computer Science 606, Springer, Berlin.

Kramer, L.

[1994] *Compact Polygons*, Ph.D. Thesis, Tübingen.

Krier, N.

[1973] The Hering classification of Möbius planes, in: Proc. Internat. Conf. Projective Planes (Washington State Univ., Pullman, Wash. 1973), pp. 157–163, Washington State Univ. Press, Pullman, Wash.

[1977] A Buekenhout oval which is not projective, *Arch. Math.* **28**, 323–324.

Kroll, H.-J.

[1970] Ordnungsfunktionen in Möbiusebenen, *Abh. Math. Sem. Univ. Hamburg* **35**, 195–214.

[1971] Ordnungsfunktionen in Miquelschen Möbiusebenen und ihre Beziehung zu algebraischen Anordnungen, *J. Geom.* **1**, 90–109.

[1977a] Anordnungsfragen in Benz-Ebenen, *Abh. Math. Sem. Univ. Hamburg* **46**, 217–255.

[1977b] Perspektivitäten in Benz-Ebenen, in: Beiträge zur geometrischen Algebra (Proc. Sympos., Duisburg, 1976), pp. 203–207. Lehrbücher u. Monographien aus dem Gebiete der Exakt. Wiss., Math. Reihe 21, Birkhäuser, Basel.

[1977c] Die Gruppe der eigentlichen Projektivitäten in Benz-Ebenen, *Geom. Dedicata* **6**, 407–413.

[1995] Eine Kennzeichnung der miquelschen Minkowski-Ebenen durch Transitivitätseigenschaften, *Geom. Dedicata* **58**, 75–77.

Kroll, H.-J. and Matraś, A.

[1997] Minkowski planes with Miquelian pairs, *Beiträge Algebra Geom.* **38**, 99–109.

Kuiper, N. H.

[1957] A real analytic non-Desarguesian plane, *Nieuw Arch. Wisk.* (3) **5**, 19–24.

Lenz, H.

[1954] Kleiner Desarguesscher Satz und Dualität in projektiven Ebenen, *Jahresber. Deutsch. Math.-Verein* **57**, 20–31.

[1992] Konvexität in Anordnungsräumen, *Abh. Math. Sem. Univ. Hamburg* **62**, 255–285.

Levenberg, L.

[1950] A class of non-Desarguesian plane geometries. *Amer. Math. Monthly* **57**, 381–387.

Lippert, M. F.

[1986] *Flat projective planes with two-dimensional non-commutative automorphism group fixing a semioval*, Dissertation, Darmstadt.

Lombardo-Radice, L.

[1955] I piani di refrazione, *Rend. Mat. V Ser.* **14**, 130–139.

Lorentz, G. G., Jetter, K., and Riemenschneider, S. D.

[1983] *Birkhoff Interpolation.* Encyclopedia of Mathematics and its Applications 19. Addison-Wesley, Reading, Mass.

Löwen, R.

[1972] *Über die Punkt- und Geradenräume ebener Ebenen*, Diplomarbeit, Tübingen

[1976] Vierdimensionale stabile Ebenen, *Geom. Dedicata* **5**, 239–294.

[1977] Schleiermachers Starrheitsbedingung für Projektivitäten in der topologischen Geometrie, *Math. Z.* **155**, 23–28.

[1979] Weakly flag homogeneous stable planes of low dimension, *Arch. Math.* **33**, 485–491.

[1981a] Homogeneous compact projective planes, *J. Reine Angew. Math.* **321**, 217–220.

[1981b] Characterization of symmetric planes in dimension at most 4, *Indag. Math.* **43**, 87–103.

[1981c] Central collineations and the parallel axiom in stable planes, *Geom. Dedicata* **10**, 283–315.

[1981d] Projectivities and the geometric structure of topological planes, in: Geometry—von Staudt's point of view (eds. P. Plaumann and K. Strambach), Bad Windsheim, 339–372.

[1982] A local 'Fundamental Theorem' for classical topological projective spaces, *Arch. Math.* **38**, 286–288.

[1983a] Topology and dimension of stable planes: on a conjecture of H. Freudenthal, *J. Reine Angew. Math.* **343**, 108–122.

[1983b] Zweidimensionale stabile Ebenen mit nicht-auflösbarer Automorphismengruppe, *Arch. Math.* **41**, 565–571.

[1984a] Compact projective planes with homogeneous ovals, *Monatsh. Math.* **97**, 55–61.

[1984b] Ebene stabile Ebenen mit vielen Zentralkollineationen, *Mitt. Math. Sem. Giessen* **165**, 63–67.

[1986a] Stable planes admitting a classical motion group, *Resultate Math.* **9**, 119–130.

[1986b] A criterion for stability of planes, *Arch. Math.* **46**, 275–278.

[1994] Topological pseudo-ovals, elation Laguerre planes, and elation generalized quadrangles, *Math. Z.* **216**, 347–369.

[1995] Ends of surface geometries, revisited, *Geom. Dedicata* **58**, 175–183.

Löwen, R. and Pfüller, U.

[1987a] Two-dimensional Laguerre planes over convex functions, *Geom. Dedicata* **23**, 73–85.

[1987b] Two-dimensional Laguerre planes with large automorphism groups, *Geom. Dedicata* **23**, 87–96.

Löwen, R. and Salzmann, H.

[1982] Collineation groups of compact connected projective planes, *Arch. Math.* **38**, 368–373.

Luksch, C.

[1987] Die Automorphismengruppe der Polynomgeometrie vom Grad n, *Mitt. Math. Sem. Giessen* **181**.

Mathsen, R. M.

[1969] $\lambda(n)$-parameter families, *Canad. Math. Bull.* **12**, 185–191.

[1972] $\lambda(n)$-convex functions, *Rocky Mountain J. Math.* **2**, 31–43.

[1982] Hereditary $\lambda(n,k)$-families and generalized convexity of functions, *Rocky Mountain J. Math.* **12**, 753–756.

Mäurer, H.

[1967] Möbius- und Laguerre-Geometrien über schwach konvexen Semiflächen, *Math. Z.* **98**, 355–386.

[1969] Die der Laguerre-Geometrie zugehörige Lie-Geometrie, *Abh. Math. Sem. Univ. Hamburg* **34**, 90–97.

[1972] Eine Kennzeichnung halbovoidaler Laguerreebenen, *J. Reine Angew. Math.* **253**, 203–213.

[1976a] Ovoide mit Symmetrien an den Punkten einer Hyperebene, *Abh. Math. Sem. Univ. Hamburg* **45**, 237–244.

[1976b] Laguerre-Ebenen mit Symmetrien an Punktepaaren und Zykeln, *J. Geom.* **8**, 79–93.

[1977] Die Bedeutung des Spiegelungsbegriffs in der Möbius- und der Laguerre-Geometrie, in: Beiträge zur geometrischen Algebra (Proc. Sympos., Duisburg, 1976), pp. 251–258, Lehrbücher und Monographien aus dem Gebiete der Exakt. Wiss., Math. Reihe 21, Birkhäuser, Basel.

[1978] Involutorische Automorphismen von Laguerre-Ebenen mit genau zwei Fixpunkten, *Monatsh. Math.* **86**, 131–142.

[1987] Eine Konstruktionsmethode für Möbiusebenen des Hering Typs VII.1, *Geom. Dedicata* **22**, 247–250.

[1991a] Die Automorphismengruppe der ebenen reellen Polynomgeometrie, *Resultate Math.* **19**, 335–340.

[1991b] Die von den harmonischen Involutionen einer euklidischen Möbiusebene erzeugte Gruppe, *Geom. Dedicata* **38**, 257–261.

Meyer, R.
[1987] *Untersuchungen über topologische Benz-Ebenen: affine Ableitungen und Anordnungstopologien*, Dissertation, Hannover.

Mohrmann, H.
[1922] Hilbertsche und Beltramische Liniensysteme, *Math. Ann.* **8**, 177–183.

Moldovan, E.
[1959] Sur une généralisation des fonctions convexes, *Mathematica (Cluj)* **1**, 49–80.
[1964] Introduction à l'étude comparative des ensembles de fonctions interpolatoires, *Mathematica (Cluj)* **6**, 145–155.
[1965] Sur les classes d'ensembles interpolatoires qui sont comparables par rapport à leur n-valence, *Mathematica (Cluj)* **7**, 319–325.
[1966] Propriétés des ensembles interpolatoires comparables par rapport à leur n-valence, *Mathematica (Cluj)* **8**, 133–136.

Montgomery, D. and Zippin, L.
[1955] *Topological Transformation Groups*, Interscience, New York.

Morozov, M. I.
[1952] On the uniform approximation of continuous functions by functions in interpolatory classes (Russian), *Izv. Akad. Nauk Ser. Mat.* **16**, 75–100.

Mostow, G. D.
[1950] The extensibility of local Lie groups of transformations and groups on surfaces, *Ann. Math.* **52**, 606–636.

Moulton, F. R.
[1902] A simple non-desarguesian plane geometry, *Trans. Amer. Math. Soc.* **3**, 192–195.

Naumann, H.
[1954] Stufen der Begründung der ebenen affinen Geometrie. *Math. Z.* **60**, 120–141.

Nemeth, A. B.
[1969] About the extension of the domain of definition of the Chebyshev systems defined on intervals of the real axis, *Mathematica (Cluj)* **11**, 307–310.

Otte, J.
[1993] *Differenzierbare Ebenen*, Dissertation, Kiel.

Payne, S. E. and Thas, J. A.

[1984] *Finite Generalized Quadrangles*, Pitman, Boston–London–Melbourne.

Percsy, N.

[1983] Les plans de Minkowski possédant des inversions sont miquelliens, *Simon Stevin* **57**, 15–35.

Petkantschin, B.

[1940] Axiomatischer Aufbau der zweidimensionalen Möbiusschen Geometrie, *Ann. Univ. Sofia Fac. Phys.-Math.* **36**, 219–233.

[1941] Über die Orientierung der Kugel in der Möbiusschen Geometrie, *Jahresber. Deutsch. Math.-Verein* **51**, 124–147.

Pfüller, U.

[1986] *Topologische Laguerreebenen*, Dissertation, Erlangen–Nürnberg.

Pickert, G.

[1956] Eine nichtdesarguessche Ebene mit einem Körper als Koordinatenbereich, *Publ. Math. Debrecen* **4**,157–160.

[1975] *Projektive Ebenen*, 2. Auflage, Springer, Berlin.

Pohl, H.-J.

[1990] Flat projective planes with 2-dimensional collineation group fixing at least two lines and more than two points, *J. Geom.* **38**, 107–157.

Polley, C.

[1968] Lokal desarguessche Salzmann-Ebenen, *Arch. Math.* **19**, 553–557.

[1972a] Lokal desarguessche Geometrien auf dem Möbiusband, *Arch. Math.* **23**, 346–347.

[1972b] Zweidimensionale topologische Geometrien, in denen lokal die dreifache Ausartung des desarguesschen Satzes gilt, *Geom. Dedicata* **1**, 124–140.

Polster, B.

[1992] The groups of projectivities of B-ovals, *Geom. Dedicata* **41**, 337–359.

[1995a] Elementary constructions of locally compact affine planes, *Geom. Dedicata* **56**, 155–175.

[1995b] Integrating and differentiating two-dimensional incidence structures, *Arch. Math.* **64**, 75–85.

[1995c] Integrating completely unisolvent functions, *J. Approx. Theory* **82**, 434–439.

[1995d] Semi-biplanes on the cylinder, *Geom. Dedicata* **58**, 145–160.

[1996a] Recycling circle planes, *Bull. Austral. Math. Soc.* **53**, 325–340.

[1996b] The piecewise projective group as group of projectivities, *Results Math.* **30**, 122–135.

[1996c] Continuous planar functions, *Abh. Math. Sem. Hamburg* **66**, 113–129.

[1997] Flat circle planes as integrals of flat affine planes, *Geom. Dedicata* **67**, 149–163.

[1998a] Invertible sharply n-transitive sets, *J. Combin. Theory Ser. A* **81**, 231–254.

[1998b] Toroidal circle planes that are not Minkowski planes, *J. Geom.* **63**, 154–167.

[1998c] Integrating n-unisolvent sets of functions, *Arch. Math.* **70**, 231–254.

[1998d] N-unisolvent sets and flat incidence structures, *Trans. Amer. Math. Soc.* **350**, 1619–1641.

[1998f] Separating sets in interpolation and geometry, *Aequationes Math.* **56**, 201–215.

[1998g] *A Geometrical Picture Book*, Universitext, Springer, New York.

[2000] Squaring the triangle, *Geom. Dedicata* **83**, 229–243.

Polster, B., Rosehr, N., and Steinke, G. F.

[1997] On the existence of topological ovals in 2-dimensional projective planes, *Arch. Math.* **68**, 418–429.

[1998] Half-ovoidal flat Laguerre planes, *J. Geom.* **60**, 113–126.

Polster, B. and Schroth, A. E.

[1997a] Semibiplanes, Kreisebenen und verallgemeinerte Vierecke, in: Überblicke Mathematik 1998, A. Beutelspacher et al. eds., pp. 128–137, Vieweg, Braunschweig.

[1997b] Split semi-biplanes in antiregular generalized quadrangles, *Bull. Belg. Math. Soc.* **4**, 625–637.

[1998a] Semi-biplanes and antiregular generalized quadrangles, *Geom. Dedicata* **96**, 207–221.

[1998b] Plane models of semi-biplanes, *Beiträge Algebra Geom.* **39**, 135–153.

[20XX] Homology semibiplanes, *Monatsh. Math.*, to appear.

Polster, B. and Steinke, G. F.

[1994] Criteria for two-dimensional circle planes, *Beiträge Algebra Geom.* **35**, 181–191.

[1995] Cut and paste in 2-dimensional projective planes and circle planes, *Canad. Math. Bull.* **38**, 469–480.

[1997] The inner and outer space of 2-dimensional Laguerre planes, *J. Austral. Math. Soc. Ser. A* **62**, 104–127.

[2000] A family of 2-dimensional Laguerre planes of generalised shear type, *Bull. Austral. Math. Soc.* **61**, 69–83.

[20XXa] Separating sets in spherical circle planes, preprint.

[20XXb] On the Kleinewillinghöfer types of flat Laguerre planes, preprint.

[20XXc] General separating sets, preprint.

[20XXd] On the automorphism groups of nested tubular circle planes, preprint.

Popoviciu, E.

[1978] Sur la convexité simultanée par rapport aux plusieurs ensembles interpolatoires, *Anal. Numér. Théor. Approx.* **7**, 101–106.

[1979] Sous-ensembles remarquables d'un ensemble interpolatoire, *Anal. Numér. Théor. Approx.* **8**, 193–202.

Prieß-Crampe, S.

[1983] *Angeordnete Strukturen*, Springer, Berlin.

Rao, R. C.

[1947] Factorial experiments derivable from combinatorial arrangements of arrays, *Suppl. J. Roy. Statist. Soc.* **9**, 128–139.

Rice, J. R.

[1964] *The Approximation of Functions I*, Addison-Wesley, Reading, Mass.

[1969] *The Approximation of Functions II*, Addison-Wesley, Reading, Mass.

Ringel, G.

[1956] Teilungen der Ebene durch Geraden oder topologische Geraden, *Math. Z.* **64**, 74–102.

Rosehr N.

[1998] *Flocks of Topological Circle Planes*, Dissertation, Würzburg.

[20XX] A note on the isotopy problem of $\mathbf{R}^2$-planes, preprint.

Salzmann, H.

[1955] Über den Zusammenhang in topologischen projektiven Ebenen, *Math. Z.* **61**, 489–494.

[1957] Topologische projektive Ebenen, *Math. Z.* **67**, 436–466.

[1958] Kompakte zweidimensionale projektive Ebenen, *Arch. Math.* **9**, 447–454.

[1959a] Homomorphismen topologischer projektiver Ebenen, *Arch. Math.* **10**, 51–55.

[1959b] Topologische Struktur zweidimensionaler projektiver Ebenen, *Math. Z.* **71**, 408–413.

[1962a] Kompakte zweidimensionale projektive Ebenen, *Math. Ann.* **145**, 401–428.

[1962b] Kompakte Ebenen mit einfacher Kollineationsgruppe, *Arch. Math.* **13**, 98–109.

[1962c] Topologische projektive Ebenen, in: Algebraical and Topological Foundations of Geometry (Proc. Colloq., Utrecht, 1959), 157–163, Pergamon, Oxford. Reprinted in: Karl Strubecker (Hrsg.), Geometrie, Wege der Forschung, 177, pp. 41–49, Wissenschaftliche Buchgesellschaft, Darmstadt, 1972.

[1963a] Characterization of the three classical plane geometries, *Illinois J. Math.* **7**, 543–547.

[1963b] Zur Klassifikation topologischer Ebenen, *Math. Ann.* **150**, 226–241.

[1964] Zur Klassifikation topologischer Ebenen II, *Abh. Math. Sem. Univ. Hamburg* **27**, 145–166.

[1965] Zur Klassifikation topologischer Ebenen III, *Abh. Math. Sem. Univ. Hamburg* **28**, 250–261.

[1966] Polaritäten von Moulton-Ebenen, *Abh. Math. Sem. Univ. Hamburg* **29**, 212–216.

[1967a] Kollineationsgruppen ebener Geometrien, *Math. Z.* **99**, 1–15.

[1967b] Topological planes, *Adv. Math.* **2**, 1–60.

[1969] Geometries on surfaces, *Pacific J. Math.* **29**, 397–402.

[1971] Zur Axiomatik der euklidischen Ebene, *Mitt. Math. Sem. Giessen* **90**, 48–50.

[1974] Compact planes of Lenz type III, *Geom. Dedicata* **3**, 399–403.

[1975a] Homogene kompakte projektive Ebenen, *Pacific J. Math.* **60**, 217–234.

[1975b] Homogene affine Ebenen, *Abh. Math. Sem. Univ. Hamburg* **43**, 216–220.

[1981] Projectivities and the topology of lines, in: P. Plaumann, K. Strambach (eds.), Geometry—von Staudt's Point of View (Proc. Bad Windsheim 1980), pp. 313–337, Reidel, Dordrecht, Netherlands.

Salzmann, H., Betten, D., Grundhöfer, T., Hähl, H., Löwen, R., and Stroppel, M.

[1995] *Compact Projective Planes*, de Gruyter, Berlin.

Schellhammer, I.

[1981] *Einige Klassen von ebenen projektiven Ebenen*, Diplomarbeit, Tübingen.

Schenkel, A.

[1980] *Topologische Minkowski-Ebenen*, Dissertation, Erlangen–Nürnberg.

Scholz, W.

[1980] *Zweidimensionale lokalkompakte zusammenhängende topologische Laguerreebenen mit Büschelsatz und die analytischen topologischen Laguerre- und Minkowskiebenen*, Dissertation, Erlangen–Nürnberg.

Schroth, A. E.

[1990] Three-dimensional quadrangles and flat Laguerre planes, *Geom. Dedicata* **36**, 365–373.

[1991] The Apollonius problem in flat Laguerre planes, *J. Geom.* **42**, 141–147.

[1992] Characterising symplectic quadrangles by their derivations, *Arch. Math.* **58**, 98–104.

[1993a] Generalized quadrangles constructed from topological Laguerre planes, *Geom. Dedicata* **46**, 339–361.

[1993b] Topological antiregular quadrangles, *Results Math.* **24**, 180–189.

[1995a] *Topological Circle Planes and Topological Quadrangles*, Pitman Lecture Notes in Mathematics 337, Longman, Harlow, Essex.

[1995b] Sisterhoods of flat Laguerre planes, *Geom. Dedicata* **58**, 185–191.

[1999a] Ovoidal Laguerre planes are weakly Miquelian. *Arch. Math.* **72**, 77–80.

[1999b] Partial circle geometries and antiregular quadrangles, *J. Geom.* **65**, 169–189.

[2000] Compact generalised quadrangles with parameter 1 and large group of automorphisms, *Geom. Dedicata* **83**, 245–272.

[20XX] The point space of compact generalised quadrangles with parameter 1, *Forum Math.*, to appear.

Schroth, A. E. and van Maldeghem, H.

[1994] Half regular and regular points in compact polygons, *Geom. Dedicata* **51**, 215–233.

Schubert, H.

[1968] *Topology*, Allyn and Bacon, Boston, Mass.

Schumaker, L. L.

[1981] *Spline Functions: Basic Theory*, Wiley, New York.

Segre, B.

[1955] Ovals in a finite projective plane, *Canad. J. Math.* **7**, 414–416.

[1956] Plans graphiques algébriques réels non desarguesiens et correspondences crémoniennes topologiques, *Rev. Math. Pures Appl.* **3**, 35–50.

Seifert, H. and Threlfall, W.

[1980] *A Textbook of Topology*, Academic Press, New York.

Sitaram, K.

[1963] A real non-Desarguesian plane, *Amer. Math. Monthly* **70**, 522–525.

Skornjakov, L. A.

[1954] Systems of curves on a plane (Russian), *Doklady Akad. Nauk SSSR* **98**, 25–26.

[1957] Systems of curves on a surface (Russian), *Trudy Moskov. Mat. Obšč.* **6**, 135–164.

Spanier, E. H.

[1966] *Algebraic Topology*, McGraw-Hill, New York–Toronto.

Steinke, G. F.

[1983a] An extension property for classical topological Benz planes, *Arch. Math.* **41**, 190–192.

[1983b] Locally classical Benz planes are classical, *Math. Z.* **183**, 217–220.

[1984a] Locally Miquelian Benz planes, *Abh. Math. Sem. Univ. Hamburg* **54**, 141–161.

[1984b] The automorphism group of locally compact connected topological Benz planes, *Geom. Dedicata* **16**, 351–357.

[1985a] Topological affine planes composed of two Desarguesian half-planes and projective planes with trivial collineation group, *Arch. Math.* **44**, 472–480.

[1985b] Some Minkowski planes with 3-dimensional automorphism group, *J. Geom.* **25**, 88–100.

[1986a] Semiclassical topological Möbius planes, *Results Math.* **9**, 166–188.

[1986b] The automorphism group of Laguerre planes, *Geom. Dedicata* **21**, 55–58.

[1987] Semiclassical topological flat Laguerre planes obtained by pasting along a circle, *Results Math.* **12**, 207–221.

[1988] Semiclassical topological flat Laguerre planes obtained by pasting along two parallel classes, *J. Geom.* **32**, 133–156.

[1989] Topological properties of locally compact connected Minkowski planes and their derived affine planes, *Geom. Dedicata* **32**, 341–351.

[1990a] On the structure of the automorphism group of 2-dimensional Laguerre planes, *Geom. Dedicata* **36**, 389–404.

[1990b] 2-dimensional Minkowski planes and Desarguesian derived affine planes, *Abh. Math. Sem. Univ. Hamburg* **60**, 61–69.

[1991a] On the structure of finite elation Laguerre planes, *J. Geom.* **41**, 162–179.

[1991b] The elation group of a 4-dimensional Laguerre plane, *Monatsh. Math.* **111**, 207–231.

[1993] 4-dimensional point-transitive groups of automorphisms of 2-dimensional Laguerre planes, *Results Math.* **24**, 326–341.

[1994] A family of 2-dimensional Minkowski planes with small automorphism groups, *Results Math.* **26**, 131–142.

[1995] Topological circle geometries, in: Buekenhout [1995], pp. 1325–1354.

[1997] Lenz–Barlotti classes of semi-classical ordered projective planes, *Australasian J. Combin.* **16**, 1–20.

[2000a] A classification of 2-dimensional Laguerre planes admitting 3-dimensional groups of automorphisms in the kernel, *Geom. Dedicata* **83**, 77-94.

[2000b] A classification of Minkowski planes over half-ordered fields, *J. Geom.* **69**, 192–214.

[20XXa] On 2-dimensional Laguerre planes of shift type, *Arch. Math.*, to appear.

[20XXb] 2-dimensional Laguerre planes admitting 4-dimensional groups of automorphisms that fix at least two parallel classes, preprint.

[20XXc] 2-dimensional Laguerre planes admitting 4-dimensional groups of automorphisms that fix a parallel class, preprint.

[20XXd] On the Klein–Kroll types of flat Minkowski planes, preprint.

Strambach, K.

[1967a] Salzmann-Ebenen mit hinreichend vielen Punkt- oder Geradenspiegelungen, *Math. Z.* **99**, 247–269.

[1967b] Eine Charakterisierung der klassischen Geometrien, *Arch. Math.* **18**, 539–544.

[1967c] Über sphärische Möbiusebenen, *Arch. Math.* **18**, 208–211.

[1968] Zur Klassifikation von Salzmann-Ebenen mit dreidimensionaler Kollineationsgruppe, *Math. Ann.* **179**, 15–30.

[1970a] Zur Klassifikation von Salzmann-Ebenen mit dreidimensionaler Kollineationsgruppe II, *Abh. Math. Sem. Univ. Hamburg* **34**, 159–169.

[1970b] Salzmann-Ebenen mit punkttransitiver dreidimensionaler Kollineationsgruppe, *Indag. Math.* **32**, 253–267.

[1970c] Zentrale und axiale Kollineationen in Salzmannebenen, *Math. Ann.* **185**, 173–190.

[1970d] Sphärische Kreisebenen, *Math. Z.* **113**, 266–292.

[1970e] Zentrale Kreisverwandtschaften und die Heringsche Klassifikation von Möbiusebenen, *Math. Z.* **117**, 41–45.

[1971] Gruppentheoretische Charakterisierungen klassischer desarguesscher und moultonscher Ebenen, *J. Reine Angew. Math.* **248**, 75–116.

[1972] Sphärische Kreisebenen mit dreidimensionaler nichteinfacher Automorphismengruppe, *Math. Z.* **124**, 289–314.

[1973] Sphärische Kreisebenen mit einfacher Automorphismengruppe, *Geom. Dedicata* **1**, 182–220.

[1974a] Kollineationsgruppen sphärischer Kreisebenen, *Abh. Math. Sem. Univ. Hamburg* **41**, 133–153.

[1974b] Der Kreisraum einer sphärischen Möbiusebene, *Monatsh. Math.* **78**, 156–163.

[1977] Der von Staudtsche Standpunkt in lokal kompakten Geometrien, *Math. Z.* **155**, 11–21.

[1986] Projektivitätengruppen in angeordneten und topologischen Ebenen, *Arch. Math.* **47**, 560–567.

Stroppel, M.

[1993a] A note on Hilbert and Beltrami systems, *Results Math.* **24**, 342–347.

[1993b] Embedding a non-embeddable plane, *Geom. Dedicata* **45**, 93–99.

[1994] Stable planes, *Discrete Math.* **129**, 181–189.

[1997] The semigroup of continuous lineations of a stable plane, *Semigroup Forum* **55**, 89–93.

[1998] Bemerkungen zur ersten nicht Desarguesschen ebenen Geometrie bei Hilbert, *J. Geom.* **63**, 183–195.

Szenthe, J.

[1974] On the topological characterization of transitive Lie group actions, *Acta Sci. Math. (Szeged)* **36**, 323–344.

Thas, J. A.

[1976] Flocks of non-singular ruled quadrics in PG$(3, q)$. *Atti Accad. Naz. Lincei Rend. Cl. Sci. Fis. Mat. Natur.* (8) **59**, 83–85.

[1990] Solution of a classical problem on finite inversive planes, in: Finite Geometries, Buildings, and Related Topics (Pingree Park, CO 1988), 145–159, Oxford Univ. Press, New York.

[1995] Generalized polygons, in: Buekenhout [1995], pp. 383–431.

Tits, J.

[1952] Sur les groupes doublement transitifs continus, *Comment. Math. Helv.* **26**, 203–224.

[1956] Sur les groupes doublement transitifs continus: correction et compléments, *Comment. Math. Helv.* **30**, 234–240.

Tornheim, L.

[1950] On n-parameter families of functions and associated convex functions, *Trans. Amer. Math. Soc.* **69**, 457–467.

Torrechante, C. E.

[1980] *Lokal miquelsche sphärische Möbiusebenen*, Dissertation, Tübingen.

Tschetweruchin, N.

[1927] Eine Bemerkung zu den Nicht-Desarguesschen Liniensystemen, *Jahresber. Deutsch. Math.-Verein* **36**, 134–136.

Umamaheswaram, S.

[1978a] $\lambda(n,k)$-parameter families and associated convex functions, *Rocky Mountain J. Math.* **8**, 491–501.

[1978b] $\lambda(n,k)$-convex functions, *Rocky Mountain J. Math.* **8**, 759–764.

Valette, G.

[1965] Structures d'ovale topologique sur le cercle. *Acad. Roy. Belg. Bull. Cl. Sci.* (5) **51**, 586–597.

van der Waerden, B. L. and Smid, L. J.

[1935] Eine Axiomatik der Kreisgeometrie und der Laguerregeometrie, *Math. Ann.* **110**, 753–776.

van Heemert, A.

[1955] Zur Kennzeichnung der Systeme der Kreise und der Kegelschnitte, *J. Reine Angew. Math.* **194**, 183–189.

Varadarajan, V. S.

[1974] *Lie Groups, Lie Algebras and Their Representations*, Prentice-Hall, Englewood Cliffs, N.J.

Warner, F. W.

[1983] *Foundations of Differentiable Manifolds and of Lie Groups*, Springer, New York.

Wefelscheid, H.

[1977] Über die Automorphismengruppen von Hyperbelstrukturen Beiträge zur geometrischen Algebra (Proc. Sympos., Duisburg, 1976), pp. 337–343. Lehrbücher u. Monographien aus dem Gebiete der Exakt. Wiss., Math. Reihe 21, Birkhäuser, Basel.

Wernicke, B.

[1987] Topologische Möbiusebenen in spiegelungsgeometrischer Darstellung, *Beiträge Algebra Geom.* **26**, 167–184.

Wilker, J. B.

[1981] Inversive geometry, in: The Geometric Vein, Coxeter Festschrift, pp. 379–442, Springer, New York–Berlin.

Wölk, D.

[1966] Topologische Möbiusebenen, *Math. Z.* **93**, 311–333.

Yaqub, J. C. D. S.

[1978] The nonexistence of Möbius planes of Hering type VI.2, *Arch. Math.* **31**, 414–416.

Zalik, R. A. and Zwick, D.

[1989] On extending the domain of definition of Chebyshev and weak Chebyshev systems, *J. Approx. Theory* **57**, 202–210.

Zielke, R.

[1973] On transforming a Tchebyshev-system into a Markov-system, *J. Approx. Theory* **9**, 357–366.

[1975] Tchebyshev systems that cannot be transformed into Markov systems, *Manuscripta Math.* **17**, 67–71.

[1979] *Discontinuous Čebyšhev Systems*, Lecture Notes in Mathematics 707, Springer, Berlin.

Index